AF356206

Recent Advances and Challenges in Emerging Power Systems

Recent Advances and Challenges in Emerging Power Systems

Guest Editor

Om P. Malik

Basel • Beijing • Wuhan • Barcelona • Belgrade • Novi Sad • Cluj • Manchester

Guest Editor
Om P. Malik
Electrical and Software
Engineering
University of Calgary
Calgary
Canada

Editorial Office
MDPI AG
Grosspeteranlage 5
4052 Basel, Switzerland

This is a reprint of the Special Issue, published open access by the journal *Inventions* (ISSN 2411-5134), freely accessible at: www.mdpi.com/journal/inventions/special_issues/Power_Sys.

For citation purposes, cite each article independently as indicated on the article page online and as indicated below:

Lastname, A.A.; Lastname, B.B. Article Title. *Journal Name* **Year**, *Volume Number*, Page Range.

ISBN 978-3-7258-3172-2 (Hbk)
ISBN 978-3-7258-3171-5 (PDF)
https://doi.org/10.3390/books978-3-7258-3171-5

Contents

Editorial

Recent Advances and Challenges in Emerging Power Systems

Om P. Malik

Department of Electrical and Software Engineering, University of Calgary, Calgary, AB T2N 1N4, Canada; maliko@ucalgary.ca

Citation: Malik, O.P. Recent Advances and Challenges in Emerging Power Systems. *Inventions* **2024**, *9*, 62. https://doi.org/10.3390/inventions9030062

Received: 29 April 2024
Accepted: 20 May 2024
Published: 23 May 2024

Diminishing fossil fuels, the continually increasing demand for energy due to rapid urbanization, pollution caused by the increased generation of electricity using fossil fuels, the consequent environmental effect, and concerns about man-made global warming have prompted a call for renewable energy solutions. This has led to the development of renewable energy sources for the generation of electricity and their integration within conventional power networks. Lack of control on the renewable energy sources results in the intermittent generation of electricity from these sources.

Differences in the characteristics of the generation of electricity from conventional and renewable energy sources offer challenges, and major restructuring in the operation, control, and protection practices of the entire power network is currently taking place to integrate the two in a seamless and efficient manner. It also offers a tremendous opportunity to rejuvenate the practices that have developed over the past 140 years and enhance the entire grid operation, making it more efficient and reliable considering the newly developed technologies.

Many investigators in all parts of the world are now working on developing new ideas and schemes from this perspective, and many advances in emerging power systems are being made. The objective of this Special Issue of *Inventions* is to provide a platform where all researchers can contribute their ideas on the new developments in electricity generation from renewable sources and the challenges encountered in the integration of conventional power systems with the electrical energy generated using various renewable energy sources with wide ranging characteristics.

The first volume of this Special Issue contains a total of 16 papers—1 review paper and 15 research papers—covering a broad range of topics related to the recent advances in power systems and the challenges faced by emerging power systems. All 16 papers are available at https://www.mdpi.com/journal/inventions/special_issues/Power_Sys (last access date 30 April 2024).

Titles of these papers and a brief description of their contents are given below.

The first article, "A Review of Perspectives on Developing Floating Wind farms", by Mohamed Maktabi and Eugen Rusu provides a review of floating wind concepts and projects around the world, which will show the reader what is going on with the projects globally. The main aim of this work is to classify floating wind concepts in terms of their number and manufacturing material, their power capacity, their number, their characteristics (if they are installed or planned), and the corresponding continents and countries where they are based. Additional available data that correspond to some of these projects, with reference to their cost, wind speeds, water depth, and distance to shore, are also classified.

The second article, "Coordinated, Centralized, and Simultaneous Control of Fast Charging Stations and Distributed Energy Resources" by Dener A. de L. Brandao et al., investigates the technical issues related to changes in the voltage profile of grid nodes and to feeder current overload in electrical power systems caused by the growing penetration of fast charging stations (FCSs) for electric vehicles and distributed energy resources (DERs). A coordinated and simultaneous control of DERs and FCSs based on a power-based control strategy, efficiently exploiting FCSs in a microgrid model, is proposed. The results show that, with the coordinated control of DERs and FCSs, the control of the power flow in a minigrid is achieved both in moments of high generation and in moments of high load, even with the maximum operation of DERs.

The third article, "IIR Shelving Filter, Support Vector Machine and k-Nearest Neighbors Algorithm Application for Voltage Transients and Short-Duration RMS Variations Analysis" by Vladislav Liubčuk et al., presents a unique and heterogeneous approach to the assessment of voltage transients and short-duration RMS variations by applying AI tools using databases of both real and synthetic data. The fundamental grid component and its harmonics filtering are investigated with an IIR shelving filter. Also, both SVM and KNN are used to classify PQ events by their primary cause in the voltage–duration plane as well as by the type of short-circuit in the three-dimensional voltage space. To avoid the difficulty in interpreting the results in the three-dimensional space, a method is developed to convert them to two-dimensional space. Based on the results of a PQ monitoring campaign in the Lithuanian distribution grid, this paper presents a unique discussion regarding PQ assessment gaps that need to be solved.

In the fourth article, "Development and Application of an Open Power Meter Suitable for NILM", the authors Carlos Rodríguez-Navarro et al. introduce an Open Multi Power Meter, an open hardware solution designed for efficient and precise electrical measurements to address the challenge of the global energy sector's increasing reliance on fossil fuels and escalating environmental concerns. The power meter, engineered around a single microcontroller architecture, features a comprehensive suite of measurement modules interconnected via an RS485 bus, to ensure high accuracy and scalability. A significant aspect of the developed meter is the integration with the Non-Intrusive Load Monitoring Toolkit, which utilizes advanced algorithms for energy disaggregation, including Combinatorial Optimization and the Finite Hidden Markov Model. The analyses performed validate its design and capabilities. Studies demonstrate the device's effectiveness, characterized by its simplicity, flexibility, and adaptability.

The fifth article, "Load Losses and Short-Circuit Resistances of Distribution Transformers According to IEEE Standard C57.110" by Vicente León-Martínez et al., describes expressions developed for the short-circuit resistances of three-phase transformers according to IEEE Standard C57.110. Considering that these resistances cause load losses of the transformer, two types of short-circuit resistance have been established: (1) the effective resistance of each phase (Rcc,z), closely related to the power loss distribution within the transformer, and (2) the effective short-circuit resistance, a mathematical parameter. Applying these to a 630 kVA oil-immersed distribution transformer, it is concluded that both types of resistances determine the total load losses of the transformer. Rcc,z accurately provides the load losses in each phase. Rcc,ef can give rise to errors depending on harmonic currents.

In the article, "Optimal Dispatch Strategy for a Distribution Network Containing High-Density Photovoltaic Power Generation and Energy Storage under Multiple Scenarios" by Langbo Hou et al., it is shown that the application of energy storage devices in the distribution network realizes peak shaving and valley filling of the load, and also relieves the pressure on the grid voltage generated by the distributed photovoltaic access. At the same time, photovoltaic power generation and energy storage cooperate and have an impact on the current distribution of the distribution network. Since photovoltaic output has uncertainty, the maximum photovoltaic output in each scenario is determined by the clustering algorithm, while the storage scheduling strategy is appropriately selected so the distribution network operates efficiently and stably. Optimization of the distribution network is carried out using an improved Particle Swarm Optimization algorithm with the objectives of minimizing network losses and voltage deviations. Studies on a 30-bus system optimally dispatched under multiple scenarios demonstrate the necessity of conducting a coordinated optimal dispatch of photovoltaics and energy storage.

In the article, "Fault Location Method for Overhead Power Line Based on a Multi-Hypothetical Sequential Analysis Using the Armitage Algorithm" by Aleksandr Kulikov et al., a method of fault location (FL) on overhead power lines (OHPLs) is proposed based on a multi-hypothetical sequential analysis using the Armitage algorithm. The inspection area of the OHPL is divided into many sections and the task of recognizing a faulted section of an OHPL is formulated as a statistical problem. The developed method makes it possible to adapt the distortions of currents and voltages on the emergency mode oscillograms to the conditions for estimating their parameters. Studies show that the implementation of the developed method has practically no effect on the speed of the FL algorithm by emergency model parameters. This ensures the uniqueness of determining the faulted section of the OHPL under the influence of random factors, which leads to a significant reduction in the inspection area of the OHPL.

In the article, "A Rank Analysis and Ensemble Machine Learning Model for Load Forecasting in the Nodes of the Central Mongolian Power System" by Tuvshin Osgonbaatar et al., a new method is proposed to predict power consumption in all nodes of the power system through the determination of rank coefficients calculated directly for the corresponding voltage level, including node substations, power supply zones, and other parts of the power system. An ensemble of decision trees is applied to construct a daily load schedule and rank coefficients are used to simulate consumption in the nodes. Initial data, obtained from daily load schedules, meteorological factors, and calendar features of the central power system, account for most of the energy consumption and generation in Mongolia for the period of 2019–2021. The daily load schedules were constructed using machine learning with a probability of 1.25%.

The contribution of the article "Inductive Compensation of an Open-Loop IPT Circuit: Analysis and Design" by Mario Ponce-Silva et al. is the inductive compensation of a wireless inductive power transmission circuit with resonant open-loop inductive coupling. Variations in the coupling coefficient k due to the misalignment of the transmitter and receiver are compensated with only one auxiliary inductance in the primary of the inductive coupling. Experiments were conducted on a low-power prototype with an input voltage of 27.5 V, output power of 10 W, switching frequency of 500 kHz, output voltage of 12 V, and transmission distance 'd' of 1.5 mm, by varying the distance "d" with several values of the compensation inductor, demonstrating the feasibility of the proposal. An efficiency of 75.10% under nominal conditions was achieved.

In the tenth article, "Performance Analysis of Harmonic-Reduced Modified PUC Multi-Level Inverter Based on an MPC Algorithm" by Umapathi Krishnamoorthy et al., a multi-level inverter for an application is selected based on a trade-off between cost, complexity, losses, and total harmonic distortion (THD). A packed U-cell (PUC) topology, composed of power switches and voltage sources connected in a series–parallel fashion, that can be extended to a greater number of output voltage levels requires fewer power switches, gate drivers, protection circuits, and capacitors. A converter is presented with a 31-level topology, switched by a variable-switching-frequency-based model predictive controller that helps in achieving optimal output with reduced harmonics. The gate driver circuit is also optimized in terms of power consumption and size complexity. A comparison of the 9-level and the 31-level PUC inverters shows that the THD for a nominal modulation index of 0.8 is 11.54% and 3.27% for the 9-level multi-level inverter and the modified 31-level multi-level inverter, respectively.

In the article, "A Decentralized Blockchain-Based Energy Market for Citizen Energy Communities" by Peyman Mousavi et al., a decentralized blockchain network based on the Hyperledger Fabric framework is introduced, enabling the formation of local energy markets of future citizen energy communities (CECs) through peer-to-peer transactions. It is designed to ensure adequate load supply and observe the network's constraints while running an optimal operation point by consensus among all the players in a CEC. The proposed framework proves its superior flexibility and proper functioning. The results show that the proposed model increases system performance, reduces costs, and reaches an operating point based on consensus among the microgrid elements.

In the article, "Robust Control and Active Vibration Suppression in Dynamics of Smart Systems" by Amalia Moutsopoulou et al., a smart structure with piezoelectric (PZT) materials is investigated for its active vibration response under dynamic disturbance. Numerical modeling with finite elements is used to achieve that. Vibration for different model values is presented considering the uncertainty of modeling. Vibration suppression was achieved with a robust controller and with a reduced-order controller. The results presented for the frequency domain and the state space domain demonstrate the advantage of robust control in the vibration suppression of smart structures.

In the article, "Compromised Vibration Isolator of Electric Power Generator Considering Self-Excitation and Basement Input" by Young Whan Park et al., two performance indices of the vibration isolator are introduced to evaluate the vibration control capability over two excitation cases, self-excitation and basement input, using the theoretical linear model of the electric power generator. The proposed strategy is devoted to enhancing the vibration control capability over the basement input, owing to the acceptable margin for self-excitation. Modification of the mechanical properties of the vibration isolator focuses on the isolator between the mass block and the surrounding building. The simulation results show that an increase in the spring coefficient and a decrease in the damping coefficient of the vibration isolator beneath the mass block could enhance the vibration reduction capability over the basement input.

In the article, "Organization of Control of the Generalized Power Quality Parameter Using Wald's Sequential Analysis Procedure" by Aleksandr Kulikov et al., considering the requirements set by industrial enterprises with respect to power quality parameters (PQPs) at the points of their connection to external distribution networks, a rationale is provided for the transition from the monitoring of a set of individual PQPs to a generalized PQP with the arrangement of the simultaneous monitoring of several parameters. The joint use of the simulation results and data from PQP monitoring systems for PQP analysis using the sampling-based procedure produces the desired effect. An example of a sequential decision-making process in the analysis of a generalized PQP based on Wald's sequential analysis procedure is provided. With this technique, it is possible to adapt the PQP monitoring procedure to the features of a specific power distribution network. The structural diagram of the device, that implements the sampling-based monitoring procedure of the generalized PQP, is also presented.

In the article, "Research of Static and Dynamic Properties of Power Semiconductor Diodes at Low and Cryogenic Temperatures" by Mikhail Ostapchuk et al., new requirements imposed on power conversion in systems with high-temperature superconductors are investigated as the main part of the losses in such systems is induced in the semiconductors of the converters. The possibility of improving the static and dynamic characteristics of power semiconductor diodes using cryogenic cooling is confirmed with a loss reduction of up to 30% achieved in some cases.

In the article, "Energy-Saving Load Control of Induction Electric Motors for Drives of Working Machines to Reduce Thermal Wear" by Tareq M. A. Al-Quraan et al., the influence of reduced voltage on the service life of an induction motor is investigated. An algorithm, developed for calculating the rate of thermal wear of induction motor insulation under a reduced supply voltage depending on the load and the mechanical characteristics of the working machine, determines the change in the rate of thermal wear under alternating external effects on the motor (supply voltage and load) and allows the forecasting of its service life. It is determined that the rate of thermal wear of the induction motor insulation increases significantly when the voltage is reduced compared to its nominal value with nominal load on the motor. The results of the experimental verification of the obtained rule for "Asynchronous Interelectro" series electric motors confirm its accuracy. Based on the obtained correlation, the rule of voltage regulation in energy-saving operation mode is derived.

Concluding remarks.

Concern about the effect of fossil fuel-based electricity generation on the environment has resulted in very significant advances in the use of renewable sources of energy to generate electricity. These advances have inevitably been accompanied with challenges, and tremendous efforts are being devoted to overcoming these. This has become evident from the interest shown in this Special Issue and the number of papers that have been received from a wide variety of geographical areas, encompassing a diverse range of research and elucidating the richness of the research field.

Considering the above, further papers are being invited for a second volume of the Special Issue "Recent Advances and Challenges in Emerging Power Systems".

Conflicts of Interest: The author declares no conflicts of interest.

Article

Coordinated, Centralized, and Simultaneous Control of Fast Charging Stations and Distributed Energy Resources

Dener A. de L. Brandao [1], **João M. S. Callegari** [1], **Danilo I. Brandao** [2] and **Igor A. Pires** [3,*]

[1] Graduate Program in Electrical Engineering, Universidade Federal de Minas Gerais, Av. Antônio Carlos 6627, Belo Horizonte 31270-901, MG, Brazil; dalbrandao@ufmg.br (D.A.d.L.B.); jmscallegari@ufmg.br (J.M.S.C.)
[2] Department of Electrical Engineering, Universidade Federal de Minas Gerais, Belo Horizonte 31270-901, MG, Brazil; dibrandao@ufmg.br
[3] Department of Electronics Engineering, Universidade Federal de Minas Gerais, Belo Horizonte 31270-901, MG, Brazil
* Correspondence: iap@ufmg.br

Abstract: The growing penetration of fast charging stations (FCSs) to electric vehicles (EVs) and distributed energy resources (DERs) in the electrical power system brings technical issue changes in the voltage profile throughout grid nodes and feeder current overload. The provision of ancillary services by DERs and FCSs arises as an appealing solution to reduce these adverse effects, enhancing the grid hosting capacity. The control of microgrids is essential for the coordinated implementation of these services. Although microgrid control is widely applied to DERs, few studies address the coordinated control of DERs and FCSs to obtain benefits for the electrical power system. This paper proposes a coordinated and simultaneous control of DERs and FCSs based on the power-based control (PBC) strategy, efficiently exploiting FCSs in a microgrid model previously unaddressed in the literature. The results show that, with the coordinated control of DERs and FCSs, the control of the power flow in a minigrid (MG) is achieved both in moments of high generation and in moments of high load, even with the maximum operation of DERs. This method allows for the maintenance of voltage levels within values considered acceptable by technical standards (above 0.93 pu). The maintenance of voltage levels is derived from reducing the overload on the point of common coupling (PCC) of the minigrid by 28%, performing the peak shaving ancillary service. Furthermore, the method allows for the control of zero power flow in the PCC of the minigrid with the upstream electric grid in periods of high generation, performing the ancillary service of valley filling. The method performs this control without compromising vehicle recharging and power dispatch by DERs.

Keywords: microgrid control; fast charging station (FCS); power-based control (PBC); distributed energy resource (DER); electric vehicle (EV)

Citation: Brandao, D.A.d.L.; Callegari, J.M.S.; Brandao, D.I.; Pires, I.A. Coordinated, Centralized, and Simultaneous Control of Fast Charging Stations and Distributed Energy Resources. *Inventions* **2024**, *9*, 35. https://doi.org/10.3390/inventions9020035

Academic Editor: Om P. Malik

Received: 23 February 2024
Revised: 18 March 2024
Accepted: 19 March 2024
Published: 25 March 2024

1. Introduction

The emission of polluting gases into the atmosphere has been the subject of debates and actions by several countries and industries. In humans, pollution can cause short-term effects such as eye, throat, and nose irritation; headaches; nausea; and can worsen cases of diseases such as bronchitis and pneumonia. Long-term effects may include heart disease, lung cancer, and pulmonary emphysema [1]. According to a study by the University of Chicago, the life expectancy of the inhabitants of South Korea is 1.4 years lower due to air pollution, as the entire South Korean population lives in areas with pollution above the levels recommended by the World Health Organization [2]. The effects on the environment are also significant, leading to soil pollution and the mortality of plants and animals thanks to polluting compounds such as sulfur dioxide and nitrogen oxides. Finally, a possible acceleration in Earth's warming is credited to carbon dioxide (due to its ability to retain heat in the atmosphere) and other gases such as methane [1].

Several governments have created laws and tax incentives to encourage solutions that reduce such emissions. Since the 1970s, the USA has followed the Clean Air Act, which regulates atmospheric emissions from stationary sources such as industries and mobile sources (i.e., combustion vehicles) [3]. Members of the European Union must meet obligations to reduce air pollution based on the National Emission Reduction Commitments Directive (NECD) [4].

Both the USA and the European Union classify the road transport sector as a high source of polluting gas emissions into the atmosphere. Combustion vehicles are responsible for 27% of polluting gas emissions in the USA [5]. One of the ways to reduce the gas emissions caused by the vehicle fleet during its use is through electrification, whether through hybrid or fully electric vehicles (EVs).

However, consumer adoption of EVs is essential. One of consumers' biggest concerns regarding EVs is recharging time and vehicle charger infrastructure [6–8]. Despite the advantage of being able to fully recharge overnight with low-power chargers, vehicles used in public and individual passenger transport require recharging at shorter periods because they cover a great distance during the day, even in urban regions. Typically, the service, policing, emergency, and cargo transportation sectors cannot rely on slow recharges.

Based on this, the Federal Highway Administration (FHWA), an agency linked to the United States Department of Transportation, proposed minimum standards for the country's road network [9] through the National Electric Vehicle Infrastructure Formula Program. The document suggests that there be four charging stations with a minimum power of 150 kW at each charging location, with a minimum distance of 80 km between the stations and less than 2 km from highways. High-power charging stations are called fast charging stations (FCSs).

Furthermore, [9] suggests the installation of 500,000 chargers by 2030. According to data from the Alternative Fuels Data Center, an organization linked to the US Department of Energy, there were around 40,000 DC fast charger ports across the country in February 2024, with approximately 27% of them being in the state of California [10]. By 2022, the number of DC fast chargers installed in the USA was around 6600, a smaller number than in countries such as Germany (12,000) and France (9000) [11], even though they have more vehicles [12] and greater territorial extension than the countries cited. China has around 760,000 fast chargers, but around 70% are installed in just 10 of the country's 22 provinces [11].

In addition to voltage disturbances [13,14] and high harmonic content [15–19], the most significant impact of inserting electric vehicle chargers is the overload of the electrical power system [20–23]. To accommodate the rise in electricity consumption, it is also necessary to generate more energy. As the energy sector is also one of the sectors that emits the most polluting gases into the atmosphere, the construction of polluting plants, such as coal or gas thermoelectric plants, has been discouraged by agreements and laws [24].

Among all the forms of energy generation, two have received the most attention: solar and wind. Due to not emitting polluting gases during their generation and the usage of renewable resources, these two sources have been receiving incentives in several countries such as the USA [25], Brazil [26], China [27], and the European Union [28].

The advantages of solar energy generation are explained by the easiness to install for small energy consumers, transforming them into prosumers. Photovoltaic modules can be installed on the roofs of houses, buildings, and condominium areas, reducing the energy costs of these consumers. Thus, the concept of distributed generation represents a transformative shift away from the conventional methods of energy production.

With distributed generation, new challenges arise for electrical power systems since the systems were originally designed to deal with generation far from large electrical energy consumption centers. Distributed generation aggravates issues related to voltage and frequency regulation, generation intermittency, and feeder overload [29,30]. Changes to electrical infrastructure through cable reconductoring and equipment replacement are alternatives, but they are complex and costly. Therefore, one of the alternatives that arise to reduce these impacts on the electrical power system is the control of microgrids capable of integrating all these new players into the electrical power system.

Despite the relevance of the subject, few studies discuss the simultaneous coordinated control of FCSs and distributed energy resources (DERs), which is highly desired in the aforementioned scenario. When it comes to providing ancillary services between FCSs and DERs, many works address vehicle-to-grid (V2G) strategies [19,31–44]. However, in fast charging applications, V2G is not an interesting alternative from the vehicle owner's point of view since the reverse power flow would increase the vehicle's total recharging time. Therefore, it is not common to find works that involve V2G and FCS.

Table 1 summarizes a comparative analysis of adherent state-of-the-art works that use FCSs to provide ancillary services, categorizing them according to voltage level, ancillary service category, control of DERs and FCSs, and the control architecture of DERs or FCSs according to [45]. None of these strategies encompass the simultaneous control of FCSs and DERs. Some use strategies with decentralized architectures that cannot compose an advanced minigrid.

Table 1. Proposals that use FCS to provide ancillary services.

Reference	Voltage Level	Ancillary Service Category	DERs Control	FCSs Control	Control Architecture [a]
[18]	low voltage (LV)	Power quality (voltage and frequency disturbances and harmonic injection)	✗	✗	✗/✗
[14]	medium voltage (MV)	Power quality (transient voltage disturbances)	✗	✗	✗/✗
[46]	MV	Reactive power support (voltage control)	✗	✓	✗/C
[13]	MV	Power quality (voltage disturbances)	✗	✗	✗/✗
[47]	MV	Active/reactive power support (voltage control)	✗	✓	✗/C
[48]	LV	Active power support (peak shaving)	✗	✓	✗/D
[49]	MV	Reactive power support (voltage support)	✗	✓	✗/D
[50]	MV	-	✗	✓	✗/D
[51]	MV	Active power support (peak shaving)	✗	✓	✗/C
[52]	LV	-	✗	✓	✗/C
[53]	MV	Active power support (load shifting)	✗	✓	✗/D
[54]	MV	Reactive power support (voltage support)	✗	✓	✗/D
[55]	MV	Active power support (load shifting)	✗	✓	✗/C
[56]	MV	-	✓	✗	C/✗
Here	LV/MV	Active power support: peak shaving and valley filling	✓	✓	C/C

[a] D = decentralized and C = centralized.

To the best of the authors' knowledge, no works were found that simultaneously control the DERs power dispatch and the FCSs power absorption. Thus, this paper proposes a simultaneous control of these two entities, allowing better grid power quality and increased operational flexibility due to controllability at different points in the network.

The main contribution of this paper is a novel method for simultaneous coordinated control between fast charging stations (FCSs) and distributed energy resources (DERs) in a minigrid (MG), categorizing it as an advanced minigrid [57]. An MG is a set composed of one or more LV microgrids with DERs connected to medium voltage through transformers, loads, generators, and FCSs also connected directly to medium voltage. This approach consists of hierarchically interconnecting smaller microgrids with the larger minigrid. Figure 1 shows the MG used in this paper, with a minigrid central controller (MGCC), microgrid central controllers (μGCCs), and fast charging station central controllers (FCSCCs). The objective of this control is to reduce or increase consumption at charging stations depending on the established boundary conditions to maintain power levels at the MG PCC that guarantee the reliability of the grid power quality parameters.

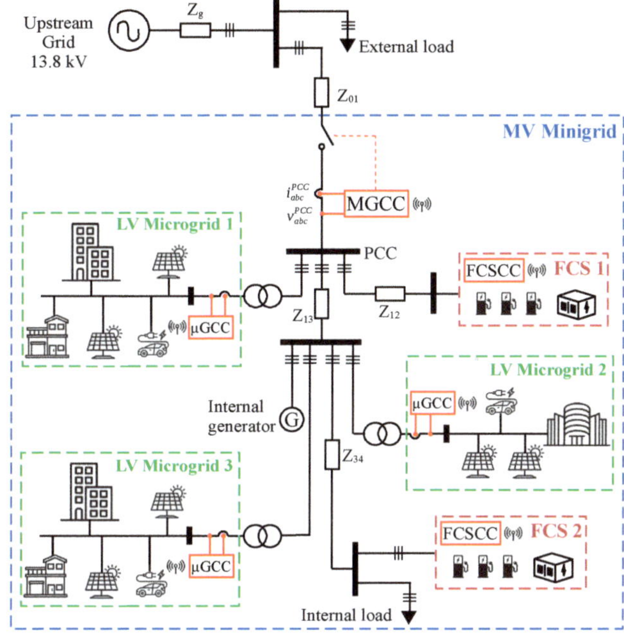

Figure 1. Representation of the medium-voltage minigrid to validate the proposal.

This paper is divided as follows: after the Introduction, Section 2 presents the literature review about the PBC method. Section 3 presents the proposed strategy and the methodology and methods used to validate the concept of the control proposal. Section 4 presents the simulation results. Section 5 presents the results and is followed by Section 6, which concludes this work.

2. Literature Review

Power-Based Control

First presented by [58] for single-phase low-voltage microgrids, other works improve the power-based control (PBC) strategy over time. Table 2 presents work involving PBC improvements, variation, or specific applications.

Figure 2 shows a schematic of the original PBC algorithm [58]. The PBC regulates, through a central controller, the dispatch of active and reactive power between a microgrid and PCC performing a proportional power sharing between DERs. To do so, the central controller uses the following measurements and parameters:

- The active power (P_{PCC}) and reactive power (Q_{PCC}) in the PCC of the microgrid in the current cycle ℓ.
- The sum of the maximum active (P_{DER}^{max}) and reactive (Q_{DER}^{max}) powers that each DER can dispatch in the current cycle ℓ.

- The sum of active (P_{DER}) and reactive (Q_{DER}) powers dispatched by DERs in the current cycle ℓ.
- The active power references (P_{PCC}^{ref}) and reactive (Q_{PCC}^{ref}) desired in the PCC in the next cycle ($\ell + 1$).

Table 2. PBC enhancements and applications.

Reference	Microgrid Configuration	Migrogrid Control Architecture	Power Quality Compensation	Active Load Control	Sharing Coefficients
[58]	1Φ to 2 wires and low voltage	Centralized	Reactive and load unbalance	✗	α_P e α_Q
[59]	1Φ and 3Φ to 4 wires and low voltage	Centralized	Reactive and load balance	✗	$\alpha_{Pa}, \alpha_{Pb}, \alpha_{Pc},$ α_{Qa}, α_{Qb} e α_{Qc}
[60]	1Φ and 3Φ to 4 wires and low voltage	Centralized	Reactive and load balance	✗	$\alpha_{Pa}, \alpha_{Pb}, \alpha_{Pc},$ $\alpha_{Qa}, \alpha_{Qb}, \alpha_{Qc},$ $\alpha_{P3\Phi}$ e $\alpha_{Q3\Phi}$
[61]	1Φ and 3Φ to 4 wires and low voltage	Centralized	Reactive and load balance	✗	$\alpha_{Pa}, \alpha_{Pb}, \alpha_{Pc},$ α_{Qa}, α_{Qb} e α_{Qc}
[62]	1Φ and 3Φ to 4 wires and low voltage	Centralized	Reactive and load balance	✗	$\alpha_{Pa}, \alpha_{Pb}, \alpha_{Pc},$ $\alpha_{Qa}, \alpha_{Qb}, \alpha_{Qc},$ $\alpha_{P3\Phi}$ e $\alpha_{Q3\Phi}$
[63]	1Φ to 2 wire and low voltage	Centralized	Reactive and load unbalance	✗	$\alpha_{P_{ESU}}, SoC_a, SoC_{Error},$ α_{Q1}, α_{Q2} e α_{Q3}
[64]	1Φ and 3Φ to 4 wires and low voltage	Distributed	Load balance	✗	$\alpha_{Pa}, \alpha_{Pb}, \alpha_{Pc}$ e $\alpha_{P3\Phi}$
[65]	1Φ and 2Φ to 4 wires and low voltage	Centralized	Reactive and load balance	✗	$\alpha_{Pa}, \alpha_{Pb}, \alpha_{Pc},$ α_{Qa}, α_{Qb} e α_{Qc}
[66]	1Φ and 2Φ to 4 wires and low voltage	Centralized	Reactive and load balance	✗	$\alpha_{Pa}, \alpha_{Pb}, \alpha_{Pc},$ α_{Qa}, α_{Qb} e α_{Qc}
[67]	1Φ and 3Φ to 4 wires and low voltage	Centralized/ decentralized	Reactive and load balance	✗	$\alpha_{Pa}, \alpha_{Pb}, \alpha_{Pc},$ α_{Qa}, α_{Qb} e α_{Qc}
Here	3Φ to 3 wires and medium voltage	Centralized	Load balance	✓	α_P^{MG}

With these measurements and parameters, it is possible to estimate the active power load (P_L) and reactive power load (Q_L) of the microgrid through (1) and (2), respectively. The ℓ cycle is the control cycle of the algorithm. The power load of the microgrid includes the entities that consume energy and generate energy that does not obey the central controller; that is, the non-dispatchable DERs:

$$P_L(\ell) = P_{PCC}(\ell) + P_{DER}(\ell) \tag{1}$$

$$Q_L(\ell) = Q_{PCC}(\ell) + Q_{DER}(\ell) \tag{2}$$

From (1) and (2), the desired power to be dispatched by the DERs in the next control cycle is defined according to (3) and (4). For instance, Equation (3) defines the active power reference for the set of DERs for the next cycle, while (4) expresses the reactive power reference:

$$P_{DER}^{ref}(\ell + 1) = P_L(\ell + 1) - P_{PCC}^{ref}(\ell + 1) \tag{3}$$

$$Q_{DER}^{ref}(\ell + 1) = Q_L(\ell + 1) - Q_{PCC}^{ref}(\ell + 1) \tag{4}$$

For there to be proportional power sharing between the DERs of the microgrid, the powers obtained in (3) and (4) are divided, respectively, by the active and reactive power maximums that each DER can dispatch at that moment. In this way, the coefficients α_P and α_Q are obtained, expressed by (5) and (6), respectively:

$$\alpha_P = \frac{P_{DER}^{ref}(\ell+1)}{P_{DER}^{max}(\ell)} \tag{5}$$

$$\alpha_Q = \frac{Q_{DER}^{ref}(\ell+1)}{Q_{DER}^{max}(\ell)} \tag{6}$$

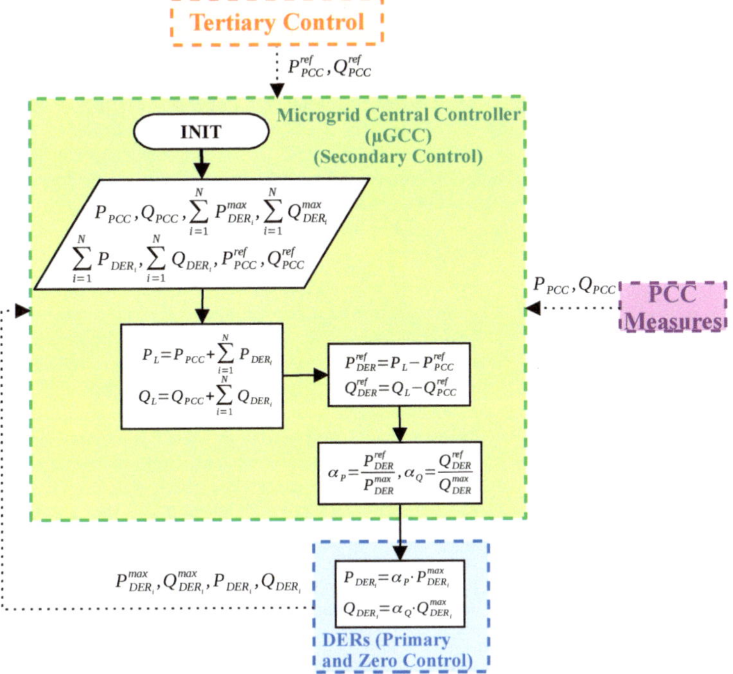

Figure 2. Schematic of the original PBC algorithm.

The lower limit for the α_P coefficient is -1, considering that the DER allows for control of the absorption of its nominal power through the central controller. The upper limit for α_P is one, which means that the DER dispatches all the available power. The coefficient α_Q also has the same lower and upper limits, with the lower limit being the maximum inductive reactive power that the DER can dispatch and the upper limit being the maximum capacitive reactive power that the DER can deliver. Current direction conventions and the position of current meters can modify the meaning of the limits of the active and reactive power coefficients.

All participant DERs of the microgrid control receive both coefficients. Each DER will carry out a dispatch proportional to its power capacity, and the central controller will be able to regulate the power dispatch in the PCC more efficiently, reducing losses in the distribution system [58].

The advantages of PBC as a microgrid control algorithm are its simplicity, good performance compared to strategies such as droop control, and its model-free approach (i.e., it is not necessary to know the grid parameters to control the microgrid, commonly required in optimal control approaches [58]). The disadvantages of PBC are mainly associated with its centralized nature, such as its difficulty in scalability and dependence on the communication link. Another disadvantage is the steady-state error in cases of non-idealities in the communication link and power measurements carried out by DERs, which can be corrected by using classical control strategies, such as proportional-integral (PI) controllers [68].

3. Methodology and Methods

3.1. Proposed Power-Based Control Applied to Fast Charging Stations

As highlighted, the PBC algorithm is limited to the maximum power that DERs can dispatch to the grid. Therefore, it is not possible to control the power dispatch in the PCC of the minigrid at the desired values if there is a violation of the lower or upper limits of the coefficients α_P and α_Q.

The proposed method herein takes advantage of the traditional PBC formulation and improves it for applications during violation of the lower and upper limits of the α_P coefficient in a minigrid (MG). The superscript index of the coefficient α_P determines whether it comes to the minigrid (MG) or microgrid (μG) controller. The proposed control will act at the tertiary level, controlling the power flow in the PCC of MGs with the upstream grid. Figure 3 shows the hierarchy between controllers. This control hierarchy has the following levels:

- Quartenary level: Composed of a distribution system operator (DSO). The electric power utility or a control entity that covers a set of minigrids controls the DSO, responsible for determining the power references for the MGCC. With the absence of the DSO, power references can be determined locally by the MGCC based on the integrity of the voltage levels (as in the work of [67]) and grid frequency or based on financial parameters involving energy consumption from the upstream grid.
- Tertiary level: Composed of an MGCC, responsible for controlling the power flow from the MG to the upstream network. Responsible for sharing and processing information from the microgrid central controller (μGCC), FCSCC, and DSO. It is responsible for sending the coefficient α_P^{MG} to the μGCC and FCSCC and receives, from the DSO, the desired power reference for the PCC of the MG.
- Secondary level: Composed of the μGCC or FCSCC, responsible for controlling the power flow from the microgrid/FCS with the upstream network. It is responsible for sharing and processing information from the MGCC and local controllers of the DER/energy managers of the FCS. Using the coefficient α_P^{MG}, you must determine your own coefficient $\alpha_P^{\mu G}$ and send it to the DERs or FCS energy managers.
- Primary level: Composed of DER local controllers and the FCS energy manager. In the case of the FCS, it may be part of the FCSCC. Responsible for sharing and processing information with μGCC/FCSCC and with the DER or FCS converters. It is he who sends the power, current, or voltage references to the DER or FCS converters.
- Level zero: Composed of the local controllers of the DER and FCS converters. They share information with the DER local controller and are responsible for dispatching/absorbing power to the grid.

In addition to the DER as actuators contributing to the injection of active power, the algorithm also uses the charging stations to control the active power at the PCC. This configuration is typical of a multiple inputs single output (MISO) system.

One of the alternatives for controlling MISO systems is to carry out proportional control between the actuators, the same way as PBC concerning DERs. As FCSs are essentially controllable system loads, participation in control will be through the reduction in power absorbed from their respective upstream connection points.

Figure 4 presents a power control diagram with the DERs and FCSs as actuators. Based on the power reference and the power measured in the PCC, the control algorithm establishes the power reference to be injected by the DER and the power reference to be absorbed by the FCS. Other non-controllable agents on the grid, whether loads or generators, are a disturbance of the controlled variable (PCC power).

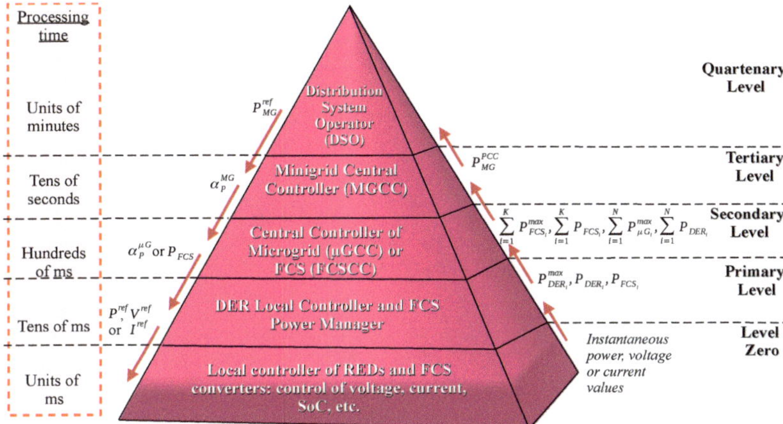

Figure 3. Control hierarchy of the proposed system.

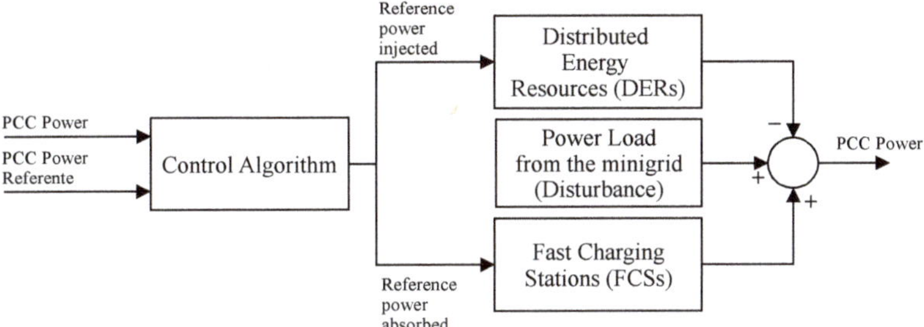

Figure 4. Simplified power control scheme in the PCC of the medium-voltage microgrid through shared control.

Therefore, the control algorithm must be able to process the measured power and power reference information at the MG PCC to send commands to the actuators (microgrids and FCS). Figure 5 shows the algorithm of the proposed system. Since the active and reactive power amounts are orthogonal to each other (i.e., decoupled), analyses can be conducted individually for each power term.

In the adaptation of PBC proposed in this work, the MGCC requires the following measurements and parameters to send commands to the actuators:

- The active power (P_{MG}^{PCC}) in the PCC of the minigrid in the current cycle ℓ. This information is collected locally by MGCC, which is connected to PCC.

- The sum of the maximum active powers ($\sum_{i=1}^{N} P_{\mu G_i}^{max}$) that each microgrid can dispatch in the current cycle ℓ. N is the number of microgrids present in the minigrid. This power is the sum of the maximum powers that DERs can dispatch. This information is sent by µGCCs to the MGCC over a low bandwidth (according to the US Federal Communications Commission [69], communication links below 25 Mbps are low-bandwidth links) and long-range communication links;

- The sum of active powers ($\sum\limits_{i=1}^{N} P_{DER_i}$) dispatched by the DERs of the microgrids in the current cycle ℓ. N is the number of microgrids present in the minigrid. This information is sent by the DERs to μGCCs through a low-bandwidth communication link, such as the radio data system (RDS). Subsequently, these data are sent to the MGCC through a low-bandwidth and long-range communication link;

- The desired active power reference in the PCC of the minigrid in the next cycle ($\ell + 1$) (P_{MG}^{ref}). This information is sent by the DSO to the MGCC over a low-bandwidth and long-range communication link;

- The sum of the maximum active powers ($\sum\limits_{i=1}^{K} P_{FCS_i}^{max}$) that each charging station can absorb in the current cycle ℓ. K is the number of FCSs present in the minigrid. This information is sent by FCSCCs to the MGCC over a low bandwidth and long-range communication link;

- The sum of active powers ($\sum\limits_{i=1}^{K} P_{FCS_i}$) absorbed by the charging stations in the current cycle ℓ. K is the number of FCSs present in the minigrid. This information is sent by FCSCCs to the MGCC over a low bandwidth and long-range communication link.

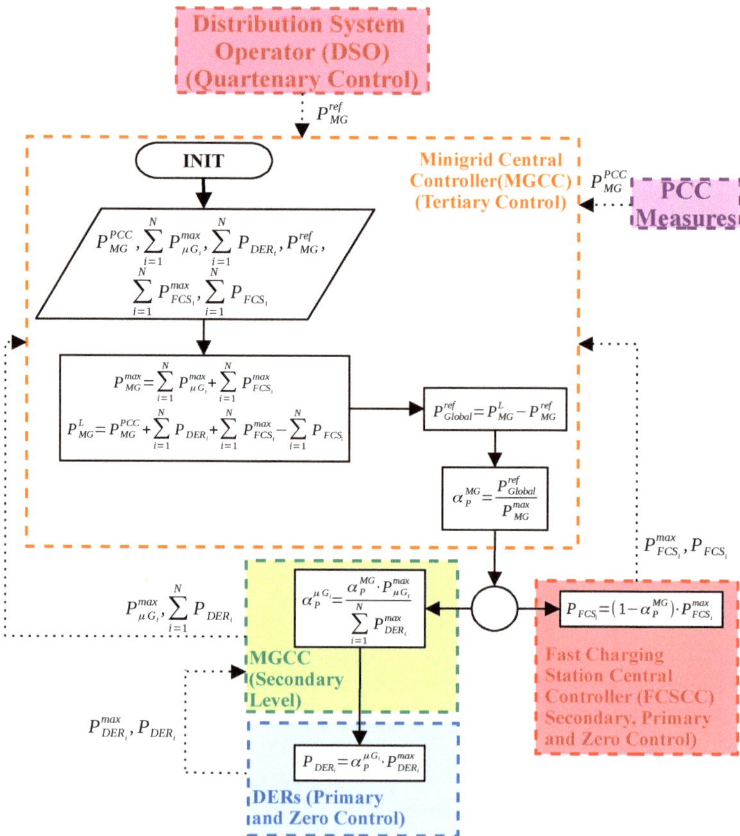

Figure 5. Algorithm of the proposed minigrid control system.

The first novelty is to calculate the total maximum power of the system, as shown in (7). With this, there will be proportionality between the power dispatch of the DERs and the increase or reduction in consumption by the FCSs:

$$P_{MG}^{max}(\ell) = \sum_{i=1}^{N} P_{\mu G_i}^{max}(\ell) + \sum_{i=1}^{K} P_{FCS_i}^{max}(\ell) \tag{7}$$

The second novelty is to calculate the power load of the minigrid, initially presented in Equation (1) for microgrids. With FCSs, it is possible to account for their contributions, as expressed in Equation (8):

$$P_{MG}^{L}(\ell) = P_{MG}^{PCC}(\ell) + \sum_{i=1}^{M} P_{DER_i}(\ell) + \left[\sum_{i=1}^{K} P_{FCS_i}^{max}(\ell) - \sum_{i=1}^{K} P_{FCS_i}(\ell) \right] \tag{8}$$

From (8), it is possible to define the desired global power to be dispatched/absorbed in the next control cycle by the set of DERs of microgrids and FCSs by using Equation (9):

$$P_{Global}^{ref}(\ell+1) = P_{MG}^{L}(\ell+1) - P_{MG}^{ref}(\ell+1) \tag{9}$$

To achieve proportional power sharing between DERs and the reduction in the active power consumed between the FCSs, the global power in (9) is divided by the maximum active power of the minigrid in (7). Thus, the coefficient α_P^{MG} expressed in (10) is obtained:

$$\alpha_P^{MG} = \frac{P_{Global}^{ref}(\ell+1)}{P_{MG}^{max}(\ell)} \tag{10}$$

Based on the α_P^{MG}, each μGCC will calculate its respective $\alpha_P^{\mu G}$ and send it to the DERs, according to (11). In turn, each DER will carry out the dispatch by (12). The FCSs defines the power to be demanded based on (13):

$$\alpha_P^{\mu G_i} = \frac{\alpha_P^{MG} \cdot P_{\mu G_i}^{max}}{\sum_{i=1}^{N} P_{DER_i}^{max}} \tag{11}$$

$$P_{DER_i} = \alpha_P^{\mu G_i} \cdot P_{DER_i}^{max} \tag{12}$$

$$P_{FCS_i} = (1 - \alpha_P^{MG}) P_{FCS_i}^{max} \tag{13}$$

Therefore, for the system to operate properly, there will be a proportionality between the power dispatched by the DERs and the power reduced by the FCSs. The charging station can operate with negative α_P^{MG} coefficients, resulting in absorbed powers superior to the maximum (for example, above the contracted demand).

3.2. Minigrid Parameters

A simulation of a medium-voltage minigrid, shown in Figure 1, containing three microgrids, two fast charging stations, a load, and a non-dispatchable distributed generator, is performed by using MATLAB/Simulink R2022b® software.

There is an external load on the PCC of the minigrid. Current sources, defined through the voltage measured at their connection point and the power standard, represent the grid elements modeled only with active power.

The MGCC can disconnect the minigrid from the upstream grid and be equipped with a grid-forming converter, maintaining the characteristics of an advanced minigrid. The grid-forming converter can also be performed by other elements, such as the internal generator (if there is a storage system) or some FCS. This work does not address the islanded mode of this minigrid nor the characterization of the grid-forming converter. Ref. [70] discusses a converter capable of operating as a grid-forming converter without injecting harmonics into the electrical grid. Ref. [71] presents an alternative to FCSs with this converter that can be used as a grid-forming unit.

Table 3 presents the characteristics of the grid elements. The FCSs are composed of a battery energy storage system (BESS) with an energy of 646.4 kWh capable of charging three 250 kW chargers simultaneously for 50 min. The power of the chargers is typical of commercial chargers such as the Tesla Supercharger V3 [72]. Table 4 presents the characteristics of the cables, represented by the grid impedances. The chosen cables operate with voltages between 8.7 kV and 15 kV. In Brazil, the typical voltage of distribution systems is 13.8 kV.

Table 3. Characteristics of the validation grid elements.

Grid Element	Internal Elements	Load Profile	Power
	Non-dispatchable DERs	-	80 kWp
LV Microgrid 1	Dispatchable DERs	-	80 kWp
	Load	Residential	200 kWp
	Non-dispatchable DERs	-	900 kWp
LV Microgrid 2	Dispatchable DERs	-	60 kWp
	Load	Commercial	450 kWp
	Non-dispatchable DERs	-	500 kWp
LV Microgrid 3	Dispatchable DERs	-	80 kWp
	Load	Residential	100 kWp
	Chargers	-	250 kW (3)
	BESS	-	750 kW
FCS 1	BESS Energy	-	646.4 kWh
	BESS state of charge (SoC)	-	20% to 100%
	Contracted demand	-	375 kW
	Chargers	-	250 kW (3)
	BESS Power	-	750 kW
FCS 2	BESS Energy	-	646.4 kWh
	BESS SoC	-	20% to 100%
	Contracted demand	-	375 kW
Internal load	-	Residential	670 kWp
External load	-	Residential	4100 kWp
Internal generator	-	-	370 kWp

Table 4. Equivalent resistances and inductances of the simulated grid cables.

Symbol	Cable Cross-Section (mm^2) [a]	Cable Length (km)	Equivalent Resistance (Ω)	Equivalent Inductance (mH)
Z_g	50	2.3	2.0	1.4
Z_{01}	50	1.7	1.2	0.9
Z_{12}	50	1.8	1.3	0.9
Z_{13}	50	1.5	1.0	0.7
Z_{34}	50	2.0	1.6	1.2

[a] Considering unipolar aluminum cables for voltages between 8.7 and 15 kV [73].

The LV Microgrid 1 (*LVμG*1) emulates a neighborhood with a predominance of houses equipped with photovoltaic systems. Among the dispatchable DERs, there is a fixed base of 30 kWp and 50 kWp of generation from solar energy, which varies throughout the day. Figure 6 shows the profiles of the dispatchable and non-dispatchable DERs and the load of the LV Microgrid 1.

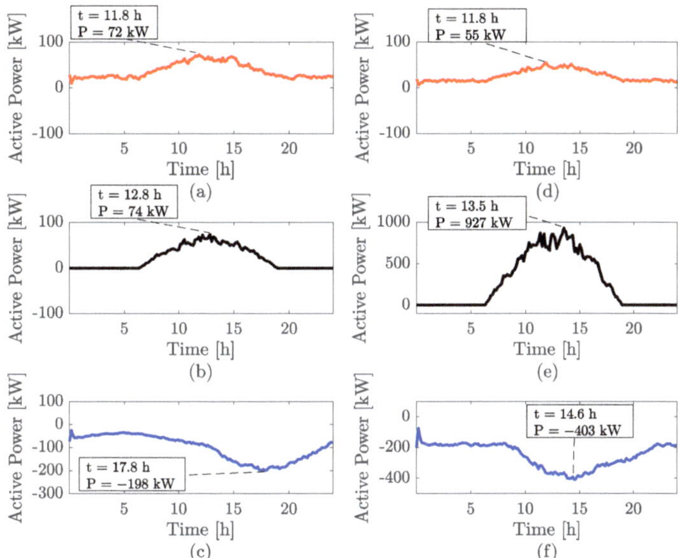

Figure 6. Power profiles: (**a**) *LVμG*1 dispatchable DERs. (**b**) *LVμG*1 non-dispatchable DERs. (**c**) *LVμG*1 load. (**d**) *LVμG*2 dispatchable DERs. (**e**) *LVμG*2 non-dispatchable DERs. (**f**) *LVμG*2 load.

A medium-sized supermarket in Brazil is the basis for the LV Microgrid 2 (*LVμG*2) data. Ref. [74] present, in their work, consumption data from this supermarket. Generation data were estimated based on its total area (5962 m^2), considering an average generation of 0.15 kWp/m^2 [75]. Among the dispatchable DERs, there is a fixed base of 20 kWp and 40 kWp of generation from variable solar energy throughout the day. Figure 6 shows the profiles of dispatchable and non-dispatchable DERs and the load of the LV Microgrid 2.

The LV Microgrid 3 (*LVμG*3) emulates a condominium with a large area dedicated to its photovoltaic plant with high generation and low consumption. Among the dispatchable DERs, there is a fixed base of 30 kWp and 50 kWp of generation from solar energy, which varies throughout the day. Figure 7 presents the profiles of the dispatchable and non-dispatchable DERs and the load of the LV Microgrid 3. Figure 7 also presents internal and external load profiles and the internal generator based on the photovoltaic energy.

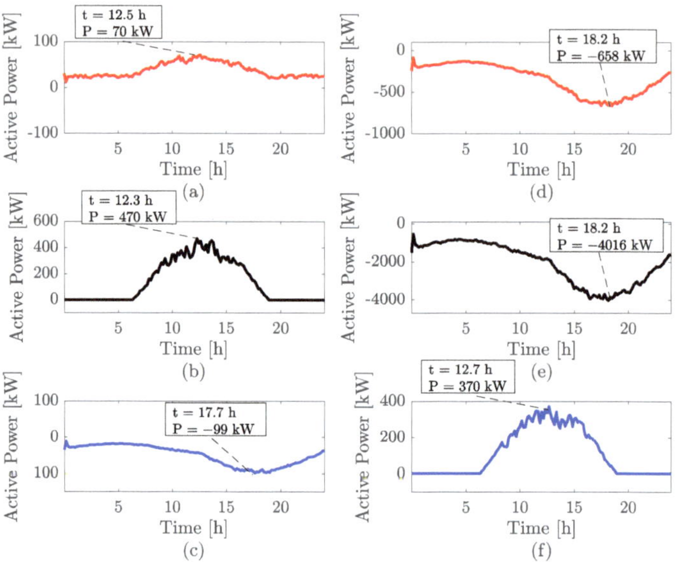

Figure 7. Power profiles: (**a**) *LVμG*3 dispatchable DERs. (**b**) *LVμG*3 non-dispatchable DERs. (**c**) *LVμG*3 load. (**d**) Internal and (**e**) external load of the minigrid and (**f**) internal generator.

3.3. Evaluated Scenarios and Metrics

There are six possibilities for the operation of the minigrid entities. They are:

1. Without FCSs: in this condition, there are no FCSs on the grid.
2. With FCSs not participating in the proposed control: Charging stations operate as constant loads due to the internal management algorithm at all times. FCSs are not controlled by the MGCC.
3. With FCSs participating in the proposed control: charging stations operate as constant loads due to the internal management algorithm. Upon receiving a control command, they start to control the energy demand according to the complement of the coefficient α_P $(1 - \alpha_P)$. Therefore, FCSs are controlled by MGCC.
4. Without DERs: in this condition, there is no type of distributed generation on the grid; that is, the grid does not have dispatchable DERs nor non-dispatchable DERs.
5. Without dispatchable DERs: All the DERs in the network dispatch all available active power. In this condition, the DERs are not controlled by the MGCC.
6. With dispatchable DERs: In this condition, the non-dispatchable DERs dispatch all available active power, and the dispatchable DERs dispatch power according to the index α_P. The MGCC commands dispatchable DERs.

It will be considered that all FCSs have their own BESS and an internal energy-management algorithm that maintains the grid power demand at a constant value within the BESS energy absorption and dispatch limits. About the DERs of each microgrid, the following conditions will be considered:

*   The non-dispatchable DERs do not have energy storage. Generation from photovoltaic modules is the basis for all energy dispatch;
*   The dispatchable DERs have energy storage;
*   For dispatchable DERs, it is considered that there is an internal energy-management system that manages the recharging of the batteries based on the generation of the photovoltaic modules. This management will guarantee a fixed installment that can

always be dispatched, even at times when there is no generation (e.g., at night). Some works suggest energy-management algorithms for DERs [63,76–78].

The possibilities for the operation of the minigrid elements allow for the evaluation of nine different scenarios, in which it is possible to evaluate the effect of the presence or absence of the proposed control. Table 5 presents the nine possible scenarios.

Table 5. Possible scenarios for the operation of minigrid elements.

Scenario	FCS?	DERs?	FCS Control?	DERs Control?
1	✗	✗	✗	✗
2	✗	✓	✗	✗
3	✗	✓	✗	✓
4	✓	✗	✗	✗
5	✓	✓	✗	✗
6	✓	✓	✗	✓
7	✓	✗	✓	✗
8	✓	✓	✓	✗
9	✓	✓	✓	✓

Scenario 1 is the base scenario, in which there are no DERs and FCSs. This scenario sketches the network with its initial design without overloads and adequate voltage levels. Scenario 2 presents the insertion of non-dispatchable DERs without power control. This scenario can increase voltage levels, especially at the DERs' connection point. Scenario 3 presents the insertion of non-dispatchable and dispatchable DERs. This scenario allows for greater operational flexibility of the grid with dispatching power from the minigrid to the upstream grid.

Scenario 4 presents the insertion of FCSs without the insertion of DERs. This scenario allows for the evaluation of a network without distributed generation and with large passive loads such as FCSs. Scenario 5 presents the insertion of non-dispatchable DERs and FCSs. This scenario relieves the grid during periods of high generation but does not relieve it during periods of low generation. Scenario 6 presents the insertion of non-dispatchable and dispatchable DERs and FCSs. This scenario relieves the grid during periods of high generation without dispatching power from the minigrid to the upstream grid.

Scenario 7 presents the insertion of FCSs controlled by the proposed algorithm without the insertion of DERs. In this scenario, the network is relieved during periods of high consumption with high use of the BESSs of the FCSs. Scenario 8 presents the insertion of FCSs controlled by the proposed algorithm with the insertion of non-dispatchable DERs. This scenario allows for the reduction in the use of BESSs from FCSs but with power dispatch from the minigrid to the upstream network.

Scenario 9 bases this work, in which both the FCS and dispatchable DERs are controlled by using the proposed algorithm. In this scenario, relief from the electrical grid is expected at times of high load with low use of the BESSs of the FCSs and with zero-flow control of the power dispatch between the minigrid and the upstream electrical grid.

The control algorithm is activated during the periods shown in Table 6. The period from 10 a.m. to 2 p.m. has the highest solar generation. The objective is to evaluate the scenarios against a zero power reference in the medium-voltage PCC. The purpose is to evaluate whether the control can not dispatch power to the upstream grid, realizing an ancillary service of valley filling.

Table 6. Microgrid control activation periods and respective reference powers.

Activation Periods	Reference Powers (kW)
10 h to 14 h	0
18 h to 22 h	1550

Another ancillary service performed by the proposed control is peak shaving. One of the metrics to establish the desired power in the PCC of a minigrid is the voltage level. A reference power of 1550 kW was defined in the activation period from 6 p.m. to 10 p.m. so that the voltage in the PCC does not exceed the values considered appropriate by Module 8 of Procedimento de Distribuição de Energia Elétrica no Sistema Elétrico Nacional (PRODIST) [79] from Agência Nacional de Energia Elétrica (ANEEL), the regulatory agency for the Brazilian electrical system. Another way to define this power is online, to always keep the voltage within the limits considered appropriate [67]. Voltage levels are also influenced by the reactive power in the PCC, which can change the active power reference levels to maintain voltage at appropriate levels. Table 7 presents the steady-state voltage rating ranges for connection points with a nominal voltage between 2.3 kV and 69 kV.

Table 7. Permanent voltage classification range for connection points with nominal voltage equal to or greater than 2.3 kV and less than 69 kV according to PRODIST.

Service Voltage	Reading Voltage Variation Range (RV) in Relation to Reference Voltage ($RefV$)
Proper	$0.93RefV \leq RV \leq 1.05RefV$
Precarious	$0.90RefV \leq RV \leq 0.93RefV$
Critical	$RV < 0.90RefV$ or $RV > 1.05RefV$

4. Simulation Results

The nine scenarios proposed in Table 5 will be evaluated. With this, it is possible to emphasize the disadvantages of each scenario and verify the performance of the control proposed in the last scenario, showing its relevance to reducing overload problems and precarious voltage levels.

4.1. Scenarios without FCS

Figure 8 presents the power terms and collective RMS voltage of the PCC considering that there are no FCSs in the minigrid (scenarios 1 to 3 of Table 5). The positive power in PCC flows from the upstream grid to the minigrid and vice versa.

In the base scenario (scenario 1 of Table 5), there is no violation of the established limits. As it only has consumer loads, there is no active power dispatch from the minigrid to the upstream network. Furthermore, the minigrid loads are not sufficient to exceed the upper limit. In this way, the voltage in the PCC remains within the limits considered appropriate, not exceeding the lower voltage limit, as shown in Figure 8b.

However, when adding only non-dispatchable DERs (scenario 2 of Table 5), the lower power limit is exceeded, reaching 1096 kW of active power dispatch in the upstream grid. With microgrid control applied to dispatchable DERs (scenario 3 of Table 5), there is a power reduction, but not enough to maintain power above the lower level.

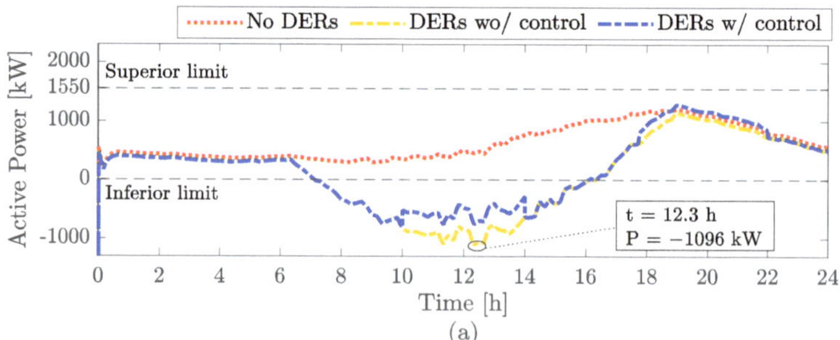

Figure 8. *Cont.*

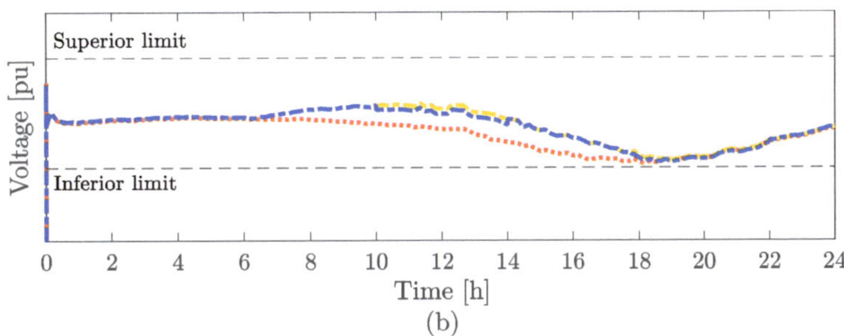

Figure 8. (**a**,**b**) Power and RMS voltage in the PCC of the microgrid considering the absence of FCS in three situations: without DERs (scenario 1), all DERs are non-dispatchable (scenario 2), and DERs dispatchable and not dispatchable (scenario 3).

4.2. Scenarios with Uncontrolled FCS

Figure 9 shows the power terms and collective RMS voltage of the PCC considering that the MGCC does not control the FCS present in the minigrid (scenarios 4 to 6 of Table 5).

Analyzing the scenario without DERs (scenario 4 of Table 5) presented in Figure 9a, a violation of the upper power limit is observed from 2 p.m. onwards, ceasing only at 10 p.m. The PCC voltage is considered precarious according to Table 7. There is also a violation of the upper limit for cases with DERs without and with control between 6 p.m. and 10 p.m. With the increase in load caused by the FCS and most of the generation concentrated at times of low load, none of the scenarios evaluated allowed for load relief from 6 p.m. to 10 p.m.

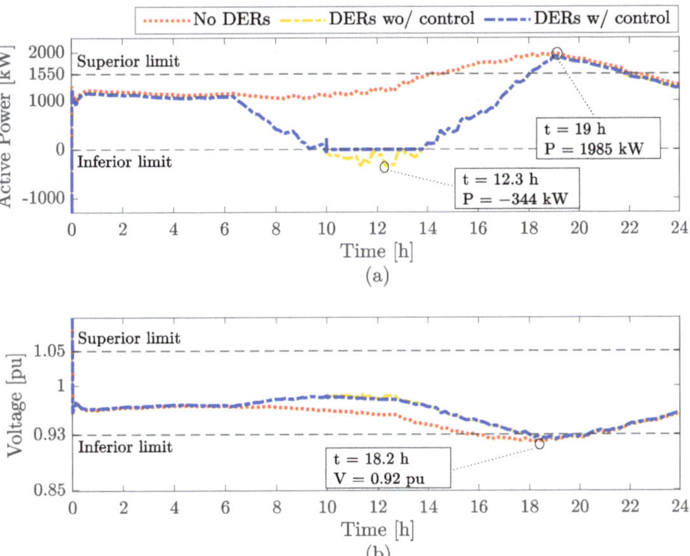

Figure 9. (**a**,**b**) Power and RMS voltage in the PCC of the microgrid with non-controllable FCS in three situations: without DERs (scenario 4), all DERs are non-dispatchable (scenario 5), and DERs are dispatchable and not dispatchable (scenario 6).

Considering the scenario with uncontrolled DERs (scenario 5 of Table 5), there was a violation of the lower limit in the highest generation interval (between 10 a.m. and

2 p.m.) with a minimum of -344 kW. For the scenario with DERs with control (scenario 6 of Table 5), there was no violation of the lower limit. It is worth mentioning that scenario 6 of Table 5 is equivalent to the original PBC since there is control of the DERs, but there is no control of the FCS.

Compared to the scenarios presented in Figure 8, it was possible, by controlling the dispatchable DERs through PBC, to regulate the active power in PCC and not dispatch active power to the upstream grid at the time of highest generation. The consumption of FCSs was essential to absorb the excess power of non-dispatchable DERs. However, the power in the PCC reached a value of 28% above the upper limit, even with the maximum power dispatch of the DERs.

4.3. Scenarios with FCS with Control

Figure 10 shows the power terms and collective RMS voltage of the PCC considering that the MGCC controls the FCS present in the microgrid (scenarios 7 to 9 from Table 5).

Analyzing the scenario without DERs (scenario 7 of Table 5) presented in Figure 10a, it is observed that there is a violation of the upper power limit from 2 p.m. onwards, as well as the scenario without DERs presented in Figure 9a. However, from 6 p.m. to 9 p.m., the power in the PCC remains below the upper limit of Table 6.

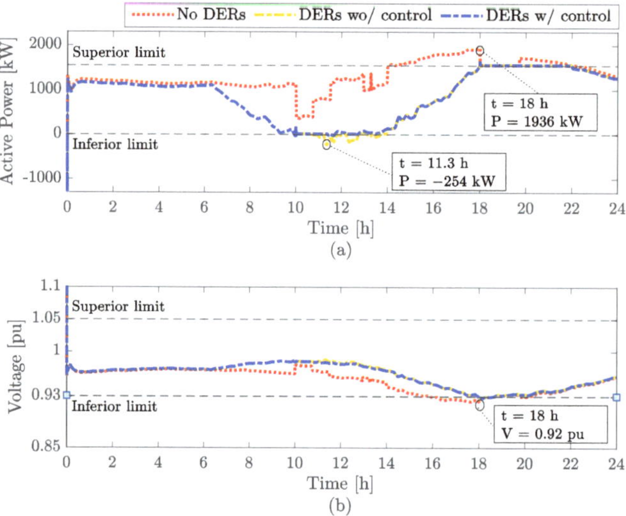

Figure 10. (**a**,**b**) Power and RMS voltage at PCC of the minigrid with FCS controllable in three situations: without DERs (scenario 7), all DERs are non-dispatchable (scenario 8), and DERs are dispatchable and not dispatchable (scenario 9).

One of the solutions to not exceed the upper limit in this scenario would be to extend the microgrid control period between 2 p.m. and 10 p.m. However, this would require a lot of energy from the BESS present at the stations since they would operate 4 h longer with reduced grid demand. Figure 11 shows the α_P^{MG} coefficients for the three scenarios evaluated in Figure 10. It is possible to observe that the index α_P^{MG} is higher for the case without DERs, reaching a maximum of one. As stations reduce their demand according to the magnitude of α_P^{MG}, the case without DERs ceases the demand from the FCS, requiring the BESS to provide all the power necessary to recharge the vehicles.

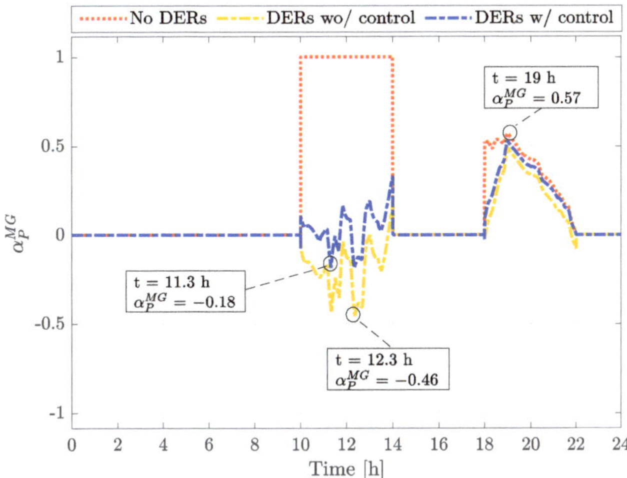

Figure 11. Coefficient values α_P^{MG} with FCS controllable in three situations: without DERs, all DERs are non-dispatchable, and DERs are dispatchable and not dispatchable.

In the scenario with uncontrolled DERs (scenario 8 of Table 5) presented in Figure 10a, there is no violation of the upper limit at any time throughout the day. Therefore, there is also no violation of the lower voltage limit in the PCC of the microgrid, which always remains within the limits considered appropriate according to Table 7. However, it was not possible to maintain the power in the PCC above the lower limit between 10 a.m. and 2 p.m.

Also, analyzing Figure 11, the coefficient α_P remained at high negative values, peaking at -0.46. As a result, dispatchable DERs must absorb energy at 46% of their maximum capacity, and charging stations must increase their demand by 46%. Considering a maximum demand increase of 20% in an FCS, the charging station limits its energy absorption and, therefore, there will be deviations from the limit reference lower than the active power in the PCC of the minigrid.

In the scenario with DERs with control (scenario 9 of Table 5), there is no violation of the lower and upper limits, maintaining the voltage within the limits considered adequate according to Table 7. Confront the values of α_P^{MG} in Figure 11 of the scenarios with DERs without control and with control, and it is observed that the minimum value of α_P^{MG} is above -0.2 (-20%), which means that FCSs can increase their demand without exceeding the imposed limits. Furthermore, at some times during the interval from 10 a.m. to 2 p.m., the α_P values become positive, which means that the FCSs reduce their active power demand while the DERs' dispatchables begin to inject active power into the grid, even in a condition of high generation by non-dispatchable DERs.

Figure 10 (DERs with control) shows that only with the simultaneous control of the DERs and FCSs was it possible to maintain the PCC power between the lower and upper limits. In this way, the PCC voltage remained at appropriate values.

By controlling only the DERs, the control is not able to keep the power in the PCC below the upper limit, as shown in Figure 9 (DERs w/control). In this condition, the PCC power exceeded the upper limit by 28%, taking the voltage in the PCC to a precarious level. By controlling only the FCS, the control is not able to maintain the power in the PCC above the lower limit, as shown in Figure 10 (DERs wo/control). In this condition, the PCC power exceeded the lower limit by 254 kW, dispatching power to the upstream network.

5. Discussion of Results

Figure 8 presents the scenarios without FCSs. It shows that the insertion of DERs without energy storages or without sufficient load at times of high generation causes an increase in voltage levels, especially at the DERs' connection points. Without controlling the DERs' power dispatch, there is a high power dispatch from the minigrid to the upstream network, which causes an increase in voltage in the PCC. For the case studied, this value was 1 MW of power in a minigrid unprepared for reverse power flow. With the control of dispatchable DERs, this value is reduced but still significant for the reverse power flow.

Increased energy consumption reduces the possibility of overvoltage on the electric grid. However, unrestrained insertion without load control can cause feeders to overload at times of high demand. Figure 9 presents scenarios with increased energy consumption through FCSs. Due to the incompatibility between generation and demand, the overload of the electrical system caused by FCSs shows that increasing consumption without coordination is not enough to avoid energy-quality problems in the electrical grid. In this case study, the insertion of FCSs into the electrical grid without coordination caused an overload of 28% in the PCCs of the minigrid during periods of high energy demand.

Figure 10 shows that the insertion of FCSs without DERs substantially increases the overload periods. Figure 11 shows that, in the case study of this work, there is an excessive use of BESSs of FCSs in a grid without distributed generation. Inserting DERs alleviates periods of overload. However, controlling only the loads (FCSs) does not prevent the reverse power flow with the upstream electrical grid, violated at 344 kW for the case study. Only the control of the DERs simultaneously and coordinated with controllable loads such as the FCSs allows ancillary services such as valley filling to be carried out, avoiding reverse power flow and peak shaving and avoiding overload on the PCC of the minigrid.

6. Conclusions

The proposed method is suitable for the simultaneous coordinated control of fast charging stations (FCSs) and distributed energy resources (DERs). By controlling only the DERs of the minigrid, there is a violation of the upper power level in the minigrid (MG) point of common coupling (PCC), taking the voltage to precarious operating levels (below 0.93 pu). Controlling only the FCSs, there is a dispatch of 254 kW between the MG and the upstream electrical grid, undesirable in a zero-flow condition. With the proposed method, it is possible to guarantee the reliability of grid power quality parameters.

Low-voltage microgrids with controllable loads, such as lower-power vehicle chargers or builds with energy storage, can also operate with the proposed control. The algorithm is interesting in microgrids with high non-dispatchable generation, being able to increase the hosting capacity with the insertion of controllable loads in the system, such as charging stations equipped with energy storages, avoiding overvoltage levels in the hosting capacity PCC of this microgrid at times of high generation without impacting voltage levels at times of greater demand. The purpose of this work performs the ancillary services of valley filling and peak shaving, maintaining voltage and power levels within appropriate values and contributing to the simultaneous control studies of DERs and FCS. This type of control allows for greater operational flexibility for the distribution system operator. With this, large cities can operate with the addition of DERs and FCSs, avoiding changes to the network infrastructure, reducing costs for the system operator and consumers, and maintaining the safety and reliability of the electrical power system.

Author Contributions: Conceptualization, all authors; methodology, all authors; software, D.A.d.L.B. and D.I.B.; validation, D.A.d.L.B.; formal analysis, D.A.d.L.B.; investigation, D.A.d.L.B.; resources, I.A.P. and D.I.B.; data curation, D.A.d.L.B.; writing—original draft preparation, D.A.d.L.B.; writing—review and editing, all authors; visualization, all authors; supervision, I.A.P. and D.I.B.; project administration, I.A.P.; funding acquisition, I.A.P. All authors have read and agreed to the published version of the manuscript.

Funding: This research received no external funding.

Institutional Review Board Statement: Not applicable.

Informed Consent Statement: Not applicable.

Data Availability Statement: Data are contained within the article.

Acknowledgments: This work was carried out with the support of the Coordination for the Improvement of Higher Education Personnel—Brazil (CAPES) through the Academic Excellence Program (PROEX), in part by Conselho Nacional de Desenvolvimento Científico e Tecnológico (CNPq), and in part by the Minas Gerais state government agency Fundação de Amparo à Pesquisa do Estado de Minas Gerais (FAPEMIG).

Conflicts of Interest: The authors declare no conflicts of interest.

Abbreviations

ANEEL	Agência Nacional de Energia Elétrica
BESS	battery energy storage system
DER	distributed energy resource
DSO	distribution system operator
EV	electric vehicle
FCS	fast charging station
FCSCC	fast charging station central controller
LV	low voltage
MG	minigrid
MGCC	minigrid central controller
MISO	multiple inputs single output
MV	medium voltage
PBC	power-based control
PCC	point of common coupling
PI	proportional-integral
PRODIST	Procedimento de Distribuição de Energia Elétrica no Sistema Elétrico Nacional
SoC	state of charge
μG	microgrid
μGCC	microgrid central controller
V2G	vehicle-to-grid

References

1. National Geographic. Air Pollution. 2023. Available online: https://education.nationalgeographic.org/resource/air-pollution/ (accessed on 12 January 2024).
2. Air Quality Life Index (AQLI). South Korea Analysis: Air Pollution Cuts Lives Short by More than a Year. 2019. Available online: https://aqli.epic.uchicago.edu/news/south-korea-analysis-air-pollution-cuts-lives-short-by-more-than-a-year/ (accessed on 12 January 2024).
3. United States Environmental Protection Agency (EPA). Summary of the Clean Air Act. 2023. Available online: https://www.epa.gov/laws-regulations/summary-clean-air-act (accessed on 13 January 2024).
4. European Environment Agency. Emissions of the Main Air Pollutants in Europe. 2023. Available online: https://eea.europa.eu/en/analysis/indicators/emissions-of-the-main-air (accessed on 12 January 2024).
5. United States Environmental Protection Agency (EPA). Carbon Pollution from Transportation. 2022. Available online: https://www.epa.gov/transportation-air-pollution-and-climate-change/carbon-pollution-transportation (accessed on 14 January 2024).
6. The New York Times. For Electric Car Owners, 'Range Anxiety' Gives Way to 'Charging Time Trauma'. 2017. Available online: https://www.nytimes.com/2017/10/05/automobiles/wheels/electric-cars-charging.html (accessed on 12 January 2024).
7. U.S. News & World Report. 11 Reasons People Don't Buy Electric Cars (and Why They're Wrong). 2023. Available online: https://cars.usnews.com/cars-trucks/advice/why-people-dont-buy-electric-cars (accessed on 16 January 2024).
8. Kantar. Understanding Consumer Attitudes Towards Electric Vehicles. 2023. Available online: https://www.kantar.com/inspiration/research-services/understanding-consumer-attitudes-towards-electric-vehicles-pf (accessed on 16 January 2024).

9. Federal Highway Administration—U.S. Departament of Transportation. National Electric Vehicle Infrastructure Formula Program. 2022. Available online: https://www.fhwa.dot.gov/environment/alternative_fuel_corridors/resources/nprm_evcharging_unofficial.pdf (accessed on 17 January 2024).
10. Alternative Fuels Data Center. Alternative Fueling Station Counts by State. 2024. Available online: https://afdc.energy.gov/stations/states (accessed on 17 January 2024).
11. International Energy Agency (IEA). Trends in Charging Infrastructure. 2023. Available online: https://www.iea.org/reports/global-ev-outlook-2023/trends-in-charging-infrastructure (accessed on 17 January 2024).
12. Statista. Estimated Number of Plug-In Electric Vehicles in Use in Selected Countries as of 2022. 2022. Available online: https://www.statista.com/statistics/244292/number-of-electric-vehicles-by-country/ (accessed on 17 January 2024).
13. Nafi, I.M.; Tabassum, S.; Hassan, Q.R.; Abid, F. Effect of Electric Vehicle Fast Charging Station on Residential Distribution Network in Bangladesh. In Proceedings of the 2021 5th International Conference on Electrical Engineering and Information Communication Technology (ICEEICT), Dhaka, Bangladesh , 18–20 November 2021; pp. 1–5. [CrossRef]
14. Alshareef, S.M. Voltage Sag Assessment, Detection, and Classification in Distribution Systems Embedded With Fast Charging Stations. *IEEE Access* **2023**, *11*, 89864–89880. [CrossRef]
15. Tovilović, D.M.; Rajaković, N.L. The simultaneous impact of photovoltaic systems and plug-in electric vehicles on the daily load and voltage profiles and the harmonic voltage distortions in urban distribution systems. *Renew. Energy* **2015**, *76*, 454–464. [CrossRef]
16. Lamedica, R.; Geri, A.; Gatta, F.M.; Sangiovanni, S.; Maccioni, M.; Ruvio, A. Integrating Electric Vehicles in Microgrids: Overview on Hosting Capacity and New Controls. *IEEE Trans. Ind. Appl.* **2019**, *55*, 7338–7346. [CrossRef]
17. Caro, L.M.; Ramos, G.; Rauma, K.; Rodriguez, D.F.C.; Martinez, D.M.; Rehtanz, C. State of Charge Influence on the Harmonic Distortion From Electric Vehicle Charging. *IEEE Trans. Ind. Appl.* **2021**, *57*, 2077–2088. [CrossRef]
18. Chudy, A.; Hołyszko, P.; Mazurek, P. Fast Charging of an Electric Bus Fleet and Its Impact on the Power Quality Based on On-Site Measurements. *Energies* **2022**, *15*, 5555. [CrossRef]
19. Nguyen, H.T.; Hosani, K.A.; Al-Sumaiti, A.S.; Nguyen, T.H.; Jaafari, K.A.A.; Alsawalhi, J.Y.; El Moursi, M.S. Fast Harmonic Rejecting Control Design to Enable Active Support of Charging Stations to Micro-Grids Under Distortion. *IEEE Trans. Transp. Electrif.* **2023**, *9*, 4132–4146. [CrossRef]
20. Steen, D.; Tuan, L.A.; Carlson, O.; Bertling, L. Assessment of Electric Vehicle Charging Scenarios Based on Demographical Data. *IEEE Trans. Smart Grid* **2012**, *3*, 1457–1468. [CrossRef]
21. Awadallah, M.A.; Singh, B.N.; Venkatesh, B. Impact of EV Charger Load on Distribution Network Capacity: A Case Study in Toronto. *Can. J. Electr. Comput. Eng.* **2016**, *39*, 268–273. [CrossRef]
22. Ramadan, H.; Ali, A.; Farkas, C. Assessment of plug-in electric vehicles charging impacts on residential low voltage distribution grid in Hungary. In Proceedings of the 2018 6th International Istanbul Smart Grids and Cities Congress and Fair (ICSG), Istanbul, Turkey, 25–26 April 2018; pp. 105–109. [CrossRef]
23. Coban, H.H.; Lewicki, W.; Sendek-Matysiak, E.; Łosiewicz, Z.; Drożdż, W.; Miśkiewicz, R. Electric Vehicles and Vehicle–Grid Interaction in the Turkish Electricity System. *Energies* **2022**, *15*, 8218. [CrossRef]
24. European Council—Council of the European Union. Climate Change Summit COP26. 2021. Available online: https://www.consilium.europa.eu/en/policies/climate-change/paris-agreement/cop26/ (accessed on 21 January 2024).
25. U.S. Department of Energy. Homeowner's Guide to the Federal Tax Credit for Solar Photovoltaics. 2023. Available online: https://www.energy.gov/eere/solar/homeowners-guide-federal-tax-credit-solar-photovoltaics (accessed on 21 January 2024).
26. Ministério do Desenvolvimento, Indústria, Comércio e Serviços—Brasil. Decreto do Governo Garante Isenção Fiscal Para Semicondutores e Inclui Energia Solar em Benefício. 2023. Available online: https://gov.br/mdic/pt-br/assuntos/noticias/2023/marco/decreto-do-governo-garante-isencao-fiscal-para-semicondutores-e-inclui-energia-solar-em-beneficio (accessed on 21 January 2024).
27. Reuters. China Sets 2022 Renewable Power Subsidy at $607 mln. 2021. Available online: https://www.reuters.com/business/energy/china-sets-2022-renewable-power-subsidy-607-mln-2021-11-16/ (accessed on 21 January 2024).
28. European Comission. EU Renewable Energy Financing Mechanism. 2023. Available online: https://energy.ec.europa.eu/topics/renewable-energy/financing/eu-renewable-energy-financing-mechanism_en (accessed on 21 January 2024).
29. Hossain, E.; Tür, M.R.; Padmanaban, S.; Ay, S.; Khan, I. Analysis and Mitigation of Power Quality Issues in Distributed Generation Systems Using Custom Power Devices. *IEEE Access* **2018**, *6*, 16816–16833. [CrossRef]
30. Zenhom, Z.M.; Aleem, S.H.E.A.; Zobaa, A.F.; Boghdady, T.A. A Comprehensive Review of Renewables and Electric Vehicles Hosting Capacity in Active Distribution Networks. *IEEE Access* **2024**, *12*, 3672–3699. [CrossRef]
31. Falahi, M.; Chou, H.M.; Ehsani, M.; Xie, L.; Butler-Purry, K.L. Potential Power Quality Benefits of Electric Vehicles. *IEEE Trans. Sustain. Energy* **2013**, *4*, 1016–1023. [CrossRef]
32. Pahasa, J.; Ngamroo, I. PHEVs Bidirectional Charging/Discharging and SoC Control for Microgrid Frequency Stabilization Using Multiple MPC. *IEEE Trans. Smart Grid* **2015**, *6*, 526–533. [CrossRef]
33. Bayat, M.; Sheshyekani, K.; Rezazadeh, A. A Unified Framework for Participation of Responsive End-User Devices in Voltage and Frequency Control of the Smart Grid. *IEEE Trans. Power Syst.* **2015**, *30*, 1369–1379. [CrossRef]
34. Weckx, S.; Driesen, J. Load Balancing with EV Chargers and PV Inverters in Unbalanced Distribution Grids. *IEEE Trans. Sustain. Energy* **2015**, *6*, 635–643. [CrossRef]

35. Akhavan-Rezai, E.; Shaaban, M.F.; El-Saadany, E.F.; Karray, F. Managing Demand for Plug-in Electric Vehicles in Unbalanced LV Systems With Photovoltaics. *IEEE Trans. Ind. Inform.* **2017**, *13*, 1057–1067. [CrossRef]

36. Rana, R.; Singh, M.; Mishra, S. Design of Modified Droop Controller for Frequency Support in Microgrid Using Fleet of Electric Vehicles. *IEEE Trans. Power Syst.* **2017**, *32*, 3627–3636. [CrossRef]

37. Datta, U.; Kalam, A.; Shi, J.; Li, J. Electric Vehicle Charging Station for Providing Primary Frequency Control in Microgrid. In Proceedings of the 2019 14th IEEE Conference on Industrial Electronics and Applications (ICIEA), Xi'an, China, 19–21 June 2019; pp. 2440–2444. [CrossRef]

38. Ahmad, F. Analysis of Microgrids for Plug-in Hybrid Electric Vehicles Charging Stations. Ph.D. Thesis, Department of Electrical Engineering, Faculty of Engineering & Technology, Aligarh Muslim University, Aligarh, India, 2019.

39. Rücker, F.; Merten, M.; Gong, J.; Villafáfila-Robles, R.; Schoeneberger, I.; Sauer, D.U. Evaluation of the Effects of Smart Charging Strategies and Frequency Restoration Reserves Market Participation of an Electric Vehicle. *Energies* **2020**, *13*, 3112. [CrossRef]

40. da Silva, E.C.; Melgar-Dominguez, O.D.; Romero, R. Simultaneous Distributed Generation and Electric Vehicles Hosting Capacity Assessment in Electric Distribution Systems. *IEEE Access* **2021**, *9*, 110927–110939. [CrossRef]

41. Lai, C.S.; Chen, D.; Xu, X.; Taylor, G.A.; Pisica, I.; Lai, L.L. Operational Challenges to Accommodate High Penetration of Electric Vehicles: A Comparison between UK and China. In Proceedings of the 2021 IEEE 15th International Conference on Compatibility, Power Electronics and Power Engineering (CPE-POWERENG), Florence, Italy, 14–16 July 2021; pp. 1–7. [CrossRef]

42. Kasturi, K.; Nayak, C.; Nayak, M. Photovoltaic and Electric Vehicle-to-Grid Strategies for Peak Load Shifting in Low Voltage Distribution System Under Time of Use Grid Pricing. *Iran. J. Sci. Technol. Trans. Electr. Eng.* **2021**, *45*, 789–801. [CrossRef]

43. Khazali, A.; Rezaei, N.; Saboori, H.; Guerrero, J.M. Using PV systems and parking lots to provide virtual inertia and frequency regulation provision in low inertia grids. *Electr. Power Syst. Res.* **2022**, *207*, 107859. [CrossRef]

44. Kiani, S.; Sheshyekani, K.; Dagdougui, H. An Extended State Space Model for Aggregation of Large-Scale EVs Considering Fast Charging. *IEEE Trans. Transp. Electrif.* **2023**, *9*, 1238–1251. [CrossRef]

45. Morstyn, T.; Hredzak, B.; Agelidis, V.G. Control Strategies for Microgrids With Distributed Energy Storage Systems: An Overview. *IEEE Trans. Smart Grid* **2018**, *9*, 3652–3666. [CrossRef]

46. Gao, X.; De Carne, G.; Andresen, M.; Brüske, S.; Pugliese, S.; Liserre, M. Voltage-Dependent Load-Leveling Approach by Means of Electric Vehicle Fast Charging Stations. *IEEE Trans. Transp. Electrif.* **2021**, *7*, 1099–1111. [CrossRef]

47. Zahedmanesh, A.; Muttaqi, K.M.; Sutanto, D. Active and Reactive Power Control of PEV Fast Charging Stations Using a Consecutive Horizon-Based Energy Management Process. *IEEE Trans. Ind. Inform.* **2021**, *17*, 6742–6753. [CrossRef]

48. Radu, A.T.; Eremia, M.; Toma, L. Use of Battery Storage Systems in EV Ultrafast Charging Stations for Load Spikes Mitigation. *Sci. Bull.-Univ. Politeh. Buchar.* **2021**, *83*, 283–294.

49. Gao, X.; De Carne, G.; Langwasser, M.; Liserre, M. Online Load Control in Medium Voltage Grid by Means of Reactive Power Modification of Fast Charging Station. In Proceedings of the 2019 IEEE Milan PowerTech, Milan, Italy, 23–27 June 2019; pp. 1–6. [CrossRef]

50. Gjelaj, M.; Hashemi, S.; Traeholt, C.; Andersen, P.B. Grid integration of DC fast-charging stations for EVs by using modular li-ion batteries. *IET Gener. Transm. Distrib.* **2018**, *12*, 4368–4376. [CrossRef]

51. Yang, H.; Pan, H.; Luo, F.; Qiu, J.; Deng, Y.; Lai, M.; Dong, Z.Y. Operational Planning of Electric Vehicles for Balancing Wind Power and Load Fluctuations in a Microgrid. *IEEE Trans. Sustain. Energy* **2017**, *8*, 592–604. [CrossRef]

52. Cheng, Q.; Chen, L.; Sun, Q.; Wang, R.; Ma, D.; Qin, D. A smart charging algorithm based on a fast charging station without energy storage system. *CSEE J. Power Energy Syst.* **2021**, *7*, 850–861. [CrossRef]

53. Lymperopoulos, I.; Qureshi, F.A.; Bitlislioglu, A.; Poland, J.; Zanarini, A.; Mercangoez, M.; Jones, C. Ancillary Services Provision Utilizing a Network of Fast-Charging Stations for Electrical Buses. *IEEE Trans. Smart Grid* **2020**, *11*, 665–672. [CrossRef]

54. Yong, J.Y.; Ramachandaramurthy, V.K.; Tan, K.M.; Mithulananthan, N. Bi-directional electric vehicle fast charging station with novel reactive power compensation for voltage regulation. *Int. J. Electr. Power Energy Syst.* **2015**, *64*, 300–310. [CrossRef]

55. Yuan, H.; Wei, G.; Zhu, L.; Zhang, X.; Zhang, H.; Luo, Z.; Hu, J. Optimal scheduling for micro-grid considering EV charging–swapping–storage integrated station. *IET Gener. Transm. Distrib.* **2020**, *14*, 1127–1137. [CrossRef]

56. Liu, Y.; Wang, Y.; Li, Y.; Gooi, H.B.; Xin, H. Multi-Agent Based Optimal Scheduling and Trading for Multi-Microgrids Integrated With Urban Transportation Networks. *IEEE Trans. Power Syst.* **2021**, *36*, 2197–2210. [CrossRef]

57. Ton, D.; Reilly, J. Microgrid Controller Initiatives: An Overview of R&D by the U.S. Department of Energy. *IEEE Power Energy Mag.* **2017**, *15*, 24–31. [CrossRef]

58. Caldognetto, T.; Buso, S.; Tenti, P.; Brandao, D.I. Power-Based Control of Low-Voltage Microgrids. *IEEE J. Emerg. Sel. Top. Power Electron.* **2015**, *3*, 1056–1066. [CrossRef]

59. Brandao, D.I.; Caldognetto, T.; Marafão, F.P.; Simões, M.G.; Pomilio, J.A.; Tenti, P. Centralized Control of Distributed Single-Phase Inverters Arbitrarily Connected to Three-Phase Four-Wire Microgrids. *IEEE Trans. Smart Grid* **2017**, *8*, 437–446. [CrossRef]

60. Brandao, D.I.; de Araújo, L.S.; Caldognetto, T.; Pomilio, J.A. Coordinated control of three- and single-phase inverters coexisting in low-voltage microgrids. *Appl. Energy* **2018**, *228*, 2050–2060. [CrossRef]

61. Araujo, L.S.; Brandao, D.I.; Silva, S.M.; Cardoso Filho, B.J. Reactive Power Support in Medium Voltage Networks by Coordinated Control of Distributed Generators in Dispatchable Low-Voltage Microgrid. In Proceedings of the 2019 IEEE 28th International Symposium on Industrial Electronics (ISIE), Vancouver, BC, Canada, 12–14 June 2019; pp. 2603–2608. [CrossRef]

62. Brandao, D.I.; Araujo, L.S.; Alonso, A.M.S.; dos Reis, G.L.; Liberado, E.V.; Marafão, F.P. Coordinated Control of Distributed Three- and Single-Phase Inverters Connected to Three-Phase Three-Wire Microgrids. *IEEE J. Emerg. Sel. Top. Power Electron.* **2020**, *8*, 3861–3877. [CrossRef]
63. Brandao, D.I.; Santos, R.P.d.; Silva, W.W.A.G.; Oliveira, T.R.; Donoso-Garcia, P.F. Model-Free Energy Management System for Hybrid Alternating Current/Direct Current Microgrids. *IEEE Trans. Ind. Electron.* **2021**, *68*, 3982–3991. [CrossRef]
64. Ferreira, D.M.; Brandao, D.I.; Bergna-Diaz, G.; Tedeschi, E.; Silva, S.M. Distributed Control Strategy for Low-Voltage Three-Phase Four-Wire Microgrids: Consensus Power-Based Control. *IEEE Trans. Smart Grid* **2021**, *12*, 3215–3231. [CrossRef]
65. dos Reis, G.; Liberado, E.; Marafão, F.; Sousa, C.; Silva, W.; Brandao, D. Model-Free Power Control for Low-Voltage AC Dispatchable Microgrids with Multiple Points of Connection. *Energies* **2021**, *14*, 6390. [CrossRef]
66. Reis, G.L.; Brandao, D.I.; Oliveira, J.H.; Araujo, L.S.; Cardoso Filho, B.J. Case Study of Single-Controllable Microgrid: A Practical Implementation. *Energies* **2022**, *15*, 6400. [CrossRef]
67. Callegari, J.M.S.; Pereira, H.A.; Brandao, D.I. Voltage Support and Selective Harmonic Current Compensation in Advanced AC Microgrids. *IEEE Trans. Ind. Appl.* **2023**, *59*, 4880–4892. [CrossRef]
68. Soares Callegari, J.M.; Araujo, L.S.; De Lisboa Brandão, D.A.; Cardoso Filho, B.J.; Brandao, D.I. Centralized Strategy Incorporating Multiple Control Actions Applied to Advanced Microgrids. In Proceedings of the 2023 IEEE 8th Southern Power Electronics Conference (SPEC), Florianopolis, Brazil, 26–29 November 2023; pp. 1–7. [CrossRef]
69. Federal Communications Commission. FCC Finds U.S. Broadband Deployment Not Keeping Pace. 2015. Available online: https://www.fcc.gov/document/fcc-finds-us-broadband-deployment-not-keeping-pace-0 (accessed on 22 January 2024).
70. Ramos, G.V.; Parreiras, T.M.; Brandão, D.A.d.L.; Silva, S.M.; Filho, B.d.J.C. A Zero Harmonic Distortion Grid-Forming Converter for Islanded Microgrids. In Proceedings of the 2023 IEEE Industry Applications Society Annual Meeting (IAS), Nashville, TN, USA, 29 October–2 November 2023; pp. 1–8. [CrossRef]
71. Brandao, D.A.L.; Parreiras, T.M.; Callegari, J.M.S.; Pires, I.A.; Brandao, D.I.; Cardoso Filho, B.D.J. Power-based Control for Microgrids and Extreme Fast Charging Stations with Low-Harmonic Current Content in the Medium Voltage Power Grid. In Proceedings of the 2023 IEEE Transportation Electrification Conference & Expo (ITEC), Detroit, MI, USA, 21–23 June 2023; pp. 1–6. [CrossRef]
72. TeslaTap. Supercharger Superguide. 2023. Available online: https://teslatap.com/articles/supercharger-superguide/ (accessed on 21 October 2023).
73. Induscabos—Condutores Elétricos. Electrical Parameters—Medium Voltage Cables. 2018. Available online: https://www.induscabos.com.br/wp-content/uploads/2018/03/induscabos-parametros-eletricos-cabos-de-media-tensao.pdf (accessed on 21 October 2023).
74. Pedrotti, R.F. Simulação Termo Energética de um Supermercado. Master's Thesis, Energy Engineering, Universidade Federal do Rio Grande do Sul, Porto Alegre, Brazil, 2015.
75. Origin. Origin1500 1.5kW Solar System Panel Specifications. 2023. Available online: https://www.originenergy.com.au/wp-content/uploads/Origin1500B_Solar_Panel_Brochure.pdf (accessed on 21 October 2023).
76. Newaz, A.; Ospina, J.; Faruque, M.O. Controller Hardware-in-the-Loop Validation of a Graph Search Based Energy Management Strategy for Grid-Connected Distributed Energy Resources. *IEEE Trans. Energy Convers.* **2020**, *35*, 520–528. [CrossRef]
77. Rafique, S.; Hossain, M.J.; Nizami, M.S.H.; Irshad, U.B.; Mukhopadhyay, S.C. Energy Management Systems for Residential Buildings With Electric Vehicles and Distributed Energy Resources. *IEEE Access* **2021**, *9*, 46997–47007. [CrossRef]
78. Ahmad, S.; Shafiullah, M.; Ahmed, C.B.; Alowaifeer, M. A Review of Microgrid Energy Management and Control Strategies. *IEEE Access* **2023**, *11*, 21729–21757. [CrossRef]
79. Agência Nacional de Energia Elétrica. Procedimentos de Distribuição de Energia Elétrica No Sistema Elétrico Nacional—PRODIST—Módulo 8—Qualidade do Fornecimento de Energia Elétrica. 2021. Available online: https://www2.aneel.gov.br/cedoc/aren2021956_2_7.pdf (accessed on 24 October 2023).

Review

A Review of Perspectives on Developing Floating Wind Farms

Mohamed Maktabi and Eugen Rusu *

Department of Mechanical Engineering, 'Dunarea de Jos' University of Galati, 800008 Galati, Romania;
mm307@student.ugal.ro
* Correspondence: eugen.rusu@ugal.ro

Abstract: Floating wind is becoming an essential part of renewable energy, and so highlighting perspectives of developing floating wind platforms is very important. In this paper, we focus on floating wind concepts and projects around the world, which will show the reader what is going on with the projects globally, and will also provide insight into the concepts and their corresponding related aspects. The main aim of this work is to classify floating wind concepts in terms of their number and manufacturing material, and to classify the floating wind projects in terms of their power capacity, their number, character (if they are installed or planned) and the corresponding continents and countries where they are based. We will classify the corresponding additional available data that corresponds to some of these projects, with reference to their costs, wind speeds, water depths, and distances to shore. In addition, the floating wind global situation and its corresponding aspects of relevance will be also covered in detail throughout the paper.

Keywords: renewable energy; floating platforms; wind; marine environment; sustainable development

Citation: Maktabi, M.; Rusu, E. A Review of Perspectives on Developing Floating Wind Farms. *Inventions* **2024**, *9*, 24. https://doi.org/10.3390/inventions9020024

Academic Editors: Om P. Malik and Alessandro Dell'Era

Received: 30 December 2023
Revised: 17 February 2024
Accepted: 19 February 2024
Published: 21 February 2024

1. Introduction

Floating wind is currently a leading candidate for renewable energy in many countries around the world, as governments and companies investing large financial resources into developing floating wind projects. The purpose of this paper is to present all the corresponding projects in the world, their implemented wind turbine types, and corresponding concepts, as this will make a very significant contribution to understanding the floating wind situation around the world.

Renewable energy has become essential to respond to the increasing world population, and its corresponding demand for energy. It is also seen as essential to stop the reliance on fuels and eliminate pollution and climate change [1].

Renewable energy is also a way to prevent countries with oil and gas resources from becoming economically and politically dominant over countries that lack these resources [2].

Unlike oil and gas energy, renewable energy is carbon-free and limitless, which makes it the perfect solution to both climate change and population growth [2].

While onshore wind energy is currently the cheapest source of renewable energy, it has weaker and more turbulent wind speeds, compared to its offshore counterpart, which is anticipated to dominate in the years to come. Floating wind projects are therefore expected to be constructed in high water-depth areas [1].

From this perspective, the European Union will need 450 GW of offshore wind by 2050 to achieve its complete decarbonization, a substantial increase on its current corresponding power capacity of 25 GW [3].

The European Union must develop 150 GW of floating wind to be carbon neutral by 2050, which is likely to happen, both due to the available financial resources and the substantial efforts of the specialized floating wind companies [4].

Europe currently has 318 MW of floating wind from 34 corresponding concepts, compared to the rest of the world, which has 32 MW power capacity from 16 concepts. Floating wind cumulative capacity is currently led by the European Union, whose future investments will facilitate its industrialization process and reduce the capital expenditures (CAPEX) of future floating wind projects [4].

In 2030, France plans to have 750 MW of floating wind power capacity, the UK plans to have 1 GW, Norway plans to have 1.5 GW (or 3 GW [5]), and Portugal plans to have 275 MW [6], as compared to current floating wind capacities of 114 MW in France, 80 MW in the UK, 95 MW in Norway, and 30 MW in Portugal. The US currently has 12 MW, and Japan has a 20 MW corresponding power capacity [4].

Floating wind projects will be implemented in areas where their offshore bottom-fixed counterparts are not feasible, due to their corresponding negative assembly impact on the marine environment and limited water-depth capacities. Floating wind projects have exceeding water-depth capacities and have less environmental impact because of their early assembly in the ports. Further, floating wind turbines are on their way toward industrialization, making them cost competitive as compared to their offshore bottom-fixed counterpart [4]. Offshore bottom-fixed turbines are generally limited to water depths of roughly 100 m, while their floating counterparts can be extended to kilometers of water depths.

The conversion of both the existing European infrastructures of oil and gas and bottom-fixed offshore wind will contribute to Europe becoming the world's floating wind leader. Europe is currently planning to take the lead in the floating wind supply chain areas, which will produce tremendous job creation in field areas that include electrical cabling, mooring, and installation. The outcome will be significant when the floating wind global market obtains 18,000 GW in the future [4].

The floating wind levelized cost of energy (LCOE) will be 250 euros/MWh when the floating wind capacity reaches 0.5 GW and will drop to 50 euros when the floating wind capacity approaches 4 GW in 2030 [7].

Romania has a current installed onshore wind capacity of 3 GW, but it lacks a corresponding electrical infrastructure in sea areas, which is currently the main obstacle to implementing floating wind projects in the country [3], although ongoing efforts are being made in this regard [8]. The solution to the lack of a corresponding offshore electrical infrastructure in Romania is possibly the implementation of Power-to-X technology, which will be used to convert the produced floating wind electrical power (mainly into hydrogen and compressed air) and eliminate the need for a tremendous electrical infrastructure in the sea region.

Figure 1 shows the most widely used bottom-fixed and floating wind turbine concepts, which are bottom-fixed monopile, floating wind spar, semi-submersible, and TLP platforms. More details about these concepts, including their advantages and disadvantages are presented in Section 4 and Table 1.

Table 1. Advantages and disadvantages of the most widely used floating wind concepts. Table data processed by the authors mainly based on the information presented in [9].

Floating Wind Turbine Types	Advantages	Disadvantages
Spar-buoy	Most simple manufacturing, convenient stability	Relatively lower water depth capacity, compared to TLP
Semi-submersible (one turbine)	Most widely used	More complex and difficult manufacturing, less stable, more expensive

Table 1. *Cont.*

Floating Wind Turbine Types	Advantages	Disadvantages
Semi-submersible (multi-turbine)	Reducing the structural materials and corresponding operation and maintenance costs	Relatively more faults, due to the interaction between the loads coming from different turbines on the same support structure, which influences each floater's operation and stability
Barge	Can be made of concrete (feasible for countries with a lack of steel material)	More complex and difficult to manufacture, less stable, more expensive
TLP	Most stable, highest water depth capacity	Most expensive, difficult to install

The following section will present collected data related to global floating wind concepts and installed and planned projects, including further classifications of some of these projects, which is required because not all projects have further available data.

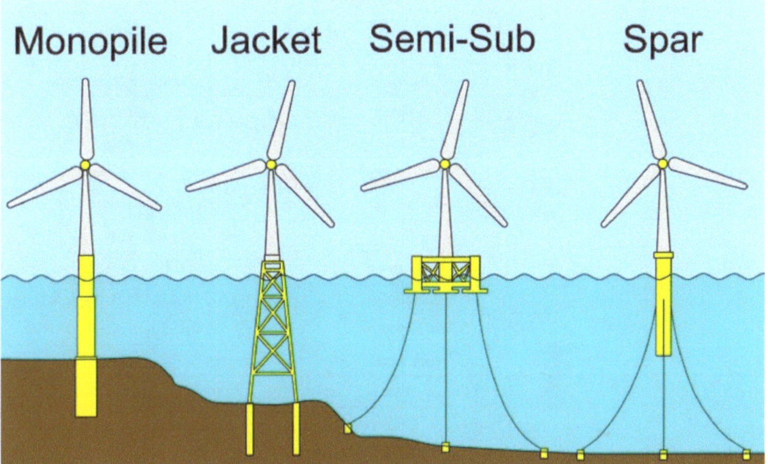

Figure 1. Most widely used bottom-fixed and floating wind support structures in the world, indicated. From left to right: monopile, jacket, semi-submersible, and spar-buoy. The authors processed the figure in accordance with the information presented in [10].

2. Materials and Methods

This section mainly presents floating wind projects and concepts from all around the world, and the data presented in this section is mainly based on the ABSG Consulting report [11] and illustrates the global floating wind situation in 2020.

The following subsection presents the world's floating wind concepts.

2.1. World's Floating Wind Concepts

This subsection presents the world's floating wind concepts.

The following Table 1 presents the main advantages and disadvantages of the world's five most common floating wind types. Refer to Section 4 for further information on each of the presented types.

Table 2 shows the four most frequently used types of floating wind turbines: spar-buoy, semi-submersible, barge, TLP, and multi-turbine type). The table also provides further information that is relevant to the most frequently used corresponding concepts of these wind turbine types, together with their corresponding details.

Table 2. Floating wind concepts applied in the world. Table data processed by the authors on the basis of information presented in [11].

Type	Concept	Designer	Hull Material
Spar-buoy	Hywind	Equinor	Steel or Concrete
	Toda Hybrid Spar	Toda	Steel and Concrete Hybrid
	Fukushima FORWARD Advanced Spar	JMU	Steel
	SeaTwirl	SeaTwirl	Steel
	Stiesdal TetraSpar	Stiesdal	Steel
Semi-submersible	WindFloat	Principle Power	Steel
	Fukushima FORWARD compact semi-submersible	MES	Steel
	Fukushima FORWARD V-shape semi-submersible	MHI	Steel
	VolturnUS	University of Maine	Concrete
	Sea Reed	Naval Energies	Steel, Concrete or Hybrid
	Cobra Semi-Spar	Cobra	Concrete
	OO-Star	Iberdrola	Concrete
	Hexafloat	Saipem	Steel
	Eolink	Eolink	Steel
	SCD nezzy	SCD Technology	Concrete
	Nautilus	NAUTILUS Floating Solutions	Steel
	Tri-Floater	GustoMSC	Steel
	TrussFloat	DOLFINES	Steel
Barge	Ideol Damping Pool Barge	Ideol	Concrete or Steel
	Saitec SATH (Swinging Around Twin Hull)	Saitec	Concrete
Tension leg platform	SBM TLP	SBM Offshore	Steel
	PivotBuoy TLP	X1 Wind	Steel
	Gicon TLP	Gicon	Concrete
	Pelastar TLP	Glosten	Steel
	TLPWind TLP	Iberdrola	Steel
Multi-turbine platform	Hexicon multi-turbine semi-submersible	Hexicon	Steel
	W2Power	EnerOcean	Steel
	Floating Power Plant	Floating Power Plant	Steel

The table shows that there are more concepts of semi-submersible than any of the other wind turbine types, followed by spar-buoy, TLP, barge, and multi-turbine platform. Most of these concepts are made of steel, and a few of concrete. Figure 2 presents an illustrative layout that addresses most of these concepts.

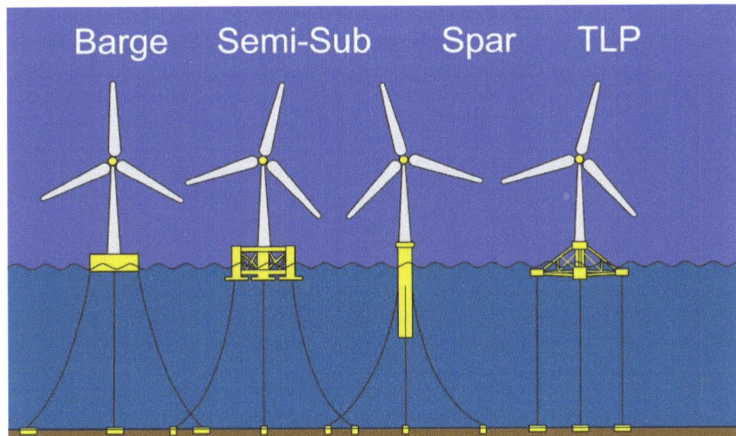

Figure 2. Most popular floating wind support structures in the world. From left to right: barge, semi-submersible, spar-buoy, and TLP. Figure processed by the authors according to the information presented in [10].

The following sub-subsection presents the world's Spar floating wind concepts.

2.1.1. World's Spar-Buoy Floating Wind Concepts

This sub-subsection presents the world's Spar floating wind concepts.

One of the most widely used floating wind spar-buoy concepts is Hywind [12], which is designed by Equinor and constructed of either steel or concrete material. Advanced Spar [13] and Sea Twirl [14], which are also well-known, are developed by JMU and Sea Twirl, respectively, and are both made of steel. Stiesdal Tetra Spar [15] and Fukushima Forward [16,17] are other worth mentioning spar concepts. They are developed by Stiesdal and JMU, respectively, and are both made of steel. Toda Hybrid Spar [18] is also a Spar floating wind concept that is developed by Toda and is a hybrid that is made of a combination of steel and concrete.

The following sub-subsection presents semi-submersible floating wind concepts used across the world.

2.1.2. World's Semi-Submersible Floating Wind Concepts

This sub-subsection presents semi-submersible floating wind concepts used across the world.

One of the most widely used floating wind semi-submersible concepts is Wind Float [19], which is designed by PRINCIPLE-POWER and made of steel. VOLTURNUS [20], OO-Star [21], and Tri-Floater [22] are also well-known floating wind semi-submersible concepts developed by the University of Maine, Iberdrola, and Gusto MSC, respectively. The first two are made of concrete, and the third is made of steel. Cobra Semi-Spar and SCD NEZZY [23] are also semi-submersibles made of concrete that have been developed by Cobra and SCD Technology, respectively. Hexa-Float [24], EOLINK, Nautilus [25], Tri-Floater, and Truss Float [11] are also floating wind semi-submersibles made of steel that have been developed by Saipem, EOLINK, Nautilus floating solutions, Gusto MSC, and DOLFINES, respectively. Sea Reed [26] is also a floating wind semi-submersible floating wind concept that is made of either steel or concrete (or both, in a hybrid) that has been developed by Naval Energies.

The following sub-subsection presents the world's barge, TLP, and multi-turbine floating wind concepts.

2.1.3. World's Barge, TLP, and Multi-Turbine Floating Wind Concepts

This sub-subsection presents the world's barge, TLP, and multi-turbine floating wind concepts.

One of the most widely used barge floating wind concepts is the IDEOL Damping Pool Barge, which was designed by IDEOL and is made of either steel or concrete. SAITEC SATH (Swinging Around Twin Hull) is a Barge floating wind concept that was developed by SAITEC and is made of concrete.

One of the most widely used floating wind TLP concepts is TLPWIND [27] which was designed by Iberdrola and is made of steel. SBM [11], Pivot Buoy [28], and PelaStar are also TLP concepts that are made of steel and were designed by SBM Offshore, X1 Wind, and GLOSTEN, respectively. GICON [29] is a TLP floating wind concept that is made of concrete and was designed by GICON.

One of the most widely used multi-turbine floating wind concepts is the HEXICON multi-turbine semi-submersible [30] which was designed by HEXICON and is made of steel. W2Power [31] and Floating Power Plant [32] are multi-turbine concepts that are made of steel and were developed by EnerOcean and Floating Power Plant, respectively.

The following subsection presents floating wind projects installed across the world in the period 2008–2020.

2.2. World's Installed Floating Wind Projects in the Period 2008–2020

This subsection presents floating wind projects installed across the world in the period 2008–2020.

Table 3 presents all the floating wind projects installed across the world in the period 2008–2020. The following will illustrate the data in this table and classify projects by referring to their contributing countries.

The following illustrates the countries that made the largest contribution to the installation of floating wind projects in the period 2008–2020.

Table 3. All the floating wind projects installed across the world in the period 2008–2020. Table data was processed by the authors on the basis of information presented in [11].

Continent	Country, Location	Year, Turbine—Power	Project Name, Designer
North America	U.S., Maine	2013, Renewegy 20 kW	VolturnUS 1:8, University of Maine
	U.S.—Oregon, WindFloat semi-submersible	2013, 5 × 6 MW	WindFloat Pacific (WFP), Principle Power
Asia	Japan, Goto	2013, Hitachi 2 MW downwind	Kabashima, Toda
	Japan, Fukue	2015, Hitachi 2 MW downwind	Sakiyama, Toda
	Japan, Fukushima	2013, 66 kV—25 MVA Floating Substation	Fukushima FORWARD Phase 1, Fukushima Offshore Wind Consortium
	Japan, Fukushima	2013, Hitachi 2 MW downwind	Fukushima FORWARD Phase 1, Fukushima Offshore Wind Consortium
	Japan, Fukushima	2015, MHI 7 MW	Fukushima FORWARD Phase 2, Fukushima Offshore Wind Consortium
	Japan, Fukushima	2016, Hitachi 5 MW downwind	Fukushima FORWARD Phase 2, Fukushima Offshore Wind Consortium
	Japan, Kitakyushu	2019, Aerodyn SCD 3 MW—2 bladed	Hibiki, Ideol

Table 3. *Cont.*

Continent	Country, Location	Year, Turbine—Power	Project Name, Designer
Europe	Denmark, Lolland	2008, 33 kW	Poseidon 37 Demonstrator [33], Floating Power Plant
	Norway, Karmøy	2009, Siemens 2.3 MW	Hywind Demo, Equinor
	Portugal, Aguçadoura	2011, Vestas 2 MW	WindFloat 1 (WF1), Principle Power
	Portugal, Viana do Castelo	2020, MHI Vestas 3 × 8.4 MW	WindFloat Atlantic (WFA), PrinciplePower
	Sweden, Lysekil	2015, 30 kW Vertical Axis Wind Turbine	SeaTwirl S1, SeaTwirl
	UK, Peterhead	2017, Siemens 5 × 6 MW	Hywind Scotland, Equinor
	UK, Dounreay	2017, N/A 2 × 5 MW	Hexicon Dounreay Trì project [34], Hexicon
	UK, Kincardineshire	2020, MHI Vestas 2 MW (former WF1) & MHI Vestas 5 × 9.5 MW	Kincardine, Principle Power
	Spain, Gran Canaria	2019, 2 × 100 kW twin-rotor	W2Power 1:6 Scale, EnerOcean
	Spain, Santander	2020, Aeolos 30 kW	BlueSATH, Saitec
	France, Le Croisic	2018, Vestas 2 MW	Floatgen, Ideol
	Germany, Baltic Sea	2017, Siemens 2.3 MW	Gicon SOF [35], GICON

Table 3 shows that the UK, Portugal, and Japan made the largest contribution to the installed floating wind projects. The table shows that the UK has a total installed power capacity of 79.5 MW, which is contributed by two floating wind projects. The first one is Kincardine [36], which was developed by Principle Power and has a power capacity of 5 × 9.5 MW. This project also contains an additional 2 MW wind turbine, which was first implemented in the WindFloat 1 (WF1) floating wind project. The UK's second floating wind project is Hywind Scotland [37], which was developed by Equinor and has a power capacity of 5 × 6 MW. The first project in the UK implemented a Vestas wind turbine brand, and the other implemented a Siemens brand.

It is seen from the table that Portugal has a total installed floating wind power capacity of 27.2 MW, which is contributed by two projects. The first project is WindFloat Atlantic (WFA) [38,39], which has a total power capacity of 3 × 8.4 MW, and the second is WindFloat 1 (WF1) [40], which has a total power capacity of 2 MW. They were both developed by Principle Power and implement wind turbines with a Vestas brand.

Japan has a total installed power capacity of 21 MW, which is contributed by seven projects. The main contributors are Fukushima FORWARD Phases I and II [41], which have a total of 14 MW power capacity and were developed by the Fukushima Offshore Wind Consortium. They are followed by the Hibiki [42], Kabashima, and Sakiyama projects. The first project was developed by Ideol, and the other two by Toda. The Hibiki project has a 2 MW power capacity and a downwind Hitachi wind turbine. The Sakiyama floating wind project also implements a 2 MW Hitachi downwind wind turbine.

Other floating wind projects in Europe include the Norwegian Hywind Demo, which has a total power capacity of 3.2 MW, implements a Siemens wind turbine brand, and is developed by Equinor.

The Spanish BlueSATH [43] and W2Power 1:6 scale projects which were developed by Saitec and EnerOcean respectively. The first project has a 30 kW power capacity, and the second has a 2 × 100 kW power capacity. The latter is accompanied by two separate wind turbines that are supported on a single multi-turbine support structure.

The Danish Poseidon 37 Demonstrator floating wind project which has a power capacity of 33 kW, and was developed by Floating Power Plant.

The French Floatgen floating wind project, which has a total power capacity of 2 MW, implements a Vestas wind turbine brand and is developed by Ideol.

The Swedish SeaTwirl S1 floating wind project [44] which is developed by SeaTwirl and has a power capacity of 30 kW. It implements a vertical-axis wind turbine (i.e., the blades rotate around the tower and not around the typical horizontal-axis wind turbine's hub, meaning their rotation axis faces the sky).

On the basis of Table 3, we conclude that Europe is currently the largest contributor to the world's installed floating wind projects.

The following subsection presents the world's planned floating wind projects in the period 2020–2027.

2.3. World's Planned Floating Wind Projects in the Period 2020–2027

This subsection presents the world's planned floating wind projects in the period 2020–2027.

Table 4 shows all the European, North American, and Asian floating wind projects in the world. Further discussions of the data presented in the table are provided in Sections 3 and 4.

Table 4. All the planned floating wind projects in the world in the period 2020–2027. Table data processed by the authors, based on the information presented in [11].

Continent	Country—Location, Floating Substructure Design—Type	Year, Turbine—Power	Project Name, Designer
Europe	Norway—Karmøy, Stiesdal TetraSpar—Spar	2020, Siemens Gamesa 3.6 MW	TetraSpar Demo [45], Stiesdal
	Norway—Haugaland, SeaTwirl Spar	2021, 1 MW Vertical Axis Wind Turbine	SeaTwirl S2 [46], SeaTwirl
	Norway—Snorre & Gullfaks offshore fields, Hywind Spar	2022, Siemens Gamesa 11×8 MW	Hywind Tampen, Equinor [47]
	Norway—Karmøy, OO-Star semi-submersible	2022, 10 MW	Flagship Demo, Iberdrola [48]
	Offshore Norway	2023, N/A	NOAKA, N/A
	Offshore UK, Ideol damping pool—barge	2021, 100 MW	Atlantis Ideol [49], Ideol
	Offshore UK, TLPWind TLP	N/A, 5 MW	TLPWind UK, Iberdrola
	Ireland—Offshore Irish west coast, Hexafloat -semi-submersible	2022, 6 MW	AFLOWT [50], Saipem
	Ireland—Offshore Kinsale, WindFloat semi-submersible	N/A, 100 MW	Emerald [51], Principle Power
	France—Gruissan, Ideol Damping Pool, barge	2021, Senvion 4×6.2 MW	EolMed [52], Ideol
	France—Offshore Napoleon Beach, SBM TLP	2021, Siemens Gamesa 3×8.4 MW	Provence Grand Large (PGL) [53], SBM Offshore
	France—Offshore Leucate-Le Barcarès, WindFloat semi-submersible	2022, MHI Vestas 3×10 MW	Golfe du Lion (EFGL) [54], Principle Power
	Spain—Offshore Canary Island, PivotBuoy TLP	2020, Vestas 200 kW	PivotBuoy 1:3 Scale [57], X1 Wind

Table 4. *Cont.*

Continent	Country—Location, Floating Substructure Design—Type	Year, Turbine—Power	Project Name, Designer
	Spain—Offshore Canary Islands, Cobra semi-spar	2020, 5 × 5 MW	FLOCAN5 [58], Cobra
	France—Offshore Brittany, Sea Reed semi-submersible	2022, MHI Vestas 3 × 9.5 MW	Groix & Belle-Ile [55], Naval Energies
	France—Offshore Le Croisic, Eolink semi-submersible	N/A, 5 MW	Eolink Demonstrator [56], Eolink
	Spain—Offshore Basque, Saitec SATH	2021, 2 MW	DemoSATH [59], Saitec
	Spain—Offshore Gran Canaria, N/A	N/A, 4 × 12.5 MW	Parque Eólico Gofio, Greenalia
	Spain—Basque, N/A	N/A, 26 MW	Balea, N/A
	Spain—Offshore Gran Canaria, N/A	N/A	WunderHexicon, Hexicon
North America	U.S.—Monhegan Island, VolturnUS semi-submersible	2023, 12 MW	New England Aqua Ventus I [11], University of Maine
	U.S.—California, WindFloat semi-submersible	2024, 100–150 MW	Red Wood Coast [60], Principle Power
	U.S.—Hawaii, WindFloat semi-submersible	2025, 400 MW	Progression South [61], Principle Power
	U.S.—California, SBM TLP/Saitec SATH	2025, 4 × 12 MW	CADEMO, SBM Offshore/ SAITEC [62]
	U.S.—California, N/A	2026, 1 GW	Castle Wind, N/A
	U.S.—Hawaii, WindFloat semi-submersible	2027, 400 MW	AWH Oahu Northwest, Principle Power
	U.S.—Hawaii, WindFloat semi-submersible	2027, 400 MW	AWH Oahu South [63], Principle Power
	U.S.—California, N/A	N/A	Diablo Canyon [64], N/A
	U.S.—Massachusetts, N/A	N/A, 10 + MW	Mayflower Wind, Atkins
Asia	Japan—Goto, Toda Hybrid spar	2021, 22 MW	Goto City [65], Toda
	Offshore Japan, Ideol Damping Pool, barge	2023, N/A	Acacia [66,67], Ideol
	Offshore Japan, SCD NEZZY Semi-Submersible	N/A, Aerodyn SCD 6 MW— 2-bladed	Nezzy Demonstrator [68], SCD Technology
	Korea—Ulsan, Hexicon multi-turbine semi-submersible	2022, 200 MW	Donghae TwinWind, Hexicon
	Korea—Ulsan, Semi-submersible	2020, 750 kW	Ulsan 750kW Floating Demonstrator, University of Ulsan
	Korea—Ulsan, N/A	2020, 5 MW	Ulsan Prototype [69,70], N/A
	Korea—Ulsan, N/A	2023, 500 MW	Gray Whale [71], N/A
	Korea—Ulsan, Hywind Spar	2024, 200 MW	KNOC (Donghae 1) [72,73], Equinor
	Korea—Ulsan, WindFloat semi-submersible	N/A, 500 MW	KFWind, Principle Power
	Korea—Ulsan, N/A	N/A, 200 MW	White Heron, N/A

This following presents the countries that made the largest contribution to planned floating wind projects in the period 2020–2027, and also outlines their corresponding power capacity:

1. The US had a power capacity of 2.45 GW, coming from nine floating wind projects in the period 2023–2027.
2. Korea had a power capacity of 1.6 GW, coming from seven floating wind projects in the period 2020–2024.
3. France had a power capacity of 113.5 MW, coming from five projects in the period 2021–2022.
4. Ireland had a power capacity of 106 MW, coming from two projects in 2022.
5. The UK had a power capacity of 105 MW, coming from two projects in 2021.
6. Spain had a power capacity of 103.2 MW, coming from six projects in the period 2020–2021.
7. Norway had a power capacity of 102.6 MW, coming from five projects in the period 2020–2023.
8. Japan had a power capacity of 28 MW, coming from three floating wind projects in the period 2020–2023.

It is worth mentioning that some other Asian countries, such as Taiwan, [74] have established a preliminary plan for their future floating wind projects. However, due to a lack of corresponding relevant details, we have eliminated them from our study (refer to the ABSG Consulting report [11] for more information about the names of the planned projects in Taiwan, for example). Paper [75] refers to large-scale offshore wind production in the Mediterranean Sea. While it does not directly relate to the present discussion, it can be taken into account in further discussions.

Figures 3 and 4 show the world's largest floating wind project (Hywind Tampen).

Figure 5 shows the world's first floating wind project (Hywind Scotland). Figures 6–8 show the world's most widely used floating multi-turbine concept (HEXICON).

The following subsection presents further details about the world's installed and planned floating wind projects in the period 2009–2026, including project costs, wind speeds, water depths, and distances to shore. The presented data does not cover all the mentioned projects in this paper, as some of them lacked corresponding data.

Figure 3. The largest installed floating wind turbine in the world (Hywind Tampen). Figure processed by the authors on the basis of information presented in [76].

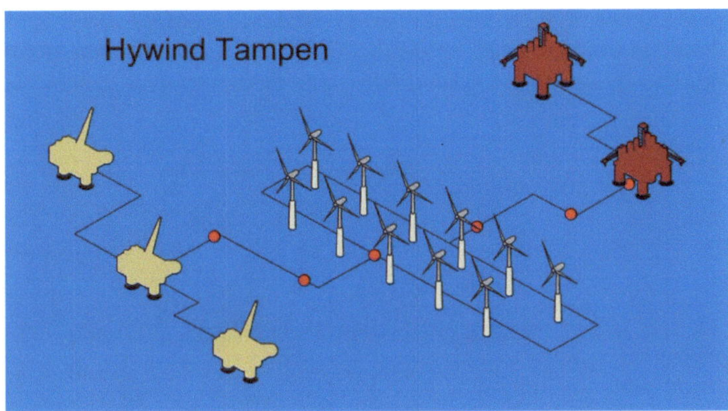

Figure 4. The largest installed floating wind project in the world (Hywind Tampen). Figure processed by the authors on the basis of information presented in [76].

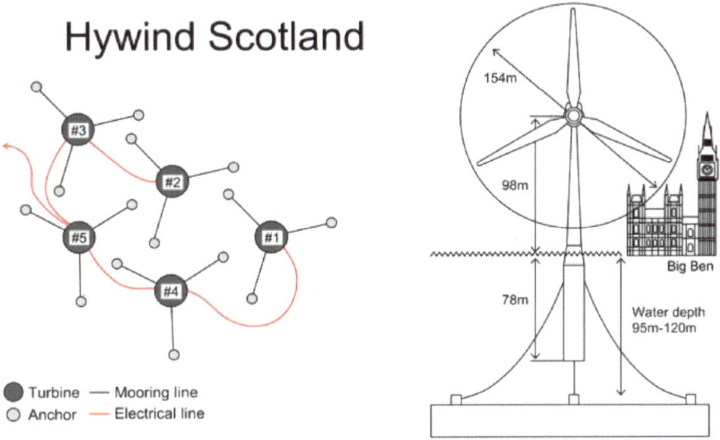

Figure 5. The first installed floating wind project in the world (Hywind Scotland). Figure processed by the authors on the basis of information presented in [77].

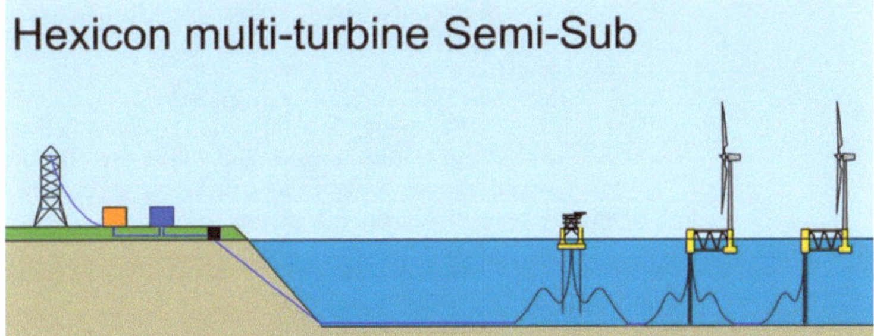

Figure 6. The most widely used multi-turbine floating wind turbine support structure in the world (HEXICON). Figure processed by the authors on the basis of information presented in [78].

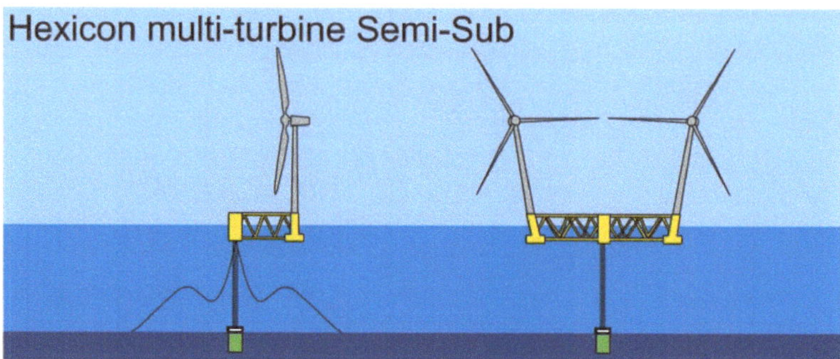

Figure 7. The most widely used multi-turbine floating wind support structure in the world (HEXICON). Figure processed by the authors on the basis of information presented in [78].

Figure 8. The most widely used multi-turbine floating wind support structure in the world (HEXICON). Figure processed by the authors on the basis of information presented in [78].

2.4. Further Details on Some of the Presented World's Installed and Planned Floating Wind Projects in the Period 2009–2026 (Based on Tables 3 and 4)

This subsection presents further details, including project costs, wind speeds, water depths, and distances to shore, which are classified on the basis of countries and their corresponding projects. This classification could not cover all the projects in this paper, as some of them lacked corresponding data. The data presented here relates to the period 2009–2026.

Table 5 presents further details of installed and planned floating wind projects in the world, which were first mentioned in their corresponding tables (Tables 3 and 4). Their corresponding mentioned data will be discussed and classified in Sections 3 and 4. In this subsection, we only present the table. Note that Table 5 contains 14/16 of the installed floating wind projects mentioned in Table 3 and 12/25 of the planned floating wind projects mentioned in Table 4.

The following section will provide results based on the collected and presented data in this section (Section 2). It will provide tables related to global floating wind concepts and installed and planned projects and will further classify some of the presented projects that have additional available data, including corresponding project costs, wind speeds, water depths, and distances to shore.

Table 5. Further details of some of the presented world's installed and planned floating wind projects in the period 2009–2026. Table data processed by the authors on the basis of information presented in [11].

Year	Project, Location, Distance to Shore	Turbine & Power, Floating Substructure Design & Type, Designer	Water Depth, Site Condition, Estimated Cost
2009	HYWIND DEMO (ZEFYROS), Offshore Karmøy Norway, 10 km	Siemens 2.3 MW, Hywind Spar, Equinor	220 m, wind speed 40 m/s & max wave height 19 m, US $71 million
2011	WINDFLOAT 1 (WF1), Offshore Aguçadoura Portugal, 5 km	Vestas 2 MW, WindFloat semi-submersible, Principle Power	49 m, wind speed 31 m/s & max wave height 17 m, US $25 million
2013	VOLTURNUS 1:8, Offshore Castine Maine US, 330 m	Renewegy 20 kW, VolturnUS, semi-submersible, University of Maine	27.4 m, 50-year wind speed 14.1 m/s & 50-year significant wave height 1.3 m, US $12 million
	SAKIYAMA, Offshore Sakiyama Fukue Island Japan, 5 km	Hitachi 2 MW downwind, Haenkaze -Toda Hybrid spar, Toda	100 m, 50-year wind speed 45.8 m/s & 50-year significant wave height 12.1 m, N/A
	FUKUSHIMA FORWARD PROJECT phase I, Offshore Fukushima Japan, 23 km	66 kV—25 MVA Floating Substation, Fukushima Kizuna—Advanced Spar, Japan Marine United Corporation (JMU)	120 m, 50-year wind speed 48.3 m/s & 50-year significant wave height 11.71 m, US $157 million for all the phases of the project
	FUKUSHIMA FORWARD PROJECT phase I, Offshore Fukushima Japan, 23 km	Hitachi 2 MW downwind, Fukushima Mira—compact semi-submersible, Mitsui Engineering & Shipbuilding Co., Ltd. (MES)	122–123 m, 50-year wind speed 48.3 m/s & 50-year significant wave height 11.71 m, US $157 million for all the phases of the project
2015	FUKUSHIMA FORWARD PROJECT, phase II, Offshore Fukushima Japan, 23 km	MHI 7 MW, Fukushima Shimpuu—V-shape Semi-Submersible, Mitsubishi Heavy Industries, Ltd. (MHI)	125 m, 50-year wind speed 48.3 m/s & 50-year significant wave height 11.71 m, US $157 million for all the phases of the project
	SEATWIRL S1, Offshore Lysekil Sweden, N/A	30 kW Vertical Axis Wind Turbine, SeaTwirl Spar, SeaTwirl	35 m, wind speed 35 m/s, N/A
2016	FUKUSHIMA FORWARD PROJECT, phase II, Offshore Fukushima Japan, 23 km	Hitachi 5 MW downwind, Fukushima Hamakaze— Advanced Spar, Japan Marine United Corporation (JMU)	110–120 m, 50-year wind peed 48.3 m/s & 50-year significant wave height 11.71 m, US $157 million for all the phases of the project
2017	HYWIND SCOTLAND, Offshore Peterhead Scotland UK, 25 km	Siemens 5 × 6 MW, Hywind Spar, Equinor	95–120 m, average wind speed 10 m/s & average wave height 1.8 m, US $210 million
2018	FLOATGEN, Offshore Le Croisic France, 20 km	Vestas 2 MW, Ideol Damping Pool-barge, Ideol	33 m, wind speed 24.2 m/s & significant wave height 5.5 m, US $22.5 million
2019	HIBIKI, Offshore Kitakyushu Japan, 15 km	Aerodyn SCD 3 MW—2 bladed, Ideol Damping Pool-barge, Ideol	55 m, typhoon-prone area, N/A
	W2POWER 1:6 SCALE, Offshore Gran Canaria Spain, N/A	2 × 100 kW twin-rotor, EnerOcean W2Power semi-submersible, EnerOcean	N/A

Table 5. *Cont.*

Year	Project, Location, Distance to Shore	Turbine & Power, Floating Substructure Design & Type, Designer	Water Depth, Site Condition, Estimated Cost
2020	WINDFLOAT ATLANTIC (WFA), Offshore Viana do Castelo Portugal, 20 km	MHI Vestas 3 × 8.4 MW, WindFloat semi-submersible, Principle Power	85–100 m, N/A, US $134 million
	KINCARDINE, Offshore Kincardineshire Scotland UK, 15 km	MHI Vestas 2 MW (former WF1)—MHI Vestas 5 × 9.5 MW, WindFloat semi-submersible, Principle Power	60–80 m, UK North Sea off the coast of Scotland, US $445 million
	BLUESATH, Offshore Santander Spain, 800 m	Aeolos 30 kW, Saitec SATH 1:6, Saitec	N/A, Abra del Sardinero, US $2.2 million
	TETRASPAR DEMO, Offshore Karmøy Norway, 10 km	Siemens Gamesa 3.6 MW, Stiesdal TetraSpar—Spar, Stiesdal	220 m, Near Zefyros (former Hywind Demo), US $20.5 million
2021	DEMOSATH, Offshore Basque Spain, 3.2 km	2 MW, Saitec SATH, Saitec	85 m, wind speed 12 m/s & significant wave height 2.8 m, $17.3 million
	EOLMED, Offshore Gruissan Mediterranean Sea France, 15 km	Senvion 4 × 6.2 MW, Ideol Damping Pool—barge, Ideol	55 m, Mediterranean Sea, US $236.2 million
	PROVENCE GRAND LARGE (PGL), Offshore Napoleon beach Mediterranean Sea France, 17 km	Siemens Gamesa 3 × 8.4 MW, SBM TLP, SBM Offshore	100 m, Mediterranean Sea, US $225 million
2022	HYWIND TAMPEN, Snorre & Gullfaks offshore fields Offshore Norway, 140 km	Siemens Gamesa 11 × 8 MW, Hywind Spar, Equinor	260–300 m, mean significant wave height 2.8 m, US $545 million
	GOLFE DU LION (EFGL), Offshore Leucate-Le Barcarès Mediterranean Sea France, 16 km	MHI Vestas 3 × 10 MW, WindFloat semi-submersible, Principle Power	65–80 m, Mediterranean Sea, US $225 million
	GROIX & BELLE-ILE, Offshore Brittany France, 22 km	MHI Vestas 3 × 9.5 MW, Sea Reed semi-submersible, Naval Energies	60 m, Atlantic Ocean off the coast of France, US $254 million
	DONGHAE TWINWIND, Offshore Ulsan Korea, 62 km	200 MW, Hexicon multi-turbine semi-submersible, Hexicon	N/A
2023	NEW ENGLAND AQUA VENTUS I, Offshore Monhegan Island in the Gulf of Maine US, 4.8 km	12 MW, VolturnUS-semi-submersible, University of Maine	100 m, 50-year wind speed of 40 m/s & 50-year significant wave height 10.2 m, US $100 million
2024	REDWOOD COAST, Offshore Humboldt County California US, 40 km	100–150 MW, WindFloat semi-submersible, Principle Power	600 m–1 km, average annual wind speed 9–10 m/s, N/A
2025	CADEMO, Offshore Vandenberg California US, 4.8 km	4 × 12 MW, SBM TLP/Saitec SATH, SBM Offshore/Saitec	85–96 m, average wind speed 8.5 m/s, N/A
2026	CASTLE WIND, Offshore Morro Bay California US, 48 km	1 GW, N/A, N/A	813 m–1.1 km, average wind speed 8.5 m/s, N/A

3. Results

This section presents the results obtained from Section 2, which are classified on the basis of global floating wind concepts and projects. Further classifications of some projects that have further available data, on corresponding countries, costs, wind speeds, water depths, and distances to shore, are also provided. The following subsections present the results from Tables 2–5 for all floating wind concepts and projects in the world in the period 2008–2027.

The following subsection presents the world's floating wind turbine concepts' results obtained from Table 2.

3.1. Results from Table 2 (World's Floating Wind Turbine Concepts—Part 1)

This subsection presents the results obtained from Table 2 for the world's floating wind turbine concepts.

The results in Table 6 show a total number of 28 floating wind turbine concepts (thirteen semi-submersibles, five spar buoys, five TLPs, three multi-turbines, and two barges).

Table 6. World's floating wind turbine concepts.

Floating Wind Turbine Types	Number of Corresponding Concepts
Spar-buoy	5
Semi-submersible	13
Barge	2
TLP	5
Multi-turbine	3
Total	28

The following subsection also presents the results obtained from Table 2 for the world's floating wind turbine concepts.

3.2. Results from Table 2 (World's Floating Wind Turbine Concepts—Part 2)

This subsection also presents the results obtained from Table 2 for the world's floating wind turbine concepts.

The results in Table 7 show a total number of 28 presented floating wind turbine concepts (eighteen are made of steel, six of concrete, and four of steel and/or concrete.

Table 7. World's floating wind turbine manufacturing materials.

Floating Wind Manufacturing Material	Number of Corresponding Concepts
Steel	18
Concrete	6
Steel and/or concrete	4
Total	28

The following subsection presents the results obtained from Table 2 for the world's floating wind turbine concepts.

3.3. Results from Table 2 (World's Floating Wind-Turbine Concepts—Part 3)

This subsection presents the results obtained from Table 2 for the world's floating wind turbine concepts.

Table 8 shows that there are a total number of 28 presented floating wind turbine concepts, thirteen semi-submersibles, eight of which are made of steel, four of concrete, and one of steel or concrete; five spar-buoys, three of which are made of steel, and two of steel and/or concrete; five TLPs, four of which are made of steel, and one of concrete; three multi-turbines, all are made of steel; two barges, one is made of concrete and one of steel or concrete.

Table 8. World's floating wind turbine concepts and their corresponding manufacturing materials.

Floating Wind Types	Number of Corresponding Concepts	Steel	Concrete	Steel and/or Concrete
Spar-buoy	5	3	-	2
Semi-submersible	13	8	4	1
Barge	2	-	1	1
TLP	5	4	1	-
Multi-turbine	3	3	-	-
Total	28	18	6	4

The following subsection presents results obtained from Table 3 for the world's installed floating wind turbine projects in the period 2008–2020.

3.4. Results from Table 3 (World's Installed Floating Wind Turbine Projects in the Period 2008–2020)

This subsection presents the results obtained from Table 3 for the world's installed floating wind turbine projects in the period 2008–2020.

The total installed floating wind capacity in Europe is 123.5 MW, which is provided by 12 projects in 8 contributing countries (the UK, Portugal, Norway, France, Spain, Denmark, Sweden, and Germany—see Table 9). Refer to Section 4 for further discussion of the contribution that each country makes to the global installed floating wind power capacity.

Table 9. World's installed floating wind turbine projects.

Continents	Total Installed Floating Wind Capacity	Corresponding Number of Projects	Corresponding Number of Countries	Corresponding Countries
Europe	123.5 MW	12	8	UK, Portugal, Norway, Germany, France, Spain, Denmark, Sweden
North America	30.2 MW	2	1	US
Asia	21 MW	4	1	Japan
Total	174.7 MW	18	10	The UK, Portugal, Norway, Germany, France, Spain Denmark, Sweden, The US and Japan

The total installed floating wind capacity in the US is 30.2 MW, which is provided by two projects.

The total installed floating wind capacity in Asia is 21 MW, which is provided by four projects in Japan.

The following subsection presents the results obtained from Table 4 for the world's planned floating wind turbine projects in the period 2020–2027.

3.5. Results from Table 4 (World's Planned Floating Wind-Turbine Projects in the Period 2020–2027)

This subsection presents results obtained from Table 4 for the world's planned floating wind turbine projects in the period 2020–2027.

Table 10 briefly classifies the power capacity of the world's planned floating wind projects on the basis of corresponding continents and countries. Further power capacity classifications regarding each of the presented countries are given in the following.

Table 10. World's planned floating wind turbine projects.

Continents	Total Planned Floating Wind Power Capacity	Corresponding Number of Projects	Corresponding Number of Countries	Corresponding Countries
Europe	525.1 MW	17	5	France, Ireland, UK, Spain, Norway
North America	2.42 GW	8	1	US
Asia	1.634 GW	9	2	Korea, Japan
Total	4.5791 GW	34	8	France, Ireland, the UK, Spain, Norway, the US, Korea and Japan

France has a planned floating wind power capacity of 113.5 MW, provided by four projects (Golfe du Lion—EFGL, GROIX & Belle-Ile, Provence Grand Large—PGL, and EOLMED).

Ireland has 106 MW, provided by two projects (Emerald and AFLOWT).

The UK has 105 MW, provided by two projects (Atlantis IDEOL and TLP Wind).

Spain has 103 MW, provided by five projects (Parque EOLICO Gofio, Balea, FLOCAN 5, Demo SATH, and Pivot Buoy 1:3 Scale).

Norway has 102.6 MW, provided by four projects (Hywind Tampen, Flagship Demo, Tetra Spar Demo, and Sea Twirl S2).

The US has 2.42 GW, provided by eight projects (Castle Wind, Progression South, AWH Oahu Northwest, AWH Oahu South, Red Wood Coast, CADEMO, New England Aqua Ventus I, and Mayflower Wind).

Korea has 1.606 GW, provided by seven projects (Gray Whale, KF Wind, DONGHAE Twin Wind, KNOC (DONGHAE 1), White Heron, Ulsan Prototype, and Ulsan 750 kW Floating Demonstrator).

Japan has 28 MW, provided by two projects (Goto City and NEZZY Demonstrator).

The following subsection presents results obtained from Table 5 for further details on the costs of some of the presented world's installed and planned floating wind projects in the period 2009–2026.

3.6. Results from Table 5 (Further Details on Costs of Some of the Presented World's Installed and Planned Floating Wind Projects in the Period 2009–2026)

This subsection presents the results obtained from Table 5 for further details on the costs of some of the presented world's installed and planned floating wind projects in the period 2009–2026.

Table 11 briefly classifies the world's planned floating wind projects' costs on the basis of corresponding continents and countries. Further cost classifications for each of the presented projects and their corresponding countries follow.

France's 962.7 million dollars of floating wind project cost is accounted for by one installed project (FLOATGEN) and four planned projects (GROIX & Belle-Ile, EOLMED, Provence Grand Large, and Golfe du Lion).

The UK's 655 million dollars is accounted for by two installed projects (Kincardine and Hywind Scotland).

Norway's 316.5 million dollars is accounted for by one installed project (Hywind Demo—ZEFYROS) and two planned projects (Hywind Tampen and Tetra Spar Demo).

Portugal's 159 million dollars is accounted for by two installed projects (Wind Float Atlantic and Wind Float 1).

Spain's 19.5 million dollars is accounted for by one installed project (Blue SATH) and one planned project (Demo SATH).

The US's 112 million dollars is accounted for by one installed project (VOLTURNUS 1:8) and one planned project (New England Aqua Ventus I).

Japan's 157 million dollars is accounted for by one installed project (Fukushima Forward Phases I & II).

Table 11. Further cost details for some of the presented world's installed and planned floating wind projects.

Continents	Corresponding Project Costs	Corresponding Number of Installed Projects	Corresponding Number of Planned Projects	Corresponding Number of Countries	Corresponding Countries
Europe	2.1127 billion dollars	7	7	5	France, UK, Norway, Portugal, Spain
North America	112 million dollars	1	1	1	US
Asia	-	-	-	-	-
Total	2.2247 billion dollars	8	8	6	France, the UK, Norway, Portugal, Spain and the US

The following subsection presents the results obtained from Table 5 for further details on wind speeds in some of the presented world's installed and planned floating wind projects in the period 2009–2026.

3.7. Results from Table 5 (Further Details about the Wind Speeds of Some of the Presented World's Installed and Planned Floating Wind Projects in the Period 2009–2026)

This subsection presents results obtained from Table 5 for further details about the wind speeds of some of the presented world's installed and planned floating wind projects in the period 2009–2026.

Table 12 briefly classifies the world's planned floating wind projects' wind speeds on the basis of corresponding continents and countries. Further wind speed classifications for each presented country follow.

Table 12. Further wind speed details for some of the presented world's installed and planned floating wind projects.

Continents	Corresponding Project Wind Speed	Corresponding Number of Installed Projects	Corresponding Number of Planned Projects	Corresponding Number of Countries	Corresponding Countries
Europe	10–40 m/s	5	1	6	Norway, Sweden, Portugal, France, Spain, UK
North America	8.5–40 m/s	1	4	1	US
Asia	45–48 m/s	2	-	1	Japan
Total	8.5–48 m/s	8	5	8	Norway, Sweden, Portugal, France, Spain, the UK, the US and Japan

Norway's 40 m/s wind speed is provided by one installed floating wind project (Hywind Demo—ZEFYROS).

Sweden's 35 m/s is provided by one installed floating wind project (Sea Twirl S1).

Portugal's 31 m/s is provided by one installed floating wind project (Wind Float 1).

France's 24.2 m/s is provided by one installed floating wind project (FLOATGEN).

Spain's 12 m/s is provided by one planned floating wind project (Demo SATH).

The UK's 10 m/s is provided by one installed floating wind project (Hywind Scotland).

The US's 8.5–40 m/s is provided by one installed floating wind project (VOLTURNUS 1:8) and four planned projects (CADEMO, Castle Wind, Red Wood Coast, and New England Aqua Ventis I).

Japan's 45–48 m/s is provided by two installed floating wind projects (Sakiyama and Fukushima Forward Phases I & II).

The following subsection presents the results obtained from Table 5 for further details about the water depths of some of the presented world's installed and planned floating wind projects in the period 2009–2026.

3.8. Results from Table 5 (Further Details about the Water Depths of Some of the Presented World's Installed and Planned Floating Wind Projects in the Period 2009–2026)

This subsection presents the results obtained from Table 5 for further details about the water depths of some of the presented world's installed and planned floating wind projects in the period 2009–2026.

Table 13 briefly classifies the world's planned floating wind projects' water depths on the basis of continents and corresponding countries. Further water depth classifications for each presented country follow.

Table 13. Further water depth details of some of the presented world's installed and planned floating wind projects.

Continents	Corresponding Projects' Water Depth	Corresponding Number of Installed Projects	Corresponding Number of Planned Projects	Corresponding Number of Countries	Corresponding Countries
Europe	33–300 m	8	6	6	Norway, UK, France, Portugal, Spain, Sweden
North America	27.4 m–1 km	1	3	1	US
Asia	55–125 m	3	-	1	Japan
Total	27.4 m–1 km	12	9	8	Norway, the UK, France, Portugal, Spain, Sweden, the US and Japan

Norway's 220–300 m water depth came from one installed floating wind project (Hywind Demo—ZEFYROS) and two planned projects (Tetra Spar Demo and Hywind Tampen.

The UK's 90–120 m came from two installed floating wind projects (Kincardine and Hywind Scotland).

France's 33–100 m came from one installed floating wind project (FLOATGEN) and four planned projects (EOLMED, GROIX & Belle-Ile, Golfe du Lion—EFGL, and Provence Grand Large—PGL).

Portugal's 49–100 m came from two installed floating wind projects (Wind Float Atlantic—WFA and Wind Float 1—WF1).

Spain's 85 m came from one installed floating wind project (Demo SATH).

Sweden's 35 m came from one installed floating wind project (Sea Twirl S1).

The US's 27.4 m–1 km came from one installed floating wind project (VOLTURNUS 1:8) and three planned projects (CADEMO, New England Aqua Ventus I, and Red Wood Coast).

Japan's 55–125 m came from three installed floating wind projects (Hibiki, Sakiyama, and Fukushima Forward Phases I & II).

The following subsection presents the results obtained from Table 5 for further details about the distance to shore of some of the presented world's installed and planned floating wind projects in the period 2009–2026.

3.9. Results from Table 5 (Further Details about the Distance to Shore of Some of the Presented World's Installed and Planned Floating Wind Projects in the Period 2009–2026)

This subsection presents the results obtained from Table 5 for further details about the distance to shore of some of the presented world's installed and planned floating wind projects in the period 2009–2026.

Table 14 briefly classifies the world's planned floating wind projects' distance to shore on the basis of corresponding continents and countries. Further distance-to-shore classifications for each presented country follow.

Table 14. Further distance-to-shore details for some of the presented world's installed and planned floating wind projects.

Continents	Corresponding Projects' Distance to Shore	Corresponding Number of Installed Projects	Corresponding Number of Planned Projects	Corresponding Number of Countries	Corresponding Countries
Europe	800 m–140 km	7	7	5	Norway, UK, France, Portugal, Spain
North America	330 m–48 km	1	4	1	US
Asia	5–62 km	3	1	2	Korea, Japan
Total	300 m–140 km	11	12	8	Norway, the UK, France, Portugal, Spain, the US, Korea and Japan

Norway's 10–140 km distance to shore came from one installed floating wind project (Hywind Demo—ZEFYROS) and two planned projects (Tetra Spar Demo and Hywind Tampen).

The UK's 15–25 km came from two installed floating wind projects (Kincardine and Hywind Scotland).

France's 15–22 km came from one installed floating wind project (FLOATGEN) and four planned projects (EOLMED, Golfe du Lion—EFGL, Provence Grand Large—PGL, and GROIX & Belle-Ile).

Portugal's 5–20 km came from two installed floating wind projects (Wind Float 1—WF1 and Wind Float Atlantic).

Spain's 800 m–3.2 km came from one installed floating wind project (Blue SATH) and one planned project (Demo SATH).

The US's 330 m–48 km came from one installed floating wind project (VOLTURNUS 1:8) and four planned projects (New England Aqua Ventus I, CADEMO, Red Wood Coast, and Castle Wind).

Korea's 62 km came from one planned floating wind project (DONGHAE Twin Wind).

Japan's 5–15 km came from three installed floating wind projects (Sakiyama, Hibiki, and Fukushima Forward Phases I & II).

The following section will primarily discuss the results presented in this section (Section 3) for global floating wind concepts and projects, and also for further classifications;

however, only some of the presented projects will be discussed with reference to their costs, wind speeds, water depths, and distances to shore, as data unavailability prevented us from engaging all of the projects presented in this paper.

4. Discussion

In this section, we will further discuss the results obtained from Section 3. This section also includes external references that are relevant to the world's floating wind situation, with particular emphasis on Europe and some related aspects. Figure 9 shows the floating wind Power-to-X technology that is used to transform the produced floating wind electrical energy (mainly into hydrogen and compressed air), eliminating the need for the construction of corresponding tremendous electrical infrastructures in countries that do not have such infrastructures in their sea region/s. Romania's floating wind feasibility will also be considered at the end of this section.

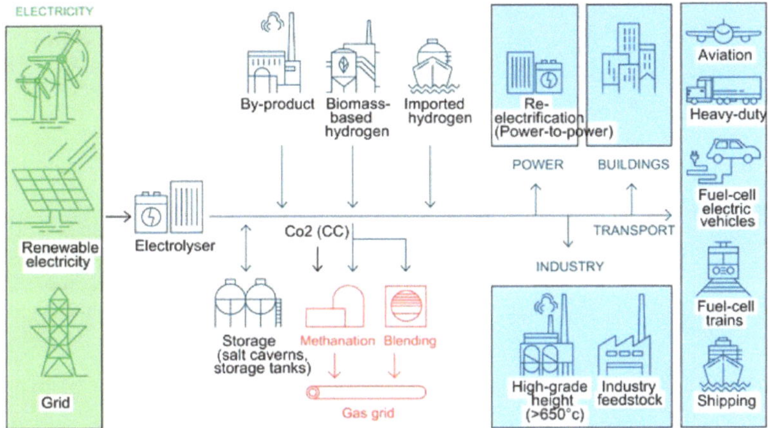

Figure 9. The floating wind Power-to-X technology, that transforms the produced floating wind electrical power (mainly into hydrogen and compressed air). Figure processed by the authors on the basis of information presented in [79].

The following subsection discusses the data presented throughout the paper, and particularly focuses on data presented in Section 3 that concerns the reliability of some of this paper's data references.

4.1. Discussions of the Data Presented in This Paper, with a Particular Focus on the Data Presented in Section 3, Which Addresses the Reliability of Some of the Paper's Data References

This subsection discusses data presented throughout the paper, with a particular focus on the data presented in Section 3 and addresses the reliability of some data references in this paper. It is worth mentioning that the reference this data is taken from is found to be the most complete and covers global floating wind concepts and projects up to 2020. The installed floating wind projects that are considered are from the period 2008–2020, and the considered planned floating wind projects are from the period 2020–2027. Our analysis is limited to these time intervals.

In the Introduction part of this paper, it was observed that Europe's floating wind plan is to achieve 150 GW by 2050 [4]. The planned power capacity is, on the basis of the results presented in Section 3, 525.1 MW, which will come from 17 projects [11]. An installed power capacity of 123.5 MW (see the results in Section 3) is also anticipated for the period 2008–2027 [11]. It is therefore obvious that there is a shortage in the presented data when compared to the overall European plan, as noted by [4]. It should also be noted that most of the projects that are planned to increase Europe's overall floating wind power capacity in Europe to 150 MW by 2050 have, according to [4], not been announced yet. It could therefore be said that our analysis is not complete because, due to the lack of corresponding published data for the planned floating wind projects (both their names and their corresponding power capacities) for the period 2020–2050, it is only able to consider floating wind projects for the period 2008–2027.

According to [4], the installed floating wind power capacity in Europe is 318 MW, which is provided by 34 floating wind concepts. However, Section 3 suggested that the installed floating wind power capacity is 123.5 MW, provided by 28 concepts (see [11]). This confirms discrepancies between the different references.

According to [4], the installed floating wind capacity is 114 MW in France, 80 MW in the UK, 95 MW in Norway, 30 MW in Portugal, 12 MW in the US, and 20 MW in Japan. According to Section 3, the installed floating wind power capacity in France is 2 MW, set against a planned capacity of 113.5 MW (the reference may be referring to this capacity, instead of the actual capacity installed in 2020). The installed capacity, according to Section 2, is 89.5 MW in the UK and 2.3 MW in Norway; however, the planned capacity is 105 MW in the UK and 102.6 MW in Norway (the reference may be referring to this capacity instead of the actual installed capacity in 2020). The installed capacity, according to Section 2, is 25.23 MW in Portugal, 30.2 MW in the US, and 21 MW in Japan. It can therefore be noted that different references have different power capacity values for the same countries (i.e., [4,11]).

According to [4], the planned floating wind capacity for 2030 is 750 MW in France, 1 GW in the UK, 1.5 MW (or 3 GW [5]) in Norway, and 275 MW in Portugal. Section 3 shows a planned power capacity of 113.5 MW in France, 105 MW in the UK, and 102.6 MW in Norway, but gives no data for Portugal, according to [11]. This corresponds to the period 2020–2027. This confirms there are discrepancies between the different references and probably a lack of detailed corresponding plans about the name of the projects and their corresponding power capacities, and this probably explains why the mentioned references did not include these details.

It can be generally concluded that the reference that the Section 2 data was taken from [11] is not 100% accurate. However, it is, to the best of our knowledge, the most complete reference existing up to 2020 that provides insight into global floating wind concepts and projects in the period 2020–2027.

The following subsection will discuss the data presented in Section 3.

4.2. Further Discussion of the Data Presented in Section 3

This subsection will discuss the data presented in Section 3. As has been reiterated throughout the paper, the plans in different countries of the world for floating wind projects do not necessarily include the projects that will cover the planned power capacities in each country. This paper's main aim was to list the announced global installed and planned floating wind projects, along with their corresponding power capacities and their details for the period 2008–2027. We can therefore assert that, according to reference [11], this paper collects all the announced installed and planned floating wind projects in the world (as of 2020) for the period 2008–2027. Their sum has a different value from the total floating wind power capacity planned by each country. Refer to Section 4.1 to see some of the discrepancies that arise between the floating wind power capacity in different references (e.g., [4,11], and possibly other references not covered in this paper).

The following seven sub-subsections will discuss the data presented in Section 3 on the global floating wind turbine concepts, installed and planned projects, and also further details, including costs, wind speeds, water depths, and distances to shore of some of these projects which contain further data.

The following sub-subsection discusses the results for global floating wind turbine concepts produced by the Section 3 data.

4.2.1. Discussion of the Results for Global Floating Wind Turbine Concepts Produced by the Section 3 Data

This sub-subsection discusses results for global floating wind turbine concepts, and specifically refers to the data presented in Section 3.

In accordance with the presented concepts in Section 3, we will discuss the following results.

The total number of the floating wind concepts is 28, coming from 5 wind-turbine types. The type that has the highest number of corresponding concepts is semi-submersible with 13 concepts (8 of which are made of steel and 5 are made of concrete or a combination of both materials). The second highest number of five concepts comes from both Spar (three of which are made of steel and two of steel and/or concrete) and TLP (four made of steel and one of concrete). The third highest number of concepts comes from the multi-turbine platform with three concepts (all made of steel). The lowest number of concepts is barge, with 2 concepts (one made of steel and one of concrete).

Semi-submersible is the most frequently used floating wind turbine type with the highest concept number; it is then followed by Spar and TLP, which have approximately half the number of concepts. The paper's data also establishes that steel is the most frequently used manufacturing material in all floating wind turbine types.

The total number of floating wind concepts is 28, 18 of which are made of steel, 6 of concrete, and 4 of steel and/or concrete.

The most frequently used manufacturing material in floating wind concepts is steel, followed by concrete or a combination of both materials.

In referring to the projects presented in Section 3, we will discuss their corresponding results in the following sub-subsections (Sections 4.2.2–4.2.7). The following sub-subsection discusses results for the global installed floating wind turbine projects in the period 2008–2020.

4.2.2. Discussion of the Results for the Global Installed Floating Wind Turbine Projects in the Period 2008–2020

This sub-subsection discusses the results obtained for the global installed floating wind turbine projects in the period 2008–2020.

With regard to installed floating wind power capacity per continent, Europe comes first, with 123.5 MW coming from the highest number of (8) corresponding countries and 12 projects. It is followed by North America, with a quarter of the power capacity (30.2 MW) coming from one country (the US) and two projects (Please note that the US project which corresponds to 30 MW power capacity had some uncertainty regarding its actual installation in the references it was taken from. The same also applies to the installed German project in this paper). Asia follows, with a sixth of Europe' power capacity, with 21 MW and four projects coming from one country (Japan). This data corresponds to the period 2008–2020, according to [11].

The overall installed floating wind power capacity in the world is 174.7 MW, which mostly comes from Europe, which is accounted for 123.5 MW (coming from 10 countries and 18 projects) in the period 2008–2020, according to [11].

With regard to Europe's installed floating wind power capacity of 123.5 MW, two countries mostly account for this value, namely the UK, with a power capacity of 89.5 MW, and Portugal, with 27.5 MW. The following six European countries make a more minor contribution to the installed European floating wind capacity: Norway with a power capacity of 2.3 MW, Germany with 2.3 MW, France with 2 MW, Spain with 230 kW, Denmark with 33 kW, and Sweden with 30 kW.

The UK is Europe's largest contributor to the installed floating wind capacity, with 89.5/123.5 MW. The second largest contributor is Portugal with 27.5/123.5 MW, followed by countries with an overall minor capacity of 6.5 MW: Norway, Germany, France, Spain, Denmark, and Sweden. According to [11], this covers the period 2008–2020, although the situation in 2023 may be different (e.g., Hywind Tampen, with 88 MW capacity, was installed in Norway in 2023, in a way that diverged from its original plan in 2022; this example corresponds to a planned project and was given merely for the sake of illustration).

The following sub-subsection discusses the results concerning the global planned floating wind turbine projects in the period 2020–2027.

4.2.3. Discussion of the Global Planned Floating Wind Turbine Projects' Results for the Period 2020–2027

This sub-subsection discusses the global planned floating wind turbine projects' results for the period 2020–2027 according to [11].

With regard to the planned floating wind power capacity per continent, North America takes the first place, with 2.42 GW power capacity coming from one corresponding country (the US) and 8 projects. Then comes Asia, with 1.634 GW coming from two countries (Korea and Japan) and nine projects. Korea (1.606 GW), contributes by far the most, followed by Japan (28 MW). Then comes Europe, with 525.1 MW produced by 5 countries and 17 projects. According to [11], this applies for the period 2020–2027.

This data corresponds to announced projects, as of 2020, for the period 2020–2027 and does not necessarily correspond to the current planned floating wind power capacities of each country.

The largest contributor to Asia's planned floating wind power capacity is Korea, followed by the much smaller contribution from Japan in the period 2020–2027, according to [11].

With regard to Europe's planned floating wind power capacity, which was 525.1 MW in the period 2020–2027, the following countries contributed: France with 113.5 MW power capacity, Ireland with 106 MW, the UK with 105 MW, Spain with 103.2 MW, and Norway with 102.6 MW.

The overall planned floating wind power capacity in the world for the period 2020–2027 is 4.5791 GW, which mostly comes from North America (the US-2.42 GW), Asia (mainly from Korea, with 1.606 GW, and Japan, with 28 MW), and Europe (525.1 MW), and specifically the following contributing countries: France (113.5 MW), Ireland (106 MW), the UK (105 MW), Spain (103.2 MW), and Norway (102.6 MW). This is for the period 2020–2027, according to [11].

The overall planned floating wind power capacity in the world is 4.5791 GW, which mostly comes from North America (the US-2.42 GW), Asia (Korea-1.606 GW and Japan-28 MW), and Europe (Total of 525.1 MW coming from France-113.5 MW, Ireland-106 MW, the UK-105 MW, Spain-103.2 MW, and Norway-102.6 MW) for the period 2020–2027, according to [11]. Note that these values correspond to the announced projects as of 2020 for the period 2020–2027 and do not necessarily correspond to the actual planned floating wind power capacity OF each country.

The following sub-subsection discusses the presented installed and planned floating wind turbine projects' results, with specific reference to countries and contributing costs in the period 2009–2026, according to [11]. Not all the presented projects throughout this paper have available corresponding data that can be classified on the basis of project cost, wind speed, water depth, and distance to shore.

4.2.4. Discussions of the Presented Installed and Planned Floating Wind Turbine Projects' Results in Terms of Their Corresponding Countries and Their Contributing Costs in the Period 2009–2026

This sub-subsection discusses some of the presented installed and planned floating wind turbine projects' results in terms of their countries and their contributing costs in the period 2009–2026, according to [11]. This is because not all the presented projects throughout the paper have available corresponding data that can be classified on the basis of project costs.

With regard to the available data of some of the presented projects throughout this paper in terms of their corresponding continents and countries, Europe is the largest contributor to floating wind projects' cost, with 2.1127 billion dollars coming from 14 projects (7 installed and 7 planned) in 5 countries (France, the UK, Norway, Portugal, and Spain). It is followed by North America, with a project cost contribution of 112 million dollars coming from two projects (one installed and one planned) in one country (the US). The reference [11], which this data was taken from, did not provide information about the contributing cost of floating wind projects in Asia.

With regard to Europe, the highest contributing countries to the floating wind project cost were, according to the available data of some of the presented projects, France (962.7 million dollars), the UK (655 million dollars), Norway (316.5 million dollars), Portugal (159 million dollars), and Spain (19.5 million dollars). This corresponds to the period 2009–2026 according to [11].

The global floating wind projects' cost contribution is, according to the available data, 2.2247 billion dollars, which mainly comes from Europe (2.1127 billion dollars, from seven installed and seven planned projects) and North America (the US, with 112 million dollars from one installed and one planned projects). This corresponds to the period 2009–2026, according to [11].

It can also be said that, according to the available data, France is the largest contributor to the floating wind projects' cost, with 962.7 million dollars (coming from 1 installed and 4 planned projects). It is followed by the UK, with 655 million dollars (coming from 2 installed projects). And then Norway, with 316.5 million dollars (coming from one installed and two planned projects). Portugal follows, with 159 million dollars (coming from two installed projects), and then Spain, with 19.5 million dollars (coming from one installed and one planned project). This corresponds to the period 2009–2026, according to [11].

It can therefore be concluded that the global contributing costs are 2.2247 billion dollars, coming from 16 projects (eight installed and eight planned) in six countries (France, the UK, Norway, Portugal, Spain, and the US). This corresponds to the period 2009–2026 according to [11]. Note that it is not claimed that this data covers all the contributing costs of floating wind projects by continents and countries, both because of a shortage of data in these classifications, and the fact that they correspond to 2020, and not necessarily to 2023, when planned projects have been installed, with further updates.

The following sub-subsection discusses the results for some of the presented installed and planned floating wind turbine projects, with specific reference to their countries and contributing wind speeds in the period 2009–2026, according to [11]. This period is selected because not all the presented projects throughout this paper have available corresponding data that can be classified on the basis of project wind speed.

4.2.5. Discussion of the Presented Installed and Planned Floating Wind Turbine Project Results in Terms of Their Corresponding Countries and Their Contributing Wind Speeds in the Period 2009–2026

This sub-subsection discusses presented installed and planned floating wind turbine projects' results in terms of their corresponding countries and their contributing wind speeds in the period 2009–2026, according to [11]. This period is selected because not all of the presented projects have available corresponding data that can be classified on the basis of project wind speed.

With regard to the available data for some of the presented projects in each of the presented continents and countries, Asia has the highest wind speed of 45–48 m/s, coming from two installed projects and one country (Japan), followed by Europe, with 10–40 m/s coming from six projects (five installed and one planned) and six countries (Norway, Sweden, Portugal, France, Spain, and the UK). And North America, with 8.5–40 m/s coming from one country (the US). This corresponds to the period 2009–2026 according to [11].

In Europe, according to the projects' available data, Norway has the highest wind speed of 40 m/s (coming from one installed project), followed by Sweden with a wind speed of 35 m/s (coming from one installed project). Portugal comes after, with a wind speed of 31 m/s (coming from one installed project), and then France, with a wind speed of

24.2 m/s (coming from one installed project). Spain, with a wind speed of 12 m/s (coming from one planned project) and the UK, with a wind speed of 10 m/s (coming from one installed project) are the last two countries. This corresponds to the period 2009–2026, according to [11].

It can therefore be concluded that the global floating wind projects' wind speed is, according to the available data, 8.5–48 m/s, with the highest wind speed coming from the following continents and countries. Asia with a wind speed of 45–48 m/s, coming from one country (Japan). Then comes Europe with a wind speed of 10–40 m/s, coming from six countries (Norway with 40 m/s, Sweden with 35 m/s, Portugal with 31 m/s, France with 24.2 m/s, Spain with 12 m/s, and the UK with 10 m/s). And North America, with a wind speed of 8.5–40 m/s coming from one country (the US). This corresponds to the period 2009–2026, according to [11].

It can be concluded that the global contributing wind speeds to the floating wind projects are 8.5–48 m/s, coming from 13 projects (8 installed and 5 planned) and 8 countries (Norway, Sweden, Portugal, France, Spain, the UK, the US, and Japan). This corresponds to the period 2009–2026, according to [11]. Note that this data is not claimed to cover all the contributing wind speeds of floating wind projects for continents and countries, as there is a shortage of data in these classifications and it corresponds to 2020, and not necessarily to 2023, where some of the planned projects have been installed and further updates have occurred.

The following sub-subsection discusses the presented installed and planned floating wind turbine projects results, with specific reference to their corresponding countries and their contributing water depths for the period 2009–2026, according to [11]. This is because not all of the presented projects throughout this paper have available corresponding data that can be classified on the basis of projects' water depths.

4.2.6. Discussion of the Presented Installed and Planned Floating Wind Turbine Project Results in Terms of Their Corresponding Countries and Their Contributing Water Depths in the Period 2009–2026

This sub-subsection discusses the presented installed and planned floating wind turbine projects' results, with specific reference to their corresponding countries and their contributing water depths in the period 2009–2026, according to [11]. This period is selected because not all of the presented projects throughout this paper have available corresponding data that can be classified on the basis of projects' water depths.

With regard to the available data for some of the presented projects in each of the presented continents and countries, North America has the highest water depth capacity, with 27.4 m–1 km coming from four projects (one installed and three planned) in one country (the US). Europe is next, with a water depth capacity of 33–330 m coming from 14 projects (eight installed and six planned) in six countries (Norway, the UK, France, Portugal, Spain, and Sweden). And then Asia, with a water depth capacity of 55–125 m coming from three installed projects and one country (Japan). This corresponds to the period 2009–2026, according to [11].

With regard to Europe, the projects' available data shows that Norway has the highest water depth capacity of 200–300 m, from one installed and two planned projects. It is followed by the UK, with a water depth capacity of 90–120 m (coming from two installed projects), and then France, with a water depth capacity of 33–100 m (coming from one installed and four planned projects). Portugal, with a water depth capacity of 49–100 m (coming from two installed projects) is then followed by Spain, with a water depth capacity of 85 m (coming from one installed project) and Sweden, with a water depth capacity of 35 m (coming from one installed project). This corresponds to the period 2009–2026 according to [11].

It can therefore be concluded that the global floating wind projects' water depth capacity is, according to the available data, 27.4 m–1 km with the highest water depth capacity in the following continents and countries. North America, with a water depth capacity of 27.4 m–1 km coming from 4 projects (1 installed and 3 planned) and one country (the US). Followed by Europe, with a water depth capacity of 33–300 m coming from 14 projects (8 installed and 6 planned) in 6 countries (Norway with 220–300 m, the UK with 90–120 m, France with 33–100 m, Portugal with 49–100 m, Spain with 85 m, and Sweden with 35 m). And, finally, Asia, with a water depth capacity of 55–125 coming from three installed projects in one country (Japan). This corresponds to the period 2009–2026, according to [11].

It can therefore be concluded that the global contributing water depths to the floating wind projects are 27.4 m–1 km, coming from 21 projects (12 installed and 9 planned) in 8 countries (Norway, the UK, France, Portugal, Spain, Sweden, the US, and Japan). This corresponds to the period 2009–2026 according to [11]. Note that this data is not claimed to cover all the contributing water depths of floating wind projects for continents and countries, both because of a shortage in data in these classifications, and because this data corresponds to 2020 and not necessarily to 2023, where some of the planned projects have been installed and further updates have occurred.

The following sub-subsection discusses the presented installed and planned floating wind turbine projects results, with specific reference to their corresponding countries and their contributing distances to shore in the period 2009–2026, according to [11]. This period is selected because not all of the presented projects throughout this paper have available corresponding data that can be classified on the basis of projects' distances to shore.

4.2.7. Discussion of the Presented Installed and Planned Floating Wind Turbine Project Results in Terms of Their Corresponding Countries and Their Contributing Distances to Shore in the Period 2009–2026

This sub-subsection discusses some of the presented installed and planned floating wind turbine projects' results, with specific reference to their corresponding countries and their contributing distances to shore in the period 2009–2026, according to [11]. This period is selected because not all the presented projects have available corresponding data that can be classified on the basis of projects' distances to shore.

With regard to the available data for some of the presented projects in each of the presented continents and countries, Europe has the highest distances to shore (of 800 m–140 km) coming from 14 projects (7 installed and 7 planned) in 5 countries (Norway, the UK, France, Portugal, and Spain). Then comes Asia, with distances to shore of 5–62 km coming from 4 projects (3 installed and 1 planned) in 2 countries (Korea and Japan). It is followed by North America, with distances to shore of 330 m–48 km coming from 5 projects (1 installed and 4 planned) in 1 country (the US). This corresponds to the period 2009–2026.

It can therefore be concluded that the global floating wind projects' distances to shore are, according to the available data, 300 m–140 km with the highest distances to shore coming in the following continents and countries. Europe, with distances to shore of 800 m–140 km coming from 14 projects (seven installed and seven planned) in 5 countries (Norway with 10–140 km, the UK with 15–25 km, France with 15–22 km, Portugal with 5–20 km, and Spain with 800 m–3.2 km). Then comes Asia, with distances to shore of 5–62 km coming from four projects (three installed and one planned) in two countries (Korea and Japan). And finally, North America, with distances to shore of 330 m–48 km coming from five projects (one installed and four planned) in one country (the US). This corresponds to the period 2009–2026.

It can therefore be concluded that the global distances to shore are 300 m–140 km, coming from 23 projects (11 installed and 12 planned) in 8 countries (Norway, the UK, France, Portugal, Spain, the US, Korea, and Japan). This data is not claimed to cover all the contributing distances to shore of floating wind projects in continents and countries, both because of a shortage in data in these classifications, and because this data corresponds to 2020, and not necessarily to 2023, where some of the planned projects have been installed and further updates have occurred.

The following three subsections will provide references for the global floating wind situation, with a specific focus on Europe, the Power-to-X technology that is relevant to floating wind farms, and the feasibility of implementing floating wind projects in Romania.

The following subsection presents references for the global floating wind situation, with a specific focus on Europe.

4.3. References Regarding the Global Floating Wind Situation with a Focus on Europe

This subsection presents references for the global floating wind situation, with a focus on Europe.

A European floating wind research project was established to support the European floating wind development, with a total cost of 50 million euros and an expected revenue of 5000 (5 billion) million euros [80]. Europe is working towards both keeping its position as the world's floating wind leader and becoming the largest floating wind manufacturer by focusing on the following aspects. It will first focus on the European pre-commercialized floating wind projects and their corresponding incentives and grants, and then turn its attention towards the European-patent floating wind concepts and collect them in a corresponding design portfolio, which it will rapidly push toward serial production. Third, it will focus on European large-scale floating wind projects and make corresponding large governmental investments before finally developing the European coastal infrastructure and making it suitable for the implementation of large-scale floating wind projects. It will also focus on financing the private sector and making European inter-governmental floating wind collaborations [4].

A typical 2 MW Spar floating wind support structure weighs 140 tons and has a draft of 100 m, a water depth of 700 m, a tower height of 70 m, and a total height of 100 m. The demonstration of a typical floating wind project takes seven years, and an additional eight years is required for its construction, as was the case with the Hywind Scotland project [81].

The overall cost of floating wind projects breaks down into the implementation of floating support structures (24%), implementation of wind turbines (33%), operation and maintenance (23%), grid connection (15%), and decommissioning (5%) [82].

Spar-buoy is the simplest floating wind support structure, and it has convenient stability. Semi-submersible is less stable because of its comparably larger water-plane area and is also relatively difficult to manufacture. TLP is the most stable floating wind support structure, but it has both the most difficult installation and an inconvenient mooring system price (See Table 1). The typical cost of a generic floating wind turbine is 8 million euros/MW [9].

Spar-buoy has both ballast and drag-embedded catenary-mooring, as well as anchor stability systems. Semi-submersible and Barge have both buoyancy and mooring stability systems. TLP has both mooring lines and suction pile anchors [82].

Romania is a feasible candidate for floating wind implementation [83,84]. However, it lacks electrical infrastructures in the sea areas, which will make it necessary to implement floating wind Power-to-X technology that will do the job of transforming the produced electrical power mainly into hydrogen or compressed air, before accordingly transporting it through ships or other means to offshore or onshore customers. This technology could also be considered to be a candidate for replacing the European gas import from other countries, by converting renewable energy's produced electricity into other chemicals, such as methanol and synthetic natural gas [85].

The following subsection presents Power-to-X technology references of relevance to the floating wind projects.

4.4. Power-to-X Technology References of Relevance to Floating Wind Projects

This subsection presents Power-to-X technology references of relevance to the floating wind projects.

Paper [86] recommends the integration of Power-to-X technology with floating wind electrical power cables. This paper states that while further Power-to-X technologies will soon come, they are currently costly, and will require some time to reduce feasible implementation and maintenance costs. It also states that there is a Power-to-X project that, through the scope of its integration with the North Sea floating wind farms, will reduce costs of billions of euros in the future. This Power-to-X project is currently proposed to be put into operation in 2029. Paper [87] presents an offshore wind weather conditions modeling, with a specific focus on maintenance aspects. Paper [88] presents a floating wind turbines' reliability approach, which is of direct relevance to their power production conditions. While the latter two references are not necessarily directly relevant to our discussion, they can be taken as a point of reference by relevant discussions in the future.

Paper [89] presents a Power-to-X project that is relevant to the discussion of floating wind projects. This project transforms the produced electrical power into hydrogen by using its integrated Power-to-X technology, which is incorporated in each of the project's corresponding floating wind turbines. The transformed produced hydrogen will then be transported to a nearby hydrogen storage subsea unit before the hydrogen power is transported to an offshore customer. This project is planned to begin operation in 2025. The Power-to-X technology in each of the corresponding wind turbines consists of fuel cells, electrolysis, HV power, and seawater treatment.

Paper [90] presents economic considerations in the use electrolysis and methanol that are relevant to the Norwegian Power-to-X technology. Both the electrolysis and methanol are cost-efficient, which makes this technology implementable. This technology has great potential to produce synthetic natural gas (SNG), which could potentially eliminate the European dependence on gas transport from other continents. The cost of this synthetic natural gas is 110–140 euros/MWh. This technology has been stated to be clean, cheap, and very feasible, especially in northern Europe.

Paper [91] presents floating wind operation and power production aspects, along with important aspects that are relevant to the floating wind Power-to-X technology.

Paper [92] presents the implementation of the Power-to-X technology as being relevant to transforming renewable electricity into green products and services. It also mentions some floating wind aspects that are relevant and shows how the use of this technology can realize the floating wind future potential of 90 GW in New Zealand by converting the produced electrical energy into green hydrogen or green ammonia. This paper also states that the offshore bottom-fixed wind future energy potential of 14.4 GW in New Zealand can be realized if this technology is implemented.

Paper [93] states that the Power-to-X technology will play a great role in achieving zero emissions by 2050 in Europe, and also suggests that this technology should be efficiently integrated into the energy system. It also presents the following detailed layout of the process through which electrical energy is converted into some chemicals, which follows. First, the electrical energy is produced from renewable energy systems, and then electrolysis, such as oxygen and heat, is used to convert the electrical energy into hydrogen, ammonia, or hydrocarbons (gas or liquid forms). This paper also states that the European hydrogen Power-to-X capacity plan is 40 GW as of 2030, with a potential increase of an additional 40 GW coming from electrolysis capacity that will potentially be shipped from Ukraine and some North African countries. The European Power-to-X capacity plan states that, after 2030, 180 GW will be generated in the North Sea.

Paper [89] states that the investment and maintenance costs of the Power-to-X technology should be reduced to enable the technology to become feasible. The implementation of such technologies will connect the future North Sea floating wind projects with each other and cut their costs by 20 billion euros (this is the "Hub-and-Spoke Project in the North Sea").

Paper [94] proposes three Power-to-X typologies for hydrogen energy production in floating wind farms. The first typology uses centralized onshore electrolysis; the second typology uses decentralized offshore electrolysis; and the third uses centralized offshore electrolysis. The first has the advantage of easier installation and lower costs; the second has the advantage of using the existing electrolysis technology that comes from offshore bottom-fixed and onshore wind industries; and the third has the advantage of reducing the maintenance of individual turbines. See the reference for further details on this.

The following subsection refers to research that provides further insight into the feasibility of building floating wind projects in Romania and also addresses relevant considerations.

4.5. Research Related to the Feasibility of Floating Wind Projects in Romania and Relevant Considerations

This subsection addresses previous research contributions that relate to the feasibility of floating wind projects in Romania and also addresses some relevant considerations.

Girleanu et al. [95] state in their paper that the north-western region of the Black Sea has a high wind power density of 500 W/m^2, which makes it feasible to implement floating wind projects with a water depth capacity of 25–125 m.

Raileanu et al. [96] state in their paper that the most feasible Romanian city for floating wind implementation is Constanta, in an offshore region that is 20 km from the shore. In referring to wind turbines available on the market, they also state there are some limitations in the Romanian wind speeds. For example, the 10 MW Vestas wind turbine has a rated wind speed of 10 m/s. However, average Romanian regions have a maximum wind speed of 8 m/s, meaning that, as a consequence, the maximum power cannot be extracted from the respective wind turbine and the levelized cost of energy (LCOE) will be higher for the corresponding case. A further remedy has been suggested, which is to only consider wind turbines with a rated wind speed that is close to the wind speed in project implementation regions.

Onea and Rusu. [97,98] state in their papers that the Black Sea has the potential to implement wind turbines with a height of 80 m and a Betz limit of 50% (i.e., wind turbines that absorb 50% of the wind they are subjected to and convert it into electricity).

In a separate paper, Raileanu et al. [99] conclude that the Romanian Black Sea is a wind energy resource, especially in the winter, between January and April. They also note that Romania has an onshore wind farm ("Fantanele & Cogealac"), which is one of the largest European onshore wind projects. This project is in an onshore area that is 20 km from the Black Sea shore. This project has a power capacity of 600 MW and an installation cost of one billion euros.

Onea et al. [100] observe that the windiest part of the Black Sea is its north-eastern region, in Ukraine. However, they add that, due to its corresponding geopolitical climate issues, the Romanian region of the Black Sea is currently the best offshore wind candidate. They also observe that this area, which has the highest wind speed in the region, has satisfactory wind and wave dynamics that will make it feasible to implement hybrid offshore wind and wave power projects.

The following section concludes the data developed throughout this paper that are directly relevant to global floating wind concepts and projects.

5. Conclusions

This section presents conclusions about the presented and developed data throughout this paper that are directly relevant to global floating wind concepts and projects.

The data presented in this paper mainly relates to floating wind projects in the period 2008–2020 and planned floating wind projects installed in the period 2020–2027.

It was found that the global installed floating wind power capacity is 174.7 MW, mainly coming from Europe (123.5 MW), North America (30.2 MW), and Asia (21 MW). Refer to Section 4 for further details about the involved countries and corresponding projects, as well as comments on the reliability of this data.

It was found that the global planned floating wind power capacity is 4.5791 GW, mainly coming from the US (2.42 GW), Asia (1.634 GW), and Europe (525.1 MW). Refer to Section 4 for further details about the involved countries and corresponding projects, as well as comments on the reliability of this data.

With regard to the installed floating wind projects in Europe, the results of this paper suggest that in the period 2008–2020, the biggest contributor was the UK, with a power capacity of 89.5/123.5 MW, followed by Portugal, with 27.5/123.5 MW, and then a range of countries with an overall minor capacity of 6.5 MW (Norway, Germany, France, Spain, Denmark, and Sweden).

With regard to the planned floating wind projects in Europe, the results of this paper suggest that in the period 2020–2027, the biggest contributor was France, with a power capacity of 113.5 MW, followed by Ireland (106 MW), the UK (105 MW), Spain (103.2), and Norway (102.6 MW). Note that these values correspond to the projects announced up to 2020, and do not necessarily correspond to the current global floating wind power capacity.

Further classifications of some of the presented installed and planned projects were made. The analysis could not cover all the presented projects throughout this paper, due to a shortage of data regarding the costs, wind speeds, water depths, and distances to shore of some of the presented projects in the paper. This data corresponds to the period 2009–2026.

According to the available data of some of the presented projects throughout this paper, the global cost contribution from floating wind projects was 2.2247 billion dollars, mainly coming from Europe (2.1127 million dollars) and North America (112 billion dollars). There was no available cost data for Asian floating wind projects.

According to the available presented data of some of the presented projects throughout this paper, the global wind speed interval from floating wind projects was 8.5–48 m/s, mainly coming from Asia (45–48 m/s), Europe (10–40 m/s), and North America (8.5–40 m/s).

According to the available presented data of some of the presented projects throughout this paper, the global water depth interval from floating wind projects was 27.4 m–1 km, mainly coming from North America (27.4 m–1 km), Europe (33–300 m), and Asia (55–125 m).

According to the available presented data of some of the presented projects throughout this paper, the global distance to shore interval from floating wind projects was 300 m–140 km, mainly coming from Europe (800 m–140 km), Asia (5–62 km), and North America (330 m–48 km). Refer to Section 4 for further details about the countries and projects involved in these classifications.

The number of global floating wind turbine concepts throughout this paper is found to be 28, coming from 5 wind turbine types as follows: semi-submersible (13), spar (5), TLP (5), multi-turbine (3), and barge (2). Eight of these wind turbine concepts are made from steel, six from concrete, and four from steel and/or concrete. Refer to Section 4, where the details and reliability of these results are discussed in further detail.

It was also mentioned in Section 4 that different researchers offer different values when considering the total number of global floating wind concepts. Refer to Section 4, where more details about the corresponding countries and references are provided.

It was also illustrated throughout the paper that different research contributions refer to different data when considering global installed and planned floating wind power capacities. The data was therefore presented as it is in the different contributions, and an effort was made to illustrate some of the differences and possibilities that arose from discrepancies and issues with reliability in the presented data. This is discussed in more detail in Section 4.

It was also hinted throughout the paper that the floating wind levelized cost of energy (LCOE) is becoming comparable with its offshore bottom-fixed and onshore counterparts and it will ultimately depend on the extent and speed of its evolution. For example, according to one finding, the floating wind LCOE will be 50 euros/MWh in 2030, when its power capacity will be 4 GW, compared to its current value of 250 euros/MWh. Other research contributions referenced throughout the paper state that there will be a floating wind power capacity of 3 GW only coming from Norway by 2030. This suggests that such sources of data on the future of floating wind capacities and corresponding LCOE costs have discrepancies and are not necessarily reliable.

This paper also considered the Power-to-X technology of relevance to floating wind projects, and it was stated that this technology eliminates the need for corresponding enormous electrical infrastructures by converting the produced electrical energy, mainly into compressed air and hydrogen, or other chemical substances (such as hydrocarbons, ammonia, methanol, and synthetic natural gas) by using electrolysis such as oxygen and heat for the countries which lack such infrastructures in their offshore regions.

This paper also considered the feasibility of floating wind projects in Romania, and concluded, on the basis of the findings and research contributions referenced throughout the paper, that the lack of corresponding electrical infrastructures in Romanian Sea regions means that it is not currently feasible to implement floating wind projects in Romania. However, the integration of the Power-to-X technology into future floating wind projects in Romania could potentially make it possible to implement such projects. It was stated throughout the paper that the European hydrogen Power-to-X technology plan is 40 GW, with a potential additional plan of an extra 40 GW, making a total of 80 GW by 2030. After 2030, there is a plan to add a further 180 GW in the North Sea region.

Author Contributions: Conceptualization, M.M. and E.R.; methodology, M.M. and E.R.; software, M.M.; resources, M.M. and E.R.; data curation, M.M.; writing—original draft preparation, M.M.; writing—review and editing, M.M. and E.R.; supervision, E.R.; project administration, E.R.; funding acquisition, E.R. All authors have read and agreed to the published version of the manuscript.

Funding: This research received no external funding.

Data Availability Statement: Data are contained within the article.

Acknowledgments: This is an improved version of the work presented at the 11th Scientific Conference organized by the Doctoral Schools of "Dunarea de Jos", which was held at the University of Galati (SCDS-UDJG), in Galati, Romania, on 8 and 9 June 2023.

Conflicts of Interest: The authors declare no conflicts of interest.

References

1. Ruiz, A.; Onea, F.; Rusu, E. Study Concerning the Expected Dynamics of the Wind Energy Resources in the Iberian Nearshore. *Energies* **2020**, *13*, 4832. [CrossRef]
2. Onea, F.; Ruiz, A.; Rusu, E. An Evaluation of the Wind Energy Resources along the Spanish Continental Nearshore. *Energies* **2020**, *13*, 3986. [CrossRef]
3. Girleanu, A.; Onea, F.; Rusu, E. Assessment of the Wind Energy Potential along the Romanian Coastal Zone. *Inventions* **2021**, *6*, 41. [CrossRef]
4. ETIP Wind. Floating Offshore Wind: Delivering Climate Neutrality. 2020. Available online: https://etipwind.eu/news/floating-offshore-wind-delivering-climate-neutrality/ (accessed on 18 February 2024).
5. ABB and ZERO. Floating Offshore Wind—Norway's Next Offshore Boom? 2018. Available online: https://new.abb.com/docs/librariesprovider50/media/tv1012-br-havvind-notat-til-zerokonferansen---engelsk.pdf (accessed on 18 February 2024).
6. Wind Europe. Scaling up Floating Offshore Wind towards Competitiveness. 2021. Available online: https://windeurope.org/policy/position-papers/scaling-up-floating-offshore-wind-towards-competitiveness/ (accessed on 18 February 2024).
7. Wind Europe. Floating Offshore Wind Energy: A Policy Blueprint for Europe. 2017. Available online: https://windeurope.org/policy/position-papers/floating-offshore-wind-energy-a-policy-blueprint-for-europe/ (accessed on 18 February 2024).
8. Onea, F.; Rusu, E.; Rusu, L. Assessment of the Offshore Wind Energy Potential in the Romanian Exclusive Economic Zone. *J. Mar. Sci. Eng.* **2021**, *9*, 531. [CrossRef]
9. Lathigara, A.; Zhao, F. *Floating Offshore Wind—A Global Opportunity*; Global Wind Energy Council: Brussels, Belgium, 2022.
10. Jakobsen, E.G.; Ironside, N. Oceans Unlocked—A Floating Wind Future. 2021. Available online: https://www.cowi.com/insights/oceans-unlocked-a-floating-wind-future (accessed on 8 June 2022).
11. BOEM. *Floating Offshore Wind Turbine Development Assessment*; ABSG Consulting Inc.: Spring, TX, USA, 2021.
12. Kubiat, U.; Lemon, M. Drivers for and Barriers to the Take-up of Floating Offshore Wind Technology: A Comparison of Scotland and South Africa. *Energies* **2020**, *13*, 5618. [CrossRef]
13. Kosasih, K.M.A.; Suzuki, H.; Niizato, H.; Okubo, S. Demonstration Experiment and Numerical Simulation Analysis of Full-Scale Barge-Type Floating Offshore Wind Turbine. *J. Mar. Sci. Eng.* **2020**, *8*, 880. [CrossRef]
14. Möllerström, E. Wind Turbines from the Swedish Wind Energy Program and the Subsequent Commercialization Attempts—A Historical Review. *Energies* **2019**, *12*, 690. [CrossRef]
15. Borg, M.; Jensen, M.W.; Urquhart, S.; Andersen, M.T.; Thomsen, J.B.; Stiesdal, H. Technical Definition of the Tetra Spar Demonstrator Floating Wind Turbine Foundation. *Energies* **2020**, *13*, 4911. [CrossRef]
16. Ishihara, T.; Liu, Y. Dynamic Response Analysis of a Semi-Submersible Floating Wind Turbine in Combined Wave and Current Conditions Using Advanced Hydrodynamic Models. *Energies* **2020**, *13*, 5820. [CrossRef]
17. Chen, J.; Kim, M.H. Review of Recent Offshore Wind Turbine Research and Optimization Methodologies in Their Design. *J. Mar. Sci. Eng.* **2022**, *10*, 28. [CrossRef]

18. Yang, R.Y.; Chuang, T.C.; Zhao, C.; Johanning, L. Dynamic Response of an Offshore Floating Wind Turbine at Accidental Limit States—Mooring Failure Event. *Appl. Sci.* **2022**, *12*, 1525. [CrossRef]
19. Gao, S.; Zhang, L.; Shi, W.; Wang, B.; Li, X. Dynamic Responses for Wind Float Floating Offshore Wind Turbine at Intermediate Water Depth Based on Local Conditions in China. *J. Mar. Sci. Eng.* **2021**, *9*, 1093. [CrossRef]
20. Liu, S.; Chuang, Z.; Wang, K.; Li, X.; Chang, X.; Hou, L. Structural Parametric Optimization of the VOLTURNUS-S Semi-Submersible Foundation for a 15 MW Floating Offshore Wind Turbine. *J. Mar. Sci. Eng.* **2022**, *10*, 1181. [CrossRef]
21. Ahn, H.; Ha, Y.J.; Cho, S.G.; Lim, C.H.; Kim, K.W. A Numerical Study on the Performance Evaluation of a Semi-Type Floating Offshore Wind Turbine System According to the Direction of the Incoming Waves. *Energies* **2022**, *15*, 5485. [CrossRef]
22. Pham, T.D.; Shin, H. The Effect of the Second-Order Wave Loads on Drift Motion of a Semi-Submersible Floating Offshore Wind Turbine. *J. Mar. Sci. Eng.* **2020**, *8*, 859. [CrossRef]
23. Desmond, C.J.; Hinrichs, J.C.; Murphy, J. Uncertainty in the Physical Testing of Floating Wind Energy Platforms' Accuracy versus Precision. *Energies* **2019**, *12*, 435. [CrossRef]
24. Ghigo, A.; Cottura, L.; Caradonna, R.; Bracco, G.; Mattiazzo, G. Platform Optimization and Cost Analysis in a Floating Offshore Wind Farm. *J. Mar. Sci. Eng.* **2020**, *8*, 835. [CrossRef]
25. Petracca, E.; Faraggiana, E.; Ghigo, A.; Sirigu, M.; Bracco, G.; Mattiazzo, G. Design and Techno-Economic Analysis of a Novel Hybrid Offshore Wind and Wave Energy System. *Energies* **2022**, *15*, 2739. [CrossRef]
26. Qu, X.; Yao, Y. Numerical and Experimental Study of Hydrodynamic Response for a Novel Buoyancy-Distributed Floating Foundation Based on the Potential Theory. *J. Mar. Sci. Eng.* **2022**, *10*, 292. [CrossRef]
27. Zhou, Y.; Ren, Y.; Shi, W.; Li, X. Investigation on a Large-Scale Braceless-TLP Floating Offshore Wind Turbine at Intermediate Water Depth. *J. Mar. Sci. Eng.* **2022**, *10*, 302. [CrossRef]
28. González, J.; Payán, M.; Santos, J.; Gonzalez, A. Optimal Micro-Siting of Weathervaning Floating Wind Turbines. *Energies* **2021**, *14*, 886. [CrossRef]
29. Walia, D.; Schünemann, P.; Hartmann, H.; Adam, F.; Großmann, J. Numerical and Physical Modeling of a Tension-Leg Platform for Offshore Wind Turbines. *Energies* **2021**, *14*, 3554. [CrossRef]
30. Lamei, A.; Hayatdavoodi, M. On motion analysis and elastic response of floating offshore wind turbines. *J. Ocean. Eng. Mar. Energy* **2020**, *6*, 71–90. [CrossRef]
31. Renzi, E.; Michele, S.; Zheng, S.; Jin, S.; Greaves, D. Niche Applications and Flexible Devices for Wave Energy Conversion: A Review. *Energies* **2021**, *14*, 6537. [CrossRef]
32. Solomin, E.; Sirotkin, E.; Cuce, E.; Selvanathan, S.P.; Kumarasamy, S. Hybrid Floating Solar Plant Designs: A Review. *Energies* **2021**, *14*, 2751. [CrossRef]
33. Haces-Fernandez, F.; Li, H.; Ramirez, D. Assessment of the Potential of Energy Extracted from Waves and Wind to Supply Offshore Oil Platforms Operating in the Gulf of Mexico. *Energies* **2018**, *11*, 1084. [CrossRef]
34. Aquatera. Dounreay-Tri Floating Wind Demonstration Project (Environmental Statement). 2016. Available online: https://marine.gov.scot/data/environmental-statement-dounreay-tri-floating-wind-demonstration-project (accessed on 18 February 2024).
35. GICON. N.d. GICON SOF. A Modular and Cost Competitive TLP Solution. Brochure. Available online: https://www.gicon.de/index.php/gicon-gruppe/infomaterial?file=files/GICON-Gruppe/Downloads/Prospekte/Englisch/GICON%C2%AE%20SOF%20(TLP)%20-%20Offshore%20substructure.pdf&cid=8716 (accessed on 18 February 2024).
36. Kyle, R.; Früh, W.G. The transitional states of a floating wind turbine during high levels of surge. *Renew. Energy* **2022**, *200*, 1469–1489. [CrossRef]
37. Ulazia, A.; Nafarrate, A.; Ibarra-Berastegi, G.; Sáenz, J.; Carreno-Madinabeitia, S. The Consequences of Air Density Variations Over Northeastern Scotland for Offshore Wind Energy Potential. *Energies* **2019**, *12*, 2635. [CrossRef]
38. Ramos, V.; Giannini, G.; Cabral, T.; López, M.; Santos, P.; Taveira-Pinto, F. Assessing the Effectiveness of a Novel WEC Concept as a Co-Located Solution for Offshore Wind Farms. *J. Mar. Sci. Eng.* **2022**, *10*, 267. [CrossRef]
39. Andersen, M.T. Floating Foundations for Offshore Wind Turbines. Ph.D. Thesis, Faculty of Engineering and Science, Aalborg University, Aalborg, Denmark, 2016.
40. Pham, T.D.; Dinh, M.C.; Kim, H.M.; Nguyen, T.T. Simplified Floating Wind Turbine for Real-Time Simulation of Large-Scale Floating Offshore Wind Farms. *Energies* **2021**, *14*, 4571. [CrossRef]
41. Dong, Y.; Chen, Y.; Liu, H.; Zhou, S.; Ni, Y.; Cai, C.; Zhou, T.; Li, Q. Review of Study on the Coupled Dynamic Performance of Floating Offshore Wind Turbines. *Energies* **2022**, *15*, 3970. [CrossRef]
42. Edwards, E.C.; Holcombe, A.; Brown, S.; Ransley, E.; Hann, M.; Greaves, D. Evolution of floating offshore wind platforms: A review of at-sea devices. *Renew. Sustain. Energy Rev.* **2023**, *183*, 113416. [CrossRef]
43. Baita-Saavedra, E.; Cordal-Iglesias, D.; Filgueira-Vizoso, A.; Morató, À.; Lamas-Galdo, I.; Álvarez-Feal, C.; Carral, L.; Castro-Santos, L. An Economic Analysis of An Innovative Floating Offshore Wind Platform Built with Concrete: The SATH® Platform. *Appl. Sci.* **2020**, *10*, 3678. [CrossRef]
44. Arredondo-Galeana, A.; Brennan, F. Floating Offshore Vertical Axis Wind Turbines: Opportunities, Challenges and Way Forward. *Energies* **2021**, *14*, 8000. [CrossRef]
45. Tamarit, F.; García, E.; Quiles, E.; Correcher, A. Model and Simulation of a Floating Hybrid Wind and Current Turbines Integrated Generator System, Part I: Kinematics and Dynamics. *J. Mar. Sci. Eng.* **2023**, *11*, 126. [CrossRef]

46. Mathern, A.; Von der Haar, C.; Marx, S. Concrete Support Structures for Offshore Wind Turbines: Current Status, Challenges, and Future Trends. *Energies* **2021**, *14*, 1995. [CrossRef]
47. Ringvej Dahl, I.; Tveitan, B.W.; Cowan, E.C. The Case for Policy in Developing Offshore Wind: Lessons from Norway. *Energies* **2022**, *15*, 1569. [CrossRef]
48. Flagship. Floating Offshore Wind Optimization for Commercialization. Brochure. 2020. Available online: https://www.flagshiproject.eu/wp-content/uploads/2021/03/003_Brochure_Cambios_Flagship.pdf (accessed on 18 February 2024).
49. Ideol. N, A. IDEOL Winning Solutions for Offshore Wind. Brochure. Available online: https://www.bw-ideol.com/sites/default/files/pdf/ideol_-_press_kit.pdf (accessed on 18 February 2024).
50. Brocklehurst, B.; Bradshaw, K. *AFLOWT Environmental Impact Assessment Scoping Report*; SEAI: Dublin, Ireland, 2022.
51. Emerald Floating Wind. Emerald Project: Foreshore License Application for Site Investigation Work—Risk Assessment for Annex IV Species. 2021. Available online: https://assets.gov.ie/240131/d557cccf-1cf4-4658-97cd-04aada10a792.pdf (accessed on 18 February 2024).
52. BOURBON. Bourbon Subsea Services Awarded by EOLMED an EPCI Contract for One of the First Floating Wind Pilot Farm in the French Mediterranean Sea. Press Release. 2022. Available online: https://www.bourbonoffshore.com/en/press-releases/bourbon-subsea-services-awarded-eolmed-epci-contract-one-first-floating-wind-pilot (accessed on 18 February 2024).
53. IGEOTEST. Provence Grand Large & Mistral Floating Offshore Wind Project. Brochure. 2013. Available online: https://www.sbmoffshore.com/newsroom/news-events/sbm-offshore-announces-successful-installation-3-floating-wind-units (accessed on 18 February 2024).
54. COPERNICUS. MET-Ocean Studies and Key Environmental Parameters for Floating Offshore Wind Technology. Brochure. 2018. Available online: https://www.copernicus.eu/en/use-cases/met-ocean-studies-and-key-environmental-parameters-floating-offshore-wind-technology (accessed on 18 February 2024).
55. EOLFI. GROIX & Belle-Ile Floating Wind Turbines, Signature of a New Phase at NAVEXPO. Press Release. 2016. Available online: https://www.eolfi.com/ressources/_pdfs/1/3d0f273-183-CP-EOLFI-signature-CDC-MERID.pdf (accessed on 18 February 2024).
56. Nair, R.J. Towing of Floating Wind Turbine Systems. Master's Thesis, UIT University, Karachi, Pakistan, 2020.
57. Voltà, L. Self-Alignment on Single Point Moored Downwind Floater—The PivotBuoy® Concept. In *EERA DeepWind'2021 Programme, Proceedings of the 18th Deep Sea Offshore Wind R&D Digital Conference, Trondheim, Norway, 13–15 January 2021*; Institute of Physics Publishing (IOP): Bristol, UK, 2021.
58. Margheritini, L.; Rialland, A.; Sperstad, I.B. The Capitalisation Potential for Ports during the Development of Marine Renewable Energy. Beppo. 2015. Available online: https://www.sintef.no/globalassets/project/beppo/beppo_wp3_report_capitalisation_potential_for_ports_in_mre.pdf (accessed on 18 February 2024).
59. RWE Offshore Wind GmbH. Demo SATH Has Achieved a Key Project Milestone with the Offshore Installation. Press Release. 2023. Available online: https://www.rwe.com/en/press/rwe-offshore-wind-gmbh/2023-08-11-demosath-has-achieved-a-key-project-milestone-with-the-offshore-installation/ (accessed on 18 February 2024).
60. Díaz, H.; Serna, J.; Nieto, J.; Guedes Soares, C. Market Needs, Opportunities, and Barriers for the Floating Wind Industry. *J. Mar. Sci. Eng.* **2022**, *10*, 934. [CrossRef]
61. Codiga, D.A. *Progression Hawaii Offshore Wind*; Schlack ITO: Honolulu, HI, USA, 2018.
62. *CADEMO Research and Demonstration Goals*; CADEMO Corporation: Palm Springs, CA, USA, 2021.
63. Cabigan, M.K. Offshore Wind Development in an Integrated Energy Market in the North Sea: Defining Regional Energy Policies in an Integrated Market and Offshore Wind Infrastructure. Master's Thesis, NTNU, Trondheim, Norway, 2022.
64. Beiter, P.; Musial, W.; Duffy, P.; Cooperman, A.; Shields, M.; Heimiller, D.; Optis, M. *The Cost of Floating Offshore Wind Energy in California Between 2019 and 2032*; NREL/TP-5000-77384; National Renewable Energy Laboratory: Golden, CO, USA, 2020.
65. *Goto Floating Wind Farm LLC Consortium Begins Offshore Wind Turbine Assembly (Towards Realization of Floating Offshore Wind Power Generator)*; Press Release; INPEX Corporation: Tokyo, Japan, 2022.
66. Belvasi, N.; Conan, B.; Schliffke, B.; Perret, L.; Desmond, C.; Murphy, J.; Aubrun, S. Far-Wake Meandering of a Wind Turbine Model with Imposed Motions: An Experimental S-PIV Analysis. *Energies* **2022**, *15*, 7757. [CrossRef]
67. Tan, L.; Ikoma, T.; Aida, Y.; Masuda, K. Mean Wave Drift Forces on a Barge-Type Floating Wind Turbine Platform with Moonpools. *J. Mar. Sci. Eng.* **2021**, *9*, 709. [CrossRef]
68. Bensalah, A.; Barakat, G.; Amara, Y. Electrical Generators for Large Wind Turbine: Trends and Challenges. *Energies* **2022**, *15*, 6700. [CrossRef]
69. Lee, J.; Xydis, G. Floating offshore wind project development in South Korea without government subsidies. *Clean Technol. Environ. Policy* **2023**. [CrossRef]
70. Ha, K.; Kim, J.B.; Yu, Y.; Seo, H. Structural Modeling and Failure Assessment of Spar-Type Substructure for 5 MW Floating Offshore Wind Turbine under Extreme Conditions in the East Sea. *Energies* **2021**, *14*, 6571. [CrossRef]
71. Technip Energies. Technip Energies, Subsea 7 and SAMKANG M&T to Perform FEED for Gray Whale 3 Floating Offshore Wind Project in South Korea. 2022. Available online: https://investors.technipenergies.com/news-releases/news-release-details/technip-energies-subsea-7-and-samkang-mt-perform-feed-gray-whale (accessed on 18 February 2024).
72. Shin, H. *Introduction to the 1.2 GW Floating Offshore Wind Farm Project in the East Sea, Ulsan, Korea*; University of Ulsan: Ulsan, Republic of Korea, 2020.

73. Strivens, S.; Northridge, E.; Evans, H.; Harvey, M.; Camp, T.; Terry, N. *Floating Wind Joint Industry Project (Phase III Summary Report)*; Carbon Trust: London, UK, 2021.
74. EOLFI. N, A. EOLFI, A Member of the Shell Group. Brochure. Available online: https://www.shell.com/energy-and-innovation/new-energies/new-energies-media-releases/shell-agrees-to-acquire-eolfi.html (accessed on 18 February 2024).
75. Pantusa, D.; Tomasicchio, G.R. Large-scale offshore wind production in the Mediterranean Sea. *Cogent Eng.* **2019**, *6*, 1661112. [CrossRef]
76. Equinor. N.d. Hywind Tampen. Available online: https://www.equinor.com/energy/hywind-tampen (accessed on 9 June 2022).
77. Equinor. N.d. Hywind Scotland. Available online: https://www.equinor.com/energy/hywind-scotland (accessed on 9 June 2022).
78. Baltscheffsky, H. Join the Future. 2021. Available online: https://www.utilityeda.com/wp-content/uploads/Wed_Session-8-a_Hexicon-AB_Henrik-Baltscheffsky_Offshore-Wind-Power.pdf (accessed on 9 June 2022).
79. IRENA. *Innovation Landscape for a Renewable-Powered Future: Solutions to Integrate Variable Renewables*; International Renewable Energy Agency: Masdar City, United Arab Emirates, 2019; ISBN 978-92-9260-111-9.
80. Tande, J.O.; Wagenaar, J.W.; Latour, M.I.; Aubrun, S.; Wingerde, A.V.; Eecen, P.; Andersson, M.; Barth, S.; McKeever, P.; Cutululis, N.A. Proposal for European Lighthouse Project: Floating Wind Energy. EU SETWind Project. 2022. Available online: https://blogg.sintef.no/wp-content/uploads/2022/03/Lighthouse_SetWind_I.pdf (accessed on 18 February 2024).
81. Nielsen, F.G. *Hywind—From Idea to the World's First Wind Farm Based upon Floaters*; University of Bergen: Bergen, Norway, 2017.
82. Butterfield, S.; Musial, W.; Jonkman, J.; Sclavounos, P. *Engineering Challenges for Floating Offshore Wind Turbines*; NREL/CP-500-38776; National Renewable Energy Laboratory: Golden, CO, USA, 2007.
83. Yildirir, V.; Rusu, E.; Onea, F. Wind Variation near the Black Sea Coastal Areas Reflected by the ERA5 Dataset. *Inventions* **2022**, *7*, 57. [CrossRef]
84. Yildirir, V.; Rusu, E.; Onea, F. Wind Energy Assessments in the Northern Romanian Coastal Environment Based on 20 Years of Data Coming from Different Sources. *Sustainability* **2022**, *14*, 4249. [CrossRef]
85. Rickert, C.; Thevar Parambil, A.M.; Leimeister, M. Conceptual Study and Development of an Autonomously Operating, Sailing Renewable Energy Conversion System. *Energies* **2022**, *15*, 4434. [CrossRef]
86. Kurniawati, I.; Beatriz, B.; Varghese, R.; Kostadinović, D.; Sokol, I.; Hemida, H.; Alevras, P.; Baniotopoulos, C. Conceptual Design of a Floating Modular Energy Island for Energy Independency: A Case Study in Crete. *Energies* **2023**, *16*, 5921. [CrossRef]
87. Skobiej, B.; Niemi, A. Validation of copula-based weather generator for maintenance model of offshore wind farm. *WMU J. Marit. Aff.* **2022**, *21*, 73–87. [CrossRef]
88. Xu, X.; Xing, Y.; Gaidai, O.; Wang, K.; Sandipkumar Patel, K.; Dou, P.; Zhang, Z. A novel multi-dimensional reliability approach for floating wind turbines under power production conditions. *Front. Mar. Sci.* **2022**, *9*, 970081. [CrossRef]
89. Fernández-Guillamón, A.; Kaushik, D.; Cutululis, N.A.; Molina-García, A. Offshore Wind Power Integration into Future Power Systems: Overview and Trends. *J. Mar. Sci. Eng.* **2019**, *7*, 399. [CrossRef]
90. Agarwal, R. Economic Analysis of Renewable Power-to-Gas in Norway. *Sustainability* **2022**, *14*, 16882. [CrossRef]
91. Connolly, P.; Crawford, C. Comparison of optimal power production and operation of unmoored floating offshore wind turbines and energy ships. *Wind. Energy Sci.* **2023**, *8*, 725–746. [CrossRef]
92. *Power to X—Transforming Renewable Electricity into Green Products and Services*; Concept Paper; Venture Taranaki: New Plymouth, New Zealand, 2021.
93. North Sea Wind Power Hub. Integration of Offshore Wind. Discussion Paper #1. 2021. Available online: https://northseawindpowerhub.eu/sites/northseawindpowerhub.eu/files/media/document/NSWPH%20Integration%20of%20offshore%20wind.pdf (accessed on 18 February 2024).
94. Ibrahim, O.S.; Singlitico, A.; Proskovics, R.; McDonagh, S.; Desmond, C.; Murphy, J.D. Dedicated large-scale floating offshore wind to hydrogen: Assessing design variables in proposed typologies. *Renew. Sustain. Energy Rev.* **2022**, *160*, 112310. [CrossRef]
95. Girleanu, A.; Onea, F.; Rusu, E. The efficiency and coastal protection provided by a floating wind farm operating in the Romanian nearshore. *Energy Rep.* **2021**, *7*, 13–18. [CrossRef]
96. Raileanu, A.B.; Onea, F.; Rusu, E. Implementation of Offshore Wind Turbines to Reduce Air Pollution in Coastal Areas—Case Study Constanta Harbor in the Black Sea. *J. Mar. Sci. Eng.* **2020**, *8*, 550. [CrossRef]
97. Onea, F.; Rusu, E. An Assessment of Wind Energy Potential in the Caspian Sea. *Energies* **2019**, *12*, 2525. [CrossRef]
98. Onea, F.; Rusu, E. Efficiency assessments for some state-of-the-art wind turbines in the coastal environments of the Black and the Caspian seas. *Energy Explor. Exploit.* **2016**, *34*, 217–234. [CrossRef]
99. Raileanu, A.; Onea, F.; Rusu, E. Assessment of the wind energy potential in the coastal environment of two enclosed seas. In Proceedings of the OCEANS 2015—Genova, Genova, Italy, 18–21 May 2015. [CrossRef]
100. Onea, F.; Raileanu, A.; Rusu, E. Evaluation of the Wind Energy Potential in the Coastal Environment of Two Enclosed Seas. *Adv. Meteorol.* **2015**, *2015*, 808617. [CrossRef]

 *inventions*

Article

IIR Shelving Filter, Support Vector Machine and k-Nearest Neighbors Algorithm Application for Voltage Transients and Short-Duration RMS Variations Analysis

Vladislav Liubčuk [1,*], Gediminas Kairaitis [1], Virginijus Radziukynas [1] and Darius Naujokaitis [1,2]

[1] Smart Grids and Renewable Energy Laboratory, Lithuanian Energy Institute, 44403 Kaunas, Lithuania
[2] Department of Applied Informatics, Kaunas University of Technology, 51368 Kaunas, Lithuania
* Correspondence: vladislav.liubcuk@lei.lt; Tel.: +370-(694)-60-558

Abstract: This paper focuses on both voltage transients and short-duration RMS variations, and presents a unique and heterogeneous approach to their assessment by applying AI tools. The database consists of both real (obtained from Lithuanian PQ monitoring campaigns) and synthetic data (obtained from the simulation and literature review). Firstly, this paper investigates the fundamental grid component and its harmonics filtering with an IIR shelving filter. Secondly, in a key part, both SVM and KNN are used to classify PQ events by their primary cause in the voltage–duration plane as well as by the type of short circuit in the three-dimensional voltage space. Thirdly, since it seemed to be difficult to interpret the results in the three-dimensional space, the new method, based on Clarke transformation, is developed to convert it to two-dimensional space. The method shows an outstanding performance by avoiding the loss of important information. In addition, a geometric analysis of the fault voltage in both two-dimensional and three-dimensional spaces revealed certain geometric patterns that are undoubtedly important for PQ classification. Finally, based on the results of a PQ monitoring campaign in the Lithuanian distribution grid, this paper presents a unique discussion regarding PQ assessment gaps that need to be solved in anticipation of a great leap forward and refers them to PQ legislation.

Keywords: power quality; shelving filter; support vector machine; k-nearest neighbors algorithm; Clarke transformation; transient; RMS variation; sag; dip; swell

Citation: Liubčuk, V.; Kairaitis, G.; Radziukynas, V.; Naujokaitis, D. IIR Shelving Filter, Support Vector Machine and k-Nearest Neighbors Algorithm Application for Voltage Transients and Short-Duration RMS Variations Analysis. *Inventions* **2024**, *9*, 12. https://doi.org/10.3390/inventions9010012

Academic Editor: Om P. Malik

Received: 12 December 2023
Revised: 28 December 2023
Accepted: 4 January 2024
Published: 9 January 2024

1. Introduction

Nowadays, greater and greater integration of artificial intelligence (AI) into the modern world is observed. Electric power systems together with power quality (PQ) application are not an exception. A state-of-the-art review [1] (2015) on both signal processing and AI techniques application in the classification of PQ disturbances distinguishes the following stages of the process:

1. Input data stage. It is obvious that the quality of machine learning is directly related to input data which is closely related to algorithms used and tasks undertaken. However, discussion on these aspects is skipped in [1]: in our opinion, one of the reasons (for that skipping) can be the deficiency of PQ monitoring systems at that time [2]. In this paper, initial data gathering is described in Section 2.1.
2. Feature extraction stage. The following taxonomy of the techniques is given in [1]: Fourier transform, Kalman filter, wavelet transform, S-transform, Hilbert–Huang transform, Gabor transform, and miscellaneous or less often used (e.g., time–frequency representation, Teager energy operator, etc.). A result of classification highly depends on both the features selection strategy and their extraction accuracy.
3. Classification stage. The following tools are highlighted in [1]: artificial neutral network, support vector machine, fuzzy expert system, and miscellaneous (e.g., k-nearest neighbors).

4. Feature selection and parameter optimization stage. In this stage, the redundant features with low recognition rate are discarded. The following tools for selection of the best suitable feature subset are highlighted in [1]: genetic algorithms, particle swarm optimization, and ant colony optimization.
5. Decision stage. Discussion on this topic is also skipped in [1]: in our opinion, one of the reasons can be that a practical significance of many research studies (algorithms) is insufficient and more comprehensive investigations (technical progress) are required for their commercialization and application in electric power systems.

The authors of [1], as can be noticed from their paper title, treat signal processing and AI algorithms as two separate fields. This is because, especially in the business world, AI is mostly associated with machine learning using the most well-known tools such as neural networks. However, if examined more deeply, the philosophy of AI is not so primitive, and the absence of a unified and simple definition accompanies it throughout the entire timeline of evolution of intelligence and thinking. Its origin dates back to the invention of writing, with the earliest known libraries on clay tablets in Mesopotamia (initially under the control of Sumerians and Akkadians, later ceded to such famous states as the Old Babylonian Empire, Assyrian Empire and New Babylonian Empire, and then followed by the defeat to Achaemenid Persian Empire) as well as in other places in particular cradles of civilization such as ancient Egypt, Indus Valley Civilization, and ancient China, later followed by Greek mythology which includes the idea of artificial beings and intelligent machines such as Talos (a giant bronze automaton made to protect Europe in Crete) [3]. At this point, we mentioned the term of 'automaton', a relatively self-operating machine, the first form of robot, which seems to be invented by ancient Egyptians: their statues of gods "spoke, moved, acted—not metaphorically, but actually", for example, replying to each question by a movement of the head [3–5]. Aristotle speculated in his *Politics* [6] (book 1, part 4) about automata as a mean for abolitionism: "For if every instrument could accomplish its own work, obeying or anticipating will of others, like the statues of Daedalus, [...] if, in like manner, the shuttle would weave and the plectrum touch the lyre without hand to guide them, chief workmen would not want servants, nor masters slaves". His syllogisms for proper reasoning are also worth mentioning, because this highly contributed to the development of logic which is an indispensable part of AI [3,7]. To continue, the prohibition against making graven images (the second commandment), taking place in the desert of Sinai, is also discussed in [3] through a prism of AI. When the discussion comes to medieval and early modern periods, the following achievements should be highlighted: (1) the flourishing of Arab mathematics and science (e.g., zairja—a device for ideas generation by mechanical means) during the Golden Age of Islam, and these ideas transferring to Europe (via Spain, Sicily, and Crusader kingdoms in the Levant); (2) appearance of mechanical clocks, the first modern measuring machine, in European towns (15th–16th centuries); (3) invention of a mechanical calculator, the Pascaline, by Blaise Pascal; (4) philosophy and other scientific contributions of René Descartes (e.g., rationalism, mind–body dualism, Cartesian coordinate system, etc.) [3]. Obviously, it is impossible to encompass (classify) all important milestones in one paragraph; hence, many contributions remain unmentioned, in particular outside Europe and the Middle East. The scientific revolution (Nicolaus Copernicus' publication on the heliocentric theory (in 1543) is often considered as its beginning, while Isaac Newton's publication on the laws of motion and universal gravitation (in 1687) is considered as its culmination) and the Age of Enlightenment (17th–18th centuries) followed by the First Industrial Revolution laid the foundation for further acceleration in the development of science in the late modern period. Moreover, science fiction did not disappear with such works as Hoffman's *The Sandman*, Goethe's *Faust*, and Mary Shelley's *Frankenstein* [3].

The history of modern AI began in the mid-20th century with such achievements as cryptanalysis of the Enigma (which enabled Allies to read considerable amounts of Morse-coded messages of Axis power), the Turing test (a method to determine whether a machine can demonstrate human intelligence), checkers (draughts) and chess engines [3,7], i.e., before AI commercialization (which, according to [3], begun after 1980) and large-scale application of artificial neural networks. Currently, the following major concepts of AI are often distinguished in the literature: (1) artificial general intelligence (or strong AI), and (2) artificial narrow intelligence (or weak AI) [8]. Also, other categories can be found such as generative AI or symbolic AI; however, detailed analysis of this field is outside the scope of this paper. According to [7], most researchers "take the weak AI hypothesis for granted", and do not care about the general intelligence; however, "all AI researchers should be concerned with the ethical implications of their work". The field of electric power systems, including its PQ application and this paper, is not an exception. In the PQ field, a wide-ranging implementation of AI may be expected; however, at the moment, the suitability of many AI algorithms still remains insufficiently explored [2]. In this paper, the small paragraph of scientific achievements of earlier times is given not without reason—all ideas (approaches) are potentially suitable (beneficial) for joint PQ and AI research. For example, the usage of logical reasoning in the field of PQ to discuss the results obtained can be found in [9]. Soft skills and philosophical insights during results interpretation are just as important as software code writing; thus, it is inseparable from the knowledge in philosophy, development of scientific ideas and their spread, etc. Ideally, for use in practice, an AI algorithm may be multicriterial and based on a complex mathematical apparatus, but simultaneously, it should have such features as simplicity, understandability, sufficient accuracy, time efficiency, cost efficiency, etc. In the case of embedded systems, algorithms must also be resource saving to avoid an undesirably fast runoff of the battery. Moreover, it is not a secret that sometimes, the conclusion about the accuracy of the developed algorithm can be limited (incorrect), because either the training data set or validation (test) data set may not represent its population sufficiently enough. In PQ papers, the risk of such or similar limitations is inevitable and cannot be fully eliminated. For example, in the case of the PQ monitoring allocation task, the size of the test schemes used is mostly up to 123 buses (IEEE test scheme); therefore, it remains unclear whether any proposed algorithm will be suitable for a large power system [2]. In addition, in our opinion, each paper should discuss in which structural part of the PQ monitoring process a proposed algorithm can be used (see [2] for the general structure of a remote PQ monitoring process).

Each PQ research should be started from the definition of target PQ events [2]. The target group of this paper consists of both transients and short-duration RMS variations. The next stage should be a selection of both features and algorithms. Let us begin the literature review with the PQ standards, because they include (reflect) universally agreed (applied) approaches. Firstly, in the case of transients, there are almost no requirements for both measurement and assessment in IEC 61000-4-30:2015 [10]. On the other hand, this standard provides ideas about transient voltage detection by the comparative method, envelope method, sliding-window method, etc. When the event is detected by either method, the following list of classification methods and parameters is given: peak value, overshoot voltage, the rate of rise (i.e., the first time derivative of either voltage or current) of the leading edge, frequency parameters (spectrum), duration, damping coefficient, frequency of occurrence, energy and power (available or conveyed), and continuity type (continuous or single-shot transient). EN 50160:2010 [11] does not provide any noteworthy information about transients nor does it establish the norms. IEEE Std 1159-2019 [12] uses the rise time, duration, spectral content and magnitude, and it also gives additional hints for the characterization of transient energy, direction, form (e.g., positive, negative, unipolar, bipolar, oscillatory, multiple-zero crossing). Secondly, in the case of voltage sags, IEEE Std 1564-2014 "Guide for Voltage Sag Indices" [13] proposes a voltage sag energy index and voltage sag severity index (using SEMI F47 curve as a reference), along with the other useful information regarding both site and system indices, which are not relevant to the scope of this paper. Meanwhile, IEC 61000-4-30:2015 states that "depending on the purpose of the measurement, other characteristics in addition to depth and duration should be considered" and presents the following ideas: voltage sag unbalance, phase angle of the beginning, phase shift, missing voltage, voltage sag distortion (e.g., estimated by total harmonic distortion—THD). EN 50160:2010 along with IEEE Std 1159-2019 use only voltage and duration. The voltage–duration plane is undoubtedly the most preferred and best understood method which, for example, has hegemony in such areas as the settings of relay protection and automation, equipment immunity testing, and fault-ride-through requirements.

PQ events classification is not an extremely recent research field. For example, [14] (2001) proposes a voltage sag classification and characterization method based on fuzzy logic: (1) voltage sags are classified into three groups—caused by grid fault, large motor starting, and interaction between motor operation and grid fault; (2) three features are extracted from the simulated voltage waveforms—phase angle shift, duration, and voltage change. As has been mentioned above, earlier achievements are a necessary prerequisite for modern research: for example, it can be highlighted that the k-nearest neighbors algorithm was firstly developed in 1951 [7], but it will still be investigated in this paper. The state-of-the-art of the joint PQ and AI field until 2015 is more or less reviewed by [1] (2015). Then, for example, [15] (2016) investigates the application of optimal multi-resolution fast S-transform along with classification and regression tree. The following PQ disturbances and their combinations have been generated by the simulation: transient, interruption, sag, swell, flicker, harmonics, and six their combinations—transient with sag, transient with swell, transient with flicker, sag with harmonics, swell with harmonics, and flicker with harmonics. The selected features—statistical parameters (maximum, minimum, standard deviation, skewness, kurtosis, etc.) and energy—are extracted from the simulated voltage waveforms. Next, [16] (2020) reviews the state-of-the-art of PQ events detection and classification until 2020. Feature extraction techniques are grouped into Fourier transform—FT (including its well-known variants: discrete-time Fourier transform—DTFT, fast Fourier transform—FFT, and short-time Fourier transform—STFT), S-transform—ST, Hilbert–Huang transform—HHT (including empirical mode decomposition—EMD), wavelet transform—WT (including tunable Q-factor wavelet transform—TQWT), and miscellaneous techniques (e.g., Gabor transform—GT, optimal feature selection based on ant colony optimization—ACO, etc.). Also, the following AI techniques, used in the PQ events detection and classification, are discussed: support vector machine—SVM (including its multiclass extension—MSVM), artificial neural network—ANN (including the following types: feedforward neural network—FNN, probabilistic neural network—PNN, recurrent neural network—RNN), fuzzy logic—FL, neuro-fuzzy system—NFS, genetic algorithm—GA, deep learning-based methods (convolutional neural network—CNN, long short-term memory—LSTM), and miscellaneous (e.g., extreme learning machine—ELM, particle swarm optimization—PSO, etc.). In addition, the paper categorizes the data used by type (synthetic or real), also—whether their data were noisy or noiseless. Many authors prefer to analyze their initial data with additionally injected noise (e.g., white noise), which can be quantified by the signal-to-noise ratio (SNR). To continue, the list of the reviewed newest papers [17–39], which are more or less relevant to the scope of this paper, is given in Table 1. In addition to the already mentioned ones, the following AI tools can also be found in the table: artificial bee colony—ABC, competitive swarm optimization—CSO, Clarke transformation—CT, decision tree—DT, ensemble methods—EM, Hilbert transform—HT, k-nearest neighbors—KNN, logistic regression—LR, multilayer perception—MLP, multiresolution analysis—MRA, ordered fuzzy decision tree—OFDT, principal component analysis—PCA, random forest—RF, variational autoencoder—VAE. Also, many simpler ideas and operations such as normalization or other signal processing techniques are left behind.

PQ classification is the main objective of the most reviewed papers, but some other interesting instances are also given in this paper. These instances are related with the following tasks: cybersecurity, localization of PQ emission source (fault), impact assessment (e.g., damage to end-user equipment), and features extraction (also including characterization and detection). For example, the localization can be determined either qualitatively or quantitatively (Figure 1a): the identification of voltage sag source side (upstream or downstream) is a qualitative approach (see [9] for more information), while the estimation of a distance to the fault node in units of length is a quantitative approach (see [2] for more information about the case of PQ monitors displacement algorithms). Please note that the classification itself often includes a wider range of tasks such as an event detection and features extraction. The proposed concept of interconnection between PQ event localization and classification is shown in Figure 1b. For example, the identification of voltage sag side can also be considered as a kind of classification by fault location (e.g., TSO grid, DSO grid, in-plant grid), and such an approach should be important in anticipation of responsibility sharing and penalties. In the most reviewed papers (especially with a wider target group), the classification is performed by a PQ event; however, it must be performed not only by a PQ event which is really more about its detection and recognition, but also by its type or its primary cause. For example, in this work, transients are classified by their primary cause (commutation or atmospheric), and short circuits are classified by their type (three-phase, two-phase-to-ground, two-phase, and single-phase). However, currently, there is a lack of research (data) about primary causes of PQ events [9].

Table 1. Highlights of the newest reviewed and relevant papers.

Reference [1]	Algorithm [2]	Task [3]	Target Group [4]					Data [5]	Parameters, Features
[17] (2023)	ST, RF	C	ImT Sw	 F	OsT H	Int N	Sg C	S, R	Maximum magnitudes (peaks) of FFT spectrum of the signal [6]
[18] (2023)	CNN-LSTM	C, S [7]	ImT Sw	 F	OsT H	Int	Sg C	S	Deep features of 1D time series signal [8]
[19] (2023)	DT, KNN, RF, SVM, VAE	C, L [9]	Sw [10]				Sg [10]	S [11]	Current, voltage, fault type and location
[20] (2023)	CNN, ST	C	ImT Sw	Sp [12] F	OsT H	Int N	Sg C	S, R	Deep features of 2D time–frequency matrix
[21] (2023)	OFDT, PCA	C	 Sw	Sp F	OsT H	Int N	Sg	S	Power spectrum estimated by Welch's method
[22] (2023)	CNN, HT	C	ImT Sw	 F [13]	OsT H	Int N	Sg C	S, R [14]	Deep features of voltage amplitude envelope
[23] (2022)	Not applicable	A	 Sw				Sg	S	Voltage, time, energy index, influence degree (severity)
[24] (2022)	Fast SVM	C	 Sw		 H	Int	Sg C	S	The input is the amplitude of one cycle of the distorted wave
[25] (2022)	2D WT, CNN [15]	C	ImT Sw		OsT	Int	Sg	S	Deep features of the colored image of the time–frequency plane
[26] (2022)	LSTM	C	ImT Sw	Sp F	OsT H	Int N	Sg C	S	The input is the raw sample with its PQ event type
[27] (2022)	DT, KNN, LR, RF, SVM	L	Sw [10]				Sg [10]	S, R	28 features extracted from voltage and current waveforms [16]
[28] (2022)	WT	F					Sg	S	Voltage, time, cross-correlation coefficients with mother wavelets
[29] (2022)	TQWT [17]	C	ImT Sw	 F	OsT H	Int N	Sg C	S	Statistical parameters, sub-band energy ratio, zero crossings, etc. [18]
[30] (2022)	LSTM	C					Sg	S, R	Voltage, fault type [19]
[31] (2021)	CT	C					Sg	S	Clarke components ellipse
[32] (2021)	FNN	C	 Sw	Sp F	OsT H	Int N	Sg C	S	Binary image of PQ signal waveform and its saliency map
[33] (2021)	ABC-PSO, MRA, PNN, WT	C	ImT Sw	Sp F	OsT H	Int N	Sg C	S	Energy, entropy, standard deviation, mean and other statistical parameters
[34] (2020)	EM, GA, SVM [20]	L	Sw [10]				Sg [10]	S	34 features extracted from voltage and current waveforms [16]
[35] (2020)	DT, ST	C	 Sw	 F	OsT H	Int	Sg C	S, R	Maximum amplitude versus both time and frequency, THD, etc.
[36] (2020)	MLP, SVM	C					Sg [21]	S	Higher-order statistics (moments and cumulants), cause
[37] (2020)	CSO, DT, GA, KNN, ST, SVM	C	ImT Sw	 F	OsT H	Int N	Sg C	S	Statistical parameters, disturbance energy ratio

Table 1. *Cont.*

Reference [1]	Algorithm [2]	Task [3]	Target Group [4]	Data [5]	Parameters, Features
[38] (2019)	ELM, WT	C	Sw Sp F OsT H Int N Sg C	S, R	Statistical parameters, wavelet coefficients
[39] (2019)	ST	F	Sg [22]	S	First and second duration times, recovery time and magnitude, THD [23]

[1] Year of publishing online (acceptance) is given in the brackets, which does not necessarily coincide with a year of scientific journal. [2] In this column, more generic names are given, which not always reflect developed (proposed) modifications by various authors. Moreover, they are not classified by purpose (e.g., for feature extraction, feature optimization, classification, etc.). [3] Type of the task: A—impact assessment, C—classification, F—feature extraction (characterization, also may be detection), L—localization, S—cybersecurity. [4] PQ events: ImT—impulsive transient, Sp—spike, OsT—oscillatory transient, Int—interruption, Sg—sag, Sw—swell, F—flicker, H—harmonics, N—notching, C—combination of multiple events. [5] Data acquisition method: S—synthetic, R—real. [6] Many not widely known challenges (issues) are present during the analysis of frequency domain. Some of them are discussed (mentioned) in both this and Section 4.3.1. [7] The defense of the classifier from an adversarial attack is investigated. [8] Deep features are complex patterns extracted from the intermediate layers of CNN. [9] The distance (location) is expressed in numbers. [10] All four types of short circuit are encompassed. [11] The original data set had been obtained by simulation, and then, it was synthetically enlarged with VAE. [12] Despite the fact that a spike is not included in the PQ legislation, authors sometimes prefer to distinguish it from impulsive transient. [13] Called as fluctuation. [14] Real data were generated by the programmable AC power source. [15] Generalized Morse, analytic Gabor and bump wavelets are investigated. [16] Papers [27,34] are kindred, including the features used by them. These features are obtained from many earlier papers and are based on various manipulations with different electrical parameters such as current, apparent power, reactive power, power factor, etc. [17] TQWT is parameterized by Q, r and J. [18] In this table, statistical parameters stand for well-known characteristics such as the mean, standard deviation, range, skewness, excess kurtosis, etc., including any manipulation (operation) with them. [19] The following parameters are required for voltage sag generation with the synthetic signal generator: signal duration, fault start time, fault duration, nominal RMS, signal frequency, sampling frequency, no-fault amplitude, phase angle offset, number of smoothing timepoints, fault severity, signal rotation, noise percentage, and the label (the type of the fault to be generated). [20] The ensemble utilizes discriminant, KNN and DT weak learners (three learners, nine methods). [21] Three causes are included: fault, motor inrush current, and transformer inrush current. [22] The eight types (causes) are included: line to line fault, self-extinguishing fault, induction motor starting, transformer energizing, multistage fault, line fault before induction motor starting, line fault with harmonics caused by a single-phase nonlinear load, and multistage fault with harmonics caused by a single-phase nonlinear load. [23] Several time parameters are required to characterize a duration of a multistage voltage sag (see Section 4.1).

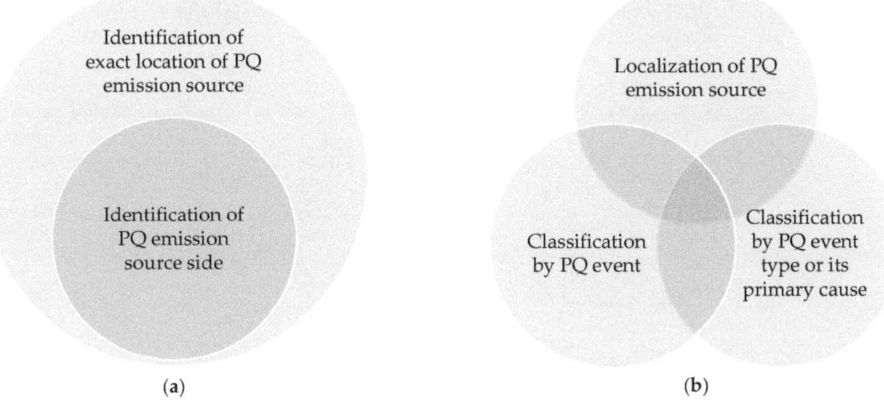

Figure 1. (**a**) Proposed classification model of the tasks of PQ pollutant localization; (**b**) Proposed interconnection model between PQ classification and pollutant localization.

To continue, in most of the papers reviewed, the main data generation method is synthetic, which sometimes is supplemented with real measurements. In these papers, the creation of a synthetic database often is based on the mathematical models of PQ disturbances (parametric equations), which can be found in [40]: impulsive transient, spike, oscillatory transient, interruption, sag, swell, flicker, harmonics, notching, interruption with harmonics, sag with harmonics, swell with harmonics, flicker with sag, and flicker with swell. This is the reason why a spike appears in the target groups (see Table 1) despite being not included in PQ legislation. It was noticed that many reviewed papers are kindred—the usage of the same PQ events models from [40] leads to a loss of uniqueness. This paper does not use these equations. Also, it remains unclear what influence the relevant aspects of Section 4.1, which is based on the PQ monitoring campaign in the Lithuanian DSO grid, have on AI performance.

It must be highlighted that many authors boast about the very high classification accuracy achieved by their algorithms: for example, under different noisy environments, the declared classification accuracy is 99.4–100% in [17], 93.3–100% in [20], etc. Nevertheless, neither these nor other PQ classifiers are currently used in practice. In our opinion, in anticipation of a great leap forward in the PQ field, authors should be more honest (transparent) and discuss the limitations of their methods and approaches (e.g., as seen in Section 4.1). The authors sometimes also highlight further development avenues: for example, [20] mentions light-weighted optimization required for deploying the algorithm in embedded systems and also online training of its PQ classification model. A great diversity of feature extraction approaches can also be noticed in Table 1, and all currently investigated techniques has a potential for wide usage in the future. Despite the remarkable contribution, this area presently is still open [2], not commercialized and not applied in practice. Various techniques are applied in the feature selection and optimization stage: for example, [19] uses a customized version of forward feature selection (an iterative method starting with no features), and [33] uses adaptive ABC-PSO for optimal feature selection. The frequency domain is often studied as well—authors must pay attention (and sometimes do) to various encountered (and not always widely known) challenges such as Heisenberg uncertainty, window function and size, spectral leakage, spectral aliasing, etc. For example, [21] uses Welch's method, which is beneficial for the noise reduction injected by the windowing of the overlapping segments of the power density spectrum. Furthermore, PQ signals can be analyzed either with or without the fundamental grid frequency component included [12]; however, despite the fact that some graphs are filtered in IEEE Std 1159-2019, other examples of such filtering were not found in the PQ literature.

The main method of analysis of this paper uses the voltage–duration plane: we will use the conventional approach and demonstrate that there are many unsolved questions. The main aim is to investigate the capabilities of both SVM and KNN to assess both transients and short-duration RMS variations, including both with the fundamental frequency component and filtered. Task No. 1 is to investigate the capabilities of the IIR shelving filter for 50 Hz and its harmonics filtering, task No. 2 is to investigate single-voltage classification in the voltage–duration plane, and task No. 3 is to investigate three-dimensional voltage classification. It goes without saying that the limitations of the research must also be discussed, and proposed ideas will be original and distinctive from the existing PQ literature as well as based on the PQ monitoring campaign in the Lithuanian DSO grid. But for now, let us begin from the first—input data—stage (Section 2.1).

2. Materials and Methods

2.1. Database

As has been mentioned in Section 1, each research on AI algorithms application should be started with a focus on the input data. The PQ data sources of this research are (1) the measurement campaigns in both the Lithuanian DSO grid and internal grids of industrial plants, (2) simulation with both MATLAB/Simulink and Siemens PSS/E software tool, and (3) literature analysis.

2.1.1. Commutation Transients

Voltage transients are classified into two categories—impulsive (non-oscillatory) and oscillatory. IEEE Std 1159-2019 divides each category further into three groups: (1) impulsive transients are classified by duration into nanosecond (less than 50 ns), microsecond (50 ns to 1 ms) and millisecond (longer than 1 ms); (2) oscillatory transients are classified by spectral content into low frequency (up to 5 kHz), medium frequency (5–500 kHz) and high frequency (0.5–5 MHz). Both categories of voltage transients are strongly interconnected because the electrical grid response to an impulsive transient can be oscillatory [12]. Meanwhile, EN 50160:2010 provides almost no information about voltage transients, and IEC 61000-4-30:2015 does not establish any measurement requirements. Moreover, measurement capabilities are limited not only due to the lack of the information but also due to technical issues related with measurement transducers (or, generally speaking, with measurement circuits)—for example, unknown magnitude and phase frequency response, or not having one voltage transformer installed in older substations (i.e., only two phase-to-ground voltages out of three are available to be measured). This is because the existing instrument transformers were designed only for fundamental frequency and electricity consumption metering—these issues have already been discussed in [2,9]. To continue, both sampling frequency and dynamic range aspects must also be considered. For example, approximately 200 MHz sampling is required to record 10 samples of 50 ns impulse, and 10 MHz sampling is required for a 5 MHz wave. The minimal requirement for sampling frequency, based on Equation (5), has been established in [2]. Dynamic range is the ratio between the largest and smallest values; thus, it becomes obvious that the equipment dedicated to atmospheric transients may be not suitable for commutation transients and vice versa. Transient occurrence probability at a single grid node is not sufficient for fast and unchallenging data gathering. Also, occurred transients can be mitigated. Considering the above, the most convenient way for data generation is simulation. Transients can be caused by various reasons; for example, instances of capacitor bank energization and ferroresonance are given in IEEE Std 1159-2019 (pp. 17–18). Among the others is transformer energization, power electronics devices, and short circuits (as it has been found in [2,9]), but let us begin with the most well-known—commutation and lightning.

The theory of commutation transients can be found in [41,42]. The connection of an RLC series circuit to a voltage source $E(t)$ is described by the second-order nonhomogeneous differential equation:

$$Ri + L\frac{di}{dt} + \frac{1}{C}\int i\,dt = E \;\Leftrightarrow\; L\frac{d^2i}{dt^2} + R\frac{di}{dt} + \frac{1}{C}i = \frac{dE}{dt}, \tag{1}$$

where R—resistance; L—inductance; C—capacitance; $i(t)$—current; t—time.

The solution of this differential equation is the sum of two components—compulsory oscillation U_1 (the general solution of the homogenous differential equation) and exponentially decaying oscillation U_2 (the particular solution of the nonhomogeneous differential equation):

$$U_1 = E_m \frac{\omega_f{}^2}{\omega_f{}^2 - \omega^2} \sin(\omega t + \varphi), \tag{2}$$

$$U_2 = -\sqrt{\sin^2\varphi + \left(\frac{\omega}{\omega_f}\cos\varphi\right)^2}\,\exp(-\delta t)\sin\left(\omega_f t + \varphi_f\right), \tag{3}$$

where E_m—grid voltage magnitude; ω—angular frequency of compulsory oscillation; ω_f—angular frequency of free oscillation; φ—phase angle of the commutation; φ_f—phase angle of free oscillation; δ—exponential decay constant.

Exponential decay constant is determined by:

$$\delta = \frac{R}{2L}. \tag{4}$$

The angular frequency of free oscillation, which determines required sampling frequency, is determined by:

$$\omega_f = \frac{1}{\sqrt{LC}}.$$

(5)

2.1.2. Atmospheric Transients

The second type to be investigated is atmospheric transients. Various lightning models are proposed in the literature; for example, one of the most popular is the Heidler function:

$$I(t) = \frac{I_m}{k} \cdot \frac{\left(\frac{t}{\tau_1}\right)^n}{1 + \left(\frac{t}{\tau_1}\right)^n} \cdot \exp\left(-\frac{t}{\tau_2}\right),$$

(6)

where I_m—current peak value; k—correction factor of current peak value; n—steepness factor (equal to 10 according to IEC 62305-1:2010 [43], but other values can also be found in the scientific literature); τ_1—rise time constant; τ_2—fall time constant [43–45].

Another example can be obtained from [46]:

$$I(t) = I_m(\exp(-C_1 t) - \exp(-C_2 t)),$$

(7)

where I_m—current peak value; C_1 and C_2—constants.

According to IEEE Std 998-2012 [47], the average peak current of the first negative return stroke is 15–20 kA. The positive strokes' current is typically higher, and its preferred limit for lightning protection is 200 kA. In nature, a 200 kA stroke occurs rarely—approximately 1% of all cases [42,47]. The selection of both the model and its parameters depends on the purpose: for example, for lightning protection design, parameters are selected with a margin. The goal of this paper is to model the situation as realistically as possible. The following function for lightning impulse modeling is selected:

$$I(t) = \begin{cases} 0, & t \in (-\infty; T_0] \\ I_m\left(1 - \exp\left(-\frac{t}{t_1}\right)\right), & t \in (T_0; T_1] \\ I_m\left(\exp\left(-\frac{t}{t_2}\right) - \exp\left(-\frac{t}{t_1}\right)\right), & t \in (T_1; +\infty) \end{cases}$$

(8)

where I_m—current peak value; t_1—rise time; t_2—fall time; T_0—moment of the beginning; T_1—moment of the peak.

When $I(t)$ is known, the electric charge of the stroke is determined by:

$$Q = \int_0^\infty I\, dt.$$

(9)

The following information about the interconnection between the lightning stroke current $I(t)$ and induced voltage is important for modeling adequacy estimation. A direct lightning strike may raise voltage up to 1000 kV, and several hundred kilovolts may be induced when the distance from the power line is greater than 5 km [42]. For better intuition, the following results, found in the literature, are given as an example:

- According to the calculations given in [42] (pp. 268–269), when the striking distance was 30 m, a 30 kA impulse induced an approximately 195 kV voltage peak, when 100 m—approximately 50 kV, and when 300 m—approximately 20 kV. The duration of these impulses is longer than 6 µs (see Section 4.1 for more details and discussion about the duration of transients). It was assumed that the height of the line (above ground) is 5 m (this assumption is not realistic), the steepness of the impulse is 20 kA/µs (see Equation (10)), the rise time of the impulse is 1.5 µs, and the speed of the return stroke is equal to 0.1 of the speed of light.

- In [45], when the striking distance was 1.16 m, 10 kA impulse induced 50 kV voltage, but the predicted result from theoretical simulation was 28 kV. The length of the single-phase overhead power line is 200 m, and the height is 10 m. The durations of induced impulses are up to 10–20 μs.
- In [46], when the striking distance was 50 m, the 12 kA impulse induced 52–80 kV voltages at various points of the single-phase overhead power line whose length is 1 km and height is 10 m. The duration of the induced impulse is up to 10 μs.

It is noteworthy that the induced overvoltage depends on lightning impulse steepness, which is defined as:

$$\alpha = \frac{dI}{dt}. \tag{10}$$

This is due to Faraday's law of induction:

$$\nabla \times \mathbf{E} = -\frac{\partial \mathbf{B}}{\partial t} \iff \oint_l \mathbf{E} \cdot d\mathbf{l} = -\int_S \frac{\partial}{\partial t} \mathbf{B} \cdot d\mathbf{S}, \tag{11}$$

where $\mathbf{E}(\mathbf{r}, t)$—electric field (generally, a function of position $\mathbf{r}$ and time t); $\mathbf{B}(\mathbf{r}, t)$—magnetic field; $d\mathbf{l}$—infinitesimal vector element of the contour; $d\mathbf{S}$—infinitesimal vector element of the surface; ∇—del operator.

2.1.3. Short-Duration RMS Variations: PQ Monitoring Campaign

IEEE Std 1159-2019 classifies RMS variations by time into two categories—short duration (shorter than 1 min) and long duration (longer than 1 min). EN 50160:2010 limits the duration up to 1 min in all recommended tables for voltage sags and swells classification. The duration of short-duration RMS events highly depends on the settings of relay protection and automation. In our opinion, RMS variations which last longer than 1 min can be excluded from the target group of this paper. The data were obtained from the measurement campaigns and theoretical simulations. Since the theoretical research of short circuits is described in detail in [9] (with the intention to be used as a database), let us continue with the PQ monitoring campaign in the Lithuanian DSO grid. The main focus of the campaign was on voltage sags; however, we have discovered that accurate calculation of the total number of events is a complicated task. Honestly, since data gathering and processing have been mainly manual to date, a tribute must be paid to those authors who at least roughly succeeded in the implementation of this (e.g., see the practical researches listed in [2]). On the other hand, in our opinion, such statistics are not important for this research, but we will present a few episodes. An absolute majority of the recorded sags were caused by grid faults; however, since the main focus was on the power supply of industrial plants, there is a small probability that at least few of them were caused by electric motor starting. The latest voltage sag in possession was recorded at the 0.4 kV point of common coupling of medical equipment manufacturer (in Kaunas municipality): the three-phase phase-to-ground voltage dropped from 230–240 V to 170–222 V (the sag's duration was 0.34 s).

In Figure 2, the statistics of the 35/10 kV substation (in Širvintos municipality) are given when a PQ analyzer was installed on the 10 kV side from January to April 2019. In the first three charts, voltage arrays are analyzed separately, i.e., the multiphase aggregation which is required in EN 50160:2010 is not applied. Such an approach will often be preferred in this work (see Section 4.1 for the argumentation). It can be noticed that the data are distributed randomly and do not follow any distribution. A trend appearance may be expected with an increase in observation period (according to the law of large numbers); however, 2–3 years may not be enough for that: the minimum monitoring period for voltage sags, swells and interruptions, recommended by IEC 61000-4-30:2015, is 1 year. Also, it should be noted that 50% and deeper sags have not occurred, which means the absence of critical situations. The shortest recorded sag duration is 0.010 s (Figure 2c), which matches the smallest possible time window defined by IEC 61000-4-30. After the multiphase aggregation, the minimal duration becomes 0.012 s (Figure 2d). The bin width of the voltage histogram is 300 V (Figure 2b), and the bin width of time histograms is 0.100 s (Figure 2c,d). The sample size of phase-to-phase voltage AB (i.e., voltage between phase A and phase B) is 11, BC—17, CA—11 (i.e., a total of 39); the means of durations—0.33 s, 0.25 s, and 0.17 s; the medians of durations—0.13 s, 0.18 s, and 0.16 s; the means of residual voltages—7383 V, 7571 V, and 7956 V; the medians of residual voltage—7789 V, 8011 V, and 7136 V (respectively).

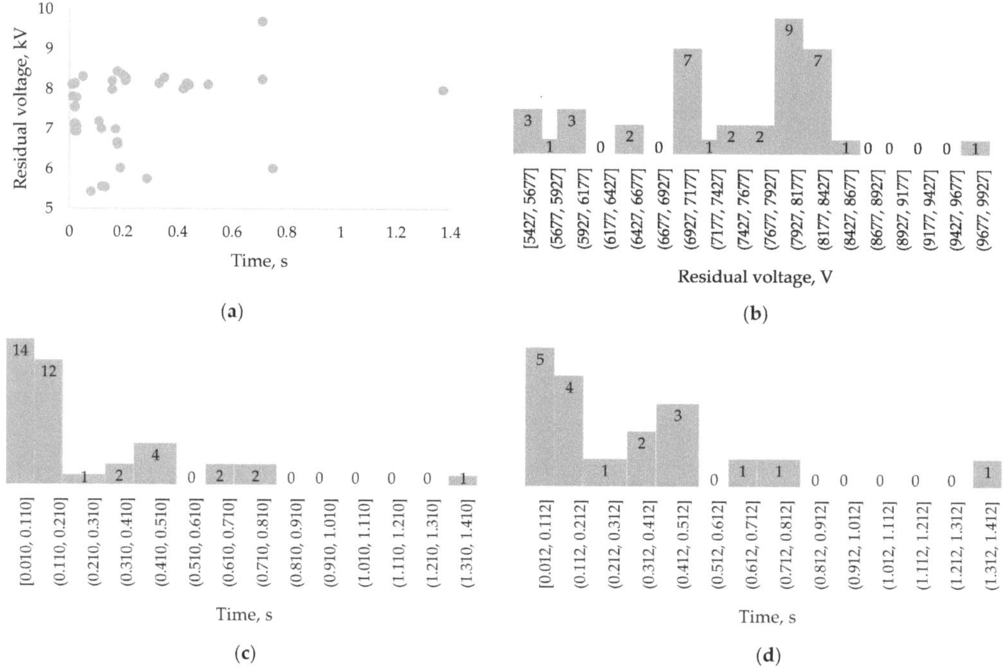

Figure 2. Voltage sags statistics of the 35/10 kV substation (located in Širvintos municipality): measurements were taken on the 10 kV side from January to April 2019. (**a**) Phase-to-phase residual voltage–duration plane; (**b**) Histogram of phase-to-phase residual voltage; (**c**) Histogram of duration; (**d**) Histogram of duration after the multiphase aggregation according to EN 50160:2010.

To continue, phase-to-ground voltage measurements in the 35 kV line (located in Rokiškis municipality) taken from 1 February 2020 to 13 February 2020 are given in Figure 3, and those taken from 15 August 2020 to 14 September 2020 are given in Figure 4. As in the previous case, there have been no sags deeper than 50%. In spite of EN 50160:2010 requirement to assess phase-to-phase voltage, phase-to-ground voltage arrays are more informative due to at least a single-phase fault identification possibility. Also, in Figure 4, the scattering of voltage phasors is given: phase unbalance can be caused by an asymmetrical fault (see [9] for examples). It is noteworthy that not every software is able to display the voltage of entire measurement period in one window (as shown in the figures); however, such a functionality is not essential for this research.

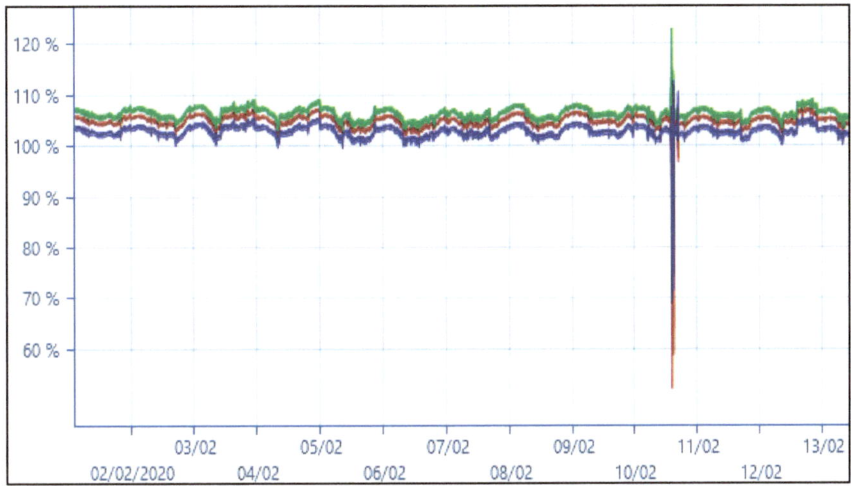

Figure 3. Phase-to-ground voltage in the 35 kV line (located in Rokiškis municipality): measurements from 1 February 2020 to 13 February 2020.

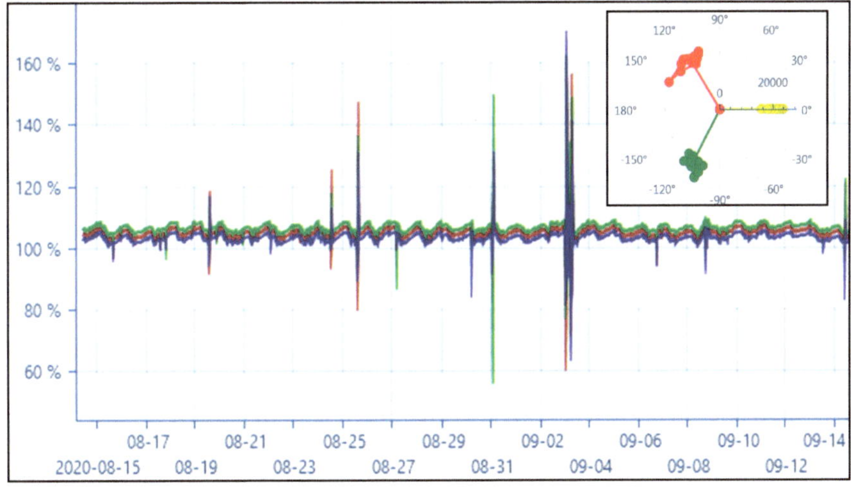

Figure 4. Phase-to-ground voltage and its phasors scattering in the 35 kV line (located in Rokiškis municipality): measurements were taken from 15 August 2020 to 14 September 2020.

Phase-to-ground voltage measurements in the 10 kV line (located in Rokiškis municipality) taken from 15 August 2020 to 15 September 2020 are given in Figure 5. Only two (of three) arrays are available due to the absence of one phase-to-ground voltage transformer. In the case of asymmetrical faults, the PQ analyzer assumes that the mode is symmetrical; thus, the estimation of the unknown voltages will be incorrect (inaccurate).

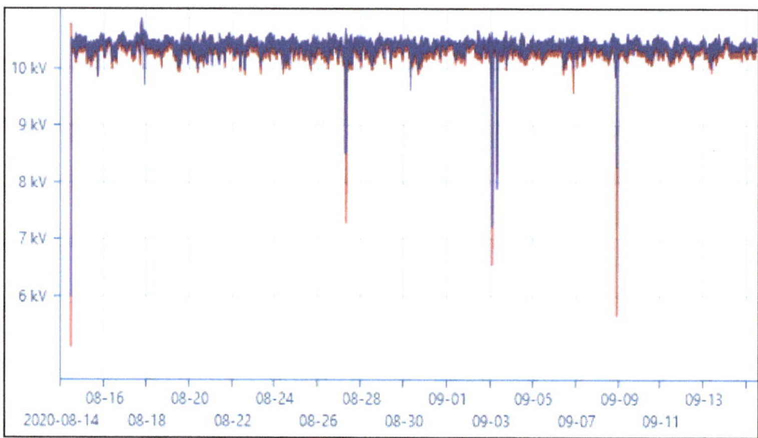

Figure 5. Phase-to-ground voltage in the 10 kV line (located in Rokiškis municipality): measurements were taken from 15 August 2020 to 15 September 2020. One voltage array is unavailable due to the absence of one voltage transformer (out of three).

2.1.4. Voltage Sags Caused by Electric Motor Starting

Another cause of voltage sag, which we would like to include, is the electric motor starting. At first glance, it seems that the data can be obtained from both practical experiments and simulations. However, the prerequisites of such an experimental work are not only the special set-up but also many different motors are required to be in possession. The simulation of an electric motor behavior requires impedance data of equivalent circuit as well as some other parameters such as rotational speed and mechanical inertia (e.g., see [48,49] for the case of a squirrel-cage induction motor). In this context, similarly as in the case of atmospheric transients (Section 2.1.2), it was decided that the most efficient way to become familiar with the topic and obtain the initial information is a literature review, postponing a more detailed investigation for the future. The following results can be given as an example:

- One figure is given in IEEE Std 1159-2019 (p. 22): the minimal residual voltage is 0.8 p.u., and the voltage sag duration is approximately 2 s. The motor type is not specified, but it is mentioned that a motor is large.
- In [14] that investigates an induction motor fed by a current controlled pulse width modulation inverter, the minimal residual voltage and duration of the simulated voltage sag is 0.88 p.u. and 0.8 s.
- In [49] (investigates an induction squirrel-cage motor), the minimal residual voltage and duration of voltage sags are as follows: (1) 0.35 p.u. and 307 ms (simulated) or 327 ms (measured) with the capacitor bank switching; (2) 0.25 p.u. and 710 ms without the capacitor bank switching. The simulation results coincide with the experimental results. It should be noticed that the applied AC motor starting method (as well as others, for example, reactor starting, autotransformer starting, Y-Δ transform, etc.) limits both inrush current and voltage sag.
- In [50] that investigates water pumps driven by the induction motors, the minimal residual voltage and duration of both simulated voltage sags are 0.85–0.87 p.u. and 0.05 s.

Please note that the given results reflect the influence of only one electric motor. In the case of a three-phase electric motor, the voltage sag is symmetrical.

To continue, in our previous work [9], similarly but not analogously to [48], the opposite case—effect of symmetrical voltage sags on electric motors—is investigated. The intermediate chain between both cases is the sags caused by a motor self-starting (see Figure 6). The higher generation of an electric motor is during a grid fault, the higher and longer-lasting is the residual voltage, and the higher is the probability of self-starting (which again reduces the voltage, thereby lowering the probability of own success) [51]. In practice, industrial companies can prioritize the self-starting of their motors; however, due to limited time, the prioritization cannot solve all problems and fully prevent the stoppages. Despite electrical machine engineering being a mature scientific field, there is still a lack of information on the topic, in particular through a prism of PQ.

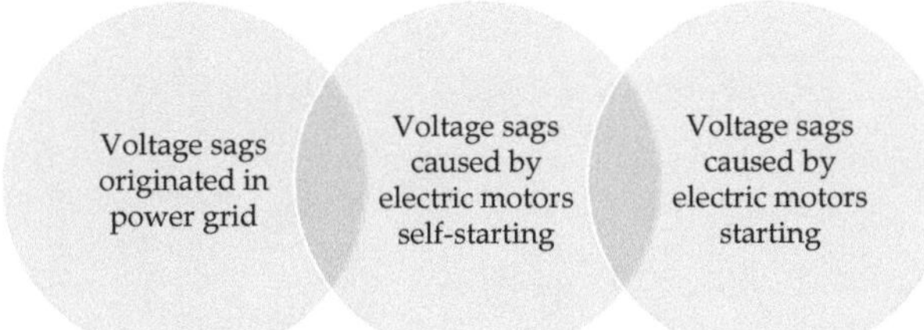

Figure 6. Proposed interconnection model of voltage sags causes from the perspective of electric motors.

2.2. Shelving Filter

As has been mentioned previously, both impulsive and oscillatory transients can be analyzed (measured) with or without the fundamental frequency component included [12]. Inspired by [52], the application of an infinite impulse response (IIR) shelving filter, obtained from [53], will be investigated for 50 Hz component removal from a PQ signal. The block diagram of the filter is given in Figure 7. It is noteworthy that any application of this filter for a PQ signal has not been found in the literature.

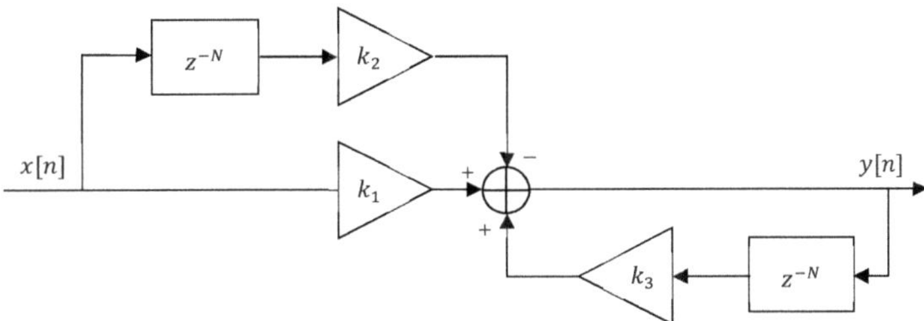

Figure 7. The block diagram of the shelving filter.

In Figure 7, it can be noticed that the implementation of the filter requires three operations of summation and three operations of multiplication. The filter has three coefficients k_1, k_2 and k_3, and its difference equation is:

$$y[n] = k_1 x[n] - k_2 x[n - N] + k_3 y[n - N], \tag{12}$$

where $x[n]$—input; $y[n]$—output; n—sample number; N—delay.

The transfer function of the filter is:

$$H(z) = \frac{Y(z)}{X(z)} = \frac{k_1 - k_2 z^{-N}}{1 - k_3 z^{-N}}, \tag{13}$$

where $X(z)$—Z-transform of $x[n]$; $Y(z)$—Z-transform of $y[n]$.

It is noteworthy that the Z-transform of a discrete-time signal $x[n]$ is defined as follows:

$$X(z) = \mathcal{Z}\{x[n]\} = \sum_{-\infty}^{\infty} x[n]z^{-n}, \tag{14}$$

where n is an integer, and z is, in general, a complex number which may be written in polar form:

$$z = re^{j\phi} = r(\cos\phi + j\sin\phi). \tag{15}$$

where r—magnitude; ϕ—complex argument; j—imaginary unit.

Coefficients k_1, k_2 and k_3 are calculated as follows:

$$k_1 = \frac{K_0 + K\beta}{1 + \beta}, \tag{16}$$

$$k_2 = \frac{K_0 - K\beta}{1 + \beta}, \tag{17}$$

$$k_3 = \frac{1 - \beta}{1 + \beta}. \tag{18}$$

Parameter β is calculated by:

$$\beta = \frac{K_r^2 - K_0^2}{K^2 + K_r^2} \tan\left(\frac{\pi \cdot N \cdot \Delta f}{2f_s}\right), \tag{19}$$

where K_0—gain of passband; K—attenuation of stopband; N—number of cuts; Δf—stopband width at K_r level, Hz; f_s—sampling frequency, Hz.

The number of cuts N is determined as follows:

$$N = \frac{f_s}{f_0}, \tag{20}$$

where f_0—frequency to be filtered and its harmonics, Hz.

Parameters K and K_r are calculated by:

$$K = K_0 \cdot 10^{-\frac{A}{20}}, \tag{21}$$

$$K_r = K_0 \cdot 10^{-\frac{L}{20}}, \tag{22}$$

where A—attenuation, dB; L—level of the cutoff frequency, dB.

2.3. Fundamental Theory of SVM

SVM is a supervised learning algorithm used for classification and regression as well as outlier detection. A large amount of information about it can be found in various literature, for example, in [7,54–56]. SVM by its origin is a binary classification algorithm, but it can be extended to any multiclass case by combining the binary classifiers using either the one-versus-one or one-versus-rest approach.

Consider the training sample:

$$\mathcal{T} = \{(\mathbf{x}_1, y_1), (\mathbf{x}_2, y_2), \ldots, (\mathbf{x}_n, y_n)\}, \tag{23}$$

where $\mathbf{x}_i$—multi-dimensional real vector; y_i—class label of $\mathbf{x}_i$. The goal of SVM is to construct a decision boundary separating the feature vectors $\mathbf{x}_i$ into two groups whose labels are either $y_i = 1$ or $y_i = -1$ in the way that ensures that the total distance between the decision boundary and the nearest points from both groups is maximized. In linear cases, the equation of a decision surface in the form of a hyperplane is as follows:

$$\mathbf{w}^{\mathrm{T}}\mathbf{x} + b = 0,$$ (24)

where $\mathbf{w}$—adjustable (and not necessarily normalized) normal vector to the hyperplane (also called the weight vector); $\mathbf{x}$—input vector; b—bias. The parameter $\frac{b}{\|\mathbf{w}\|}$ determines the offset of the hyperplane from its initial position along the vector $\mathbf{w}$. Therefore, the Euclidian norm $\|\mathbf{w}\|$ needs to be minimized in order to maximize the margin of separation. In the case of a hard margin, the principle of SVM linear separation takes the following form:

$$\begin{cases} \mathbf{w}^{\mathrm{T}}\mathbf{x}_i + b \geq 1, & \text{if } y_i = 1 \\ \mathbf{w}^{\mathrm{T}}\mathbf{x}_i + b \leq -1, & \text{if } y_i = -1 \end{cases} \Rightarrow y_i\left(\mathbf{w}^{\mathrm{T}}\mathbf{x}_i + b\right) \geq 1, \forall i \in \{1, 2, \ldots, n\}.$$ (25)

In case of overlapping data, when a line cannot separate groups without any mistake, SVM classification is extended by establishing the soft margin concept. It allows misclassification to happen, thereby violating the condition of Equation (25) and raising the need to minimize classification error ξ. Thus, Equation (25) becomes:

$$y_i\left(\mathbf{w}^{\mathrm{T}}\mathbf{x}_i + b\right) \geq 1 - \xi_i, \ \xi_i \geq 0, \ \forall i \in \{1, 2, \ldots, n\},$$ (26)

where ξ_i—slack variable, which is a measure of the deviation from the ideal points separation. If $\xi_i > 1$, the point falls on the wrong side of the margin. The goal of optimization is based on the hinge loss, which is a loss function used in training classifiers (in particular SVM). Let us skip the intermediate steps with the hinge loss (since they are beyond the scope of this paper) and give the final formulation of the optimization problem:

$$\min\left(\frac{1}{2}\|\mathbf{w}\|^2 + C\sum_{i=1}^{N}\xi_i\right),$$ (27)

where C—regularization parameter. The parameter C controls the width of the gap (margin size) in exchange of misclassification, and it has to be set by the user. Larger values of C correspond to the harder margin.

Nonlinear classification can also be efficiently performed with SVM. Two nonlinear kernels are used in this paper. The radial basis function (RBF) kernel is defined as follows:

$$k\left(\mathbf{x}_j, \mathbf{x}_k\right) = \exp\left(-\frac{\|\mathbf{x}_j - \mathbf{x}_k\|^2}{2\sigma^2}\right),$$ (28)

where $\|\mathbf{x}_j - \mathbf{x}_k\|^2$—squared Euclidian distance between the two feature vectors; σ—width, which is often expressed as:

$$\gamma = \frac{1}{2\sigma^2}.$$ (29)

The polynomial kernel function takes the following form:

$$k\left(\mathbf{x}_j, \mathbf{x}_k\right) = \left(\mathbf{x}_j^{\mathrm{T}}\mathbf{x}_k + c\right)^p,$$ (30)

where c—non-negative free parameter; p—degree of the polynomial.

It is noteworthy that $k\left(\mathbf{x}_j, \mathbf{x}_k\right) = k\left(\mathbf{x}_k, \mathbf{x}_j\right)$ (for both kernels explored), and the maximum value is reached when $\mathbf{x}_j = \mathbf{x}_k$ (but there is no such need of its existence) [54].

2.4. Fundamental Theory of KNN

KNN is a supervised learning algorithm which is also used for classification and regression. The main principle of label assignment is based on majority rule involving a selected number k of nearest neighbors. This number is selected freely (but reasonably) by a user. Also, in a binary classification, it would be wise to choose an odd number since it allows avoiding tied votes [7]. Various methods can be used to estimate the distance (e.g., Chebyshev distance), but the already mentioned Euclidian distance, which is a special case of Minkowski distance, will be preferred in this paper. In n-dimensional space, the formula of Euclidian distance is:

$$d(p,q) = \|p - q\| = \sqrt{(p_1 - q_1)^2 + (p_2 - q_2)^2 + \ldots + (p_n - q_n)^2}, \tag{31}$$

where p_i and q_i are Cartesian coordinates of points p and q respectively.

One majority voting disadvantage occurs when the class distribution is skewed. This can be solved by the distance weighting. In this paper, the weights are inversely proportional to the distance, i.e., $1/d$. Therefore, closer points became more important than more distant ones. Also, it is noteworthy that KNN does not work well in high-dimensional spaces; this problem is called the curse of dimensionality, which is a phenomenon where feature space becomes increasingly sparse [7]. Intuitively, it can be understood that even the closest points can indeed be too distant in a high-dimensional space.

2.5. Additional Methods

2.5.1. Geometric Analysis

In Section 3.2.2, data dispersion areas and their boundaries in the feature space are analyzed geometrically. For this purpose, in this section, a relevant theoretical background is given for both spaces $\mathbb{R}^2$ and $\mathbb{R}^3$.

When a line passes different points (x_1, y_1, z_1) and (x_2, y_2, z_2), the canonical equation of this line is as follows:

$$\frac{x - x_1}{x_2 - x_1} = \frac{y - y_1}{y_2 - y_1} = \frac{z - z_1}{z_2 - z_1}. \tag{32}$$

The general equation of a plane is:

$$ax + by + cz + d = 0, \tag{33}$$

where a, b and c are the components of the normal vector $\mathbf{n} = \{a, b, c\}$ which is perpendicular to the plane.

The coplanarity condition of points (x_1, y_1) and (x_2, y_2) is given by:

$$\begin{vmatrix} x & y & 1 \\ x_1 & y_1 & 1 \\ x_2 & y_2 & 1 \end{vmatrix} = 0. \tag{34}$$

The canonical equation of an ellipsoid is:

$$\frac{(x - x_0)^2}{a^2} + \frac{(y - y_0)^2}{b^2} + \frac{(z - z_0)^2}{c^2} = 1, \tag{35}$$

where (x_0, y_0, z_0)—center coordinates; a, b and c—the lengths of the semi-axes. In a two-dimensional case, when $z = 0$, then Equation (35) represents an ellipse.

An elliptic paraboloid is represented by the following equation:

$$z = \frac{(x - x_0)^2}{a^2} + \frac{(y - y_0)^2}{b^2} + z_0, \tag{36}$$

where (x_0, y_0, z_0)—center coordinates; a and b—constants.

The equation of a parabola, whose vertex coordinates are (x_0, c), is given by a univariate quadratic function:

$$a(x - x_0)^2 + b(x - x_0) + c = 0. \tag{37}$$

Rotation in two-dimensional Euclidian space is performed with the rotation matrix:

$$R = \begin{bmatrix} \cos\theta & -\sin\theta \\ \sin\theta & \cos\theta \end{bmatrix}, \tag{38}$$

where θ—rotation angle about the origin of a two-dimensional Cartesian coordinate system. The positive angle corresponds to a counterclockwise rotation, and negative corresponds to a clockwise rotation. In the case of three-dimensional Euclidian space, the rotation matrix takes the following form:

$$R = \begin{bmatrix} \cos\alpha & -\sin\alpha & 0 \\ \sin\alpha & \cos\alpha & 0 \\ 0 & 0 & 1 \end{bmatrix} \begin{bmatrix} \cos\beta & 0 & \sin\beta \\ 0 & 1 & 0 \\ -\sin\beta & 0 & \cos\beta \end{bmatrix} \begin{bmatrix} 1 & 0 & 0 \\ 0 & \cos\gamma & -\sin\gamma \\ 0 & \cos\gamma & \cos\gamma \end{bmatrix}, \tag{39}$$

where α, β and γ are Euler angles about the abscissa, ordinate and applicate, respectively.

2.5.2. Clarke Transformation

In Section 3.2.2, the idea to transform three-dimensional space into two-dimensional is raised. For this task, we will attempt to apply the alpha–beta ($\alpha\beta\gamma$) transformation, also known as Clarke transformation, which transforms the time domain components of a three-phase system to two components in an orthogonal stationary frame [57]. For a three-phase voltage signal, the mathematical expression of power invariant Clarke transformation takes the following form:

$$\begin{bmatrix} u_\alpha(t) \\ u_\beta(t) \\ u_\gamma(t) \end{bmatrix} = \sqrt{\frac{2}{3}} \begin{bmatrix} 1 & -\frac{1}{2} & -\frac{1}{2} \\ 0 & \frac{\sqrt{3}}{2} & -\frac{\sqrt{3}}{2} \\ \frac{1}{\sqrt{2}} & \frac{1}{\sqrt{2}} & \frac{1}{\sqrt{2}} \end{bmatrix} \begin{bmatrix} u_a(t) \\ u_b(t) \\ u_c(t) \end{bmatrix}, \tag{40}$$

where $\begin{bmatrix} u_a(t) & u_b(t) & u_c(t) \end{bmatrix}^{\mathrm{T}}$—three-phase voltage $u_{abc}(t)$; $\begin{bmatrix} u_\alpha(t) & u_\beta(t) & u_\gamma(t) \end{bmatrix}^{\mathrm{T}}$—corresponding voltage in the $\alpha\beta\gamma$ reference plane after the transformation. If the input is symmetrical:

$$\begin{bmatrix} u_a(t) \\ u_b(t) \\ u_c(t) \end{bmatrix} = \begin{bmatrix} \sqrt{2}U\cos\theta(t) \\ \sqrt{2}U\cos(\theta(t) - \frac{2}{3}\pi) \\ \sqrt{2}U\cos(\theta(t) + \frac{2}{3}\pi) \end{bmatrix}, \tag{41}$$

then the output of the transformation is as follows:

$$\begin{bmatrix} u_\alpha(t) \\ u_\beta(t) \\ u_\gamma(t) \end{bmatrix} = \begin{bmatrix} \sqrt{3}U\cos\theta(t) \\ \sqrt{3}U\sin\theta(t) \\ 0 \end{bmatrix}, \tag{42}$$

where U—rms voltage; and $\theta(t)$—generic time-varying angle (can also be set to ωt). Therefore, in the case of a balanced signal, the third component is equal to zero. This is the main reason of our choice, which makes it potentially promising. However, in the case of asymmetry, the third component is not equal to zero (see Section 3.2.2 and Appendix A for more information about results interpretation and decisions made).

We also noticed that the result of the power invariant Clarke transformation (Equation (40)) is similar to a power invariant direct-quadrature-zero (DQZ) transformation when the α-axis is aligned to the d-axis and the angle θ is equal to zero:

$$\begin{bmatrix} u_d(t) \\ u_q(t) \\ 0 \end{bmatrix} = \sqrt{\frac{2}{3}} \begin{bmatrix} \cos\theta & \cos(\theta - \frac{2\pi}{3}) & \cos(\theta + \frac{2\pi}{3}) \\ -\sin\theta & -\sin(\theta - \frac{2\pi}{3}) & -\sin(\theta + \frac{2\pi}{3}) \\ \frac{1}{\sqrt{2}} & \frac{1}{\sqrt{2}} & \frac{1}{\sqrt{2}} \end{bmatrix} \begin{bmatrix} u_a(t) \\ u_b(t) \\ u_c(t) \end{bmatrix}, \tag{43}$$

where θ—angle between the α-axis and d-axis (equal to zero); $\begin{bmatrix} u_a(t) & u_b(t) & u_c(t) \end{bmatrix}^{\mathrm{T}}$—three-phase voltage $u_{abc}(t)$; $\begin{bmatrix} u_d(t) & u_q(t) & 0 \end{bmatrix}^{\mathrm{T}}$—corresponding voltage in DQZ reference plane after the transformation. This transformation is the product of Clarke transformation and Park transformation. The latter converts the components of the orthogonal stationary $\alpha\beta\gamma$ frame to an orthogonal rotating DQZ reference frame [57,58]. The rotation angular velocity of both the d-axis and q-axis is equal to ω, and the angle θ is equal to ωt: in this particular case, when θ is equal to zero, then ω is also zero.

3. Results

3.1. Shelving Filter

Let us begin with the examination of a filter transient response to a 50 Hz signal of 10 kV phase-to-ground voltage. When the stopband width Δf is set to 1 Hz, the filter's transient process lasts a relatively long time—longer than 0.5 s (Figure 8a), which may result in a high risk to conceal an overall majority of PQ events. Therefore, in an iterative way, it has been figured out that the shortest process is observed when Δf is equal to 25 Hz (Figure 8b). An analogous tendency is also observed in the filter impulse response (Figure 9).

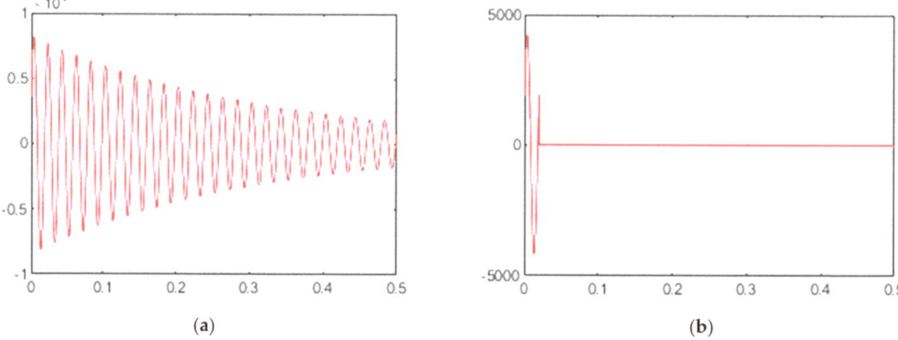

(a)   (b)

Figure 8. (**a**) Transient response to 50 Hz signal (phase-to-ground voltage of 10 kV grid) when Δf is equal to 1 Hz; (**b**) Transient response to 50 Hz signal (phase-to-ground voltage of 10 kV grid) when Δf is equal to 25 Hz.

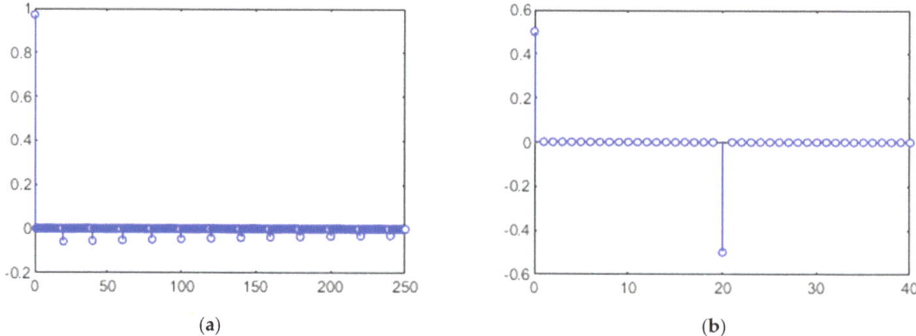

(a)   (b)

Figure 9. (**a**) Impulse response when Δf is equal to 1 Hz; (**b**) Impulse response when Δf is equal to 25 Hz.

In the pole–zero plot (Figure 10), a stopband width Δf corresponds with the distance between the pole and zero across the respective radius of the unit circle. The system remains stable in the both cases (i.e., impulse response approaches zero), because all poles are inside the unit circle. Since all zeros are on the boundary of the unit circle, maximal attenuation is achieved.

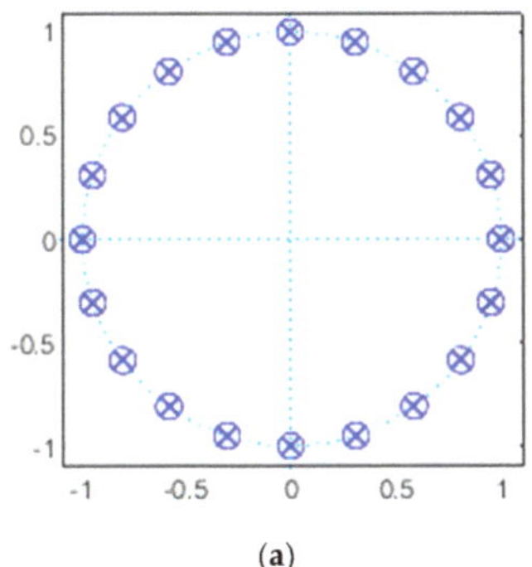

(a)

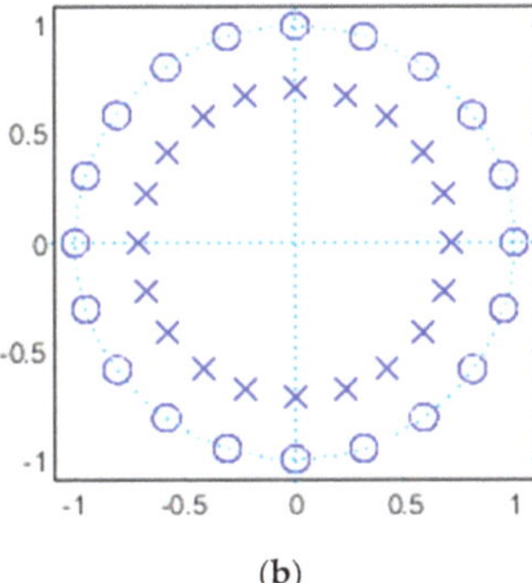

(b)

Figure 10. (**a**) The pole–zero plot when Δf is equal to 1 Hz; (**b**) The pole–zero plot when Δf is equal to 25 Hz.

In Figure 11, both types of voltage transients, described in Sections 2.1.1 and 2.1.2, and their filtering results are given. The first is a commutation transient caused by a 10 kV line switching on (Figure 11a), which is described by Equations (1)–(5). The second is an atmospheric transient which propagated from a 110 kV grid to a 10 kV grid (Figure 11b). This oscillation is the result of the grid response to the lightning impulse, modeled by Equation (8), when the current peak value I_m is 205 kA, rise time t_1 is 21 µs, and fall time t_2 is 300 µs. The occurrence frequency of such an impulse is lower than 0.01. Its electric charge is 87 C, which is lower than (but close to) the 100 C used in IEC 62305-1:2010 for the highest lightning protection level from the first positive impulse. The filtering results of the transients in phase A (red curve) are given in Figure 11c–f. When Δf is 1 Hz, in the case of the commutation transient, it is clearly seen (examining both magnitude and time) that the true signal is hidden beyond the heavy transient response (Figure 11c), but in the case of the atmospheric transient—the difference between true and output signals is not visually clear (Figure 11d). On the other hand, when Δf is 25 Hz and the transient response is the lightest (shortest), the signal distortions are clearly seen in Figure 11e,f, which cannot be recognized as a desirable result.

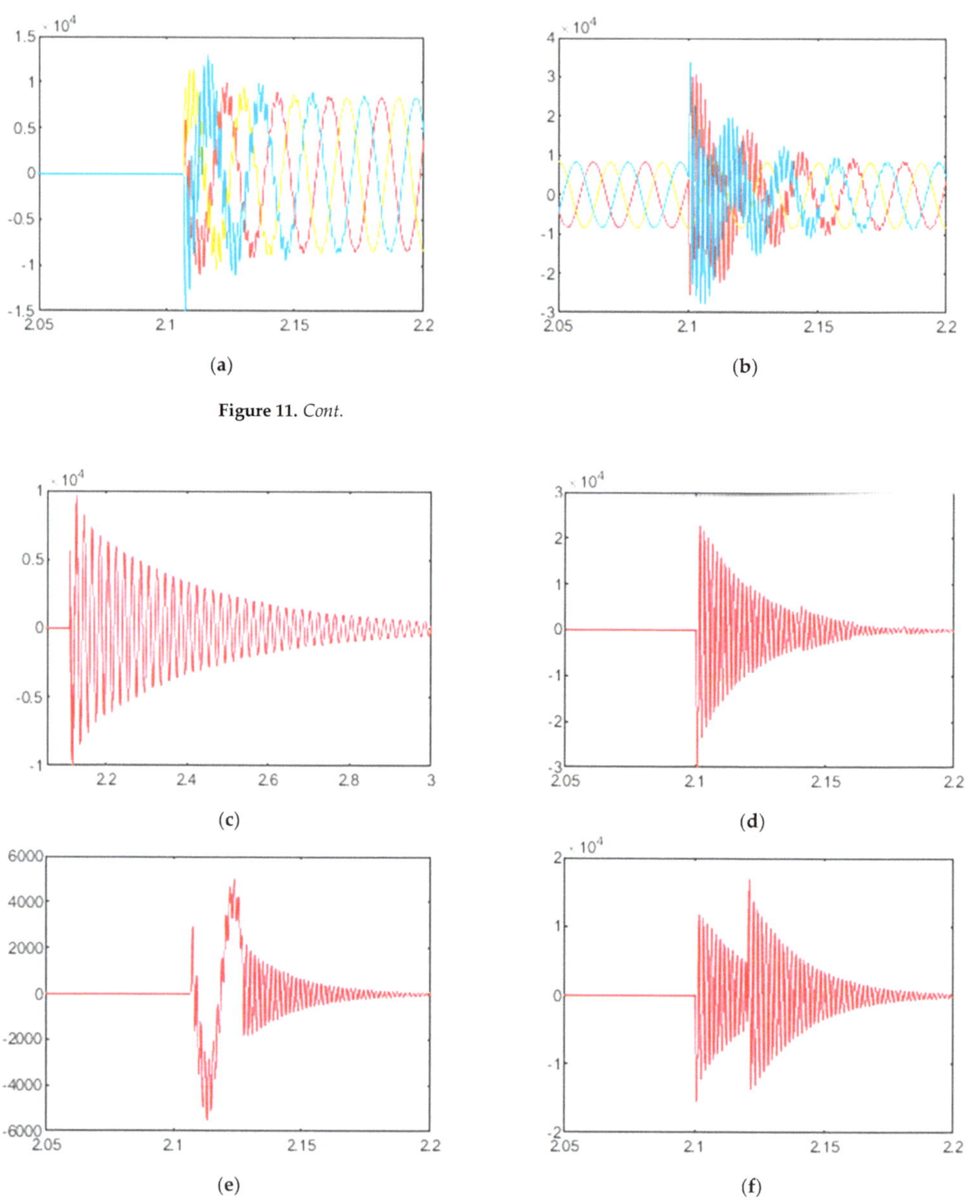

(**a**)

(**b**)

Figure 11. *Cont.*

(**c**)

(**d**)

(**e**)

(**f**)

Figure 11. (**a**) Phase-to-ground voltage transient caused by 10 kV line commutation (switching on); (**b**) Phase-to-ground voltage transient in a 10 kV line caused by lightning stroke; (**c**) Filter output of phase A (red curve) of a commutation transient when Δf is equal to 1 Hz; (**d**) Filter output of phase A (red curve) of an atmospheric transient when Δf is equal to 1 Hz; (**e**) Filter output of phase A (red curve) of a commutation transient when Δf is equal to 25 Hz; (**f**) Filter output of phase A (red curve) of an atmospheric transient when Δf is equal to 25 Hz.

A signal's distortion can be evaluated from the magnitude and phase frequency response of the digital filter (Figure 12): the magnitude and phase of signal frequency components are less distorted when Δf is equal to 1 Hz (especially phase); however, as it has been shown in Figure 8, its transient response is heavier than in the case when Δf is equal to 25 Hz.

In the case of RMS variations, the filter is not useful because the frequency of voltage sag/swell is 50 Hz. The example of a single-phase fault is given in Figure 13. Considering the above, it is decided to continue the research with the fundamental component included.

3.2. Classification

3.2.1. Single-Voltage Classification

The investigation is started with single-voltage array analysis, which is part of a three-phase system. In this case, the feature vector has the following form:

$$\mathbf{x}_i = \begin{bmatrix} U_i & t_i \end{bmatrix}^{\mathrm{T}}, \tag{44}$$

where U—voltage (either phase-to-phase or phase-to-ground), p.u.; t—duration, ms. In Table 2, the features are given exactly in this form. Four PQ event groups are labeled (atmospheric transient, commutation transient, single-phase short circuit, and electric motor starting), and the fifth group is for uncategorized events without any label assigned (gray zone). The results of both SVM and KNN, obtained in Python, are given in Figures 14–20. The characteristic points are marked with the outline.

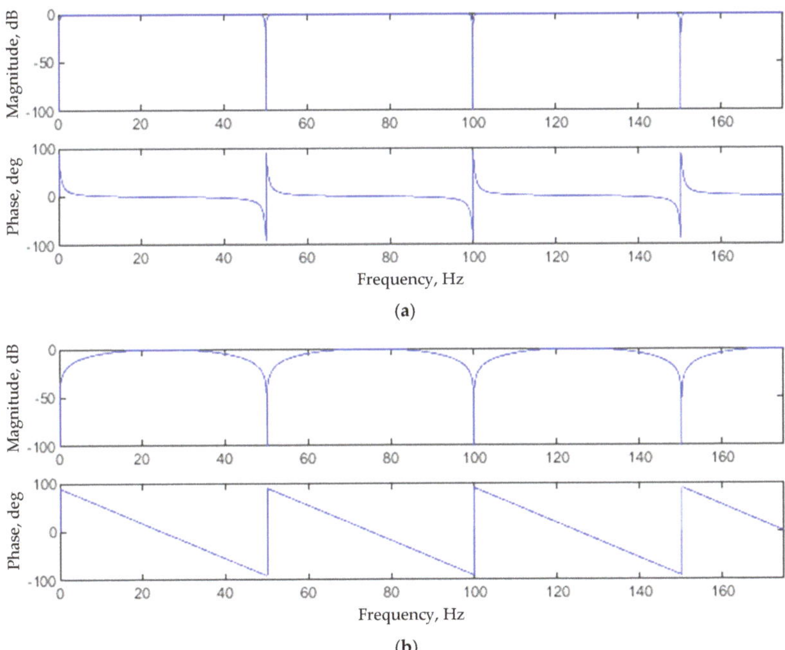

Figure 12. (a) Frequency response when Δf is equal to 1 Hz; (b) Frequency response when Δf is equal to 25 Hz.

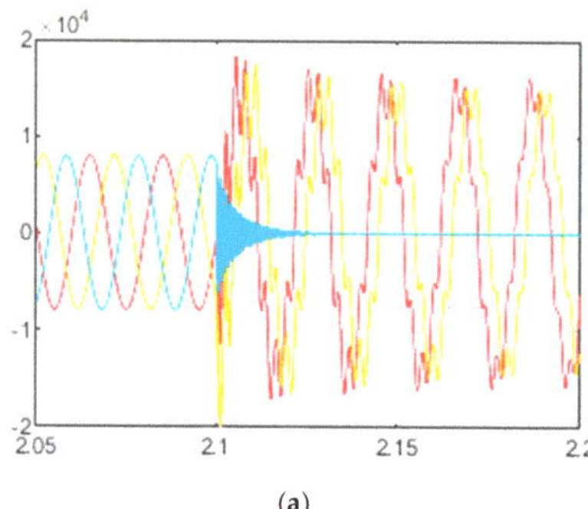 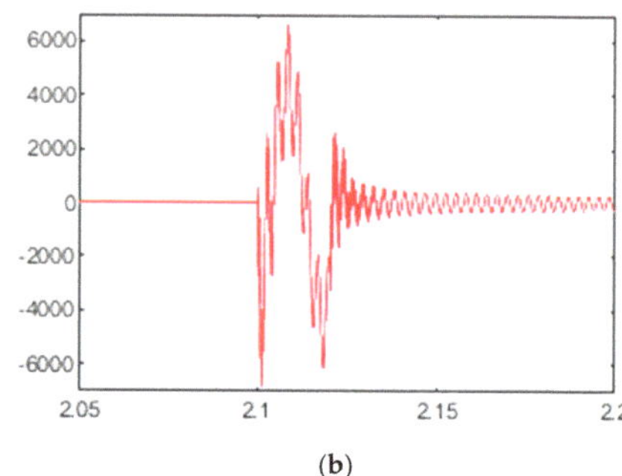

(a)   (b)

Figure 13. (**a**) Phase-to-ground voltage of a single-phase short circuit in a 10 kV line; (**b**) Filter output of phase A (red curve) when Δf is equal to 25 Hz.

Table 2. Characteristic data points of the training set.

Group	Data									
Atmospheric transient	8 p.u. 0.02 ms	5 p.u. 0.02 ms	4 p.u. 50 ms		4 p.u. 0.3 ms	4 p.u. 0.005 ms		3.2 p.u. 55 ms	3.1 p.u. 55 ms	3 p.u. 0.01 ms
Commutation transient	2 p.u. 31 ms	2 p.u. 20 ms	1.88 p.u. 30 ms	1.8 p.u. 10 ms	1.5 p.u. 25 ms	1.3 p.u. 10 ms	1.29 p.u. 10 ms	1.25 p.u. 5 ms	1.05 p.u. 33 ms	1 p.u. 30 ms
Single-phase short circuit	2 p.u. 55 s	1.8 p.u. 19 s	1.7 p.u. 60 s	1.55 p.u. 56 s	1.5 p.u. 16 s	1.4 p.u. 46 s	1.3 p.u. 1000 ms	1.2 p.u. 27 s	1.1 p.u. 60 s	1.1 p.u. 35 s
Electric motor starting	0.88 p.u. 800 ms	0.87 p.u. 50 ms		0.85 p.u. 50 ms	0.8 p.u. 2000 ms		0.35 p.u. 327 ms	0.35 p.u. 307 ms		0.25 p.u. 710 ms
Not categorized	2 p.u. 1000 ms	1.8 p.u. 1600 ms	1.8 p.u. 600 ms	1.8 p.u. 10 ms	1.78 p.u. 1780 ms	1.4 p.u. 3000 ms	1.4 p.u. 600 ms	1.3 p.u. 60 s	1.2 p.u. 3000 ms	1.2 p.u. 2300 ms
	1.2 p.u. 2000 ms	1.1 p.u. 60 s	1.1 p.u. 3000 ms	1.1 p.u. 1800 ms	1.1 p.u. 600 ms	1.1 p.u. 10 ms	0.84 p.u. 200 ms	0.82 p.u. 100 ms	0.8 p.u. 150 ms	0.74 p.u. 340 ms
	0.7 p.u. 25 ms		0.68 p.u. 20 ms		0.6 p.u. 750 ms		0.57 p.u. 290 ms	0.55 p.u. 80 ms		0.4 p.u. 60 ms

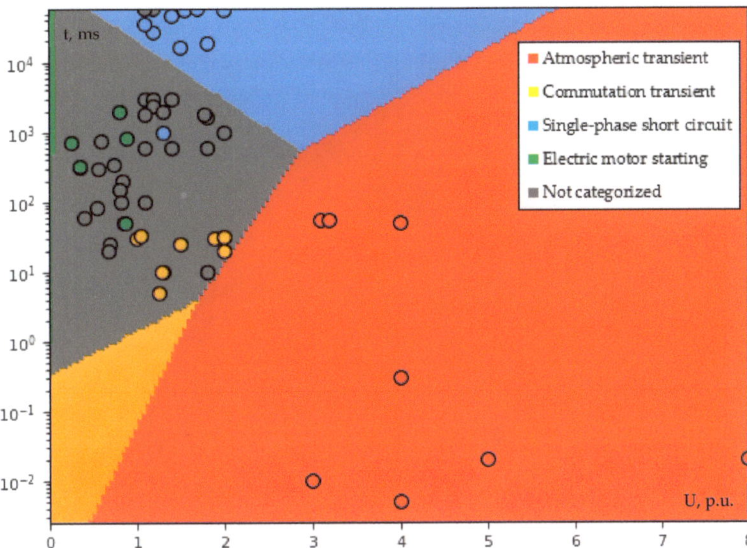

Figure 14. Single-voltage classification according to the type of event (primary cause) by using SVM with linear kernel.

Despite the background in Section 2.1, an additional explanation should be given about single-phase short circuits. The Lithuanian MV grid operates in either isolated or compensated neutral mode; thus, two phase-to-ground voltage swells along with one interruption are inherent to the single-phase fault. It can last several hours, because even under these circumstances, the grid is still able to ensure power supply to end-users (see [2,9] for more information). Since the size of our time window is limited to 1 min (according to the definition of short-duration RMS variation given by IEEE Std 1159-2019), during a single-phase fault, each point will be at 1 min, and after that, it will be once at a random time (which depends on the moment of the disconnection). Thus, the magnitude of voltage swell is given according to the simulation results obtained in [9], but its duration is determined with a random number generator. It should be noted that such circumstances tighten the dynamic range requirements.

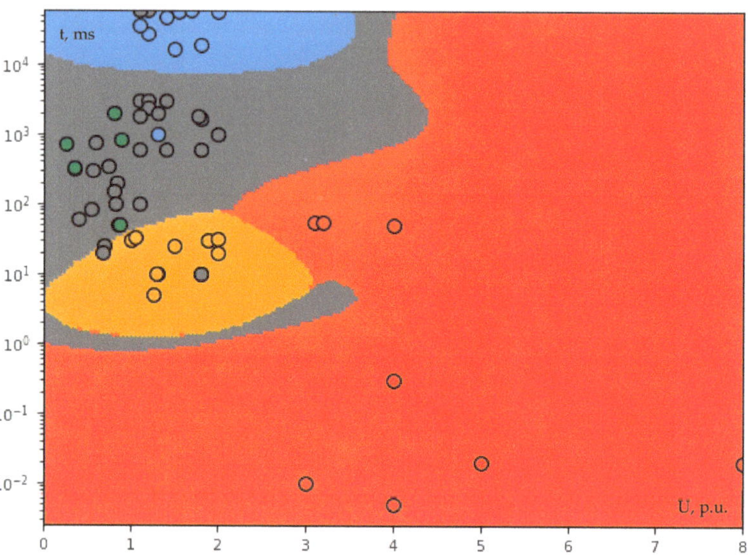

Figure 15. Single-voltage classification according to the type of event (primary cause) by using SVM with RBF kernel when the width parameter γ is equal to 0.35.

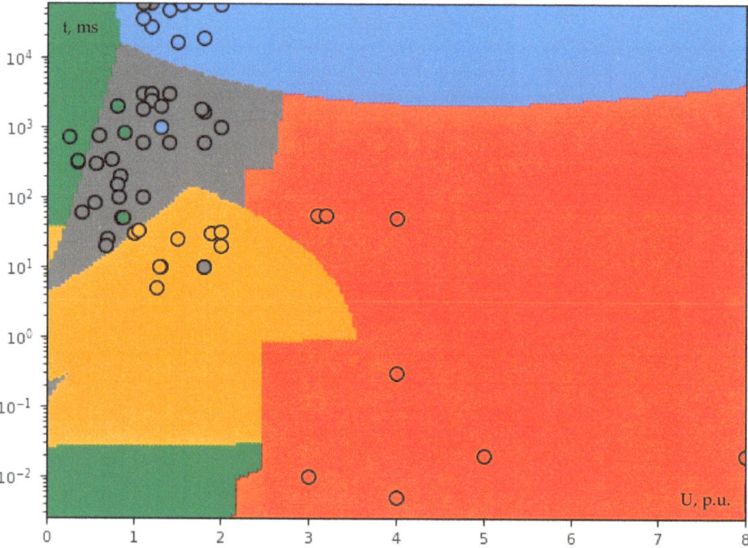

Figure 16. Single-voltage classification according to the type of event (primary cause) by using SVM with the second-order polynomial kernel.

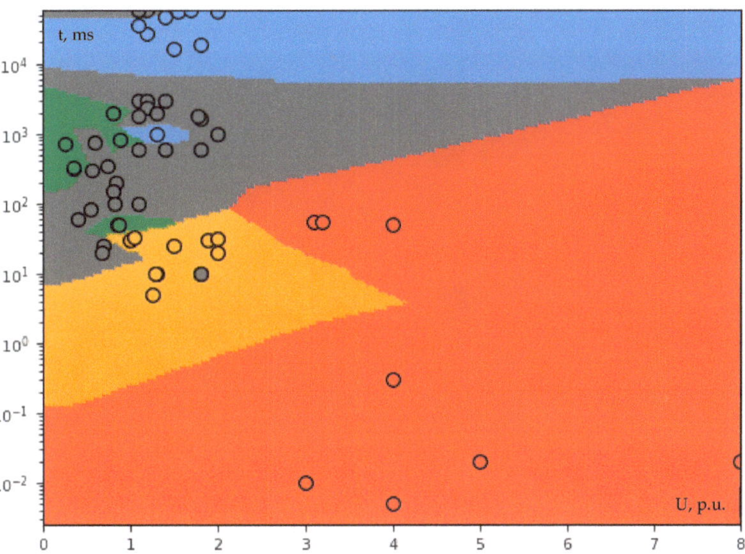

Figure 17. Single-voltage classification according to the type of event (primary cause) by using KNN when *k* is 1.

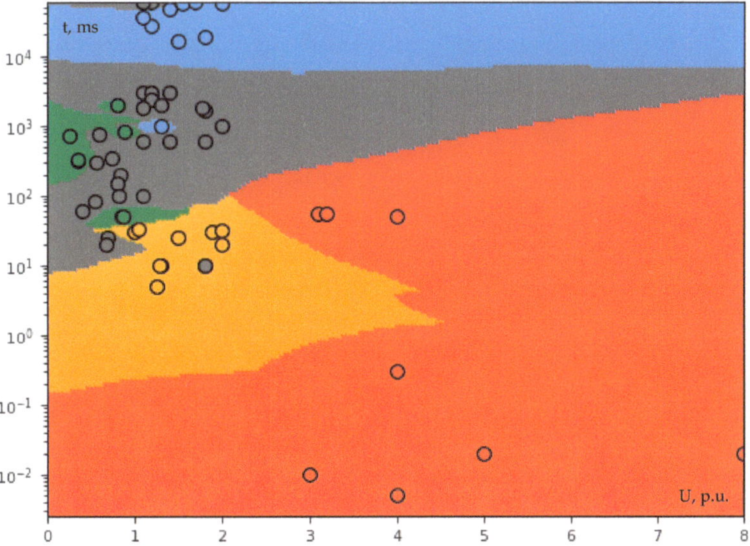

Figure 18. Single-voltage classification according to the type of event (primary cause) by using KNN when *k* is 4.

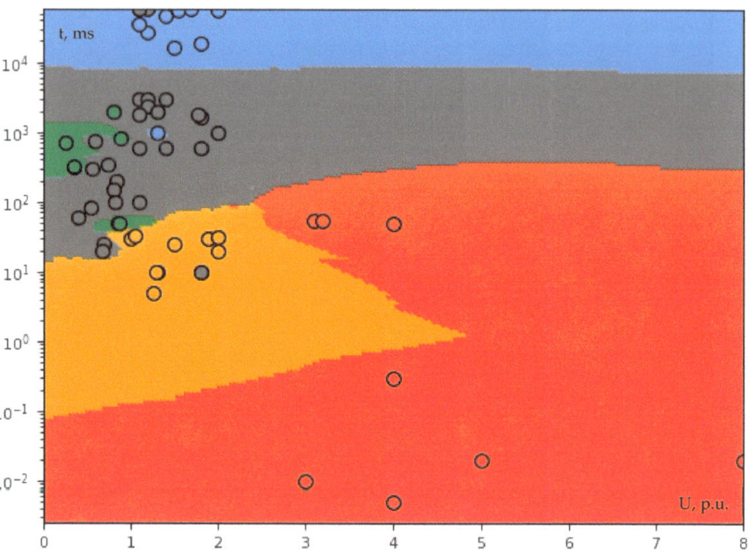

Figure 19. Single-voltage classification according to the type of event (primary cause) by using KNN when k is 7.

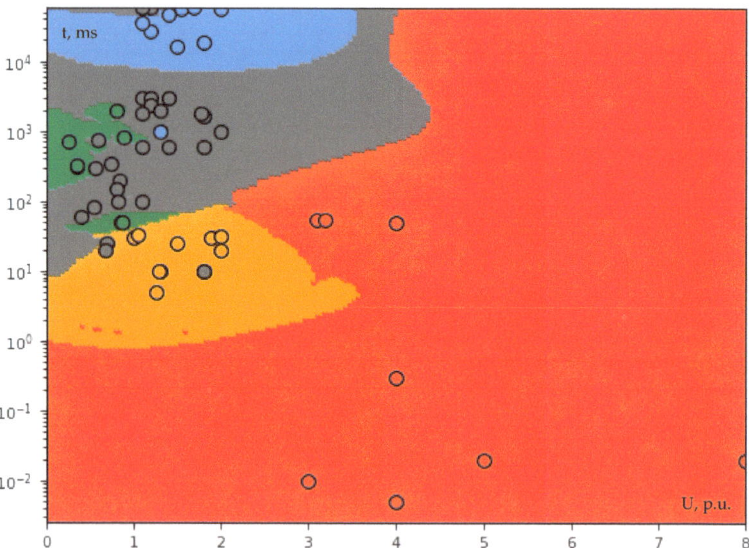

Figure 20. Single-voltage classification according to the type of event (primary cause) by using the ensemble of both SVM with RBF kernel when the width parameter γ is equal to 0.35 and KNN when k is 3.

Typically, machine learning requires a large database; however, it is not essential to this task because the PQ database can be enlarged with a random number generator (see Section 4.3 for more details). Since one of the goals is to avoid both abnormal range and extreme values during data generation (see Section 4.2 for more information about data quality), competency and experience remain much more important, and these skills can be acquired mostly during only PQ monitoring campaigns. It is noteworthy that the information about the typical range of the characteristics of PQ events can be found only in few sources, and one of them is IEEE Std 1159-2019.

Firstly, it seems that KNN performs better than SVM with a soft margin (in all SVM cases, the regularization parameter C is equal to 1). For example, in Figure 14 (it shows SVM results with linear kernel), both all green and all yellow points are outside their zones, which does not mean that these zones are inappropriate but indicates insufficient coverage (span). On the contrary, this problem is not encountered in the case of KNN, but that does not necessarily mean that the best possible solution was found. Secondly, it is clear that a single either SVM or KNN cannot deal with the overlapping in the voltage–duration plane, for example: (1) an atmospheric transient (especially damped) can have the same characteristics as a commutation transient (but not vice versa); however, the grid operator must be able to identify such a discrepancy by itself; (2) no events are expected (possible) in the upper right corner of Figures 14–20; (3) despite the fact that a voltage swell is inherent not only to a single-phase fault but also to a two-phase-to-ground fault, the overlapping zone is not large because two-phase-to-ground faults are usually disconnected as soon as possible (say up to 2–3 s but typically within a few tens of milliseconds). Also, it is obvious that the approach can be expanded by adding more groups, but at present, this is hardly implemented due to insufficient knowledge and experience in the field. Thirdly, a few words must be written about classification accuracy. As mentioned in Section 1, most authors affirm that their classification accuracy is close to perfect. However, in the field of PQ, it actually depends on either the validation or test set: since each zone (category) covers at least some part of the correct (expected) range, it is not problematic to either adjust or select a test set similar to the training set with maintained realism and solid argumentation and thereby achieve desired (high) accuracy. Fourthly, in spite of the fact that each zone covers its correct range, it also can fall in the regions where it is unexpected; however, this issue does not necessarily have a significant effect on the performance (accuracy): for example, currently, it is not possible to confirm nor deny that the amplitude of any commutation transient can be higher than 4 p.u. (contrary to Figures 17–19), and it can be confirmed that the amplitude of any voltage swell during a single-phase fault is not expected to be higher than 3 p.u. and lower than 1 p.u. (contrary to Figures 14–20). Lastly, in order to implement the task, the EN 50160:2010 requirement for the multiphase aggregation had to be rejected (see Section 4.1 for more details). Also, currently, the data of the entire power system (consisting of HV, MV and LV grids) is plotted on one plane: on the one hand, such an approach does not have serious drawbacks and, on the other hand, the data segregation by, for example, either voltage lever or neutral mode could (slightly) improve the performance.

3.2.2. Three-Dimensional Voltage Classification

An asymmetrical short circuit cannot be described with any single voltage value; therefore, the approach must be based on three-dimensional voltage classification. In this case, the feature vector has the following form:

$$\mathbf{x}_i = \begin{bmatrix} U_{1i} & U_{2i} & U_{3i} \end{bmatrix}^{\mathrm{T}}, \tag{45}$$

where U_1, U_2 and U_3 are either phase-to-phase or phase-to-ground voltages at a fault node arranged in ascending order, p.u. On the one hand, the order of numbers in the feature array is not important; on the other hand, in this paper, the arrangement in ascending order is preferred for research purposes. Analogously to Section 3.2.1, for this case, data quality is more important than quantity.

It is noteworthy that no three-dimensional voltage classification was found in the existing PQ literature, and most likely the reason for this is the absence of comprehensive investigations of asymmetrical faults (and, respectively, database), as has been argued by [9]. Both training sets (Tables 3 and 4) are based on the results of this paper with all its limitations that are inherent to the inductive reasoning method, including the fact that the results were obtained in the BRELL-based test scheme (different network properties in other countries may have different impacts on the patterns of PQ events). Also, it is clear that phase-to-phase and phase-to-ground cases must be investigated separately except for the case of a symmetrical fault due to the following axiom established by [9]:

Axiom 1. *In the case of a three-phase fault, both phase-to-phase and phase-to-ground voltage sag depths are equal and independent of neutral mode.*

The training set in the form of Equation (45) is given in Tables 3 and 4. All four types of short circuits are included. Please note that the group label is assigned only to the voltage of the fault node (regardless of the scenario), while the voltages of the rest nodes fall in the gray zone. In Table 4, the same phase-to-ground voltage swells of a single-phase short circuit can be noticed as in Table 2. In addition, each feature vector after Clarke transformation is given in parentheses. Please note that the transformation output depends on the order of elements in the input vector, and this is important for this research because the third element of the output is discarded.

Table 3. Phase-to-phase voltage characteristic data points of the training set.

Group	Data									
Three-phase short circuit	0.8 (0.98)	0.5 (0.61)	0.4 (0.49)	0.3 (0.37)	0.15 (0.18)	0.1 (0.12)	0.05 (0.06)	0.0 (0.00)		
	0.8 (0.98)	0.5 (0.61)	0.4 (0.49)	0.3 (0.37)	0.15 (0.18)	0.1 (0.12)	0.05 (0.06)	0.0 (0.00)		
	0.8 (0.00)	0.5 (0.00)	0.4 (0.00)	0.3 (0.00)	0.15 (0.00)	0.1 (0.00)	0.05 (0.00)	0.0 (0.00)		
Two-phase-to-ground short circuit	0.0 (0.97)	0.0 (0.77)	0.0 (0.76)	0.0 (0.76)	0.0 (0.75)	0.0 (0.73)	0.0 (0.65)	0.0 (0.54)		
	0.9 (0.64)	0.7 (0.49)	0.7 (0.49)	0.7 (0.49)	0.69 (0.49)	0.68 (0.48)	0.6 (0.42)	0.5 (0.35)		
	0.9 (0.52)	0.72 (0.41)	0.71 (0.41)	0.7 (0.40)	0.7 (0.40)	0.68 (0.39)	0.6 (0.35)	0.5 (0.29)		
Two-phase short circuit	0.0 (0.97)	0.0 (0.93)	0.0 (0.86)	0.0 (0.85)	0.0 (0.77)	0.0 (0.76)	0.0 (0.70)	0.0 (0.65)		
	0.9 (0.64)	0.86 (0.61)	0.8 (0.56)	0.79 (0.59)	0.72 (0.51)	0.7 (0.49)	0.65 (0.46)	0.6 (0.42)		
	0.9 (0.52)	0.86 (0.49)	0.8 (0.46)	0.79 (0.46)	0.72 (0.42)	0.7 (0.40)	0.65 (0.38)	0.6 (0.35)		
Single-phase short circuit	1.0 (1.23)	0.9 (1.11)	0.82 (1.00)	0.8 (0.98)	0.7 (1.06)	0.7 (1.12)	0.7 (1.02)	0.6 (1.06)		
	1.0 (1.22)	0.9 (1.10)	0.82 (1.00)	0.8 (0.98)	0.9 (0.98)	0.8 (0.91)	0.7 (0.86)	0.6 (0.73)		
	1.0 (0.00)	0.9 (0.00)	0.82 (0.00)	0.8 (0.00)	0.9 (0.12)	1.0 (0.15)	0.9 (0.12)	1.0 (0.23)		
Not categorized	0.7 (1.12)	0.6 (0.94)	0.6 (1.08)	0.6 (0.90)	0.52 (0.95)	0.5 (1.02)	0.5 (0.94)	0.4 (1.02)	0.4 (0.82)	0.4 (0.73)
	0.8 (0.92)	0.8 (0.86)	0.7 (0.80)	0.6 (0.74)	0.52 (0.64)	0.5 (0.61)	0.5 (0.61)	0.6 (0.62)	0.4 (0.49)	0.4 (0.49)
	1.0 (0.15)	0.8 (0.12)	1.0 (0.21)	0.8 (0.12)	0.9 (0.22)	1.0 (0.29)	0.9 (0.23)	1.0 (0.31)	0.8 (0.23)	0.7 (0.17)
	0.3 (0.76)	0.3 (0.82)	0.3 (0.53)	0.2 (0.49)	0.1 (0.95)	0.1 (0.74)	0.0 (0.86)	0.0 (0.77)	0.0 (0.65)	0.0 (0.54)
	0.6 (0.56)	0.5 (0.49)	0.3 (0.37)	0.2 (0.24)	0.8 (0.60)	0.6 (0.46)	0.8 (0.57)	0.72 (0.51)	0.6 (0.42)	0.5 (0.35)
	0.7 (0.21)	0.8 (0.25)	0.5 (0.12)	0.5 (0.17)	0.9 (0.44)	0.7 (0.32)	0.8 (0.46)	0.72 (0.42)	0.6 (0.35)	0.5 (0.29)

Table 4. Phase-to-ground voltage characteristic data points of the training set.

Group	Data									
Three-phase short circuit	0.8 (0.98)	0.5 (0.61)	0.4 (0.49)	0.3 (0.37)	0.15 (0.18)	0.1 (0.12)	0.05 (0.06)	0.0 (0.00)		
	0.8 (0.98)	0.5 (0.61)	0.4 (0.49)	0.3 (0.37)	0.15 (0.18)	0.1 (0.12)	0.05 (0.06)	0.0 (0.00)		
	0.8 (0.00)	0.5 (0.00)	0.4 (0.00)	0.3 (0.00)	0.15 (0.00)	0.1 (0.00)	0.05 (0.06)	0.0 (0.00)		
Two-phase-to-ground short circuit	0.0 (1.23)	0.0 (1.08)	0.0 (1.08)	0.0 (1.02)	0.0 (0.99)	0.0 (0.98)	0.0 (0.90)	0.0 (0.73)		
	0.0 (0.00)	0.0 (0.00)	0.0 (0.00)	0.0 (0.00)	0.0 (0.00)	0.0 (0.00)	0.0 (0.00)	0.0 (0.00)		
	1.5 (0.87)	1.32 (0.76)	1.3 (0.76)	1.25 (0.72)	1.22 (0.70)	1.2 (0.69)	1.1 (0.64)	0.9 (0.52)		
Two-phase short circuit	0.5 (1.02)	0.5 (1.02)	0.5 (1.02)	0.5 (0.94)	0.48 (0.93)	0.43 (0.74)	0.42 (0.78)	0.4 (0.78)		
	0.5 (0.61)	0.5 (0.61)	0.5 (0.61)	0.5 (0.61)	0.48 (0.59)	0.43 (0.53)	0.42 (0.51)	0.4 (0.51)		
	1.0 (0.29)	1.0 (0.29)	1.0 (0.29)	0.9 (0.23)	0.9 (0.24)	0.69 (0.15)	0.75 (0.19)	0.8 (0.19)		
Single-phase short circuit	0.0 (1.94)	0.0 (1.59)	0.0 (1.51)	0.0 (1.40)	0.0 (1.53)	0.0 (1.37)	0.0 (1.27)	0.0 (1.05)		
	1.8 (1.27)	1.4 (0.99)	1.4 (0.99)	1.3 (0.92)	1.2 (0.85)	1.2 (0.85)	1.1 (0.78)	0.9 (0.64)		
	1.8 (1.04)	1.5 (0.84)	1.4 (0.81)	1.3 (0.75)	1.5 (0.79)	1.3 (0.72)	1.2 (0.69)	1.0 (0.55)		
Not categorized	1.0 (2.27)	0.8 (2.01)	0.8 (1.96)	0.8 (1.06)	0.6 (1.44)	0.6 (1.06)	0.58 (0.89)	0.52 (1.03)	0.5 (0.80)	0.4 (0.92)
	2.0 (1.87)	1.7 (1.56)	0.8 (0.98)	0.8 (0.98)	1.2 (1.12)	0.6 (0.73)	0.58 (0.71)	0.52 (0.64)	0.6 (0.67)	0.5 (0.55)
	2.0 (0.58)	1.8 (0.55)	2.0 (0.69)	0.9 (0.06)	1.3 (0.38)	1.0 (0.23)	0.8 (0.13)	1.0 (0.28)	0.7 (0.10)	0.9 (0.26)
	0.4 (0.76)	0.4 (1.23)	0.4 (0.82)	0.4 (0.82)	0.3 (1.30)	0.3 (0.82)	0.2 (1.14)	0.1 (0.95)	0.0 (0.76)	0.0 (0.69)
	0.5 (0.55)	0.4 (0.49)	0.4 (0.49)	0.4 (0.49)	1.1 (0.90)	0.5 (0.50)	0.2 (0.24)	0.8 (0.60)	0.7 (0.49)	0.62 (0.44)
	0.7 (0.15)	1.3 (0.52)	0.8 (0.23)	0.7 (0.23)	1.2 (0.49)	0.8 (0.25)	1.3 (0.64)	0.9 (0.44)	0.7 (0.40)	0.62 (0.36)

The results in three-dimensional space are given in Figures 21 and 22. As in the previous case, the characteristic points from Tables 3 and 4 have the outlines.

Calculations in three-dimensional space were carried out only with KNN because it turned out to be more suitable in Section 3.2.1. This also will be realized later (in this section) after transforming it to two-dimensional space. The range of phase-to-phase voltage is up to 1 p.u., while the range of phase-to-ground voltage is up to 2 p.u. In Python, each calculation took up to 5 min in three-dimensional space (including both training and display steps) and up to 1 min in two-dimensional space (the base frequency of the 4-core processor is 3.50 GHz, RAM—16 GB).

It is difficult to analyze and interpret Figures 21 and 22; thus, let us examine the zones of the characteristic points separately. The results of both phase-to-phase and phase-to-ground cases, obtained in MATLAB, are shown in Figure 23. The theoretical background of the used equations along with their parameters is given in Section 2.5.1: planes that do not coincide with a coordinate hyperplane are colored; a counterclockwise rotation is represented by a positive rotation angle, while a clockwise rotation is represented by a negative rotation angle. The findings convince that the idea of using the ascending order was superb, enabling the observation of many interesting tendencies:

1. In the case of a three-phase fault (Figure 23a), all points lie on the line belonging to the plane which forms the angles of 45° with both abscissa and ordinate.
2. In the case of a two-phase-to-ground fault (Figure 23b), all points (of both phase-to-phase and phase-to-ground voltages) belong to the same coordinate hyperplane, i.e., are coplanar (see Equation (34)), which is defined by the pair of ordinate and applicate.
3. In the case of a two-phase fault (Figure 23c), the phase-to-phase pattern is identical to the case of a two-phase-to-ground fault, while phase-to-ground points are encircled with the ellipse lying in the plane which forms the angles of 45° with both the abscissa and ordinate (analogously to the case of a three-phase fault).
4. In the case of a single-phase fault (Figure 23d), phase-to-phase points are covered with one half of the paraboloid, while phase-to-ground points are encircled with the ellipse lying in the coordinate hyperplane defined by the pair of ordinate and applicate.

As in the previous case, please note that the data are not segregated by grid voltage or neutral mode, and despite this, the trends still appeared. Such analysis and findings would be not be possible with the multiphase aggregation required by EN 50160:2010. The next stage is the elimination of inductive reasoning limitations, which should be performed by foreign authors. Then, two outcomes are possible: (1) if it turns out that the findings of this paper are universal, it would mean a great leap forward toward the successful completion of the classification task; (2) if the results are different, the geometry of clusters' boundaries should be modified accordingly as long as it will be required for the universalization. Please note that there is an option to alter the currently set parameters of the used ellipses and paraboloid, and the applied approach is not the only way to define a feature cluster's border (see Section 4.3.1).

As has been mentioned above, it is obvious that the three-dimensional feature space of Figures 21 and 22 is difficult to analyze (especially visually). Respectively, the analysis of larger n-dimensional spaces will possibly be even more complicated. For this purpose, we create a new method to transform a three-dimensional feature space into a two-dimensional. The method is based on Clarke transformation (see Section 2.5.2) and implemented in MATLAB/Simulink. Transformation output is given in the brackets of Tables 3 and 4, whereas the results in the $\alpha\beta$ plane, which is obtained by discarding component γ, are given in Figures 24–28 (where the colors are kept the same as in the legend of Figure 21). More information about the transformation procedure can be found in Appendix A. It is noteworthy that the proposed method is completely appropriate for the analysis of short circuits; however, this does not indicate the suitability for universal application to various tasks (but it may be). In Figures 24–26, similarly as in Section 3.2.1, it is clearly seen that the performance of SVM is not satisfying enough with both regularization parameter C values—1 (which corresponds to a soft margin) and 10 (which corresponds to a hard margin). Luckily, KNN works much better (Figures 27 and 28). In addition, similarly as in Figure 23, the geometric analysis of the clusters in the $\alpha\beta$ plane is given in Figures 29 and 30.

It can be noticed that phase-to-ground voltage is more informative than the phase-to-phase voltage due to smaller overlapping; nevertheless, in order to enhance the performance, both voltages should be assessed regardless of EN 50160:2010. Moreover, it would be greatly beneficial to identify areas where no data points are expected (e.g., the upper left corner).

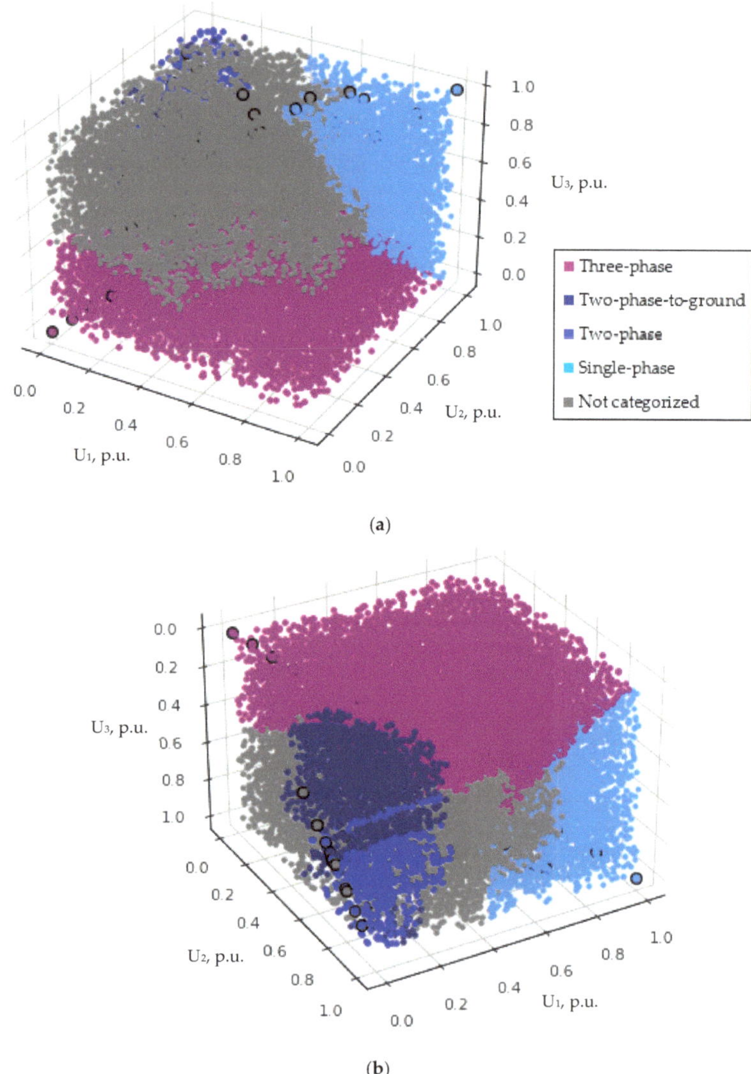

Figure 21. Phase-to-phase voltage three-dimensional classification according to the type of short circuit by using KNN when *k* is 3. (**a**) Front view; (**b**) Back view.

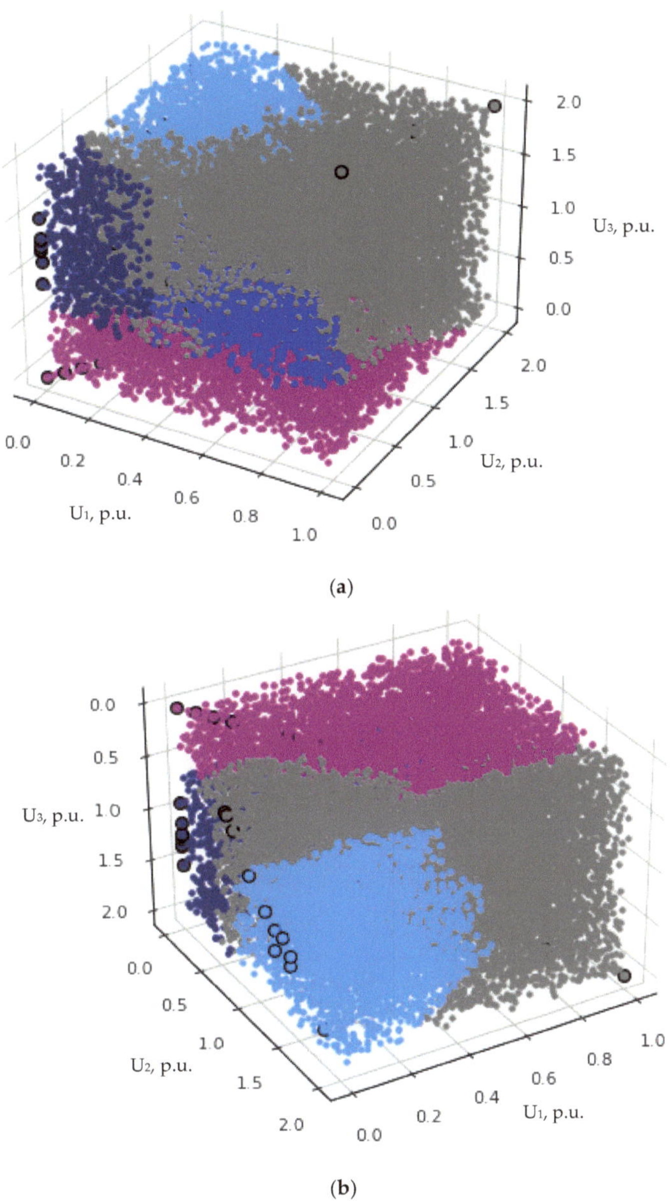

Figure 22. Phase-to-ground voltage three-dimensional classification according to the type of short circuit by using KNN when *k* is 3. (**a**) Front view; (**b**) Back view.

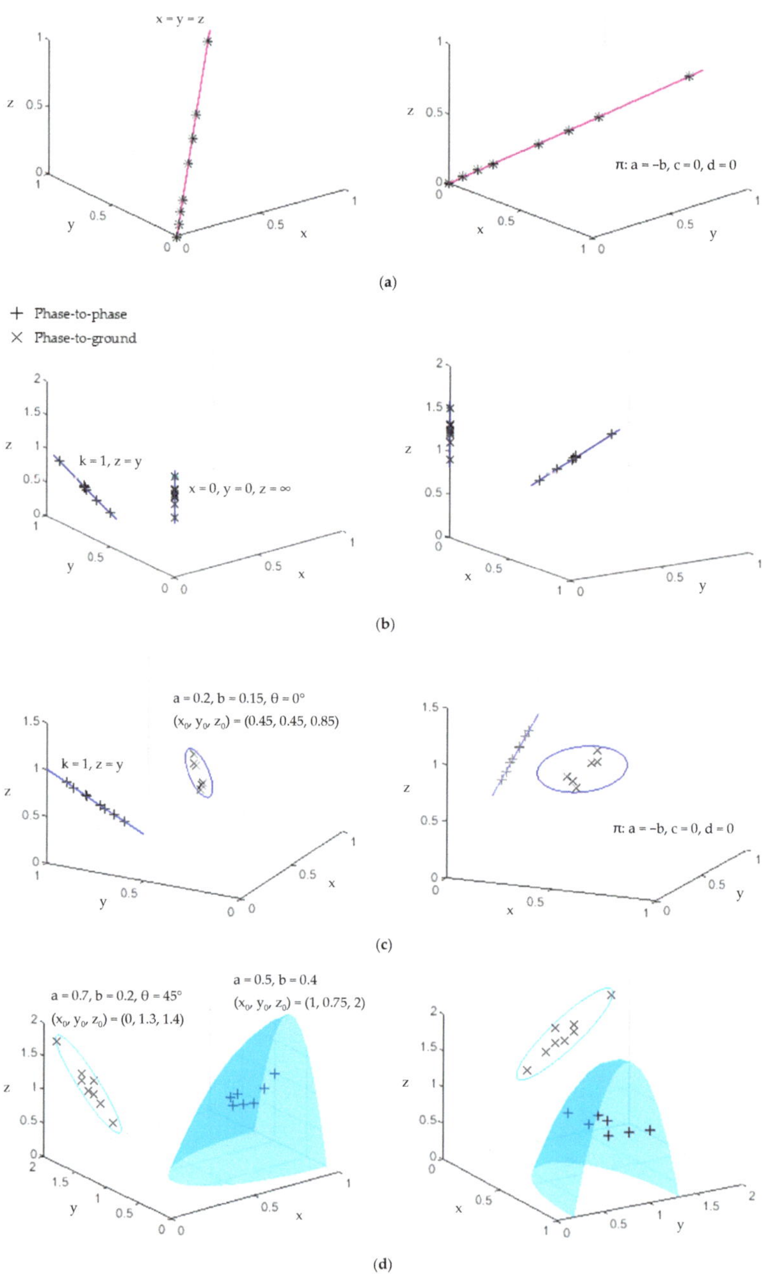

Figure 23. Three-dimensional geometric analysis of the characteristic features. (**a**) Three-phase fault; (**b**) Two-phase-to-ground fault; (**c**) Two-phase fault; (**d**) Single-phase fault.

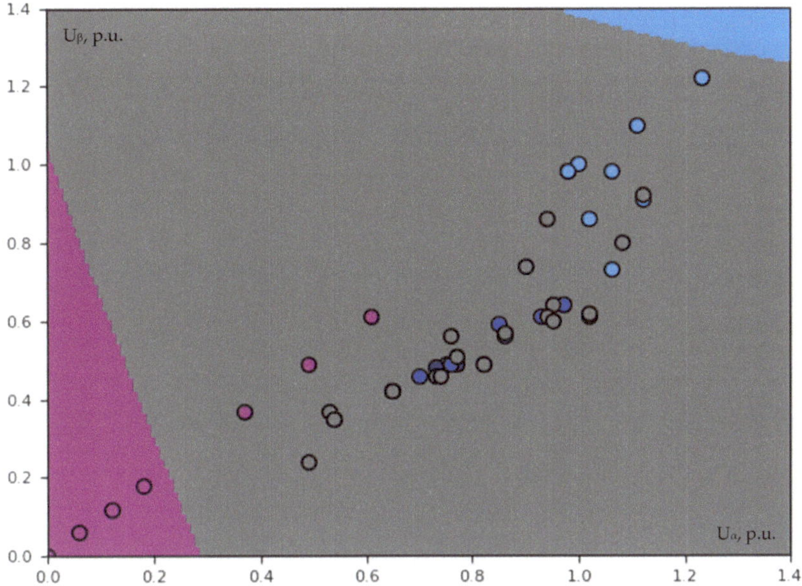

Figure 24. Phase-to-phase voltage classification in the αβ plane according to the type of short circuit by using SVM with RBF kernel when the regularization parameter *C* is equal to 1.

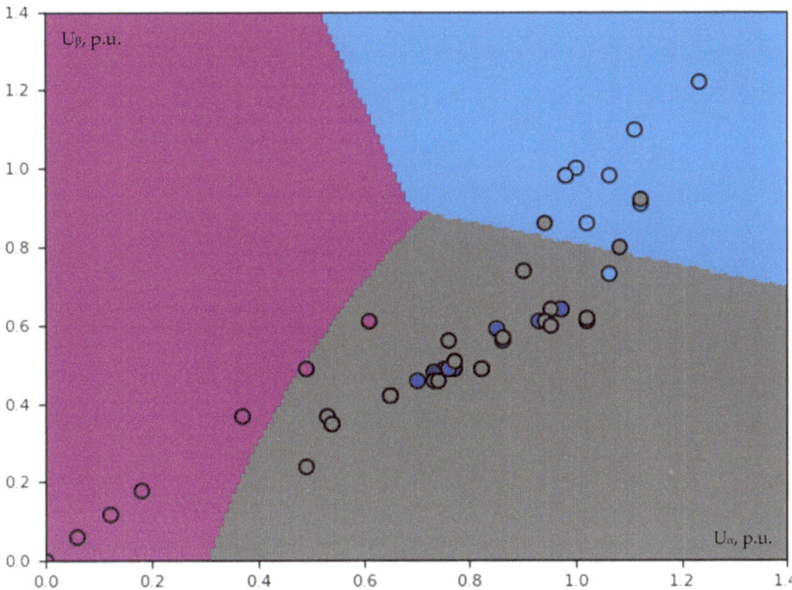

Figure 25. Phase-to-phase voltage classification in the αβ plane according to the type of short circuit by using SVM with RBF kernel when the regularization parameter *C* is equal to 10.

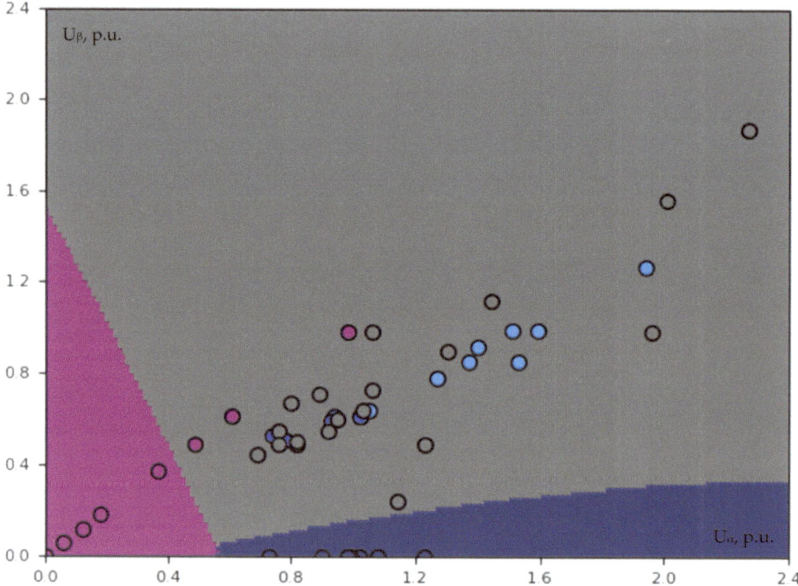

Figure 26. Phase-to-ground voltage classification in the $\alpha\beta$ plane according to the type of short circuit by using SVM with RBF kernel when the regularization parameter C is equal to 1.

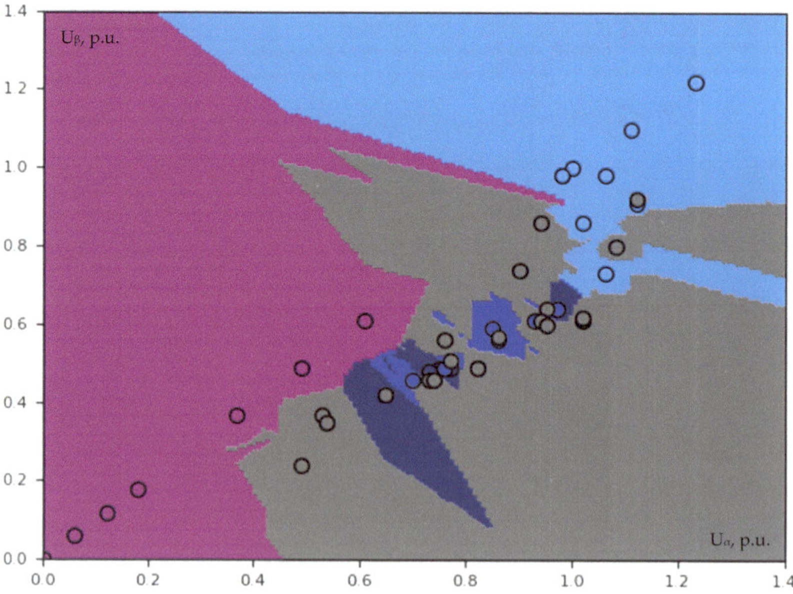

Figure 27. Phase-to-phase voltage classification in the $\alpha\beta$ plane according to the type of short circuit by using KNN when k is 3.

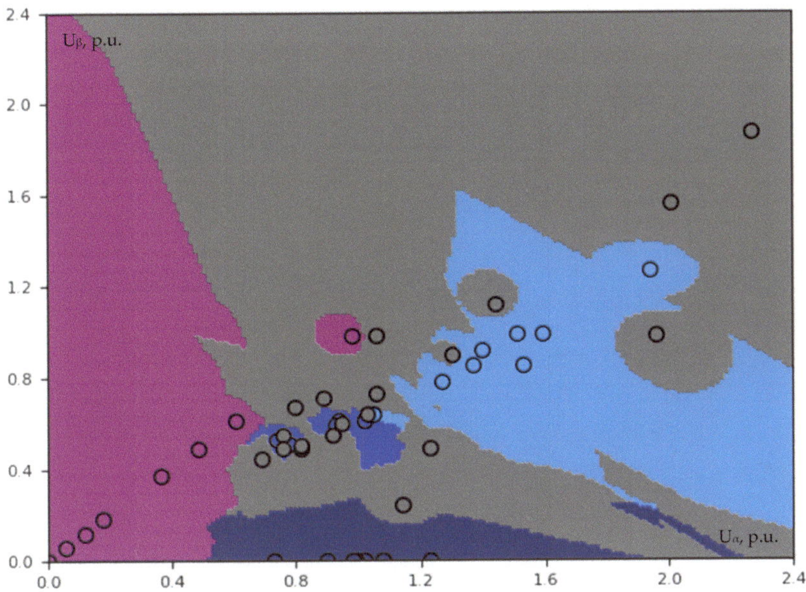

Figure 28. Phase-to-ground voltage classification in the $\alpha\beta$ plane according to the type of short circuit by using KNN when *k* is 3.

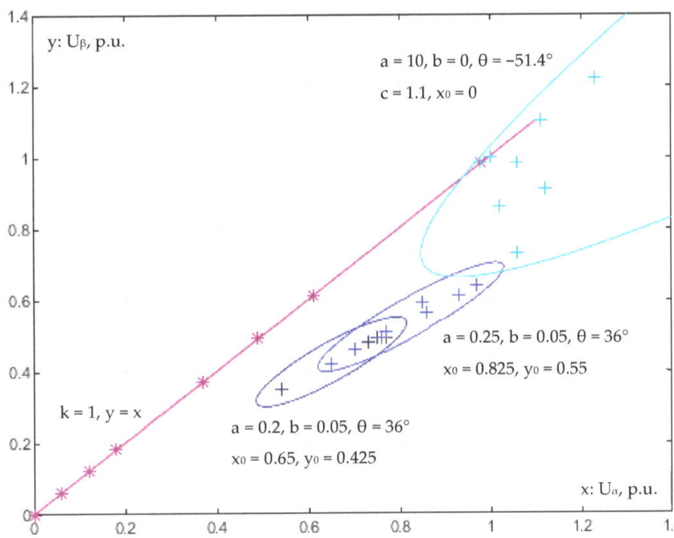

Figure 29. Geometric analysis of characteristic features of phase-to-phase voltage in $\alpha\beta$ plane.

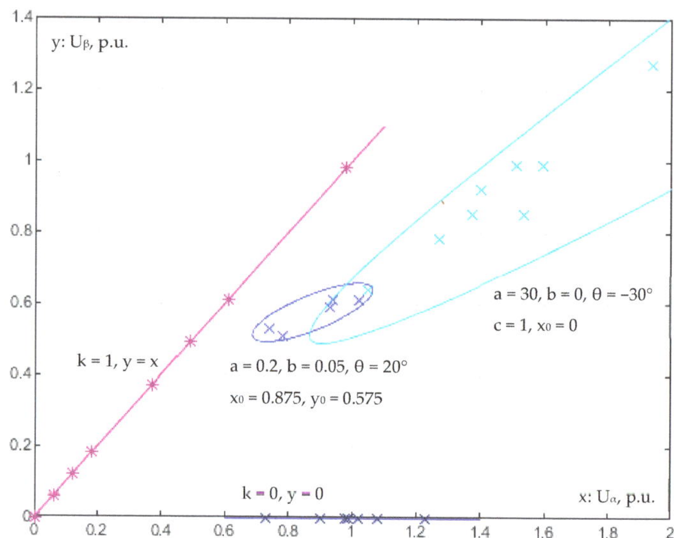

Figure 30. Geometric analysis of characteristic features of phase-to-ground voltage in $\alpha\beta$ plane.

4. Discussion

4.1. PQ Events Assessment

A lot of universally unsolved issues in the field of PQ events assessment are considered to be highly limiting factors for the further research and application of AI algorithms [2]. Some of these issues, particularly those regarding voltage sag assessment, which were noticed during the PQ monitoring campaign in the Lithuanian DSO grid (see Section 2.1.3), are discussed in this section. Many of them can be treated as PQ legislation gaps. Appropriate amendments in EN 50160:2010, IEC 61000-4-30:2015, IEEE Std 1159-2019, IEEE Std 1564-2014 and other relevant standards (documents) are essential to accelerate the development and usefulness of PQ monitoring systems. At the moment, there is no similar discussion to this in the existing PQ literature.

Firstly, it has been noticed that the RMS voltage of some voltage sags is time-varying (multistage), i.e., they have a stepped shape. The example is given in Figure 31: a voltage sag event was recorded on 6 March 2019 on the 10 kV side of the 35/10 kV substation (located in Širvintos municipality). Therefore, although the stepped shape can be characterized according to EN 50160:2010, i.e., with a single residual voltage and duration (as in Section 2.1.3), such an approach seems to be inaccurate and inappropriate because it does not reflect any information about the voltage sag profile. IEC 61000-4-30:2015 also briefly mentions this aspect, i.e., that the envelopes of both voltage sag and swell are not necessarily rectangular. This is one of the main drawbacks of currently the most popular method of the voltage–duration plane (e.g., as in Figure 2a). In our opinion, the problem can be solved by using the voltage sag energy characteristic, which is defined in IEEE Std 1564-2014 as follows:

$$E = \int_{0}^{T} \left(1 - \frac{U^2(t)}{U_N{}^2}\right) dt, \tag{46}$$

where $U(t)$—RMS voltage during the event; U_N—nominal voltage. Despite the fact that the energy index considers the RMS voltage profile, it is more difficult to understand (interpret) its result (especially for specialists having lower competence). Therefore, in our opinion, the feature vector should be at least three-dimensional and contain the values of voltage, duration and energy. Also, please note that the terminology used to describe the magnitude of a voltage sag is often confusing, and reference voltage selection for voltage sags assessment currently remains an open question [9,12]: the residual (remaining) voltage is preferred by both EN 50160:2010 and IEEE Std 1159-2019, while IEEE Std 1564-2014 does not reject the idea of using either pre-event or nominal voltage (e.g., see Equation (46)).

Secondly, during voltage sag feature extraction, another question arises regarding the multiphase aggregation required by EN 50160:2010. It is obvious that the multiphase aggregation leads to information loss: for example, from Figure 31, it is much more informative to extract all three phase-to-phase residual voltages (5.8 kV, 7.0 kV, 7.1 kV) than only one (5.8 kV). The same also applies to the time domain. Therefore, the multiphase aggregation increases the entropy of a message, but in machine learning, the goal is to minimize uncertainty. Considering the above, in our opinion, the approach depends on the task: the multiphase aggregation is completely not preferred in this paper.

Thirdly, incorrect marking of the voltage sag event start can be clearly seen in Figure 31. If we refer to the definitions given in [39] (see Table 1), it would probably be the second duration time needed for multistage voltage sag characterization. This is exactly what the software does; however, such a feature extraction technique does not match the definition in PQ legislation. It is obvious that it is impossible to recheck all data manually; thus, software errors during feature extraction and other stages will reduce the benefit gained from AI algorithms. According to EN 50160:2010, a voltage sag begins when at least one voltage out of three drops below a 90% value, and it ends when all three voltages rise above the end threshold, which is equal to same 90% plus 2% hysteresis as recommended in IEC 61000-4-30:2015. The standard argues that the hysteresis—the difference between the start and end thresholds—is required to avoid the counting multiple events when the voltage magnitude oscillates about the threshold level. However, in our opinion, based on the experience gained during the PQ monitoring campaign in the Lithuanian DSO grid, the usefulness and necessity of this parameter should be reviewed and rethought. Please note that the mentioned hysteresis has nothing in common with well-known iron core saturation.

Fourthly, after capturing the voltage sag bursts (Figures 32–34), a lack of literature on the topic was encountered [2]. For example, a 3 s time interval separates the voltage sags of Figure 32; thus, it is highly probable that these events are dependent. Currently, there are no investigations directly relating voltage sags with their primary causes [9,59]. According to IEC 61000-4-30:2015, multiple voltage sags "may occur [. . .] during a failed attempt to auto-reclose and re-energize a faulty line section". Also, it may be caused during windstorms as well as by certain sequences of operation of relay protection and automation under both normal and emergency operating conditions. However, more detailed research is needed to substantiate these thoughts. Analysis of the primary causes plays an important role in preventive measures planning [9] as well as in the determination of typical time intervals between the events, which will be beneficial for both legal regulation and machine learning. Presently, IEC 61000-4-30:2015 allows counting the number of sags that "occurs at approximately the same time" as a single event. According to IEEE Std 1564-2014, recommendations regarding the time interval cannot be made at the present stage, since "the discussion is ongoing on which aggregation time to use".

Fifthly, a voltage sag followed by a power supply interruption (Figure 33) should be treated more seriously/heavily (i.e., as more dangerous) than casual voltage sag. Power supply interruptions can be identified from electric current measurements. Moreover, the current direction coincides with the power-flow direction; hence, electric current measurements should be highly beneficial for the determination of voltage sag source location. Also, current measurements provide useful information about electric energy consumption; thus, it can be beneficial for impact assessment on end-user equipment. After the detailed investigation of the situation from the primary causes to consequences (e.g., economic loss, damaged equipment), the concept of responsibility sharing (considering the obligation to pay penalties, compensations) should be introduced. For example, the information about voltage sags regulation in Sweden can be found in the CEER 5th Benchmarking

Report [60] (2011, p. 67); however, the up-to-date situation is unknown. In the document, two voltage–duration planes are given (for up to and including 45 kV and above 45 kV), and each of them is divided into three areas as follows: (1) voltage sags in Area A are treated as rapid voltage change (RVC) events; (2) the grid operator has the responsibility to mitigate voltage sags in Area B; (3) there shall not be any voltage sags in Area C. After deeper examination, it can be noted that the events in Area C can occur only in case of inappropriate (e.g., wrong logic, too high delay) grid relay protection and automation response to a fault: duration threshold for up to and including 45 kV starts from 1 s, for above 45 kV—from 0.6 s. For example, no event of Figure 2a falls in Area C.

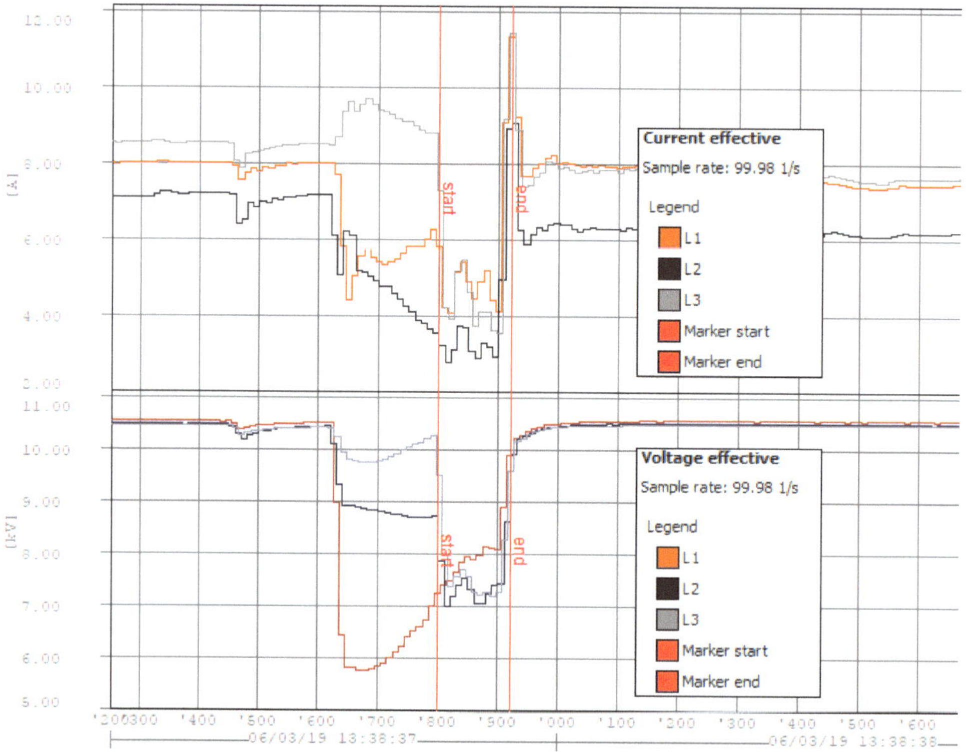

Figure 31. Time-varying voltage sag depth with incorrectly determined moment of its beginning. Phase-to-phase voltage and current measurements during a voltage sag event (occurred on 6 March 2019) on the 10 kV side of the 35/10 kV substation (located in Širvintos municipality).

The three dependent voltage sags are shown in Figure 34. Similar to Figure 32, both time intervals between them are also approximately 3 s. The power supply interruption, followed after the second voltage sag, is eliminated after 3 s (probably by an automatic circuit recloser, or if this reclosing is unsuccessful—by an automatic transfer switch) causing both the third voltage sag and current transient during the commutation. The third voltage sag probably occurs due to transformers energizing (magnetizing current is greater than 150 A) because it is followed by the smooth operation of the power grid afterwards. In addition, two time-varying phase-to-phase voltage sags are seen during the second event (similar to Figure 31). Moreover, it can be noticed that the voltage sag events' currents decrease in Figures 32 and 33, and they increase in Figure 34.

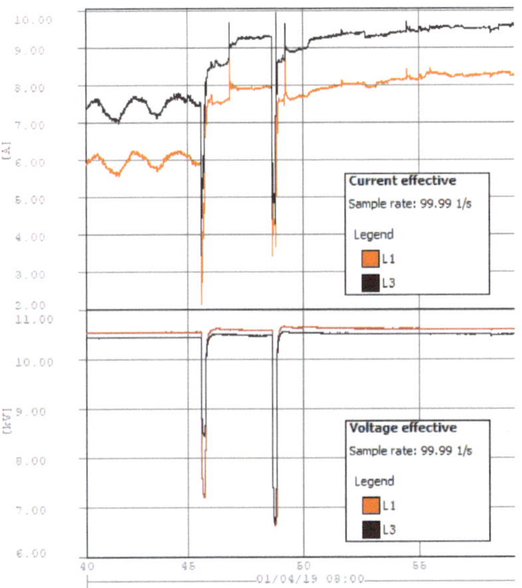

Figure 32. Two probably dependent voltage sags separated by 3 s time interval. Phase-to-phase voltage and current measurements during voltage sag events (occurred on 1 April 2019) on the 10 kV side of the 35/10 kV substation (located in Širvintos municipality).

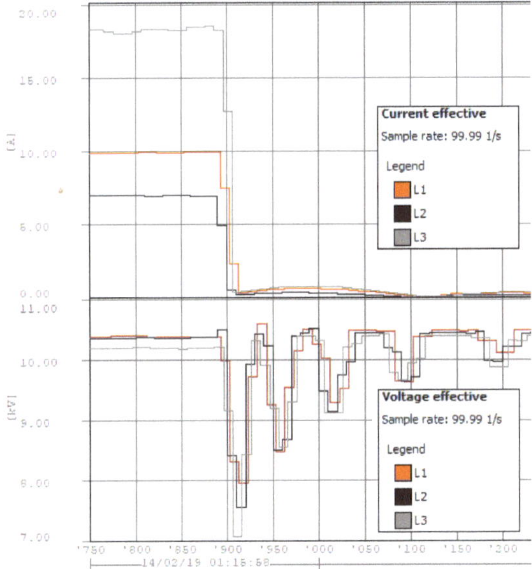

Figure 33. Voltage sag burst followed by power supply interruption. Phase-to-phase voltage and current measurements during voltage sag events (occurred on 14 February 2019) on the 10 kV side of the 35/10 kV substation (located in Širvintos municipality).

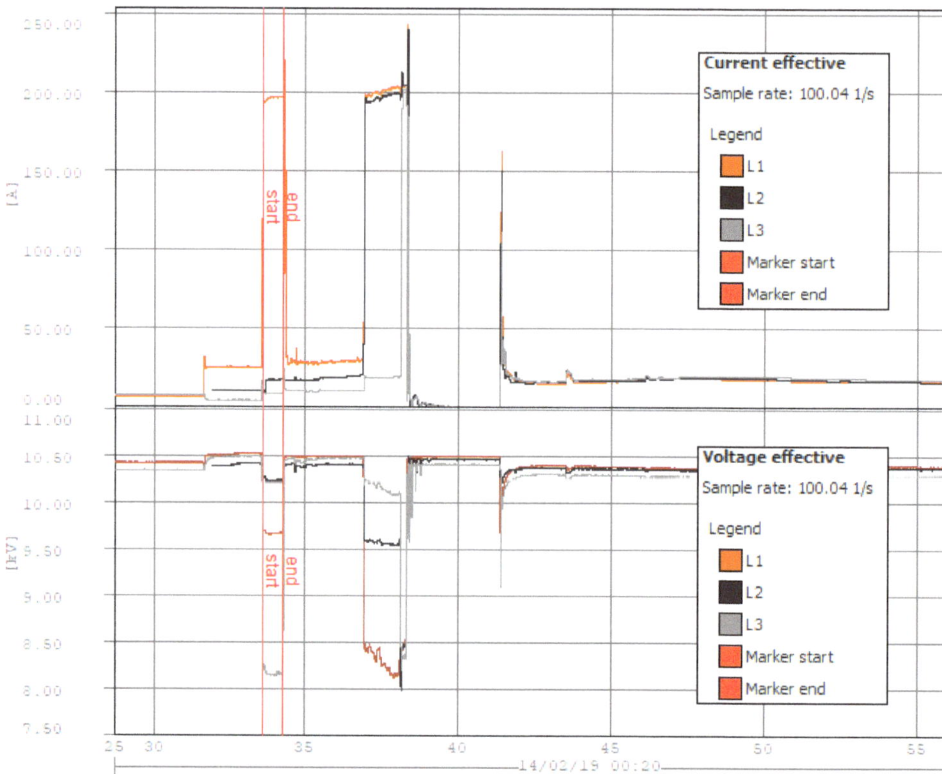

Figure 34. Phase-to-phase voltage and current measurements during voltage sag events (occurred on 14 February 2019) on the 10 kV side of the 35/10 kV substation (located in Širvintos municipality). Power supply, interrupted after the second voltage sag, is restored after slightly more than 3 s, causing both the third voltage sag and current transient.

Sixthly, there is a lack of discussion regarding how to assess PQ events when they are recorded by two or more measuring devices. This issue is important mainly for central data processing. In general, if a voltage sag is observed by n monitors, the number of alternatives to select k monitors for the assessment, when the order of the selection is not important, is equal to:

$$C_n^k = \frac{A_n^k}{P_k} = \frac{n!}{k!(n-k)!}, \quad 1 \le k \le n, \tag{47}$$

where n—total number of monitors that recorded the event; k—number of selected monitors for the assessment; A_n^k—partial permutation; P_k—permutation. Probably, in many particular cases, only extreme cases of k, i.e., either one or all of n, will be considered. On the other hand, a slight redundancy in measurements is sometimes desirable [2]; thus, other values of k can also be considered. In case the order of selection is important (e.g., for sorting the array by distance from short circuit location), the total number of possible combinations is equal to:

$$A_n^k = \frac{n!}{(n-k)!}, \quad 1 \le k \le n. \tag{48}$$

Seventhly, it is understandable that the accuracy of PQ algorithms can be affected by power consumption and transformers' tap changers (or, generally speaking, by long-duration RMS variations). The example is given in Figure 35: the maximal absolute error is determined by the voltage profile range (in normal mode, i.e., excluding RVC and others short-duration events) which in the given case is approximately equal to 2 kV. This error will be summed with other absolute errors such as the instrument error. This aspect is irrelevant in the case of the reference voltage being either pre-fault (also known as sliding reference voltage) or residual, however: (1) the implementation of pre-fault voltage tracking is much more complex; (2) in our opinion, residual voltage is important for impact assessment on a customer and his equipment, but it is not appropriate enough for voltage sag magnitude (depth) assessment. Since all possible nuances regarding this are not described in PQ standards as a universally agreed approach, questions may arise during the development of algorithms, which can also lead to data interpretation errors.

To continue, the information about ohmic losses in power lines can be found in the classical literature such as [61,62]. The relationship between the vector components of the voltage drop ΔU along the line and its loading takes the following form:

$$\begin{cases} \Delta U_x = \frac{PR+QX}{U_2}, \\ \Delta U_y = \frac{PX+QR}{U_2}, \end{cases} \tag{49}$$

where R—line resistance; X—line reactance; P—active power flow; Q—reactive power flow; U_2—voltage at the end of the line.

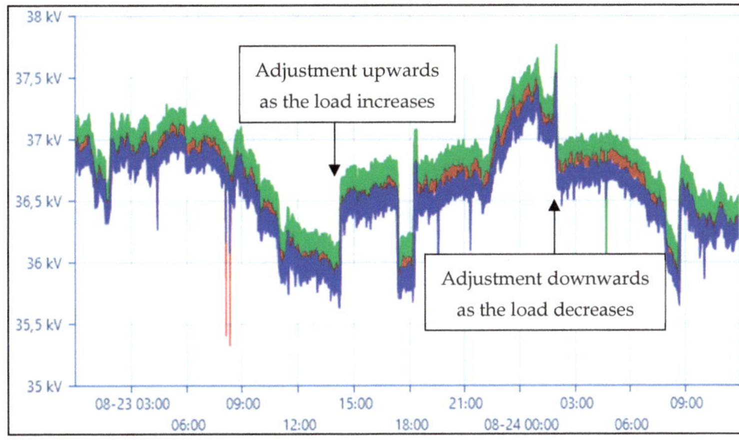

Figure 35. End-users and transformer's on-load tap changer influence on voltage profile of the 35 kV node (located in Rokiškis municipality).

Eighthly, let us briefly discuss the measurement of waveform. As has been mentioned in Section 2.1.1, the magnitude and phase frequency response of a measurement circuit must be known in order to evaluate distortions and the validity of results. For example, since this characteristic is unknown in the case of Figure 36, it cannot be determined whether the observed waveform distortion is caused by the frequency response, instrument internal reasons, insufficient sampling frequency or other reasons. For example, in IEEE Std 1159-2019 (pp. 68–69), the two figures of unrealistic voltage waveforms are given, which most likely resulted from instrument error. As has been figured out by [2], currently, the best achievements in determining the frequency response of measurement circuits are (1) the "Primary Networks PQ Analysis" study [63] carried out by one of the British DSOs, and (2) the three examples given by IEEE Std 1159-2019 (pp. 37–38). However, the spectrum of the first work is limited up to 5 kHz (i.e., only for harmonics and inter-harmonics), and the second work does not provide any information about the phase frequency response. It should be noted that the results of both of the aforementioned research studies, in the overlapping frequency range where they can be compared, are not similar. It should be noted that when dealing with the higher spectral components, the relationship between impedance and frequency must be taken into consideration: inductive reactance is directly proportional and capacitive reactance is inversely proportional to the signal frequency.

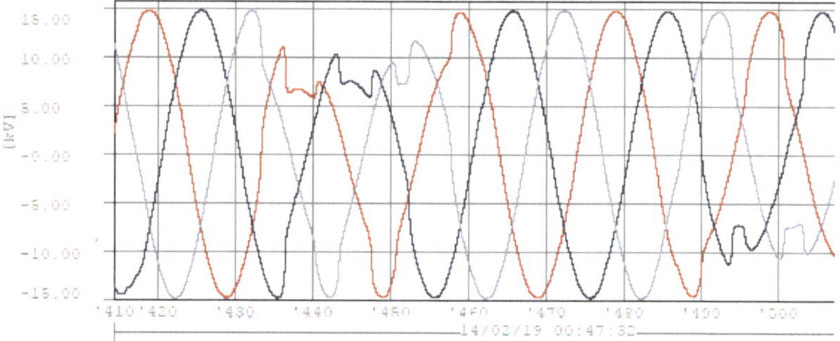

Figure 36. Measured 10 kV voltage waveform during the voltage sag events.

Ninthly, there are some gaps in the definitions of transients' duration, especially oscillatory ones. IEC 61000-4-30:2015 states that this duration "is a difficult parameter to define due to damping, irregularity of waveforms, etc.". In spite of this, IEEE Std 1159-2019 provides the typical durations of oscillatory transients—however, without a clear definition of them. On the other hand, the nomenclature of impulsive transients seems to be more or less clear. For example, in the case of a surge waveform of 1.2/50 (which perhaps is the most commonly used), it is clearly defined that 1.2 stands for the rise time in microseconds (from 10–90% peak), and 50 stands for the duration. The definition of this duration is different, for example: (1) according to IEEE Std 1159-2019, it starts from the beginning of the event and lasts to a 50% peak in the decay phase, but (2) in IEC 61000-4-5:2014 [64], this definition is based on full-width at half-maximum (which can be multiplied by a constant). At first glance, it may appear that the discussion on the issue can be omitted, since expected errors are small. However, it depends on the application, but the issue will be relevant at least during the code writing. Obviously, the definition may vary by scope (e.g., substation shielding, equipment immunity, predictive maintenance, PQ compliance). For example, in the case of predictive maintenance, it may be desirable not to lose the time of the decaying tail of an impulse. In this paper, such errors are not very important at the moment due to the insufficient state-of-the-art of the field.

Tenthly, during both the PQ measurement and assessment stages, questions regarding the RMS measurement method may arise. The standard RMS method is appropriate for a pure sine wave, while the true-RMS method calculates the correct RMS value even if such a waveform is distorted. This aspect is also important for smart meters, since a high level of harmonics is typically expected in an LV grid. The difference between these methods is best seen from the comparison of their formulas. An average responding meter assumes a sinusoidal wave and calculates the RMS value according to the following equation:

$$u_{RMS} = \frac{U_p}{\sqrt{2}},$$

(50)

while the true-RMS value is calculated by:

$$U_{RMS} = \sqrt{\frac{1}{N}\sum_{i=1}^{N} U_i^2},$$ (51)

where U_p—peak value; U_i—instantaneous value; N—window size [65,66].

Eleventhly, despite the fact that DSOs cannot influence the frequency, the requirement for this parameter is given in EN 50160:2010. On the other hand, a frequency array can be useful as additional information: for example, if the PQ monitor either does not measure the frequency or does it incorrectly, concerns about the validity of both voltage and current data must be raised. The example of a frequency measurement result under the condition of power supply interruption, which lasted approximately 1 h, is given in Figure 37. It can be understood that the frequency value has a certain random behavior in the normal operation mode. However, in the emergency mode of this case, the constant value is outputted. Although this constant has no meaning, it can be beneficial for both checking the operation of the device and power supply interruption detection. To conclude, many encountered issues are described, but this list is definitely not full. These aspects in the form of legal gaps inhibit the progress in many areas such as PQ classification, impact assessment, compliance verification, etc.; therefore, a great leap forward is impossible without universally agreed answers (solutions) to the questions raised [2]. However, solid and competent solutions cannot be made without enough experience, which inevitably requires more measurement data and analysis of various situations. This once again emphasizes the need for PQ monitoring.

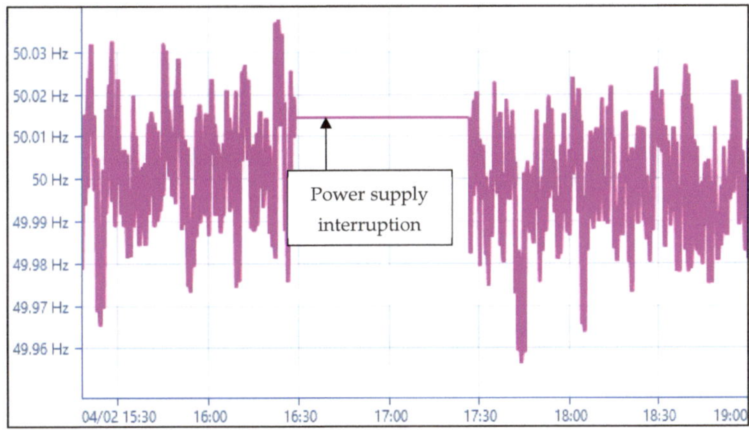

Figure 37. Frequency measurement during power supply interruption.

4.2. PQ—Part of Smart Grid and Its Communication Network

The development of traditional electric power systems started from such events as the war of the currents, invention of polyphase system, standardization of both frequency and voltage levels, and electrification (which in North America went more smoothly than in Europe due to the world wars, different national interests and the Cold War; for example, the Soviet Union refusal to accept the Marshall Plan) [62,67]. Now, it is already in the past, and the new challenges related to information and communication technologies (ICT) are emerging in anticipation of smart grids. Problems that will negatively impact PQ data assessment can occur during data transmission, i.e., in between local data processing and central data processing blocks. Moreover, it should be understood that not only pure PQ data packets may be used but also mixed ones containing data packets of other applications. Packet aggregation approaches can be applied due to technical convenience and comprehensive assessment (e.g., PQ interoperability with the following smart grid applications has been identified as beneficial in [2,9]: advanced metering infrastructure—AMI, outage management—OM, predictive maintenance—PM, substation automation—SA). A smart grid should be understood in a broader context through a prism of Power-to-X concept, i.e., when the electricity sector is integrated with other sectors such as gas, heat, fuel, chemical, etc. An example of such a concept is shown in Figure 38. Surplus electric power (which typically occurs due to fluctuating renewable energy generation) can be used in many sectors, for example, for hydrogen production by the electrolysis of water. Hydrogen is required for fuel cells (including hydrogen transport) as well as ammonia NH_3 production or methanation process. Since the enthalpy change in methanation of both carbon monoxide CO and dioxide CO_2 is negative (i.e., the reaction is exothermic), released heat can be used for the anaerobic digestion of biogas as well as for biogas upgrading to biomethane (i.e., the separation of methane CH_4 from CO_2). Every monitoring system in each sector will have its own sensors and data packets, and it will be a part of a smart grid's communication network, which currently remains to be a further research avenue.

In a noisy communication channel, the amount (probability) of altered messages can be diminished by selecting a proper modulation scheme. The theoretical bit error rate (BER) of some common modulations (phase-shift keying—PSK, quadrature amplitude modulation—QAM, frequency-shift keying—FSK) dependence on the energy per bit to noise power spectral density ratio E_b/N_0 (also known as normalized SNR) in an additive white Gaussian noise (AWGN) channel is given in Figure 39. The AWGN model accurately imitates real data transmission mediums (wired, optical, wireless); thus, it is often used [68]. The highest BER can be expected with 32-PSK modulation, and the lowest can be expected with 32-FSK. Lower-order PSK and QAM schemes are more robust to errors. Also, it is obvious that a lower BER is achieved with a stronger signal, but more energy is required to generate it. In all cases, demodulation is coherent, i.e., both the phase and frequency in local oscillators of the transmitter and receiver are ideally synchronized. This condition reduces the BER. Influences of both differential encoding and channel coding are not considered. Differential encoding reduces error probability, but it increases a number of errors if an error occurs [68]. BER highly correlates with the channel quality index (CQI), which can be negatively influenced, for example, by rainy weather. The lowest CQI corresponds to the poorest quality of a signal, for which more robust (against interference, noise) modulations must be used (e.g., 4-PSK). When the CQI is high, 32-QAM or 64-QAM schemes can be used, which are more efficient (i.e., have a higher ratio of average information per symbol to average code length) but less resistant to errors.

To continue, let us take a deeper look at the influence of AWGN noise through a prism of modulation order. Constellations in the AWGN channel output are given in Figure 40 when the energy per bit to noise power spectral density ratio is equal to 4 dB. When the size of the data packet was equal to 100 symbols, 4 of them were decoded incorrectly by a receiver in the case of 4-QAM, and 20 were decoded incorrectly in the case of 16-QAM. In addition, the data transfer performance is be evaluated with error vector magnitude (EVM), which is 41.0% in the case of 4-QAM and 24.5% in the case of 16-QAM. EVM is calculated by the following equation:

$$EVM_{RMS}[\%] = 100 \cdot \sqrt{\frac{1}{P_{avg}} \cdot \frac{1}{N} \sum_{i=1}^{N} (I_i - I_i^*)^2 + (Q_i - Q_i^*)^2}, \qquad (52)$$

where P_{avg}—average power of reference constellation (equal to 1 W); I_i—in-phase component of reference constellation; I_i^*—in-phase component of received signal; Q_i—quadrature component of reference constellation; Q_i^*—quadrature component of received signal.

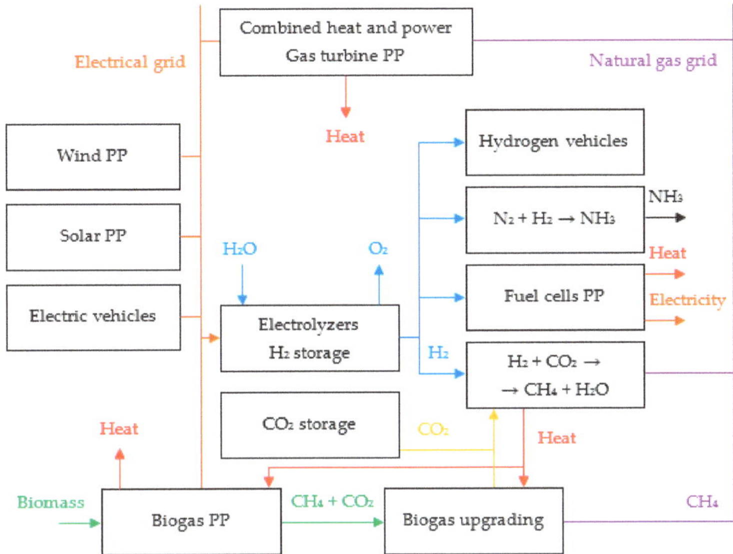

Figure 38. Example of Power-to-X concept.

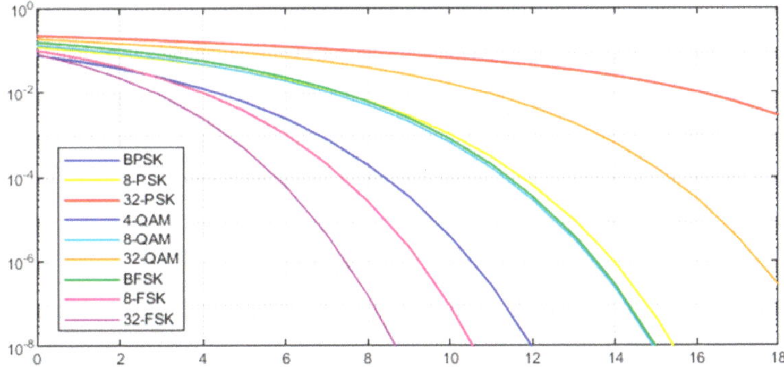

Figure 39. BER dependence on energy per bit to noise power spectral density ratio (from 0 to 18 dB) and modulation type in AWGN channel. Curves of BPSK and 4-QAM coincide.

Another issue can be an insufficient level of received power. Communication system coverage is evaluated with reference signal received power (RSRP) and reference signal received quality (RSRQ). For example, according to the Communications Regulatory Authority of the Republic of Lithuania [69], the RSRP of 2G GSM ranges from −95 to −75 dBm, while the RSRP of 4G LTE ranges from −115 to −95 dBm (at 1.5 m height when the receiver's antenna gain is equal to 0 dBi). It is obvious that base stations located at a higher location will provide a larger coverage. Also, a higher RSRP can be reached by installing mode base stations, but then, both the cost and complexity of a smart grid's ICT system will be increased.

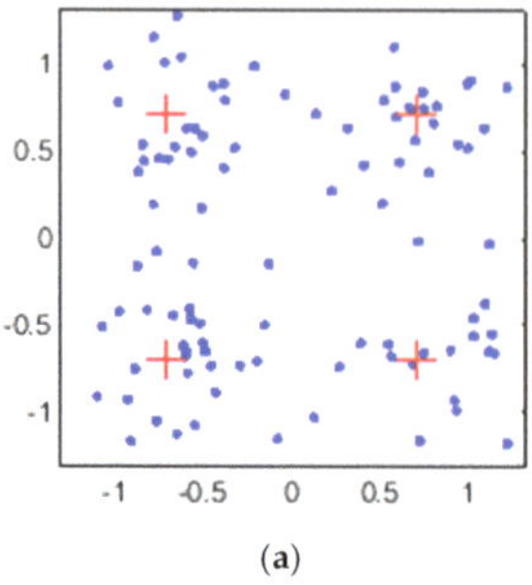 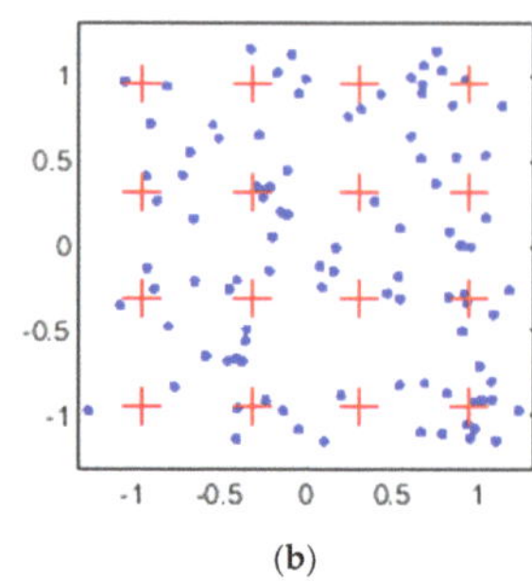

(a)　　　　　　　　　　　　　　(b)

Figure 40. Constellations of received signal in AWGN channel output. Data packet consists of 100 symbols. Energy per bit to noise power spectral density ratio is equal to 4 dB. (**a**) Case of 4-QAM (EVM is 41.0%, 4 symbols out of 100 were decoded incorrectly); (**b**) Case of 16-QAM (EVM is 24.5%, 20 symbols out of 100 were decoded incorrectly).

Two examples of the coverage calculation in EDX SignalPro software are given in Figure 41: 4G LTE technology is chosen, but the ICT grid designing principles are similar to the case of other wireless wide area network technologies such as GPRS and WiMAX. The total area of the site is 902 km^2 (in comparison, the area of Lithuania is 65,300 km^2): one half is covered by the city, and the rest is covered by forests, mountains, and suburbs. The highest altitude is 600 m, which is a potential place for one of the 4G LTE communication towers. The height of all the antenna towers is 40 m. The maximal gain of the antennas is 17.4 dBi, the output power maximum is 60 W (47.7 dBm), polarization is horizontal, and MIMO technology is not used. Each three-sector antenna broadcasts 1 Gbps on average (i.e., 334 Mbps per sector on average), and the channel bandwidth is up to 20 MHz. In Figure 41, it has been tried to reach sufficient coverage with a minimum number of LTE base stations, and it is clearly seen that 1920 MHz communication (Figure 41a) ensures a higher RSRP in comparison with the 800 MHz case (Figure 41b): in the case of 1920 MHz, the RSRP in 61% (550 km^2) of the area is between −90 and −80 dBm (green area). Moreover, different coverage patterns are clearly seen in Figure 41, and the reason for this is not only different frequencies but also different origins (and, respectively, limitations) of the propagation models: COST 231—the empirical—model (the extension of the Okumura–Hata model for frequencies up to 2 GHz) is used for 1920 MHz frequency, and "free space + RMD"—the theoretical—model (assessing the mutual effect of both free space path loss and reflection plus multiple diffraction, i.e., terrain and obstacles influence) is used for 800 MHz frequency.

In addition, four wireless point-to-point microwave links are shown in Figure 41b. The designing example of such a link is given in Figure 42: the distance between the base stations is approximately 11 km, the maximum elevation is approximately 600 m, and the chosen frequency is 7.8 GHz. The strongest signals of point-to-point communication propagate within the first Fresnel zone: the theoretical allowable limit of the coverage is 40% of the zone. In Figure 42, there are no obstacles in the zone; thus, the communication reliability is greater than 99.999%; however, such a value is not always the best solution because, for example, a 99.997% level can be significantly cheaper.

To continue, minimum quality of service (QoS) requirements must be established for each smart grid application. The example in the case of security, bandwidth, reliability, and latency can be found in [70]; however, PQ application has been skipped. In addition, IEEE Std 2030-2011 [71] with its smart grid interoperability reference methodology should be considered, as it has been done by [2]. QoS requirements can be quantitative and qualitative [70]. These requirements are discussed through a prism of PQ in Table 5.

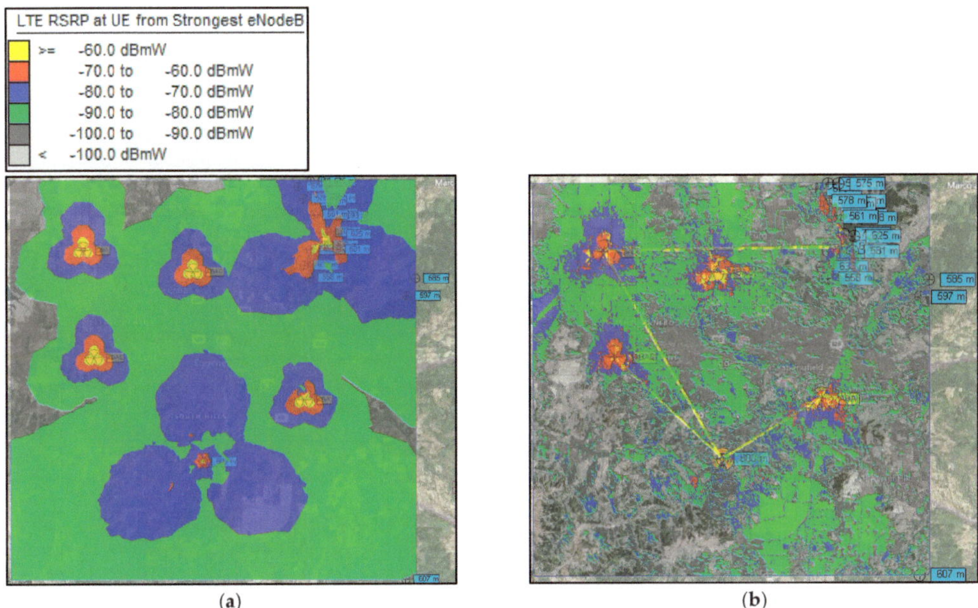

(**a**)   (**b**)

Figure 41. (**a**) RSRP in the case of 1920 MHz communication (COST 231 model); (**b**) RSRP in the case of 800 MHz communication ("free space + RMD" model).

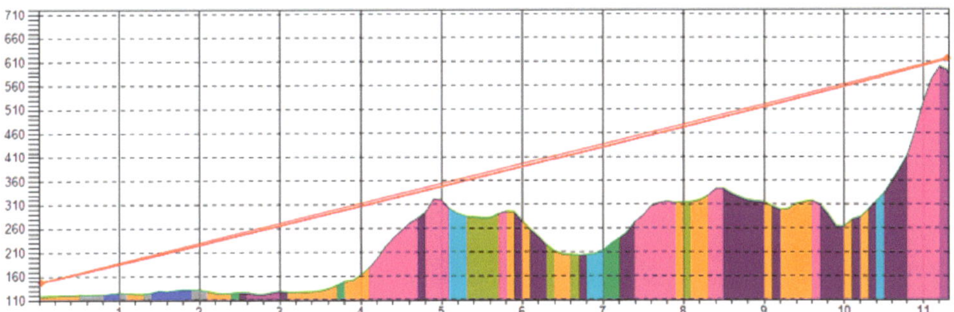

Figure 42. Example of the point-to-point link: the line-of-sight and the first Fresnel zone in the terrain cross-section plot.

Finally, the importance of this section can be supported with the already mentioned PQ study [63] carried out by the "Western Power Distribution" company, which is one of the British DSOs. This research has been already cited in this and our previous paper [2], but in this section, we would like to focus on ICT problems which were faced with less than 50 monitors when both 4G LTE and IEC 61850 were used in the communication network. Currently, the project is one of the few best examples of a remote PQ monitoring [2]. From the following, outcomes must be learned in anticipation of a massive PQ monitoring system:

- Trial sites can have poor communication, hence: (1) this criterion should be considered during trial sites selection; (2) alternative sites with better communication should be considered; and (3) the availability of communication alternatives should be ensured. Surveys revealed that no single cellular network operator is capable of covering all sites, in particular with 4G LTE. In this case, roaming SIM cards can be used; thus, the communication hub will be able to utilize an available provider at each site.

- IEC 61850 can be implemented differently (particularly in terms of file transfer mechanisms), and this is not essentially desirable. Moreover, in the case of IEC 61850 usage, the monitor sometimes is unable to reply to all requests. This can lead to (small) data loss, which subsequently cannot be retrieved.
- The episodic instability of one monitor was solved by developing a method of remote triggering. This allows avoiding site visits when a reset is needed.
- Since PQ data do not need to be transmitted continuously, file transfer is preferred because it can be carried out asynchronously. This approach requires less resources and is more robust (resilient) to a temporary loss of communication.
- Monitors installation has been sped up by pre-configuring and pre-commissioning them and communication hubs prior to traveling to site.
- In comparison with more commonly used CSV and JSON file formats, the HDF format offers several advantages such as faster data retrieval, storage and memory saving.

Table 5. QoS requirements for PQ application.

Requirement	General Information	Nuances in PQ Monitoring
Latency	Data transmission delay between smart grid components	PQ monitoring is not a time critical or real-time application; thus, the delay requirement can be not very strict [2]
Bandwidth	Wireless communication frequency determines its coverage and bandwidth (data rate) [70]; hence, low, medium and high frequencies will have their specific roles in a smart grid. Therefore, a detailed examination of data rate, transmission distance and other features is essential in order to select appropriate technology [1]	PQ monitoring is a wide area network application whose end-nodes are static (hence, handoff regions are not relevant). The required bandwidth can be diminished by feature extraction which, however, not always can be applied, for example, in the case of transient waveforms
Data rate	Various types of data (text, pictures, audio, video, etc.) will be generated by smart grid applications at different rates [70]. Hence, it is important to select appropriate ICT, considering its technical characteristics and cost efficiency	Data rate depends on feature extraction techniques. Probably, PQ application will be a part of a smart grid's ICT network, and there are numerous possible ways to implement this task, including technology selection (e.g., see [2] for the list of potential ICTs)
Throughput	The sum of data transferred between smart grid components in a specific time interval [70]	PQ measurements must be carried out continuously. This is not essential for PQ data transmission: it could be scheduled considering an ICT network's traffic profile and the memory of the monitor with an obvious exception in case of a high-priority request. Also, the throughput will depend on the monitors' quantity optimization (including smart meters and other relevant devices), data redundancy factor, feature extraction, etc.
Reliability	Quantified success of proper data transferring. In [70], the reliability requirement is higher than 98% for all smart grid applications	Currently, the PQ system reliability can reach 98%; however, it must be significantly increased after the integration with SA. Also, perhaps, a reliability requirement could be slightly lowered with an increase in the data redundancy factor. The transferring of high-entropy messages is more valuable and thus must be more reliable than low-entropy messages [2]

Table 5. *Cont.*

Requirement	General Information	Nuances in PQ Monitoring
Accuracy andprecision	Accuracy is the difference between a measurement and a true (accepted) value, while precision characterizes how close the measurements are to each other [3]	As already mentioned, along with well-known aspects such as the instrument error, the magnitude and phase frequency response of the measurement circuit also play an important role. In the case of frequency domain analysis, which is an integral part of PQ, even more less understood nuances emerge: for example, Heisenberg uncertainty, resolution bandwidth, spectrum aliasing, spectral leakage, etc. [2]
Validity	Characterizes information usefulness, strength, accuracy and other relevant features which are required to efficiently achieve the desired goal. It is closely related to feature extraction strategies: useful data must be retrieved	Validity depends directly on feature extraction. This aspect is important for every PQ task regardless of whether it is technical, economic or political
Accessibility	Access rules must be established for interested parties such as TSO, DSO, regulator, industry, and households. Equal opportunities must be offered for the members of each group without discrimination [70]	Probably, each user will have access to simplified and understandable to him information restricted to his narrow area of responsibility. Meanwhile, grid operators must possess full data
Interoperability	Various protocols and ICTs are expected in smart grids. Thus, a proper protocol conversion must be ensured to achieve the best possible interoperability	Various protocols and ICTs can be used for PQ data and its smooth interactions with other applications (AMI, OM, PM, SA, etc.) must be ensured. Along with popular general-purpose protocols and data formats, more specialized protocol could also be used such as PQDIF specified in IEEE Std 1159.3-2019 [72]
Security	Protection from both physical threats and cyberattacks	Currently, PQ data are not as critical as they will become in the future, especially after the integration with SA. All structural parts of the PQ monitoring process are potentially vulnerable, in particular local data assessment, data transmission and central data assessment. For example, [18] investigates classification defense from an adversarial attack. The question regarding minimum storage time of PQ history currently also remains unanswered [2]

[1] In general, on one side, lower frequencies have lower bandwidths, but they can be used for long-distance communication. On the other side, the propagation path of higher frequencies is shorter, but they have a higher bandwidth, resulting in a higher data rate. [2] In information theory, the entropy of a message describes its informational value which depends on the average level of surprise (uncertainty). It is directly analogous to the entropy which is central to the second law of thermodynamics. [3] Errors can be systematic (e.g., faulty calibration, parallax error) and random (e.g., mistakes during measurement reading). Systematic errors are one-sided, cannot be easily analyzed by statistical analysis and cannot be eliminated by performing additional experiments. Random errors are two-sided, can be easily detected by statistical analysis and reduced by increasing a number of measurements (according to the law of large numbers).

Despite various problems and challenges, remote PQ monitoring has many advantages. One of them is a fast detection of data loss, which can occur, for example, due to the loss of synchronization between a PQ monitor and the grid. Many synchronization loss flags were found during the PQ monitoring campaign in the Lithuanian DSO grid (Section 2.1.3) [2]. The data quality topic is slightly covered by IEEE Std 1159-2019, which discusses issues of missing data points, data latching, abnormal range, and extreme value. The example of data loss is also given by [2]. Moreover, data latching is clearly seen in Figure 37; however, this particular case is not an anomaly because randomness in the frequency measurement disappears during the power supply interruption. In the case of classification tasks, it is important to detect both abnormal ranges and extreme values of PQ phenomena in order to reduce the probability of large errors as well as to select the appropriate dynamic range of a measuring device. To conclude, in a remote PQ monitoring system, it is much easier to immediately detect data quality issues and deal with them.

4.3. Further Development of the Algorithms

In this section, we will discuss some ideas regarding a further expansion of the classification: Section 4.3.1 reviews some mathematical concepts, while Section 4.3.2 reviews the integration with PM application focusing on electrical grid insulation.

4.3.1. Mathematical Methods

To begin with, an upgrade of the current approach to probabilistic classification seems to be promising and realizable. However, the challenge lies in finding probability values due to insufficient experience in PQ monitoring [2,9]. Three interrelated groups of PQ probabilistic analysis can be highlighted—event occurrence frequency, probability of values of parameters characterizing the event, and probability of its primary cause (see Figure 43a). After that, this model can be further connected with both probabilistic analysis in other closely related fields (see [9] for the examples) and PQ classification (see Figure 43b). In addition, it can be stated that such labels as "not categorized", "not identified" and "others" should be absent in an ideal PQ classification algorithm; however, data aggregation and grouping processes may also have their own limitations: for example, the dependence on an expert's competence and subjectivity. Moreover, since PQ monitoring capabilities are limited and may not meet expectations, probability and statistics should play an important role in synthetic data generation, for example, by Monte Carlo methods. Despite the fact that this paper demonstrated the unnecessity of a large-sized training set for PQ machine learning (in contrast to biometrics), such data abundance will be essential for probabilistic and statistical analysis (according to both law of large numbers and central limit theorem), and synthetic data are not suitable for this purpose at all. Also, please note that biometric data are much easier to obtain, while PQ events occurrence frequency is not sufficient to quickly create a database.

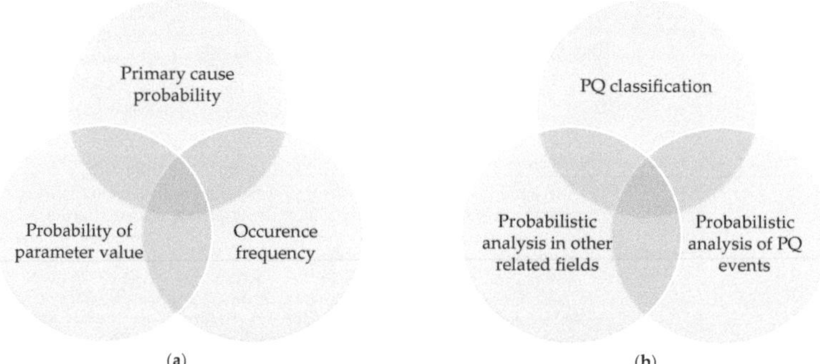

Figure 43. (**a**) Proposed structure of PQ events probabilistic analysis; (**b**) Proposed interconnection model between PQ classification and probabilistic analysis of both PQ and related fields.

Let us continue the discussion about decision boundaries of a classification. In this paper, a simplified geometrical method has been used, when an analytic expression of PQ data cluster's border is found manually (see Section 3.2.2). However, there are other ways to obtain these equations. As an example, let us examine one area of two-phase fault of Figure 25. The result is given in Figure 44: the closed boundary is split into the upper and lower parts (which is a common practice in classical mathematics) and then approximated with the polynomials.

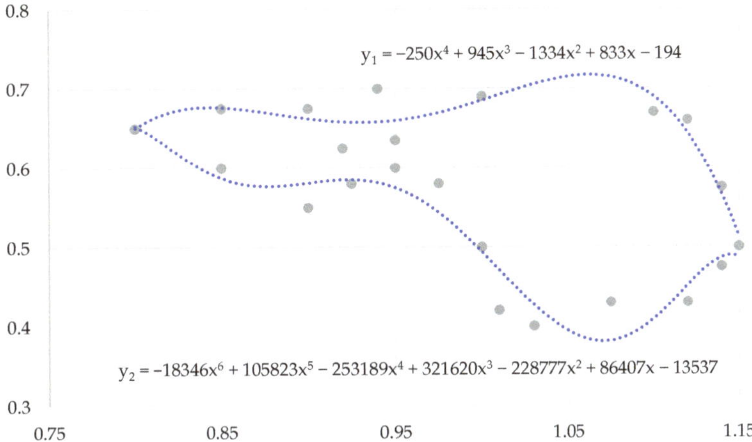

Figure 44. Example of polynomial regression application to find an analytical expression of a curve. The closed curve is divided into the upper and lower arcs, and both coefficients of determination are higher than 0.85.

At first glance, the similarity with Taylor series (or its special case—Maclaurin series) may be noticed in Figure 44, because they are among the best known; but indeed, it is a polynomial regression which solves the curve-fitting task focusing on all data points, while Taylor series approximates the function around the point x_0. However, the accuracy of both methods improves with the increasing order of their polynomials. For real-valued continuous function $f(x)$ that is differentiable n times at x_0, the Taylor series is given by:

$$f(x) = f(x_0) + \frac{f'(x_0)}{1!}(x - x_0) + \frac{f''(x_0)}{2!}(x - x_0)^2 + \ldots + \frac{f^{(n)}(x_0)}{n!}(x - x_0)^n + \frac{f^{(n+1)}(\zeta)}{(n+1)!}(x - x_0)^{n+1}, \tag{53}$$

where n—order of the polynomial; ζ—remainder in the Lagrange form.

Obviously, there are many more series expansions in addition to the already mentioned, and after determining the analytic expression of a closed curve in any way, it will be possible to estimate such characteristics as arc length, area, overlapping area, etc., which can upgrade PQ analytics to the next level. The area (volume) and arc length in $\mathbb{R}^n$ can be estimated by well-known either surface or line integrals, whose interrelationship is defined by either Green's or Stokes' theorem, which are also well known. Nonetheless, let us take a deeper look at the extension of real numbers when PQ features are plotted on the complex plane. It is entirely possible to encounter such a situation because complex numbers are widely used in electrical engineering. In this case, the line integral of a positively oriented closed contour γ, which is described by a complex function $f(z)$, can be evaluated using the residue theorem:

$$\oint_\gamma f(z)dz = 2\pi i \sum_{k=1}^{n} \operatorname*{Res}_{z=z_k} f(z), \tag{54}$$

where i—imaginary unit. Please note that function $f(z)$ is not fully holomorphic, i.e., is not holomorphic at singularities z_k; otherwise, the line integral over a closed path in a complex plane will be equal to zero (according to Cauchy's integral theorem). Also, it is noteworthy that the geometric intuition of complex integrals is a much more complicated question than in the case of real valued functions (e.g., see [73] for more information). One of several ways to calculate residues is the series method: the residue of $f(z)$ at the point z_0 is the coefficient c_{-1} in the Laurent series of $f(z)$ around z_0. In other words, this series around z_0 is given by:

$$\sum_{n=-\infty}^{\infty} c_n(z-z_0)^n = \sum_{n=-\infty}^{-1} c_n(z-z_0)^n + \sum_{n=0}^{\infty} c_n(z-z_0)^n = \frac{c_k}{(z-z_0)^k} + \ldots + \frac{c_{-1}}{z-z_0} + \sum_{n=0}^{\infty} c_n(z-z_0)^n, \tag{55}$$

where the residue lies in the principal part of the series as follows:

$$\operatorname*{Res}_{z=z_0} f(z) = c_{-1}. \tag{56}$$

Next, the limitation of overlapping areas can probably be eliminated (mitigated) with both a combination (ensemble) of different algorithms and more dimensions in a feature space. On the other hand, if both ideas are implemented poorly, it will bring about confusion and worsen the result. Another dimension, supplementing the classification in either voltage–duration or other investigated planes, can obviously be the frequency domain. It is relevant not only to transients, harmonics and inter-harmonics but also to voltage sags (due to discovered interconnection between sag and transient by [9]). Many problematic aspects of this domain are listed in both this and our previous paper [2]: for example, Heisenberg uncertainty, which is an important concept not only in quantum mechanics but also in signal processing. Nonetheless, in this paragraph, let us briefly present one of these challenges—spectral leakage caused by windowing. It is important to select appropriate window function and be aware of its characteristics. For example, both the time and frequency domains of a Bohman window are shown in Figure 45. In addition, in a similar manner to the already mentioned (investigated) digital filters and measurement transducers, windowing also may have undesirable phase response (such a characteristic is not given). Moreover, it goes without saying that a phenomenon of spectral leakage is very important in the assessment of both harmonics and inter-harmonics.

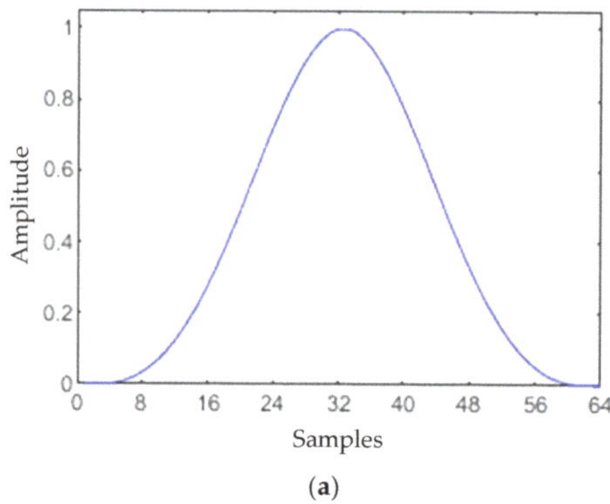
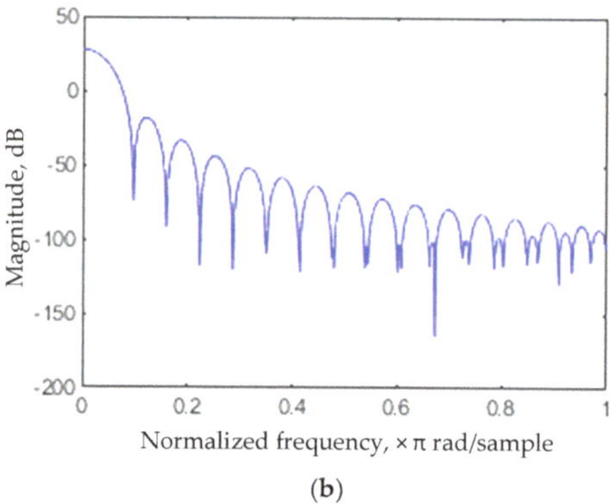

(a)   (b)

Figure 45. (**a**) Time domain of Bohman window; (**b**) Frequency domain of Bohman window.

4.3.2. Integration with Other Applications: Case of PM of Grid Insulation

The PQ and PM interconnection idea has already been mentioned several times in this and our previous papers [2,9]. PM can be treated as a partial case of PQ impact assessment. The latter can be divided into two large and mutually related groups, as shown in Figure 46: (1) influence on the equipment and its aging (of both grid infrastructure and end-user), and (2) influence on grid reliability in the form of power supply interruptions and outages. One of the best examples regarding PQ impact assessment is methods for voltage sag severity quantification proposed by IEEE Std 1564-2014. In this section, let us take a deeper look at electrical insulation, which is one of the main parts of a power grid's infrastructure. In electric power systems, a large variety of insulation is used—gas, liquid, solid, and combined. Each material has its own advantages and disadvantages as well as different characteristics which should be stored in a database. Moreover, transients monitoring is not only a prerequisite for impact assessment on insulation, but it also is a prerequisite for a better understanding of these phenomena's propagation, reflection and mitigation, which still remains poorly understood.

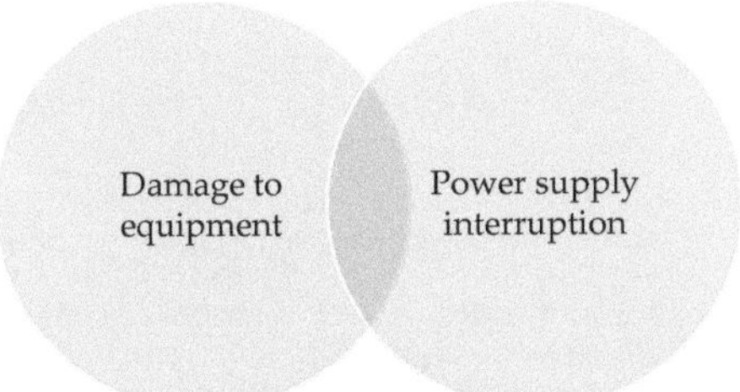

Figure 46. Proposed classification model of PQ impact.

It is clear that the most popular used dielectric gases are air (nitrogen) and sulfur hexafluoride SF_6. The latter has an octahedral geometry: a central sulfur atom is surrounded by six fluorine atoms, the bond angles are right angles, and the dipole moment is zero. This gas has high electronegativity and density as well as a property of "self-healing" after the exposure to an electric arc. Its main drawback is a very high global warming potential value as well as a long lifespan. It is noteworthy that both these characteristics are even higher than refrigerant trichlorofluoromethane $CFCl_3$ (widely used earlier), which is an ozone-depleting chlorofluorocarbon also known as freon-11 or R-11. The destruction principle is the following (see Figure 47): (1) a chlorine atom $Cl^\bullet$ is removed from the $CFCl_3$ molecule by ultraviolet radiation; (2) chlorine reacts with ozone O_3, converts it to an oxygen O_2, and forms chlorine monoxide $ClO^\bullet$ (by removing an oxygen atom); (3) $ClO^\bullet$ reacts with another ozone molecule and splits it into two oxygen molecules, thereby regenerating $Cl^\bullet$ and enabling the cycle to repeat. It should be noted that environmentally friendly materials will become more and more important in smart grids due to strict environmental policy and sustainable development.

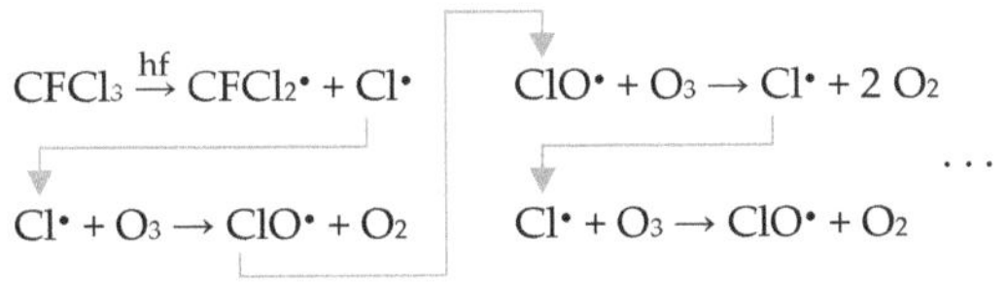

$$CFCl_3 \xrightarrow{hf} CFCl_2^\bullet + Cl^\bullet \qquad ClO^\bullet + O_3 \rightarrow Cl^\bullet + 2\,O_2$$

$$Cl^\bullet + O_3 \rightarrow ClO^\bullet + O_2 \qquad Cl^\bullet + O_3 \rightarrow ClO^\bullet + O_2 \quad \cdots$$

Figure 47. Principle of the ozone depletion caused by trichlorofluoromethane.

To continue, the most well-known liquid insulation undoubtedly is an insulating oil. Most often, it is a mixture of higher alkanes C_nH_{2n+2} produced during the petroleum refining process—mineral oil (which is distinct from both vegetable oil and synthetic oil)—which also serves as a coolant. Properties of liquid insulation depend on impurities, in particular water and gases: water can be dissolved, in an emulsion state (including colloidal solution), or accumulated at the bottom of the tank, while gases can be either dissolved or dispersed [42,74]. Obviously, there are many more factors such as temperature and voltage transient frequency; therefore, various sensors and monitoring systems will be essential to PM implementation. Formerly, the insulating fluid of both transformers and capacitors consisted of polychlorinated biphenyls $C_{12}H_{10-n}Cl_n$ (Figure 48a), whose production is currently internationally banned due to carcinogenicity and other dangerous properties. In addition to the mentioned materials, there are many other liquid dielectrics:

for example, 1-phenyl-1-xylyl ethane $C_{16}H_{18}$ (Figure 48b), also called PXE, is used as an impregnating agent in the capacitors manufacturing process.

(a)

(b)

Figure 48. (**a**) The isomer of polychlorinated biphenyl; (**b**) The isomer of 1-phenyl-1-xylyl ethane.

Some of the most widely known materials of solid insulators are porcelain (ceramic) and glass, because they can be easily spotted with the naked eye in overhead power lines. However, there are many more materials used: these are various rocks (minerals) such as mica and soapstone, and there are also a large variety of polymers—both duroplasts (e.g., various resins, polyurethane (Figure 49a), polyamide (Figure 49b), etc.) and thermoplastics (e.g., polyethylene $(C_2H_4)_n$, polypropylene $(C_3H_6)_n$ (Figure 49c), polytetrafluoroethylene $(C_2F_4)_n$ (Figure 49d), etc.). Furthermore, layered solid insulation is generally more electrically resistant than monolithic insulation of the same length: generally, its resistance depends on partial discharges in the gaps between layers, which is prevented by both impregnation and coverage with either gas or liquid [42]. Such a type of insulation can be called combined. Electrotechnical paper, a cellulosic material (Figure 49e), is often included in such combinations (especially in HV equipment), for example, paper–oil insulation or paper–polypropylene film–synthetic liquid (e.g., already mentioned PXE) insulation.

Undoubtedly, the breakdown voltage and loss tangent are among the most important parameters of an insulator. The loss tangent of a solid insulator consists of three components:

$$\tan \delta = \tan \delta_1 + \tan \delta_2 + \tan \delta_3, \tag{57}$$

where δ_1—loss angle due to a flow of free charges; δ_2—loss angle due to polarization; δ_3—loss angle due to both partial discharges and ionization processes [42]. The loss tangent also depends on various external factors. In addition to the already mentioned water and gaseous impurities in liquids, insulation must be resistant to various mechanical, thermal, physical, chemical and other effects, and it must meet many requirements regarding acidity, viscosity, wettability, immunity against radiation, etc. For example, acidity can be characterized by pH scale along with the potassium hydroxide KOH amount needed to neutralize acids, wettability—by the contact angle of a liquid droplet on a solid surface (when the zero angle corresponds to a perfect wetting), and the resistance against radiation consists of both electromagnetic radiation (including full spectrum, in particular both X-rays and gamma) and particle radiation (e.g., helium ions $_2^4\mathrm{He}^{2+}$, electrons e^-, positrons e^+, neutrons n^0). In addition, an uneven distribution of the electric field gradient along the insulator along with the influence of both corona and grading rings must also be taken into consideration. To sum up, a heterogeneity in materials, their properties and external factors complicates PM research and development, delays the implementation, and can also reduce its accuracy. Moreover, as has been mentioned many times, real data capture remains challenging—not only due to the insufficient occurrence frequency of voltage transients but also due to their damping, a lack of measuring devices and limitations of their technical capabilities [2,9]. Perhaps progress can be accelerated by laboratory tests, but they require complex and expensive set-up.

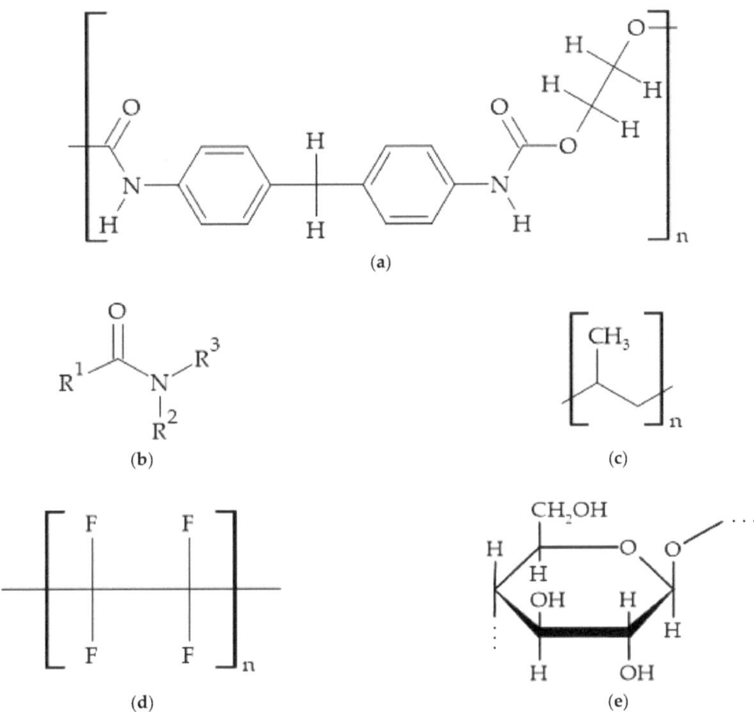

Figure 49. (**a**) Polyurethane; (**b**) General structure of an amide; (**c**) Polypropylene; (**d**) Polytetrafluo-roethylene; (**e**) Structural unit of a linear chain of cellulose is β-D-glucose [74].

5. Conclusions

1. Research on the fundamental grid component removal from a PQ signal has not been found in the existing literature. However, data processing with an IIR shelving filter showed that such filtering distorts PQ signals and converts them to the unsuitable form for a further analysis: (1) when Δf is set to 1 Hz, a filter's transient process lasts a relatively long time, concealing the signal, and (2) when Δf is set to 25 Hz, the duration of the transient process is minimized but unfortunately in exchange for frequency response quality.

2. Contrary to biometrics, a large database is not always needed for PQ machine learning. On the other hand, it is more difficult to create a high-quality PQ database due to insufficient PQ events occurrence frequency (especially at a single grid node), which highly correlates with the difficulty in covering all possible scenarios (situations). The results provided by KNN are much more satisfying than SVM; however, it does not mean that KNN alone can successfully cope with all arising classification challenges, for example, with clusters overlapping which perhaps could be solved by both AI ensemble and adding more additional dimensions to the feature vector.

3. The proposed unique approach of three-dimensional voltage classification is a successful outcome achieved by using the simulation results presented in [9], and the geometric analysis of the feature clusters highlighted certain patterns, which has the potential to bring the problem significantly closer to successful completion. Moreover, developed the Clarke transformation-based method showed outstanding results; thus, it can be considered as a promising tool suitable for the simplification of short

circuit data analysis in three-dimensional space. In order to successfully implement all proposed mathematical operations, it is important to maintain the same order of elements in the vectors.

4. At present, many methodological gaps in PQ assessment inhibit AI application. Some of them have been noticed during the analysis of PQ data measured in the Lithuanian DSO grid: for example, regarding a multistage voltage sag, voltage sag burst, voltage sag followed by a power line disconnection, etc. It should be underlined that the majority of these solutions must be universally agreed upon in the form of PQ law amendments. It is noteworthy that such a discussion as that given in Section 4.1 is presented for the first time in the scientific literature.

Author Contributions: Conceptualization, V.L.; methodology, V.L., G.K., V.R. and D.N.; software, V.L. and G.K.; validation, V.L., G.K., V.R. and D.N.; formal analysis, V.L.; investigation, V.L. and G.K.; resources, V.L. and V.R.; data curation, V.L.; writing—original draft preparation, V.L. and G.K.; writing—review and editing, V.L., G.K. and V.R.; visualization, V.L.; supervision, V.R. and D.N.; project administration, V.L. and V.R. All authors have read and agreed to the published version of the manuscript.

Funding: This research was funded by the Lithuanian Energy Institute.

Data Availability Statement: Data are contained within the article.

Conflicts of Interest: The authors declare no conflict of interest.

Appendix A

Clarke transformation of a balanced signal is shown in Figure A1: in a symmetrical case, as has been mentioned in Section 2.5.2, component γ is equal to zero and therefore can be neglected. In the opposite case, when a signal is asymmetrical, γ is not equal to zero; therefore, it can be treated as a kind of unbalance evaluating index. Usually, its value is the lowest, but an exceptional case—phase-to-ground voltage of two-phase-to-ground fault—is given in Figure A2 (see Tables 3 and 4). Obviously, such a feature is beneficial because it even more distinguishes the group. It is noteworthy that a phase-shift, a characteristic feature of asymmetrical faults (e.g., see [2,9] for more details), is not assessed in the calculations; however, fortunately, neither this aspect nor γ discarding significantly impact the result. Moreover, the information loss effect is minimized even more since the approach is applied to all cases equally.

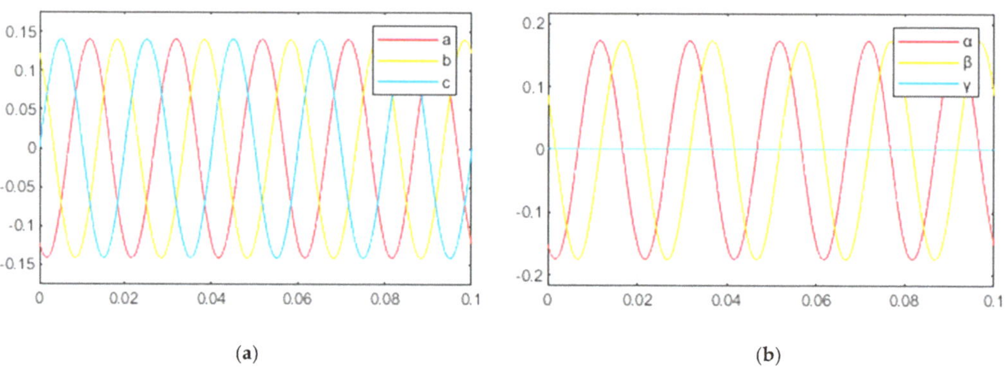

(a)   (b)

Figure A1. (**a**) Balanced signal; (**b**) Balanced signal after the Clarke transformation.

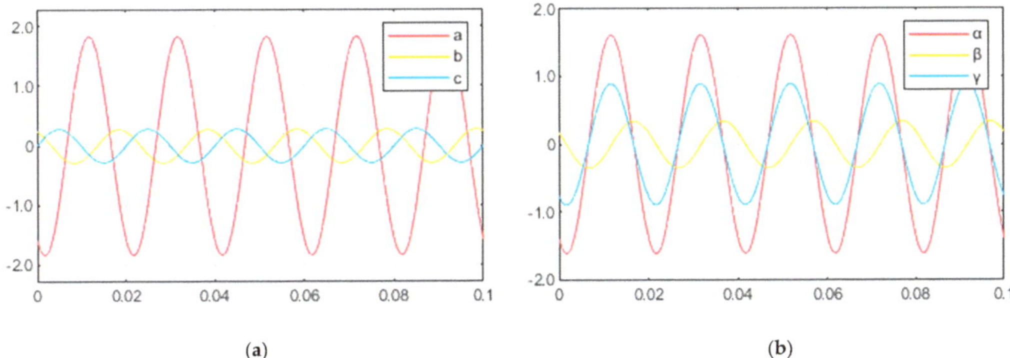

Figure A2. (**a**) Unbalanced signal; (**b**) Unbalanced signal after the Clarke transformation. Note that in the given case, contrary to the most asymmetrical cases, component β is lower than γ.

References

1. Khokhar, S.; Mohd Zin, A.A.B.; Mokhtar, A.S.B.; Pesaran, M. A Comprehensive Overview on Signal Processing and Artificial Intelligence Techniques Applications in Classification of Power Quality Disturbances. *Renew. Sustain. Energy Rev.* **2015**, *51*, 1650–1663. [CrossRef]
2. Liubčuk, V.; Radziukynas, V.; Naujokaitis, D.; Kairaitis, G. Grid Nodes Selection Strategies for Power Quality Monitoring. *Appl. Sci.* **2023**, *13*, 6048. [CrossRef]
3. McCorduk, P. *Machines Who Think: A Personal Inquiry into the History and Prospects of Artificial Intelligence*, 2nd ed.; A K Peters: Natick, MA, USA, 2004.
4. Egypt Independent. Ancient Egyptians Invented First Robot 4000 Years Ago: Study. Available online: https://egyptindependent.com/ancient-egyptians-invented-first-robot-4000-years-ago-study (accessed on 30 September 2023).
5. Maspero, G. *Manual of Egyptian Archaeology and Guide to the Study of Antiquities in Egypt*; Cambridge University Press: Cambridge, UK, 2010.
6. Aristotle. *Aristotle's Politics*; Clarendon Press: Oxford, UK, 1885.
7. Russell, S.J.; Norvig, P. *Artificial Intelligence: A Modern Approach*, 3rd ed.; Pearson Education: London, UK, 2016.
8. Hasija, Y. Artificial Intelligence and Digital Pathology Synergy: For Detailed, Accurate and Predictive Analysis of WSIs. In Proceedings of the 8th ICACCS, Coimbatore, India, 25–26 March 2022.
9. Liubčuk, V.; Radziukynas, V.; Kairaitis, G.; Naujokaitis, D. Power Quality Monitors Displacement Based on Voltage Sags Propagation Mechanism and Grid Reliability Indexes. *Appl. Sci.* **2023**, *13*, 11778. [CrossRef]
10. *LST EN 61000-4-30:2015*; Electromagnetic Compatibility (EMC)—Part 4–30: Testing and Measurement Techniques—PQ Measurement Methods (IEC 61000-4-30:2015). Lithuanian Standards Board: Vilnius, Lithuania, 2016.
11. *EN 50160:2010*; Voltage Characteristics of Electricity Supplied by Public Electricity Networks. CENELEC: Brussels, Belgium, 2010.
12. *IEEE Std 1159-2019*; IEEE Recommended Practice for Monitoring Electric Power Quality. IEEE: New York, NY, USA, 2019.
13. *IEEE Std 1564-2014*; IEEE Guide for Voltage Sags Indices. IEEE: New York, NY, USA, 2014.
14. Kezunovic, M.; Liao, Y. A New Method for Classification and Characterization of Voltage Sags. *Electr. Power Syst. Res.* **2001**, *58*, 27–35. [CrossRef]
15. Huang, N.; Peng, H.; Cai, G.; Chen, J. Power Quality Disturbances Feature Selection and Recognition Using Optimal Multi-Resolution Fast S-Transform and CART Algorithm. *Energies* **2016**, *9*, 927. [CrossRef]
16. Khetarpal, P.; Tripathi, M.M. A Critical and Comprehensive Review on Power Quality Disturbance Detection and Classification. *Sustain. Comput. Inform. Syst.* **2020**, *28*, 100417. [CrossRef]
17. Jiang, Z.; Wang, Y.; Li, Y.; Cao, H. A New Method for Recognition and Classification of Power Quality Disturbances Based on IAST and RF. *Electr. Power Syst. Res.* **2024**, *226*, 109939. [CrossRef]
18. Zhang, L.; Jiang, C.; Chai, Z.; He, Y. Adversarial Attack and Training for Deep Neural Network Based Power Quality Disturbance Classification. *Eng. Appl. Artif. Intell.* **2024**, *127*, 107245. [CrossRef]
19. Khan, M.A.; Asad, B.; Vaimann, T.; Kallaste, A.; Pomarnacki, R.; Hyunh, V.K. Improved Fault Classification and Localization in Power Transmission Networks Using VAE-Generated Synthetic Data and Machine Learning Algorithms. *Machines* **2023**, *11*, 963. [CrossRef]
20. Fu, L.; Deng, X.; Chai, H.; Ma, Z.; Xu, F.; Zhu, T. PQEventCog: Classification of Power Quality Disturbances Based on Optimized S-Transform and CNNs with Noisy Labeled Datasets. *Electr. Power Syst. Res.* **2023**, *220*, 109369. [CrossRef]

21. Khetarpal, P.; Tripathi, M.M. Power Quality Disturbance Classification Taking into Consideration the Loss of Data during Pre-Processing of Disturbance Signal. *Electr. Power Syst. Res.* **2023**, *220*, 109372. [CrossRef]
22. Liu, Y.; Jin, T.; Mohamed, M.A. A Novel Dual-Attention Optimization Model for Points Classification of Power Quality Disturbances. *Appl. Energy* **2023**, *339*, 121011. [CrossRef]
23. Li, M.; Li, Y.; Tian, D.; Lou, J.; Chen, Y.; Liu, X. Assessment of Voltage Sag/Swell in the Distribution Network Based on Energy Index and Influence Degree Function. *Electr. Power Syst. Res.* **2023**, *216*, 109072.
24. Lin, W.-M.; Wu, C.-H. Fast Support Vector Machine for Power Quality Disturbance Classification. *Appl. Sci.* **2022**, *12*, 11649. [CrossRef]
25. Salles, R.S.; Ribeiro, P.F. The Use of Deep Learning and 2-D Wavelet Scalograms for Power Quality Disturbances Classification. *Electr. Power Syst. Res.* **2023**, *214*, 108834. [CrossRef]
26. Wang, J.; Zhang, D.; Zhou, Y. Ensemble Deep Learning for Automated Classification of Power Quality Disturbances Signals. *Electr. Power Syst. Res.* **2022**, *213*, 108695. [CrossRef]
27. Mohammadi, Y.; Miraftabzadeh, S.M.; Bollen, M.H.J.; Longo, M. Voltage-Sag Source Detection: Developing Supervised Methods and Proposing a New Unsupervised Learning. *Sustain. Energy Grids Netw.* **2022**, *32*, 100855. [CrossRef]
28. Upadhya, M.; Singh, A.K.; Thakur, P.; Nagata, E.A.; Ferreira, D.D. Mother Wavelet Selection Method for Voltage Sag Characterization and Detection. *Electr. Power Syst. Res.* **2022**, *211*, 108246. [CrossRef]
29. Yang, L.; Guo, L.; Zhang, W.; Yang, X. Classification of Multiple Power Quality Disturbances by Tunable-Q Wavelet Transform with Parameter Selection. *Energies* **2022**, *15*, 3428. [CrossRef]
30. Turonović, R.; Dragan, D.; Gojić, G.; Petrović, V.B.; Gajić, D.B.; Stanisavljević, A.M.; Katić, V.A. An End-to-End Deep Learning Method for Voltage Sag Classification. *Energies* **2022**, *15*, 2898. [CrossRef]
31. Mohstaham, M.B.; Jalilian, A. Classification of Multi-Stage Voltage Sags and Calculation of Phase Angle Jump Based on Clarke Components Ellipse. *Electr. Power Syst. Res.* **2022**, *205*, 107725.
32. Zhang, Y.; Zhang, Y.; Zhou, X. Classification of Power Quality Disturbances Using Visual Attention Mechanism and Feed-Forward Neural Network. *Measurements* **2022**, *188*, 110390. [CrossRef]
33. Chamchuen, S.; Siritaratiwat, A.; Fuangfoo, P.; Suthisopapan, P.; Khunkitti, P. High-Accuracy Power Quality Disturbance Classification Using the Adaptive ABC-PSO As Optimal Feature Selection Algorithm. *Energies* **2021**, *14*, 1238. [CrossRef]
34. Mohammadi, Y.; Salarpour, A.; Leborgne, R.C. Comprehensive Strategy for Classification of Voltage Sags Source Location Using Optimal Feature Selection Applied to Support Vector Machine and Ensemble Techniques. *Electr. Power Energy Syst.* **2021**, *124*, 106363. [CrossRef]
35. Khoa, N.M.; Dai, L.V. Detection and Classification of Power Quality Disturbances in Power System Using Modified-Combination between the Stockwell Transform and Decision Tree Methods. *Energies* **2020**, *13*, 3623. [CrossRef]
36. Nagata, E.A.; Ferreira, D.D.; Bollen, M.H.J.; Barbosa, B.H.C.; Ribeiro, E.G.; Duque, C.A.; Ribeiro, P.F. Real-Time Voltage Sag Detection and Classification for Power Quality Diagnostics. *Measurements* **2020**, *164*, 108097. [CrossRef]
37. Bravo-Rodríguez, J.C.; Torres, F.J.; Borrás, M.D. Hybrid Machine Learning Models for Classifying Power Quality Disturbances: A Comparative Study. *Energies* **2020**, *13*, 2761. [CrossRef]
38. Wang, J.; Xu, Z.; Che, Y. Power Quality Disturbances Classification Based on DWT and Multilayer Perceptron Extreme Learning Machine. *Appl. Sci.* **2019**, *9*, 2315. [CrossRef]
39. Li, P.; Zhang, Q.S.; Zhang, G.L.; Liu, W.; Chen, F.R. Adaptive S Transform for Feature Extraction in Voltage Sags. *Appl. Soft Comput.* **2019**, *80*, 438–449. [CrossRef]
40. Wang, J.; Xu, Z.; Che, Y. Power Quality Disturbance Classification Based on Compressed Sensing and Deep Convolution Neural Networks. *IEEE Access* **2019**, *7*, 78336–78346. [CrossRef]
41. Pukys, P. *Teorinė Elektrotechnika II*, 4th ed.; Bartkevičius, S., Lazauskas, V., Pukys, P., Stonys, J., Virbalis, A., Eds.; Kaunas University of Technology—Publishing House "Technologija": Kaunas, Lithuania, 2011.
42. Baublys, J.; Jankauskas, P.; Marckevičius, L.A.; Morkvėnas, A. *Izoliacija ir Viršįtampiai*; Kaunas University of Technology—Publishing House "Technologija": Kaunas, Lithuania, 2008.
43. *LST EN 62305-1:2011*; Protection Against Lightning—Part 1: General Principles (IEC 62305-1:2010). Lithuanian Standards Board: Vilnius, Lithuania, 2011.
44. Vejuvić, S.; Lovrić, D. Exponential Approximation of the Heidler Function for the Reproduction of Lightning Current Waveshapes. *Electr. Power Syst. Res.* **2010**, *80*, 1293–1298. [CrossRef]
45. Yu, Z.; Zhu, T.; Wang, Z.; Lu, G.; Zeng, R.; Liu, Y.; Luo, J.; Wang, Y.; He, J.; Zhuang, C. Calculation and Experiment of Induced Lightning Overvoltage on Power Distribution Line. *Electr. Power Syst. Res.* **2016**, *139*, 52–59. [CrossRef]
46. Evdokunin, G.A.; Petrov, N.N. Lightning-Induced Voltage on Overheard Lines. In Proceedings of the 2016 ElConRusNW, St. Petersburg, Russia, 2–3 February 2016.
47. *IEEE Std 998-2012*; IEEE Guide for Direct Lightning Stroke Shielding of Substations. IEEE: New York, NY, USA, 2013.
48. Pedra, J.; Sainz, L.; Córdoles, F. Effects of Symmetrical Voltage Sag on Squirrel-Cage Induction Motors. *Electr. Power Syst. Res.* **2007**, *77*, 1672–1680. [CrossRef]
49. Silva, B.F.; Vanço, W.E.; Da Silva Gonçalves, F.A.; Bissochi, C.A., Jr.; De Carvalho, D.P.; Guimarães, G.C. A Proposal for the Study of Voltage Sag in Isolated Synchronous Generators Caused by Induction Motor Start-up. *Electr. Power Syst. Res.* **2016**, *140*, 776–785. [CrossRef]

50. Hashem, M.; Abdel-Salam, M.; Nayel, M.; El-Mohandes, M.T. Mitigation of Voltage Sag in a Distribution System During Start-up of Water-Pumping Motors Using Superconducting Magnetic Energy Storage: A Case Study. *J. Energy Storage* **2022**, *55*, 105441. [CrossRef]
51. Вольдек, А.И. Электрические Машины, 3rd ed.; Толвинская, Е.В., Ed.; «Энергия»: Leningrad, Russia, 1978.
52. Marozas, V. (Biomedical Engineering Institute of Kaunas University of Technology, Kaunas, Lithuania). *Personal communication*, 2018.
53. Orfanidis, S.J. *Introduction to Signal Processing*; Prentice Hall: Hoboken, NJ, USA, 1996.
54. Haykin, S. (Ed.) *Neural Networks and Learning Machines*, 3rd ed.; Person Education: Upper Saddle River, NJ, USA, 2009.
55. Using a Hard Margin vs Soft Margin in Support Vector Machines. Available online: https://www.section.io/engineering-education/using-a-hard-margin-vs-soft-margin-in-support-vector-machines (accessed on 17 November 2023).
56. Using a Hard Margin vs. Soft Margin in SVM. Available online: https://www.baeldung.com/cs/svm-hard-margin-vs-soft-margin (accessed on 17 November 2023).
57. MathWorks. Clarke and Park Transforms. Available online: https://se.mathworks.com/solutions/electrification/clarke-and-park-transforms.html (accessed on 15 November 2023).
58. MathWorks. Park Transform. Available online: https://se.mathworks.com/help/sps/ref/parktransform.html (accessed on 15 November 2023).
59. De Santis, M.; Di Stasio, L.; Noce, C.; Varilone, P.; Verde, P. Indices of Intermittence to Improve the Forecasting of the Voltage Sags Measured in Real Systems. *IEEE Trans. Power Deliv.* **2021**, *37*, 1252–1263. [CrossRef]
60. Council of European Energy Regulators. *5th CEER Benchmarking Report on the Quality of Electricity and Gas Supply*; CEER: Brussels, Belgium, 2011.
61. Svinkūnas, G.; Navickas, A. *Elektros Energetikos Pagrindai*, 2nd ed.; Kaunas University of Technology—Publishing House "Technologija": Kaunas, Lithuania, 2013.
62. Kundur, P. *Power Systems Stability and Control*, 1st ed.; Balu, N.J., Lauby, M.G., Eds.; McGraw-Hill: New York, NY, USA, 1994.
63. Western Power Distribution. Primary Networks Power Quality Analysis. Six Monthly Project Progress Reports, 2018–2021. Available online: https://www.westernpower.co.uk/projects/primary-networks-power-quality-analysis-pnpqa (accessed on 5 May 2021).
64. *LST EN 61000-4-5:2014*; Electromagnetic Compatibility (EMC)—Part 4–5: Testing and Measurement Techniques—Surge Immunity Test (IEC 61000-4-5:2014). Lithuanian Standards Board: Vilnius, Lithuania, 2014.
65. What Is True RMS Measurement? Available online: https://www.electricalvolt.com/2018/09/what-is-true-rms-measurement/?utm_content=cmp-true (accessed on 22 October 2023).
66. What Do RMS and True RMS Stand for? Available online: https://www.promaxelectronics.com/ing/news/561/what-do-rms-and-true-rms-stand-for-here-we-explain-you-the-differences (accessed on 22 October 2023).
67. Ponce-Jara, M.A.; Ruiz, E.; Gil, R.; Sancristóbal, E.; Pérez-Molina, C.; Castro, M. Smart Grid: Assessment of the Past and Present in Developed and Developing Countries. *Energy Strat. Rev.* **2017**, *18*, 38–58. [CrossRef]
68. Kašėta, S. *Telekomunikacijų Teorija*; Kaunas University of Technology—Publishing House "Technologija": Kaunas, Lithuania, 2012.
69. Communications Regulatory Authority of the Republic of Lithuania. Expected Coverage Areas of Cellular Networks. Available online: https://www.rrt.lt/judriojo-rysio-tinklu-tiketinos-aprepties-zonos/?highlight=RSRP (accessed on 10 October 2023).
70. Faheem, M.; Shah, S.B.H.; Butt, R.A.; Raza, B.; Anwar, M.; Ashraf, M.W.; Ngadi, M.A.; Gungor, V.C. Smart Grid Communication and Information Technologies in the Perspective of Industry 4.0: Opportunities and Challenges. *Comput. Sci. Rev.* **2018**, *30*, 1–30. [CrossRef]
71. *IEEE Std 2030-2011*; IEEE Guide for Smart Grid Interoperability of Energy Technology and Information Technology Operation with the Electric Power System (EPS), End-Use Applications, and Loads. IEEE: New York, NY, USA, 2011.
72. *IEEE Std 1159.3-2019*; IEEE Recommended Practice for Power Quality Data Interchange Format (PQDIF). IEEE: New York, NY, USA, 2019.
73. Youtube. Complex Integration, Cauchy and Residue Theorems [Video]. Available online: https://www.youtube.com/watch?v=EyBDtUtyshk (accessed on 23 November 2023).
74. Kurienė, A. Chemija. In *Trumpas Chemijos Kursas*; "Gimtinė": Vilnius, Lithuania, 1999.

 *inventions*

Article

Development and Application of an Open Power Meter Suitable for NILM

Carlos Rodríguez-Navarro, Francisco Portillo , Fernando Martínez, Francisco Manzano-Agugliaro * and Alfredo Alcayde

Department of Engineering, University of Almeria, ceiA3, 04120 Almeria, Spain;
crn565@inlumine.ual.es (C.R.-N.); portillo@ual.es (F.P.); fmg714@ual.es (F.M.); aalcayde@ual.es (A.A.)
* Correspondence: fmanzano@ual.es

Abstract: In the context of the global energy sector's increasing reliance on fossil fuels and escalating environmental concerns, there is an urgent need for advancements in energy monitoring and optimization. Addressing this challenge, the present study introduces the Open Multi Power Meter, a novel open hardware solution designed for efficient and precise electrical measurements. This device is engineered around a single microcontroller architecture, featuring a comprehensive suite of measurement modules interconnected via an RS485 bus, which ensures high accuracy and scalability. A significant aspect of this development is the integration with the Non-Intrusive Load Monitoring Toolkit, which utilizes advanced algorithms for energy disaggregation, including Combinatorial Optimization and the Finite Hidden Markov Model. Comparative analyses were performed using public datasets alongside commercial and open hardware monitors to validate the design and capabilities of this device. These studies demonstrate the device's notable effectiveness, characterized by its simplicity, flexibility, and adaptability in various energy monitoring scenarios. The introduction of this cost-effective and scalable tool marks a contribution to the field of energy research, enhancing energy efficiency practices. This research provides a practical solution for energy management and opens advancements in the field, highlighting its potential impact on academic research and real-world applications.

Keywords: open hardware energy monitoring; non-intrusive load monitoring; innovative metering technology; electrical measurement accuracy

Citation: Rodríguez-Navarro, C.;
Portillo, F.; Martínez, F.;
Manzano-Agugliaro, F.; Alcayde, A.
Development and Application of an
Open Power Meter Suitable for
NILM. *Inventions* **2024**, *9*, 2.
https://doi.org/10.3390/
inventions9010002

Academic Editor: Om P. Malik

Received: 1 December 2023
Revised: 14 December 2023
Accepted: 19 December 2023
Published: 21 December 2023

1. Introduction and Literature Review

Efficient and accurate electrical consumption monitoring has never been more critical in the evolving energy management and sustainability landscape [1]. As the world grapples with the challenges posed by its dependence on fossil fuels and the pressing need to transition towards more sustainable energy sources, innovations in energy monitoring technologies have emerged as a critical area of focus [2]. Energy monitoring is also considered a crucial component of the upcoming smart power grid infrastructure [3] since integrating widely fluctuating distributed generation sources presents a challenge to the stability of power generation and distribution networks [4].

A simple approach to gauging the power usage of separate devices in a home is by employing smart appliances that track their energy consumption. Nevertheless, this strategy is complicated and expensive [5]. Alternatively, Non-intrusive Load Monitoring (NILM) [6] offers a more practical and cost-effective means to estimate the energy consumption of individual devices. NILM has emerged as a crucial approach in this domain, leveraging advanced computational capabilities to estimate individual electrical device consumption using a single smart meter sensor [7]. In the literature, there are various categories of methods used in NILM. These categories can be grouped into four main categories:

- Methods of optimization: these methods use optimization techniques to conduct load disaggregation. Examples of these methods are Vector Support Machines (SVMs) [8], Bird Swarm Algorithms (BSAs) [9], Genetic Algorithms [10], and Particle Swarm Optimization (PSO) [11], among others;
- Supervised methods: these methods use tagged training datasets where individual exposures are known. Some examples of supervised methods are Bayesian [12], Vector Support Machines (SVM) [13], the algorithm of Discriminative Disaggregation Sparse Coding (DDSC) [14], and Artificial Neural Networks (ANN) [15], as well as their extensions;
- Unsupervised methods: use clustering techniques and statistical models for pattern recognition and load segmentation. Examples of unsupervised methods include Combinatorial Optimization (CO) [16], Hidden Markov Models (HMM) and their extensions, such as the FHMM (Factorial Hidden Markov Model) [17];
- Other approaches: in addition to the above categories, other approaches and techniques are used in NILM. Especially interesting is the processing of transient active power responses, measured when powered on and sampled at 100 Hz [18], so that using three stages (adaptive threshold event detection, convolutional neural network, and k-nearest neighbors' classifier), new devices can be automatically identified without the need for additional retraining or modeling for future expansions. Other ways can include semi-supervised learning methods, methods based on signal decomposition, approaches based on change detection, and different approaches proposed in the literature.

In terms of datasets available for energy disaggregation, some of the most commonly used are the following:

- AMPds16 (Anomaly detection in the network traffic dataset of 2016, Canada) [19]: provides detailed readings, such as voltage, current, frequency, and power for an overall meter and 19 individual circuits with 20 Hz of sampling;
- BERDS (Berkeley Energy Disaggregation Dataset, USA) [20]: provides active, reactive, and apparent power measurements at 20″ increments;
- BLOND (Technical University of Munich, Germany) [21]: contains voltage and current readings in two versions (BLOND-50 and BLOND-250) with different sample rates (50 kHz for aggregated circuits and 6.4 kHz for individual appliances);
- BLUED (Building-Level Fully Labeled Electricity Disaggregation Dataset, USA) [22]: includes high-frequency data (with 12 kHz of sampling) at the household level for approximately eight days, with events recorded whenever an appliance changes state;
- COOLL (Controlled On/Off Loads Library–University of Orleans, USA) [23]: Provides current and voltage data at a sampling rate of 100 kHz for 12 distinct types of appliances;
- DEPS (Higher Polytechnic School of the University of Seville, Spain) [24]: power, voltage, and current readings at the frequency of 1 Hz on six devices present in a classroom taken during a month;
- iAWE (Indian Ambient Water and Energy, India) [25]: it provides comprehensive real-time electricity and gas consumption data from 33 household sensors in an apartment in Delhi, covering both aggregate and individual appliance consumption patterns.

Various commercial and research meters exist in the current energy management and monitoring field, offering capabilities for measuring electricity consumption and power quality [26]. These include sophisticated power quality analyzers that professionals use for diagnostic purposes, identifying energy waste, and preventing energy-related issues [27]. However, such devices are often expensive and complex, making them less accessible to non-expert users [28]. They are primarily utilized for advanced energy audits and network analysis tasks. The software accompanying these devices is typically proprietary, but a trend toward open-source solutions is emerging, as seen in various domains [29]. Open-source software, already transformative in sectors like telecommunications and cloud computing [30], is now making significant inroads into the energy sector [31]. It offers benefits like accelerated development, reduced costs, and enhanced stability and interoperability [32].

Among open-source developments based on Arduino, several projects stand out:

- OpenEnergyMonitor [33]: this system was designed for home energy monitoring, providing real-time analysis of energy usage. It supports active power, root mean square (RMS) voltage, and RMS current measurements at a high sampling rate and features an HTML5 interface, Wi-Fi and ethernet support, and an API. However, it lacks capabilities for measuring reactive power and power factor.
- Arduino Energy Monitor: this open-source project leverages an Arduino board and a non-invasive current sensor, displaying measurements on an LCD screen or a web interface. It offers real-time consumption data, storage, and communication capabilities, making it suitable for home monitoring and energy efficiency projects.
- EmonTx: aimed at energy efficiency, renewable energy, and building monitoring projects, EmonTx is an open-source system that measures and records electricity consumption in real time. It includes hardware that connects to electrical circuits and uses sensors to measure energy consumption. The data are transmitted via radio frequency or wires to a receiver that sends it to a computer or cloud platform for visualization and analysis. The software associated with EmonTx v4 allows the system to be configured, calibrated, and visualize the collected data. It also offers logging and long-term data storage functions, allowing detailed energy consumption monitoring and usage pattern detection.
- There are also commercial Arduino-based projects that are not open-source:
- IoTaWatt [34] is an IoT device based on an ESP32 microcontroller [35] that monitors energy consumption in real-time, recording data and transmitting it to the cloud for analysis. It also measures energy generated by renewable sources and adapts to different monitoring needs.
- Smappee [36] is a commercial energy monitor that offers a variety of devices to measure and monitor electrical energy consumption. It provides a user-friendly interface and provides detailed information about real-time energy consumption. It also offers logging and analysis capabilities through its online platform.

Furthermore, platforms based on other boards like Raspberry Pi [37] have led to the development of devices like Wattson, which uses a non-invasive current sensor and an LED screen; emonPi, a device for energy monitoring and data logging, providing real-time consumption information and online access for analysis; and RPICT, a hardware project for energy monitoring that uses current transformers to measure and monitor electrical energy consumption, offering a cost-effective and customizable solution for real-time energy monitoring and analysis.

Another energy meter, the Open Z Meter (oZm) [38], developed by the Universities of Granada and Almeria, stands out as an energy quality analyzer and an open-source, open-hardware device with IoT capabilities. It can record and process extensive data, measuring various electrical variables such as voltage, intensity, active power, reactive power, Total Harmonic Distortion (THD), power factor, and harmonics of intensity and voltage up to 50 at a high sampling frequency [39]. The latest version allows the analysis of three-phase systems [40].

Despite the plethora of available options, there is a significant disparity in the performance and accuracy of energy monitors on the market, with some offering essential functions and others, like the one above, providing high accuracy but needing more scalability and expandability.

In this context, this research introduces the Open Multi Power Meter (OMPM), a solution to address these gaps, particularly in the NILM field, offering a balance of accuracy, scalability, and user-friendliness. The OMPM is an open-hardware solution with firmware developed in open source [41]. The device's open hardware nature not only makes it accessible to a broader range of users but also encourages innovation and customization, allowing it to be tailored to specific research or operational needs.

Central to the OMPM's utility is its compatibility with the Non-Intrusive Load Monitoring Toolkit (NILMTK) [42], which employs advanced algorithms for energy disaggregation, a method that uses computational techniques to estimate the power usage of individual appliances from a single meter reading that records the total power demand [43]. The NILMTK's Combinatorial Optimization (CO) and Finite Hidden Markov Model (FHMM) algorithms [44] are particularly adept at dissecting complex energy usage patterns, making them ideal for assessing the OMPM's performance. By leveraging these tools, the OMPM can provide detailed insights into electricity consumption, leading to more informed energy management decisions and efficiency optimization [45], since without direct feedback, expecting consumers to actively participate in a sustainable and efficient energy system is unrealistic [46].

The article is organized into several sections, each focusing on distinct aspects of OMPM development and its application. The sections cover the materials and methods used in creating the OMPM, the measurement module, the sequencer module, and the metrics and process of disaggregation. Results and discussion are presented, highlighting the performance and effectiveness of the OMPM in various scenarios. The article concludes with a summary of the key findings, implications of the research, and suggestions for future research.

2. Materials and Methods

The OMPM stands out for integrating a single microcontroller architecture with a suite of measurement modules interconnected through an RS485 bus system, enabling the integration of multiple low-cost measurement modules as needed. This design ensures a balance between high accuracy in electrical measurements and the required flexibility for wide-scale implementation.

The core components of this new hardware, aimed at the acquisition and recording of electrical measurements, are as follows:

- ESP32 nodeMCU: the central processing unit that manages the hardware's operations and data processing;
- PZEM-004 modules (one for measure module): these modules are crucial for measuring various electrical parameters since, in a single device, we obtain the voltage, current, power, and power factor,
- SD card reader: for reading data stored on SD cards;
- SD card: used for data storage and retrieval,
- Schottky diodes BAT54SW (one for measure module): essential for preventing reverse current flow,
- I2C screen (16 × 2, optional): this screen displays system information and measurements,
- Power supply (5 V/800 mA): provides the necessary power to the system,
- Additional components: including a simple switch, a resistor, an enclosure box, etc., for the complete hardware setup.

Furthermore, a primary sequencer circuit has been selected to automate the measurement process. This circuit is designed to manage various combinations of application activations and deactivations. The components for this optional hardware include:

- Arduino One: serves as the primary controller for the sequencer circuit;
- Optoisolated relay module (8×, compatible with Arduino): these relays enable controlled switching operations,
- Power Supply (12 V, 1 A): powers the sequencer system.
- Adding the price of all the components, the budget of the control unit with the display, the SD card reader, one 8 GB memory card, and the power supply to power the entire assembly is around EUR 22, to which EUR 5 would have to be added for each measurement channel, which would mean a total of EUR 52 at most for a 6-channel acquisition unit (5 measurement channels for applications plus one for the aggregate). It should be noted that each additional measurement channel, thanks to the expandable design using an RS485 bus, only needed a measurement module and a Schottky diode,

removing about EUR 5 from the budget. In summary, the cost of this simple optional unit would be around EUR 13.

The following subsections will detail the measurement system based on the PZEM-004 modules, data acquisition, and sequencer systems. The explanation will provide insights into the functionality and capabilities of each component within the system, illustrating how they collectively contribute to an efficient and scalable energy measurement solution.

2.1. PZEM-004 Module

The PZEM-004 module, developed by Peacefair [47], is a highly popular and cost-effective real-time power consumption monitoring tool. It stands as the cornerstone of the proposed solution, given its ability to measure five essential electrical characteristics of a circuit: RMS voltage, RMS current, active power, frequency, and power factor. This module's versatility and affordability make it a key component in energy monitoring applications. Key features of the PZEM-004 module include self-powering capability, optocoupled outputs for TTL level serial communication, and the use of Rogowski coils for current measurement, enhancing the accuracy and reliability of the readings.

The heart of the PZEM-004 module is the Vango Tec 9881 microcontroller. This ARM Cortex-M0-based controller boasts a 32-bit architecture and is equipped with 32 kb of flash memory and 8 kb of RAM. It is specifically designed for control and monitoring applications in the electrical energy sector, with built-in protection against over-current, over-voltage, and short-circuit scenarios. Additionally, the module features communication interfaces such as UART, SPI, and I2C, facilitating seamless integration with other electronic devices. Figure 1 presents the block diagram of the PZEM-004, showcasing its internal configuration and connectivity.

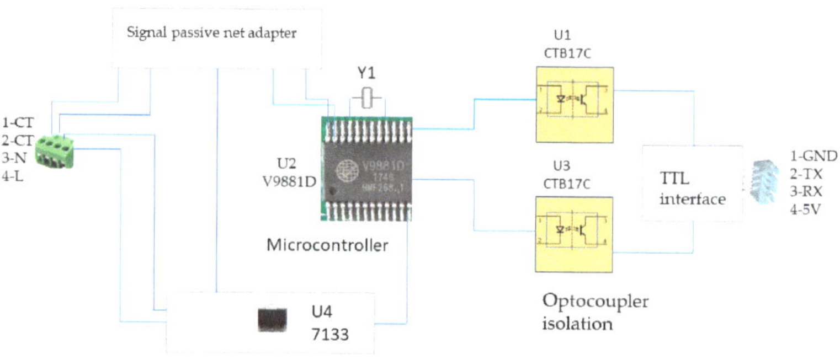

Figure 1. Block diagram of the PZEM-004 module (source: own elaboration).

Regarding the precision of the PZEM-004, each module is equipped with a calibration function. This feature allows for offset and gain adjustments, ensuring accurate and reliable readings. The electrical specifications for measurements with the PZEM-004 T-100A are as follows:

- Voltage: 80–260 V; Resolution: 0.1 V; Accuracy: 0.5%.
- Current: measuring range: 0–100 A; Initial measuring current: 0.024; Resolution: 0.001; Accuracy: 0.5%.
- Active power: measuring range: 0–23 kW; Initial power: 0.4 W; Resolution: 0.1 W; Display format: <1000 W (e.g., 999.9 W) and ≥1000 W (e.g., 1000 W); Accuracy: 0.5%.
- Power factor: measurement range: 0.00–1.00; Resolution: 0.01; Accuracy: 1%.
- Frequency: Measuring range: 45 Hz–65 Hz; Resolution: 0.1 Hz; Accuracy: 0.5%.
- Active energy: measuring range: 0–9999.99 kWh; Resolution: 1 Wh; Accuracy: 0.5%; Display format: <10 kWh (Wh unit) and ≥10 kWh (kWh unit).

- The PZEM module is a versatile tool that can be used in a variety of industrial automation projects. However, in most cases, it is used in isolation. A solution has been developed using multiple PZEM modules connected to an RS485 bus. The RS485 bus is a physical layer standard widely used in industrial automation. It is known for its noise resistance, extended data transmission range, and ability to support up to 127 devices on a single network. OMPM's solution takes advantage of the RS485 bus to enable communication between multiple PZEM modules. This allows users to collect data from a variety of sources and perform more complex analyses.

2.2. Measurement Module

The OMPM employs an ESP32 NodeMCU microcontroller central to the operation, managing the data collection from the measurement modules. Connectivity with the SD card adapter and the optional I2C display is achieved through the MISO/MOSI, CS, SCK, SCL, and SDA/SCL lines of the ESP32. The SD card serves as the primary storage medium for measurement data, formatted in CSV for each meter and connected to the SPI bus as follows:

- CS: GPIO 5;
- MOSI: GPIO 23;
- MISO: GPIO 19;
- SCK: GPIO 18.

The system incorporates six PZEM-004 modules, each with Rogowski coils for current measurement. Voltage measurements are conducted through parallel wiring, which powers the measurement modules. This setup allows for recording intensity measurements for six different electrical devices.

The implementation of an RS485 bus enhances the scalability of the system. This setup enables the transmission of voltage, current, power, frequency, and power factor measurements from each module to the central controller via the RX (GPIO 16) and TX lines (GPIO 17). The measurement acquisition frequency is above 10 Hz.

Additionally, the design includes a 2×16 LCD connected to the microcontroller via I2C, with the following wiring:

- SDA: GPIO 13;
- SCL: GPIO 14.

The entire assembly is powered by a 5 V DC supply from the controller's USB bus. This is feasible due to the low power consumption of the RX/TX part of each PZEM-004 module, which is primarily required to power the optocouplers in each module's transmission part. A small switch connected to GPIO15 is incorporated to activate the recording of measurements on the SD card.

Figure 2 illustrates the wiring diagram of the OMPM solution, highlighting the optional but convenient 16×2 LCD screen.

The RS485 bus implementation is non-standard, utilizing Schottky diodes to block reverse current and prevent interference and a standard 10 K resistor connected between the positive and line to limit current through the diodes. This setup also helps maintain the RX line voltage, facilitating signal detection on the TX line.

Regarding the firmware of the ESP32, it is essential to program a unique address for each PZEM-004 module to ensure univocal identification. The addresses used in OMPM are as follows:

- 0×110: aggregate consumption;
- 0×120: plug 1;
- 0×130: plug 2;
- 0×140: plug 3,
- 0×150: plug 4;
- 0×160: plug 5.

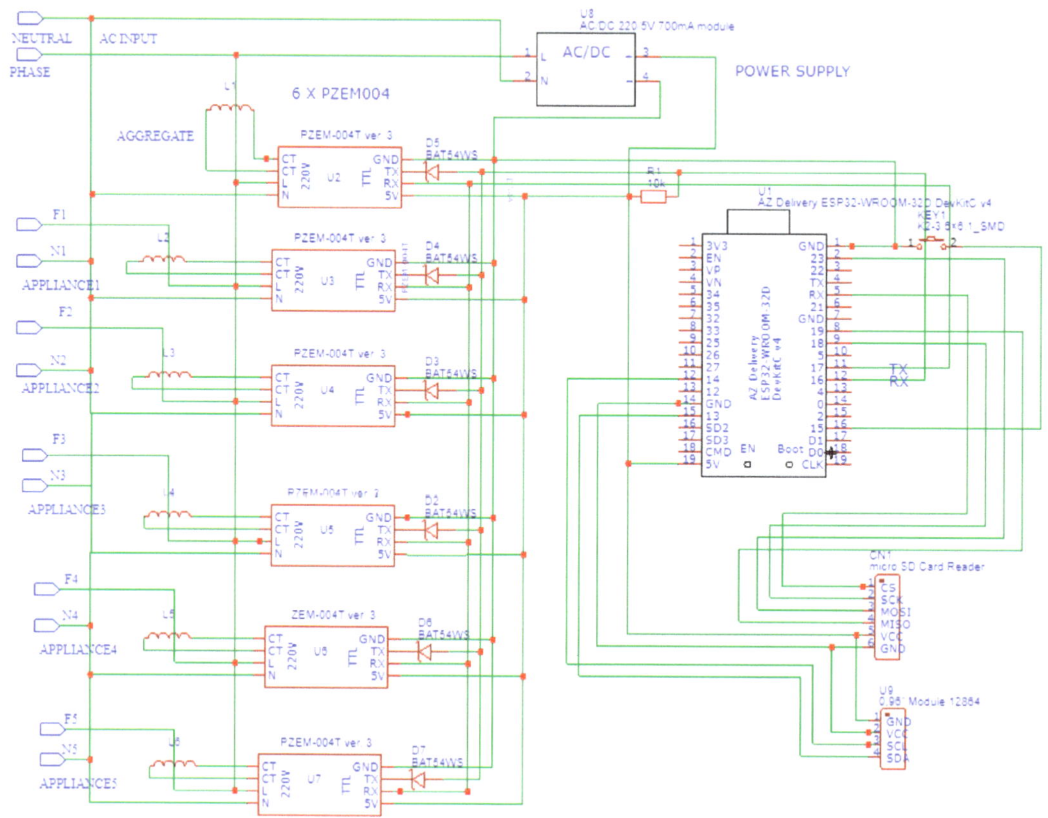

Figure 2. Wiring diagram of the OMPM.

The acquisition firmware developed for the microcontroller involves initializing the SD card, capturing the current date and time via STP using a network connection, and creating six files for each application, differentiated by the counter number concatenated with the first capture date. The files' headers are formatted in NILMTK style: "timestamp, VLN, A, W, F, PF", corresponding to various measurement parameters like date and time stamp (timestamp), nominal voltage (VLN), current (A), power (W), frequency (F), and power factor (PF).

The main program periodically records all readings from each meter, ensuring each is active and accessible. Each set of measurements is logged in its corresponding file, along with the timestamp value, as shown in the flowchart in Figure 3.

Figure 4 shows a photograph of the final circuit in operation, displaying measurements on the LCD, the ESP32 nodeMCU capturing data on an SD card, and the PZEM-004 modules in operation.

The experiment involves connecting target devices to each PZEM-004 module for measurement and analysis using NILMTK [42]. The devices include a meter for total consumption, a fan, a laptop computer, a light bulb, an LED light, and an electric welder.

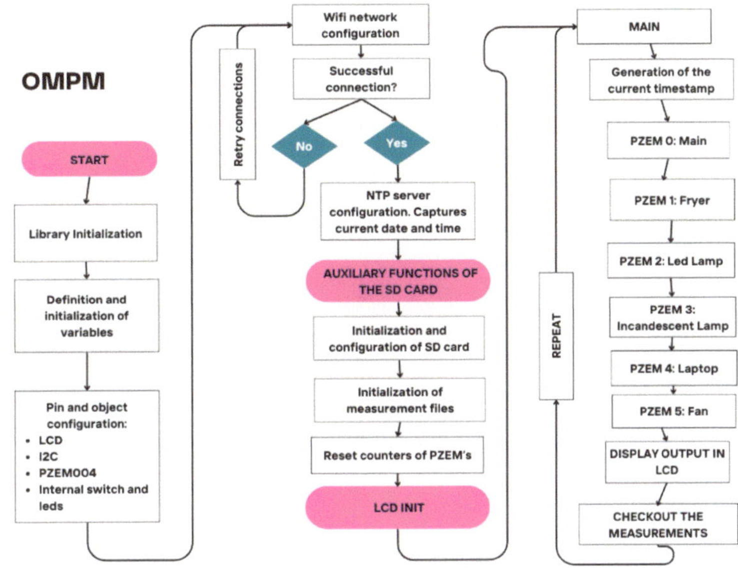

Figure 3. OMPM firmware flow chart.

Figure 4. Photograph of the final assembly.

2.3. Sequencer Module

A streamlined approach has been adopted to accurately simulate the real-world behavior of an installation using the measurement module described. This involves using an Arduino Uno board, which is interfaced through six digital bits with a relay board equipped with opto-coupled inputs. A 12 V DC voltage source powers the setup.

In this configuration, the normally open contacts of each relay on the board are wired in parallel to the manual switches of the five different applications used in this setup. These manual switches are kept in the open position. This arrangement allows for the automated control of each application, replicating the operational conditions typically found in electrical installations.

2.4. Metrics Used in This Work Available in NILMTK

To evaluate the quality of the results obtained in the disaggregation using NILMTK, four essential metrics will be used in this work:

- Error in total Allocated Energy (*EAE*) [48] quantifies the mean absolute error in energy estimation, calculated by Equation (1) as follows:

$$EAE = \left| \sum_t y_t^{(n)} - \sum_t \hat{y}_t^{(n)} \right| \qquad (1)$$

where $\hat{y}_t^{(n)}$ is the assigned power of appliance n at each time interval t, and $y_t^{(n)}$ is the real power of the same appliance. This metric effectively quantifies the discrepancies between estimated and actual energy usage, indicating algorithm precision. A lower *EAE* means greater accuracy of the algorithm;

- Mean Normalized Error in Assigned Power (*MNEAP*) [48] is a metric that evaluates the average absolute error in a normalized form, expressed as a percentage. It is articulated as follows in Equation (2):

$$MNEAP = \frac{\sum_t \left| y_t^{(n)} - \hat{y}_t^{(n)} \right|}{\sum \left| y_t^{(n)} \right|} \qquad (2)$$

where $y_t^{(n)}$ is the assigned power of appliance n at each time interval t, and $y\,\hat{y}_t^{(n)}$ is the real power of the same appliance. A lower *MNEAP* value signifies enhanced accuracy of the algorithm;

- Root Mean Square Error (*RMSE*) [49] is a standard metric that quantifies the magnitude of deviation in energy estimations, providing insight into the variance between energy consumption values predicted by the model and the real figures, as depicted in Equation (3).

$$RMSE = \sqrt{\frac{1}{T} \sum \left(y_t^{(n)} - \hat{y}_t^{(n)} \right)^2} \qquad (3)$$

where $\hat{y}_t^{(n)}$ is the assigned power of appliance n at each time interval t, $y_t^{(n)}$ is the real power of the same appliance, and T represents the total number of observations or time intervals over which the energy consumption is recorded;

- F-score. Known as the F1-score [50], this critical metric in machine learning evaluates the balance between model precision and recall. Derived from the confusion matrix within NILMTK, it embodies an amalgamation of precision and recall. *Precision* (positive predictive values), given in Equation (4), is concerned with the accurate prediction of 'ON' states. At the same time, *Recall* (Sensitivity), calculated as per Equation (5), focuses on correctly identifying actual appliance activations.

$$Precision = \frac{TP}{TP + FP} \qquad (4)$$

$$Recall = \frac{TP}{TP + FN} \qquad (5)$$

TP represents true positives, *FP* false positives, and *FN* false negatives.

The F1-score synthesizes these aspects, offering a composite measure that signals robust accuracy in identifying and predicting appliance states, as expressed in Equation (6). This metric, expressed as one, is desirable to be as close as possible to unity.

$$F1 = \frac{2 \times Precision \times Recall}{Precision + Recall} \qquad (6)$$

2.5. Disaggregation with NILMTK

For the disaggregation process utilizing the OMPM, the NILMTK [42] is employed. This toolkit facilitates analyzing and processing energy consumption data, as illustrated in Figure 5.

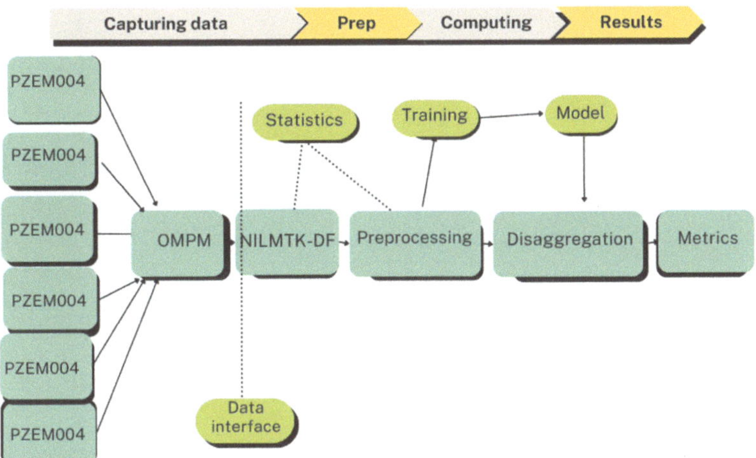

Figure 5. Implementation of NILMTK in the OMPM setup.

A new converter, UALM2, has been developed specifically for use with the OMPM and adapted from the DSUAL [51,52] converter (based on the iAWE [53] converter). This adaptation is necessary because the PZEM modules do not measure apparent or reactive power, resulting in only five available measurements. The UALM2 converter is designed to generate a dataset consistent with the measurements from the OMPM, which is compiled into six files. Each file contains a timestamp and five key electrical measurements: RMS voltage, RMS current, real power, frequency, and power factor.

Upon the generation of this new dataset, it becomes possible to graphically display various parameters such as active power, power factor, voltage, and current for all applications, including the aggregate meter, as shown in Figure 6a. Moreover, the recording of all measurements is visually represented in Figure 6b.

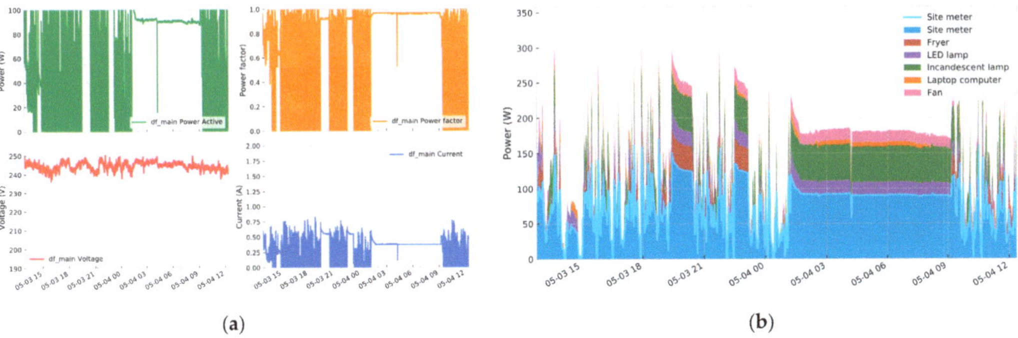

(a)   (b)

Figure 6. (**a**) Critical measurements for the aggregate meter; (**b**) consumption of individual applications.

The dataset is divided into training, validation, and testing. The NILMTK's two implemented algorithms, CO and FHMM, are executed with various methods of filling (mean, median, and first) and different sampling periods (ranging from 1 s to 30 min). This process aims to identify the optimal combination for generating the disaggregation model and calculating all available metrics in NILMTK, such as F1-score, EAE, and MNEAP.

3. Results and Discussion

Following the execution of the two algorithms under consideration, CO and FHMM, an initial assessment reveals insights into the optimal combination of sampling times and filling methods. Table 1 below presents the first estimations, comparing the performance of the CO and FHMM algorithms with various filling methods (mean and median) across different sampling periods. This preliminary analysis suggests that a 60-s sampling interval may be the most effective choice.

Table 1. First estimations.

	CO (Mean)	FHMM (Mean)	CO (Median)	FHMM (Median)
1 s	6.45	8.47	7.21	11.07
10 s	5.68	5.81	5.61	6.62
30 s	4.84	4.62	4.61	4.83
60 s	3.94	4.27	3.90	4.33
5 min	6.46	9.66	5.71	8.39
15 min	7.49	11.95	7.74	14.72

A more comprehensive set of experiments was conducted based on this initial understanding. The CO and FHMM algorithms were run using all three filling methods (mean, median, and first) over an expanded range of sampling periods. The results of this extensive testing are summarized in Table 2.

Table 2. Final results.

	CO (Mean)	FHMM (Mean)	CO (Median)	FHMM (Median)	CO (First)	FHMM (First)
1 s	7.74	14.65	12.56	13.18	10.12	13.96
10 s	8.57	7.68	9.73	8.10	5.38	7.11
30 s	4.00	5.12	4.22	5.02	4.16	5.60
60 s	3.70	5.57	3.77	4.84	4.25	5.47
5 min	7.78	10.28	7.49	11.42	13.31	12.41
10 min	8.73	13.49	8.95	13.10	10.88	14.54
15 min	9.18	14.61	8.95	15.43	12.60	16.38
30 min	9.34	14.46	9.13	13.74	9.69	14.29

Table 2 shows that a 60-s sampling interval consistently yields the most favorable results across all combinations. Notably, combining the CO algorithm with the mean filling method emerges as the most effective, underscoring the potential of this system in the context of NILM.

A notable difference emerges in the optimal sampling times when comparing the results achieved using NILMTK on the OMPM with those obtained from the iAWE dataset [54]. The best results for the iAWE dataset were obtained with a significantly more extended sampling period, exceeding 10 min, as seen in Table 3, which presents the results obtained for the iAWE dataset, showing the performance of the CO and FHMM algorithms with different filling methods across various sampling intervals.

Table 3. Results for the iAWE dataset.

	CO (Mean)	FHMM (Mean)	CO (Median)	FHMM (Median)	CO (First)	FHMM (First)
1 s	11.01	124.36	12.65	117.12	11.46	112.09
10 s	11.02	23.09	10.43	22.02	10.31	21.79
30 s	10.23	15.35	10.29	15.65	10.24	15.41
60 s	9.93	12.88	9.93	12.83	9.81	12.45
5 min	9.94	10.38	9.47	10.41	9.48	10.27
10 min	9.23	10.02	9.33	10.05	9.27	10.03

The analysis reveals that the most efficient algorithm for the iAWE dataset is the CO using the mean method as the filling method and a sampling period of 10 min. This contrasts with the 60-s sampling requirement for the OMPM.

DEPS [24] is a three-phase consumption dataset with active power, reactive power, voltage, and current measurements. It comprises ten industrial meters whose characteristics are described in Table 4.

Table 4. Summary of measurements recorded in the DEPS dataset.

Meter	Registered Measures	Sampling Period
1 × Three-phase main meter (RST)	P, Q	1 s
3 × Phase meters (R, S y T)	P, Q, V, I	1 s
6 × Device Meters	P, Q, V, I	1 s

The main meter (Main_RST) measures the aggregate active (P) and reactive (Q) power of the system. It also functions as a phase-based meter, allowing P, Q, voltage (V), and current (I) to be recorded for each phase. The devices are divided into two lighting groups (Lights_1 and Lights_2), three air conditioners (HVAC_1, HVAC_2, and HVAC_4), and a computer rack (Rack). Lighting data include only active power. Air conditioning equipment data have active power, reactive power, voltage, and current. Rack data include active power, reactive power, voltage, and current.

NILMTK also enables the calculation of evaluation metrics using the MeterGroup to validate results via the validation set. The performance of the models can be assessed using different metrics such as FEAC, F1-score, EAE, MNEAP, and RMSE, which provide insights into the accuracy and reliability of the disaggregation process. To illustrate the effectiveness of the approach, the primary metrics obtained for various applications on OMPM are presented in Tables 5 and 6, which display the results of the same metrics in the DEPS dataset [24].

Table 5. Main metrics obtained for applications.

	Fryer	LED Lamp	Bulb Lamp	Laptop	Fan
F1-score	0.420	0.789	0.756	0.453	0.741
EAE	0.002	0.001	0.011	0.002	0.012
MNEAP	1.138	0.349	0.484	1.150	0.502
RMSE	17.417	7.339	22.688	13.816	12.651

Table 6. Results of the main metrics for DEPS.

	Lights_1	Lights_2	HVAC_1	HVAC_2	HVAC_4	Rack
F1-score	0.915	0.860	0.968	0.972	0.463	0.945
EAE	0.61	0.59	1.62	2.56	0.49	0.49
MNEAP	0.16	0.26	0.59	0.94	1.23	0.12
RMSE	108.8	88.9	165.9	194.0	72.5	36.0

The F1-score metric shows satisfactory accuracy in the OMPM setup, with values close to or exceeding 50% even in the worst-case scenarios (laptop, fryer). Compared to results from the DEPS dataset, except for excellent values for HVAC_1 and HVAC_2, the OMPM yields better results overall.

The EAE metric shows excellent results for the OMPM, with almost negligible discrepancies for all appliances. The results for DEPS in terms of EAE are equally excellent.

The MNEAP metric yields good values, especially for the LED lamp, followed by the halogen lamp and fan. The laptop and fryer also exhibit impressive results, with the arithmetic mean for MNEAP being 0.724. The results for MNEAP with DEPS are remarkably like those obtained with OMPM, with an arithmetic mean of 0.73.

The RMSE metric shows exceptionally favorable results for all applications with the OMPM, particularly when compared to the DEPS dataset. For DEPS, significantly higher RMSE values are observed, except for the rack, indicating a substantial difference in performance.

Figure 7 presents the index correspondence for the OMPM dataset, illustrating the performance of the combinatorial algorithm across various sampling methods for different appliances.

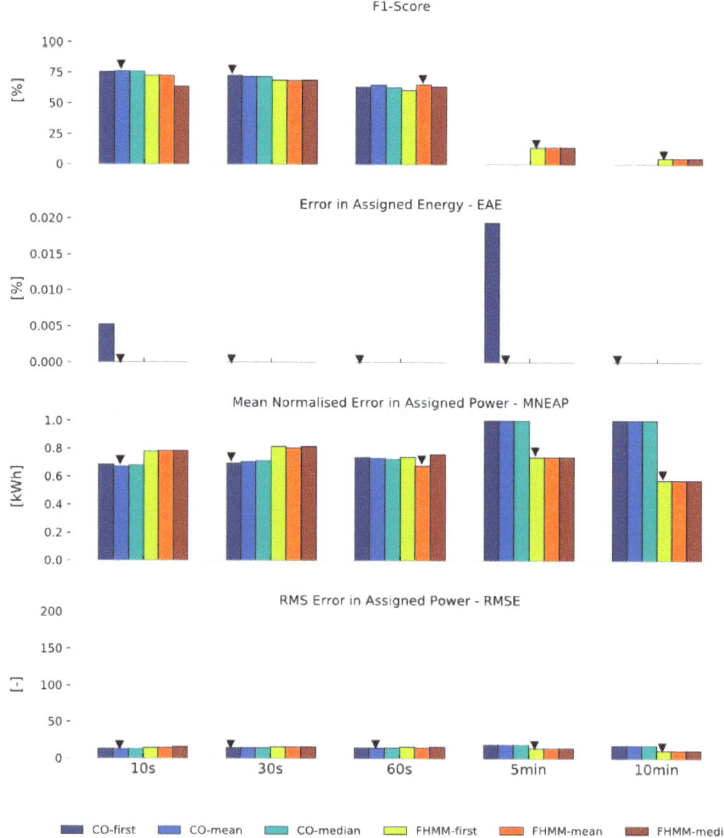

Figure 7. OMPM index correspondence.

The analysis shows varied yields across different metrics and sampling times. For the F1-score metric, the results are similar for less than one-minute sampling times with both algorithms and other fill methods. As for the SEA metric, zero error is observed in almost all cases, which means a very accurate energy estimate in both datasets. As for the MNEAP metric, a noticeable increase in error is observed when the sampling time exceeds one minute. This increase is more pronounced for the combinatorial model, while it decreases for the FHMM. For the RMSE metric, even with sampling times extended to 10 min, the values are still very good (low), so no algorithm, sampling time, or filling method stands out as significantly superior.

Since the best behavior offered is with the CO algorithm, Figure 8a showcases the OMPM results, while Figure 8b shows the results of the DEPS dataset, whose best performance was with the FHMM algorithm.

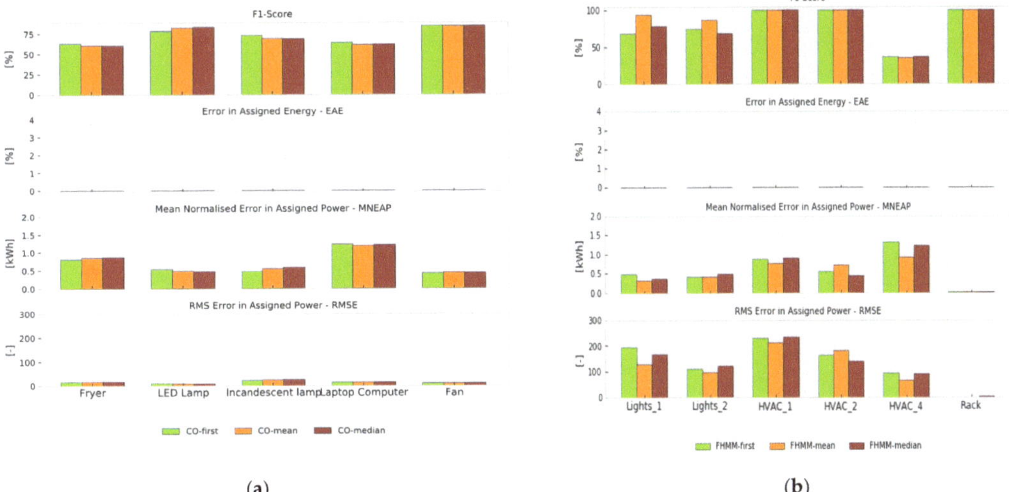

Figure 8. (**a**) OMPM results with CO algorithm; (**b**) DEPS results with FHMM algorithm.

The comparison reveals that while the F1-score metric excels for some devices with DEPS, it performs poorly for others (notably HVAC_4), indicating that the OMPM results are more consistent across different devices. The EAE metric results for DEPS are equally excellent, mirroring those obtained with OMPM. The MNEAP metric for DEPS notably excels for the rack application but is otherwise like OMPM. Finally, regarding the RMSE metric, OMPM's results are notably worse when compared to DEPS, highlighting a significant disparity in performance between the two datasets in this metric.

In the final stage of analysis, the median method of the CO algorithm is applied to assess its efficacy on the OMPM dataset. The results are presented in Figure 9a, providing a comprehensive view of the algorithm's performance across various metrics. For a comparative perspective, the outcomes of employing the same model on the DEPS dataset are also examined (Figure 9b).

The analysis indicates a notable variability in the F1-score metric within the DEPS dataset across different devices. While specific devices exhibit high performance, others, such as HVAC_4 and Lights_2, display comparatively lower scores. This variability contrasts with the more uniform results observed across the OMPM dataset, suggesting a higher degree of consistency and homogeneity in its performance.

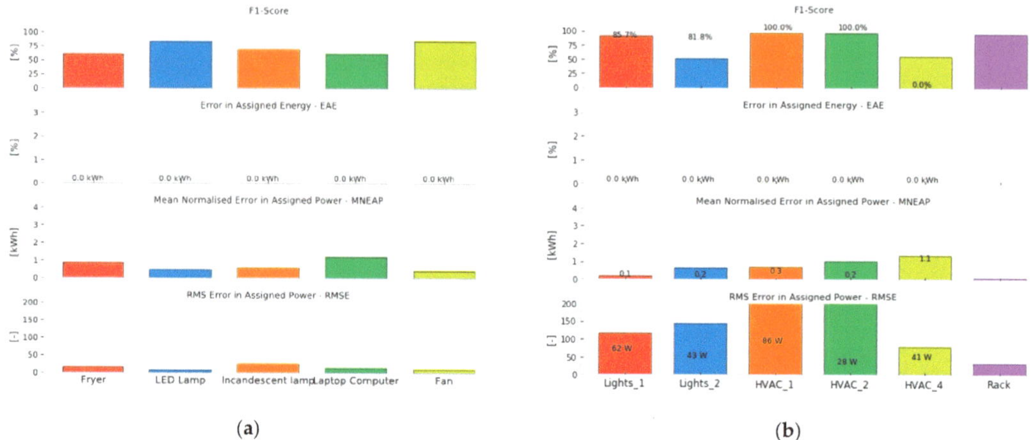

Figure 9. (**a**) Evaluation results of the optimal model using the OMPM dataset; (**b**) evaluation results of the optimal model using the DEPS dataset.

Regarding the EAE metric, the DEPS and OMPM datasets yield exceptional results, indicating high accuracy in energy consumption estimations.

In the context of the MNEAP metric for the DEPS dataset, standout performance is noted for the rack application. At the same time, the results for other appliances align closely with those obtained from the OMPM dataset.

Lastly, the RMSE metric reveals significant disparities, with the OMPM dataset demonstrating markedly superior accuracy in energy consumption estimations compared to the DEPS dataset, which exhibits less favorable outcomes. This distinction highlights the robustness of the OMPM dataset in providing reliable and precise energy consumption data.

In summary, comparing the results of the metrics obtained with OMPM concerning DEPS, for the F1-score, quite similar results are obtained (giving a specific slight advantage to DEPS). However, for EAE and MNEAP, similar results are obtained (now giving a specific slight advantage to OMPM). Notably, the RMSE metric stands out very clearly, where much better results are undoubtedly obtained for OMPM. Given the proposed solution's low cost and the metrics obtained, especially for RMSE, OMPM is an extremely interesting solution in the field of NILM.

4. Conclusions

This research has presented an open hardware-based solution that stands out for its scalability, affordability, and replicability while upholding the exceptional precision synonymous with professional-grade solutions. With a modest budget of approximately EUR 52 or less, this open-source solution delivers measurements of six simultaneous channels, encompassing voltage, current, power, and power factor. These channels can be effortlessly expanded up to a maximum of 127 by adding as many measurement modules to the bus as needed, each costing around EUR 5.

In addition to its remarkable scalability, this open system can be used as a self-scaling multi-system for acquiring electrical measurements and implementing the NILM task. The system utilizes open-source software, encompassing the microcontroller's firmware responsible for capturing measurements and storing them in files and the post-processing phase for NILM, which relies on the NILMTK toolkit specifically tailored to incorporate the dataset generated by this innovative hardware.

A noteworthy aspect of this work is the development of a new converter tailored to the OMPM measurement files. This converter creates a new dataset supporting a 13-digit timestamp, facilitating the application of NILMTK's various phases, including validation, training, and metrics evaluation.

Significant differences have emerged when comparing the results obtained from applying NILMTK metrics to the OMPM dataset with those derived from the DEPS dataset (generated using professional hardware). Notably, the OMPM dataset requires shorter sampling times and exhibits a remarkable 200% difference in the RMSE metric compared to the DEPS dataset. The comparative analysis of the OMPM and other public datasets, including measurements from commercial and open hardware monitors, underlines the device's accuracy and scalability. The results from these comparisons validate the OMPM's effectiveness and highlight its simplicity and adaptability, making it a valuable tool for a wide range of applications, from academic research to practical energy management solutions.

The promising results achieved with the OMPM dataset using NILMTK metrics open new possibilities for researchers to generate their datasets and further enhance NILM research. The scalability of the proposed solution, facilitated by the implementation of an RS485 bus, allows for the use of multiple channels with a single microcontroller. This scalability ensures the capture of all fundamental electrical measurements with commendable accuracy. Having been successfully assessed with six modules and the number of circuits in a typical household, the system holds the potential for future expansion to accommodate even more modules.

To evaluate this new hardware, applications with low power consumption were chosen to increase the complexity of disaggregation tasks. The hardware yielded highly satisfactory results across various metrics, suggesting its potential utility in ongoing NILM research.

As an improvement, it is worth mentioning that currently, each module is fed directly from the mains voltage using a simple RC circuit, a rectifier diode, and a Zener diode, with a U3 regulator (7133) at the output. An improvement could be achieved by feeding the regulator from an isolated, independent source, such as an R05P125, which offers a promising direction for further research and development.

Future work could focus on enhancing the accuracy of the measurement modules to create a system for disaggregating energy consumption in real-time, for example, by sending the measurement files to a Raspberry Pi running NILMTK every few seconds.

This research has demonstrated the potential of a low-cost, open-source hardware solution for NILM tasks. The proposed system's scalability, affordability, replicability, and remarkable accuracy make it a valuable tool for researchers and practitioners. Future work should focus on refining the hardware design and exploring its potential applications in various domains, including energy management, smart homes, and grid monitoring.

Author Contributions: Conceptualization, C.R.-N. and A.A.; methodology, C.R.-N. and A.A.; software, C.R.-N.; validation, C.R.-N., F.P., F.M.-A., F.M. and A.A.; formal analysis, C.R.-N. and A.A.; investigation, C.R.-N., F.P., F.M.-A., F.M. and A.A.; resources, C.R.-N., F.P., F.M.-A., F.M. and A.A.; data curation, C.R.-N. and A.A.; writing—original draft preparation, C.R.-N., F.P. and A.A.; writing—review and editing, C.R.-N., F.P., F.M.-A., F.M. and A.A.; visualization, C.R.-N., F.P., F.M.-A., F.M. and A.A.; supervision, F.M. and A.A. All authors have read and agreed to the published version of the manuscript.

Funding: This research received no external funding.

Data Availability Statement: The data will be made available upon request to the corresponding author.

Conflicts of Interest: The authors declare no conflict of interest.

References

1. Lian, J.; Zhang, Y.; Ma, C.; Yang, Y.; Chaima, E. A review on recent sizing methodologies of hybrid renewable energy systems. *Energy Convers. Manag.* **2019**, *199*, 112027. [CrossRef]
2. Juaidi, A.; AlFaris, F.; Saeed, F.; Salmeron-Manzano, E.; Manzano-Agugliaro, F. Urban Design to Achieving the Sustainable Energy of Residential Neighbourhoods in Arid Climate. *J. Clean. Prod.* **2019**, *228*, 135–152. [CrossRef]
3. Alcayde, A.; Robalo, I.; Montoya, F.G.; Manzano-Agugliaro, F. SCADA System for Online Electrical Engineering Education. *Inventions* **2022**, *7*, 115. [CrossRef]

4. Meliani, M.; Barkany, A.E.; Abbassi, I.E.; Darcherif, A.M.; Mahmoudi, M. Energy Management in the Smart Grid: State-of-the-Art and Future Trends. *Int. J. Eng. Bus. Manag.* **2021**, *13*, 18479790211032920. [CrossRef]
5. Machlev, R.; Belikov, J.; Beck, Y.; Levron, Y. MO-NILM: A Multi-Objective Evolutionary Algorithm for NILM Classification. *Energy Build.* **2019**, *199*, 134–144. [CrossRef]
6. Hart, G.W. Nonintrusive Appliance Load Monitoring. *Proc. IEEE* **1992**, *80*, 1870–1891. [CrossRef]
7. Fortuna, L.; Buscarino, A. Non-Intrusive Load Monitoring. *Sensors* **2022**, *22*, 6675. [CrossRef]
8. Papageorgiou, P.G.; Christoforidis, G.C.; Bouhouras, A.S. Odd Harmonic Distortion Contribution on a Support Vector Machine NILM Approach. In Proceedings of the SyNERGY MED 2022—2nd International Conference on Energy Transition in the Mediterranean Area, Thessaloniki, Greece, 17–19 October 2022.
9. Melin, P.; Miramontes, I.; Carvajal, O.; Prado-Arechiga, G. Fuzzy Dynamic Parameter Adaptation in the Bird Swarm Algorithm for Neural Network Optimization. *Soft Comput.* **2022**, *26*, 9497–9514. [CrossRef] [PubMed]
10. Chui, K.T.; Lytras, M.D.; Visvizi, A. Energy Sustainability in Smart Cities: Artificial Intelligence, Smart Monitoring, and Optimization of Energy Consumption. *Energies* **2018**, *11*, 2869. [CrossRef]
11. Gong, F.; Liu, C.; Jiang, L.; Li, H.; Lin, J.Y.; Yin, B. Load Disaggregation in Non-Intrusive Load Monitoring Based on Random Forest Optimized by Particle Swarm Optimization. In Proceedings of the 2017 IEEE Conference on Energy Internet and Energy System Integration, Beijing, China, 26–28 November 2017.
12. Bing, Q.; Liya, L.; Xin, W. Low-Rate Non-Intrusive Load Disaggregation with Graph Shift Quadratic Form Constraint. *Appl. Sci.* **2018**, *4*, 554. [CrossRef]
13. Quanbo, Y.; Huijuan, W.; Botao, W.; Yaodong, S.; Hejia, W. A Fusion Load Disaggregation Method Based on Clustering Algorithm and Support Vector Regression Optimization for Low Sampling Data. *Future Internet* **2019**, *11*, 51. [CrossRef]
14. Albert, F. Algorithms for Energy Disaggregation. Master's Thesis, Universitat Politècnica de Catalunya, Barcelona, Spain, 2016. Available online: https://upcommons.upc.edu/handle/2117/89937 (accessed on 30 November 2023).
15. Lei, J.; Jiaming, L.; Suhuai, L.; Sam, W.; Glenn, P. Power Load Event Detection and Classification Based on Edge Symbol Analysis and Support Vector Machine. *Appl. Comput. Intell. Soft Comput.* **2012**, *2012*, 742461. [CrossRef]
16. Berrettoni, G.; Bourelly, C.; Capriglione, D.; Ferrigno, L.; Miele, G. Preliminary Sensitivity Analysis of Combinatorial Optimization (CO) for NILM Applications: Effect of the Meter Accuracy. In Proceedings of the 2021 IEEE 6th International Forum on Research and Technology for Society and Industry (RTSI), Naples, Italy, 6–9 September 2021. [CrossRef]
17. Hui, L. *Non-Intrusive Load Monitoring*; Liu, H., Ed.; Science Press: Beijing, China; Central South University: Chansha, China, 2020; Volume 1, ISBN 978-9811518591.
18. Athanasiadis, C.L.; Papadopoulos, T.A.; Doukas, D.I. Real-Time Non-Intrusive Load Monitoring: A Light-Weight and Scalable Approach. *Energy Build.* **2021**, *253*, 111523. [CrossRef]
19. Makonin, S.; Popowich, F.; Bartram, L.; Gill, B.; Bajić, I.V. AMPds: A Public Dataset for Load Disaggregation and Eco-Feedback Research. In Proceedings of the 2013 IEEE Electrical Power & Energy Conference, Halifax, NS, Canada, 21–23 August 2013.
20. Maasoumy, M.; Sanandaji, B.M.; Poolla, K.; Vincentelli, A.S. BERDS-BERkeley EneRgy Disaggregation Data Set. In Proceedings of the Workshop on Big Learning at the Conference on Neural Information Processing Systems (NIPS), Lake Tahoe, NV, USA, 9 December 2013.
21. Kriechbaumer, T.; Jacobsen, H.A. BLOND, a Building-Level Office Environment Dataset of Typical Electrical Appliances. *Sci. Data* **2018**, *5*, 180048. [CrossRef] [PubMed]
22. Cannas, B.; Carcangiu, S.; Carta, D.; Fanni, A.; Muscas, C.; Sias, G.; Canetto, B.; Fresi, L.; Porcu, P. NILM Techniques Applied to a Real-Time Monitoring System of the Electricity Consumption. *Acta IMEKO* **2021**, *10*, 139–146. [CrossRef]
23. Yu, M.; Wang, B.; Lu, L.; Bao, Z.; Qi, D. Non-Intrusive Adaptive Load Identification Based on Siamese Network. *IEEE Access* **2022**, *10*, 11564–11573. [CrossRef]
24. Arias, S. DEPS: Dataset de La Escuela Politécnica Superior. Available online: https://github.com/AriasSilva/DEPS_NILM_Dataset (accessed on 3 June 2022).
25. Batra, N.; Gulati, M.; Singh, A.; Srivastava, M.B. It's Different: Insights into Home Energy Consumption in India. In Proceedings of the BuildSys 2013—5th ACM Workshop on Embedded Systems For Energy-Efficient Buildings, New York, NY, USA, 11–15 November 2013.
26. Montoya, F.G.; García-Cruz, A.; Montoya, M.G.; Manzano-Agugliaro, F. Power Quality Techniques Research Worldwide: A Review. *Renew. Sustain. Energy Rev.* **2016**, *54*, 846–856. [CrossRef]
27. Montoya, F.G.; Manzano-Agugliaro, F.; López, J.G.; Alguacil, P.S. Power Quality Research Techniques: Advantages and Disadvantages. *Dyna* **2012**, *79*, 66–74.
28. Montoya, F.; Baños, R.; Alcayde, A.; Montoya, M.; Manzano-Agugliaro, F. Power Quality: Scientific Collaboration Networks and Research Trends. *Energies* **2018**, *11*, 2067. [CrossRef]
29. Biansoongnern, S.; Plangklang, B. Nonintrusive Load Monitoring (NILM) Using an Artificial Neural Network in Embedded System with Low Sampling Rate. In Proceedings of the 2016 13th International Conference on Electrical Engineering/Electronics, Computer, Telecommunications and Information Technology (ECTI-CON), Chiang Mai, Thailand, 28 June–1 July 2016; pp. 1–4.
30. Kromanis, R.; Forbes, C. A Low-Cost Robotic Camera System for Accurate Collection of Structural Response. *Inventions* **2019**, *4*, 47. [CrossRef]

31. Mousavi, P.; Ghazizadeh, M.S.; Vahidinasab, V. A Decentralized Blockchain-Based Energy Market for Citizen Energy Communities. *Inventions* **2023**, *8*, 86. [CrossRef]
32. Klimt, J.; Eiling, N.; Wege, F.; Baude, J.; Monti, A. The Role of Open-Source Software in the Energy Sector. *Energies* **2023**, *16*, 5855. [CrossRef]
33. OpenEnergyMonitor. Available online: https://openenergymonitor.org/ (accessed on 30 November 2023).
34. IoTaWatt. Available online: https://iotawatt.com/ (accessed on 30 November 2023).
35. ESP32. Available online: http://esp32.net/ (accessed on 30 November 2023).
36. Smappee. Available online: https://www.smappee.com/ (accessed on 30 November 2023).
37. Raspberry Pi. Available online: https://www.raspberrypi.com/ (accessed on 30 November 2023).
38. OpenZmeter. Available online: https://openzmeter.com/ (accessed on 30 November 2023).
39. Viciana, E.; Alcayde, A.; Montoya, F.; Baños, R.; Arrabal-Campos, F.; Zapata-Sierra, A.; Manzano-Agugliaro, F. OpenZmeter: An Efficient Low-Cost Energy Smart Meter and Power Quality Analyzer. *Sustainability* **2018**, *10*, 4038. [CrossRef]
40. Viciana, E.; Arrabal-Campos, F.M.; Alcayde, A.; Baños, R.; Montoya, F.G. All-in-One Three-Phase Smart Meter and Power Quality Analyzer with Extended IoT Capabilities. *Measurement* **2023**, *206*, 112309. [CrossRef]
41. OMPM. Available online: https://github.com/crn565/OMPM (accessed on 30 November 2023).
42. Batra, N.; Kelly, J.; Parson, O.; Dutta, H.; Knottenbelt, W.; Rogers, A.; Singh, A.; Srivastava, M. NILMTK: An Open Source Toolkit for Non-Intrusive Load Monitoring. In Proceedings of the e-Energy 2014—5th ACM International Conference on Future Energy Systems, Cambridge, UK, 11–13 June 2014.
43. Ebrahim, A.F.; Mohammed, O.A. Pre-Processing of Energy Demand Disaggregation Based Data Mining Techniques for Household Load Demand Forecasting. *Inventions* **2018**, *3*, 45. [CrossRef]
44. Bonfigli, R.; Principi, E.; Fagiani, M.; Severini, M.; Squartini, S.; Piazza, F. Non-intrusive load monitoring by using active and reactive power in additive Factorial Hidden Markov Models. *Appl. Energy* **2017**, *208*, 1590–1607. [CrossRef]
45. Carrie Armel, K.; Gupta, A.; Shrimali, G.; Albert, A. Is Disaggregation the Holy Grail of Energy Efficiency? The Case of Electricity. *Energy Policy* **2013**, *52*, 213–234. [CrossRef]
46. Zoha, A.; Gluhak, A.; Imran, M.; Rajasegarar, S. Non-Intrusive Load Monitoring Approaches for Disaggregated Energy Sensing: A Survey. *Sensors* **2012**, *12*, 16838–16866. [CrossRef]
47. Peacefair Peacefair Site. Available online: https://www.peacefairmeter.com/ (accessed on 11 December 2023).
48. Celi Peñafiel, C.E.; Guartan Castro, F.E. Aplicación de Técnicas de Machine Learning Para La Desagregación y Pronóstico Del Perfil de Carga En El Sector Industrial. Bachelor's Thesis, Universidad Politécnica Salesiana, Cuenca, Ecuador, 2021. Available online: http://dspace.ups.edu.ec/handle/123456789/21273 (accessed on 30 November 2023).
49. Klemenjak, C.; Makonin, S.; Elmenreich, W. Towards Comparability in Non-Intrusive Load Monitoring: On Data and Performance Evaluation. In Proceedings of the 2020 IEEE Power and Energy Society Innovative Smart Grid Technologies Conference, ISGT 2020, Washington, DC, USA, 17–20 February 2020.
50. Kelly, J.; Knottenbelt, W. Neural NILM: Deep Neural Networks Applied to Energy Disaggregation. In Proceedings of the BuildSys 2015—Proceedings of the 2nd ACM International Conference on Embedded Systems for Energy-Efficient Built, Seoul, Republic of Korea, 4–5 November 2015; pp. 55–64.
51. DSUAL. Available online: https://github.com/crn565/DSUAL_without-armonics (accessed on 30 November 2023).
52. Rodriguez-Navarro, C.; Alcayde, A.; Isanbaev, V.; Castro-Santos, L.; Filgueira-Vizoso, A.; Montoya, F.G. DSUALMH-A New High-Resolution Dataset for NILM. *Renew. Energy Power Qual. J.* **2023**, *21*, 238–243. [CrossRef]
53. IAWE. Available online: https://iawe.github.io/ (accessed on 30 November 2023).
54. OZm vs IAWE. Available online: https://github.com/crn565/NILMTK_ozm_vs_iawe (accessed on 30 November 2023).

 *inventions*

Article

Load Losses and Short-Circuit Resistances of Distribution Transformers According to IEEE Standard C57.110

Vicente León-Martínez *, Elisa Peñalvo-López, Clara Andrada-Monrós and Juan Ángel Sáiz-Jiménez

Electrical Engineering Department, Universitat Politècnica de València, Camino de Vera, 14, 46022 València, Spain; elpealpe@upvnet.upv.es (E.P.-L.); claanmon@etsii.upv.es (C.A.-M.); jasaiz@die.upv.es (J.Á.S.-J.)
* Correspondence: vleon@die.upv.es

Abstract: Load losses determine transformers' efficiency and life, which are limited by overheating and deterioration of their elements. Since these losses can be characterized by short-circuit resistances, in this article, we have developed expressions for the short-circuit resistances of three-phase transformers according to IEEE Standard C57.110. Imposing the condition that these resistances must cause load losses of the transformer, two types of short-circuit resistance have been established: (1) the effective resistance of each phase ($R_{cc,z}$) and (2) the effective short-circuit resistance of the transformer ($R_{cc,ef}$). The first is closely related to the power loss distribution within the transformer. The second is just a mathematical parameter. Applying these resistances to the 630 kVA oil-immersed distribution transformer of a residential network, we have concluded that both types of resistances determine the total load losses of the transformer. However, only $R_{cc,z}$ accurately provides the load losses in each phase. $R_{cc,ef}$ can give rise to errors more significant than 16% in calculating these losses, depending on imbalances in the harmonic currents.

Keywords: distribution transformers; short-circuit resistances; load losses; harmonics; efficiency

1. Introduction

Three-phase transformers are essential machines for the operation and stability of power systems. They interconnect electrical networks with different voltage levels, transferring electrical energy from generation centers (large power transformers) to consumption points (distribution transformers), and their use is growing due to the strong demand for energy in today's societies. However, in their operation, these machines waste energy in the core and windings [1,2], which reached worldwide values of 1181 TWh in 2020 and could be higher than 1845 TWh in 2040 according to estimates provided by the organization United for Efficiency (U4E) [2], patronized by the United Nations. These energy losses raise the temperature of the transformers and cause the following adverse effects:

- emission of greenhouse gases [2–5],
- deterioration in the properties of the core material and insulation, reducing the transformer's life [6,7],
- decrease in power transmission capacity [6,7].

Core losses (P_0) are little affected by voltage harmonics [8], and their values are usually considered the same as those provided by the manufacturers. Winding losses or load losses (P_{cc}) are caused by the circulation of currents through the primary and secondary windings of the transformer [9].

When the transformers supply non-linear loads, the load losses can be calculated by applying IEEE Standard C57.110-2018 [10–13] and other well-known standards [14–20]. Alternatively, the load losses of three-phase transformers can be determined with the use of short-circuit resistances (R_{cc}) according to the well-known equation [21,22]:

$$P_{cc} = R_{cc} \cdot \left(I_A^2 + I_B^2 + I_C^2 \right) \tag{1}$$

Citation: León-Martínez, V.; Peñalvo-López, E.; Andrada-Monrós, C.; Sáiz-Jiménez, J.Á. Load Losses and Short-Circuit Resistances of Distribution Transformers According to IEEE Standard C57.110. *Inventions* **2023**, *8*, 154. https://doi.org/10.3390/inventions8060154

Academic Editor: Om P. Malik

Received: 31 October 2023
Revised: 1 December 2023
Accepted: 6 December 2023
Published: 8 December 2023

in which (I_A, I_B, I_C) are the RMS values of the currents measured in each phase ($z = A, B, C$) of the primary or secondary winding depending on R_{cc}, which is the short-circuit resistance referred to the primary or secondary winding.

The short-circuit resistances determine not only the load losses but also efficiency and proper operation of the transformers. These parameters, jointly with the short-circuit reactance, determine the short-circuit impedances that limit the values of short-circuit currents and their adverse effects on the transformer windings due to vibrations [23–26] and electromagnetic forces [27–30].

References [31–35] establish different methods for determining a transformer's short-circuit impedances, but none specifies expressions for the short-circuit resistances as a function of the harmonic frequencies.

In 2023, L. Sima et al. [36] established expressions for the short-circuit resistance of three-phase transformers feeding non-linear loads. L. Sima et al. worked with a transformer model like the one depicted in Figure 1, which is implicit in IEEE Standard C57.110-2018 [10]. As we know, the paper published by L. Sima et al. is the only one in the technical literature that develops short-circuit resistances with values depending on harmonic frequencies.

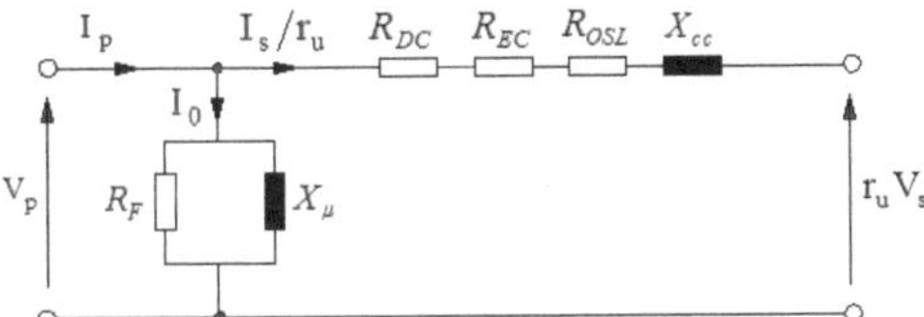

Figure 1. Single-phase equivalent circuit of three-phase transformers proposed by L. Sima.

The short-circuit resistance referred to as the primary winding (R_k) developed by L. Sima et al. by direct application of IEEE Standard C57.110-2018 [10] is expressed as follows:

$$R_k = \frac{P_K}{I_p^2} = R_{DC} + R_{EC} + R_{OSL} \tag{2}$$

where P_K represents the total load losses of the transformer defined by this standard, and I_p is the combined RMS value of the three currents of the primary winding, that is, $I_p^2 = I_{pA}^2 + I_{pB}^2 + I_{pC}^2$.

The short-circuit resistance of L. Sima et al. (R_k), as indicated by Equation (2), is the sum of three resistances: R_{DC}, R_{EC}, and R_{OSL} (Figure 1). Each characterizes the load losses caused by the three power phenomena present in the operation of three-phase transformers, as established by IEEE Standard C57.110-2018 [10].

In Equation (2), R_{DC} [37] is the combined direct current (or ohmic) resistance of the primary (R_{DCp}) and secondary (R_{DCs}) windings of the transformer:

$$R_{DC} = R_{DCp} + r_u^2 R_{DCs} = \frac{P_{DC}}{I_p^2} \tag{3}$$

where $r_u \cong V_p/V_{s0}$ is the transformation ratio, approximately defined by the quotient between the primary (V_p) and the no-load secondary (V_{s0}) voltages. The resistance R_{DC} characterizes the load losses (P_{DC}) when direct currents circulate through the transformer windings.

The resistance R_{EC} due to the combined skin effects of the primary (R_{ECp}) and secondary (R_{ECs}) windings of the transformer

$$R_{EC} = R_{ECp} + r_u^2 R_{ECs} = \frac{P_{EC}}{I_p^2} \tag{4}$$

causes the eddy current losses (P_{EC}), which are expressed by IEEE Standard C57.110-2018 as follows:

$$P_{EC} = P_{ECN} \cdot \sum_{h=1}^{\infty} h^2 \left(\frac{I_{ph}}{I_{pN}} \right)^2 \tag{5}$$

where P_{ECN} is the value of the eddy current losses at the nominal frequency ($f_N = 50 - 60$ Hz), I_{ph} is the RMS value of the harmonic of order $h = f_h / f_1$ of the transformer's primary currents, and I_{pN} is the rated RMS value of these currents.

The short-circuit resistance R_{OSL} is obtained by L. Sima et al. from Equation (2) as:

$$R_{OSL} = R_k - R_{DC} - R_{EC} \tag{6}$$

As explained by L. Sima et al. [36], R_{OSL} is not defined in the primary or secondary windings, unlike the previous two (R_{DC} and R_{EC}), since its losses do not cause additional heating in the windings, but in the tank and other metallic parts of the transformers.

Reference [36] constitutes the first approximation in the technical literature in the study of short-circuit resistances of transformers feeding non-linear loads according to IEEE Standard C57.110-2018.

The direct relationship between the short-circuit resistances and the load losses of the three-phase transformers, implicit in Equations (1) and (2), allows the use of these resistances as indicators for monitoring the operating status of these machines, as well as being sufficient to determine deterioration caused by overheating. To do this, the short-circuit resistance expressions must provide the correct values of the load losses in each phase of the windings, regardless of the type of currents and their RMS values in the three phases.

The short-circuit resistances referred to as "primary" by L. Sima et al. (R_k) can determine the total load losses of three-phase transformers according to IEEE Standard C57.110-2018; however, they do not correctly determine the load losses in each phase of the transformer, as will be demonstrated in this article. This is because the expressions for the short-circuit resistances of L. Sima et al. are established using both the total load losses included in IEEE Standard C57.110-2018 [10] and the combined RMS values of the primary currents (I_p).

To avoid the errors that these two technologies introduce in the calculation of the load losses of each phase of the three-phase transformers, the expressions for the effective short-circuit resistances of each phase ($R_{cc,z}$) have been developed in the second section of this article (Materials and Methods) referring to the secondary winding of the transformer. These resistances constitute the main novelty of this paper and are unpublished in the technical literature. The expressions of the resistances $R_{cc,z}$ have been established by imposing the condition that they must dissipate the actual load losses of each phase of the transformer according to IEEE Standard C57.110-2018 ($P_{cc,z}$), when each phase's secondary currents (I_{sz}) flow through them. Because IEEE Standard C57.110-2018 [10] does not include the expressions for $P_{cc,z}$, before developing the resistances $R_{cc,z}$, the expressions for the phase load losses ($P_{cc,z}$) have been obtained by adapting the expressions of the total load losses included in this standard [5,10] as another novelty of the article.

Likewise, in the second section of the article, the expressions for the effective short-circuit resistance of the transformer ($R_{cc,ef}$) have been developed, referring to the secondary winding. These short-circuit resistances have been defined following the same procedure used by L. Sima et al. in [36], but using the secondary combined currents (I_s) instead of the primary currents (I_p). That is, $R_{cc,ef}$ are the short-circuit resistances of L. Sima et al. (R_k) referred to as secondary (R_k'). Thus, $R_{cc,ef}$ and R_k have the same properties and disadvantages.

In the third section of the article (Results), the short-circuit resistances referred to as secondary ($R_{cc,z}$ and $R_{cc,ef}$) developed in Section 2, are calculated from the measurements carried out by a Fluke 435 Series II analyzer on the secondary of a 630 kVA, immersed in oil, and three-phase distribution transformer, which supplies a moderately distorted residential area. In the fourth section (Discussion), the effective short-circuit resistances $R_{cc,z}$, and $R_{cc,ef}$ are used to calculate the load losses in each phase ($P_{cc,z}$) and total (P_{cc}) in the analyzed distribution transformer of Section 3. The results demonstrate that both short-circuit resistances determine the same values of total load losses (P_{cc}) as IEEE Standard C57.110-2018; however, only the load losses calculated with $R_{cc,z}$ match the load losses determined in each phase applying the standard. The use of $R_{cc,ef}$ and R_k resistances determine errors in the values of $P_{cc,z}$ greater than 16% for the analyzed distribution transformer. The fifth section summarizes the main conclusions.

2. Materials and Methods

The expressions for the total load losses of three-phase transformers (P_{cc}) with non-linear loads that were obtained in [5] by adaptation of IEEE Standard C57.110-2018 are used in this section to establish the transformer load losses for each phase ($P_{cc,z}$). Based on these losses, the expressions for the effective short-circuit resistances of each phase ($R_{cc,z}$) and the effective short-circuit resistance of the transformer ($R_{cc,ef}$) are developed.

2.1. Load Losses of Three-Phase Transformers Adapted from IEEE Standard C57.110-2018

According to IEEE Standard C57.110-2018 [10], the load losses (P_{cc}) of three-phase transformers are expressed as the sum of three losses:

$$P_{cc} = P_{DC} + P_{EC} + P_{OSL} \tag{7}$$

Each of these losses is due to three different phenomena, which occur because of the circulation of currents through the transformer windings:

- P_{DC} are the power losses by Joule effect that would be produced when direct currents circulate through the transformer windings [37].
- P_{EC} are the eddy current losses due to the skin phenomenon in the conductors of the coils.
- P_{OSL} are the other stray losses that originate in the tank and other metallic parts of the transformer due to electromagnetic induction.

The direct current losses (P_{DC}) are determined, in general, as follows [5]:

$$P_{DC} = \frac{1}{3} \cdot P_{DCN} \sum_{z=A,B,C} \sum_{h_z=1}^{h_{z,max}} \left(\frac{I_{hz}}{I_{sN}} \right)^2 \tag{8}$$

where

- P_{DCN} are the nominal losses in direct current.
- I_{sN} is the rated RMS value of the secondary currents of the transformer.
- I_{hz} is the RMS value of the harmonic of order $h_z = f_{hz}/f_1$ (f_{hz} = harmonic frequency, f_1 = 50–60 Hz is the fundamental frequency) of each phase ($z = A, B, C$) of the secondary currents.
- $h_{z,max}$ is the order of the highest-frequency harmonic used in the calculation.

The losses caused by the skin effect (P_{EC}) are proportional to the square of the RMS values of each phase harmonic current (I_{hz}) and frequency according to the following expression [5]:

$$P_{EC} = \frac{1}{3} \cdot P_{ECN} \sum_{z=A,B,C} \sum_{h_z=1}^{h_{z,max}} h_z^2 \cdot \left(\frac{I_{hz}}{I_{sN}} \right)^2 \tag{9}$$

where I_{sN}, I_{hz}, h_z and $h_{z,max}$ have the meanings previously indicated and P_{ECN} represents the nominal losses due to the skin effect, measured with the nominal secondary currents (I_{sN}), at the fundamental frequency.

The losses produced by the eddy currents induced in the metallic parts of the transformer (P_{OSL}) are obtained as follows [5]:

$$P_{OSL} = \frac{1}{3} \cdot P_{OSLN} \sum_{z=A,B,C} \sum_{h_z=1}^{h_{z,max}} h_z^{0.8} \cdot \left(\frac{I_{hz}}{I_{sN}} \right)^2 \tag{10}$$

where P_{OSLN} is the nominal value of these losses, measured when the transformers operate with the nominal secondary currents (I_{sN}), at the fundamental frequency.

Transformer manufacturers usually provide nominal load losses (P_{DCN}, P_{ECN}, P_{OSLN}). Their sum is equal to the nominal load losses of the transformer ($P_{ccN} = P_{DCN} + P_{ECN} + P_{OSLN}$) included in the manufacturer catalogues. However, sometimes manufacturers do not provide the individual values of P_{ECN} and P_{OSLN}, but rather provide them together ($P_{SLN} = P_{ECN} + P_{OSLN}$). On these occasions, IEEE Standard C57.90™ [38] establishes that $P_{ECN} = 0.33 \cdot P_{SLN}$ and $P_{OSLN} = 0.66 \cdot P_{SLN}$ in oil-immersed transformers, $P_{ECN} = 0.66 \cdot P_{SLN}$ and $P_{OSLN} = 0.33 \cdot P_{SLN}$ in dry-type transformers.

Substituting expressions (8)–(10) in Equation (7), the total load losses (P_{cc}) adapted from IEEE Standard C57.110-2018 [10] for three-phase transformers with unbalanced and non-linear loads can be calculated as [5]:

$$P_{cc} - \sum_{z=A,B,C} P_{cc,z} = \frac{1}{3} \cdot \sum_{z=A,B,C} \sum_{h_z=1}^{h_{z,max}} \left(P_{DCN} + P_{ECN} \cdot h_z^2 + P_{OSN} \cdot h_z^{0.8} \right) \cdot \left(\frac{I_{hz}}{I_{sN}} \right)^2 \tag{11}$$

From the above expression, the load losses of each phase ($z = A, B, C$) can be determined as follows:

$$P_{cc,z} = \frac{1}{3} \cdot \sum_{h_z=1}^{h_{z,max}} \left(P_{DCN} + P_{ECN} \cdot h_z^2 + P_{OSLN} \cdot h_z^{0.8} \right) \cdot \left(\frac{I_{hz}}{I_{sN}} \right)^2 \tag{12}$$

2.2. Short-Circuit Resistances Referred to Secondary of Three-Phase Transformers
2.2.1. Effective Short-Circuit Resistance for Each Phase of the Transformer

The effective short-circuit resistances ($R_{cc,z}$) referred to as the secondary could be defined for each phase ($z = A, B, C$) of the transformer as those that would produce the same load losses in each phase ($P_{cc,z}$), that is:

$$P_{cc,z} = R_{cc,z} \cdot \sum_{h_z=1}^{h_{z,max}} I_{hz}^2 \tag{13}$$

Matching the last equation and (12), it turns out that:

$$R_{cc,z} = R_{DCN} + R_{ECN} \frac{\sum_{h_z=1}^{h_{z,max}} h_z^2 \cdot I_{hz}^2}{\sum_{h_z=1}^{h_{z,max}} I_{hz}^2} + R_{OSLN} \frac{\sum_{h_z=1}^{h_{z,max}} h_z^{0.8} \cdot I_{hz}^2}{\sum_{h_z=1}^{h_{z,max}} I_{hz}^2} \tag{14}$$

being the loss coefficients of each phase ($z = A, B, C$) caused by the eddy currents in the windings (F_{HL}^z) and in other parts of the transformer (F_{HL-STR}^z), respectively,

$$F_{HL}^z = \frac{\sum_{h_z=1}^{h_{z,max}} h_z^2 \cdot I_{hz}^2}{\sum_{h_z=1}^{h_{z,max}} I_{hz}^2} \qquad F_{HL-STR}^z = \frac{\sum_{h_z=1}^{h_{z,max}} h_z^{0.8} \cdot I_{hz}^2}{\sum_{h_z=1}^{h_{z,max}} I_{hz}^2} \tag{15}$$

and R_{DCN}, R_{ECN}, and R_{OLSN} are the nominal short-circuit resistances, which characterize the nominal direct current losses (P_{DCN}), the nominal eddy current losses (P_{ECN}), and the rated other stray losses (P_{OSLN}), according to the following expressions.

$$R_{DCN} = \frac{P_{DCN}}{3 \cdot I_{sN}^2} \quad R_{ECN} = \frac{P_{ECN}}{3 \cdot I_{sN}^2} \quad R_{OSLN} = \frac{P_{OSLN}}{3 \cdot I_{sN}^2} \tag{16}$$

Therefore,

$$R_{cc,z} = R_{DCN} + R_{ECN} \cdot F_{HL}^z + R_{OSLN} \cdot F_{HL-STR}^z = R_{DCN} + R_{EC,z} + R_{OSL,z} \tag{17}$$

where

$$R_{EC,z} = R_{ECN} \cdot F_{HL}^z \qquad R_{OSL,z} = R_{OSLN} \cdot F_{HL-STR}^z \tag{18}$$

are the short-circuit resistances due to the phenomena of the eddy current ($R_{EC,z}$) and other stray losses ($R_{OSL,z}$) in each phase of the transformer. The values depend on the order (h_z) and the RMS values (I_{hz}) of the current harmonics of each secondary phase, as observed in Equation (14). Therefore, the short-circuit resistances $R_{EC,z}$ and $R_{OSL,z}$ are responsible for:

- the increase in the short-circuit resistance of the transformer feeding non-linear loads, and
- the different values of the short-circuit resistances in each phase of the transformer with non-linear loads, not foreseen by L. Sima et al. in Equation (2).

From Equations (14) and (17), it can also be seen that the direct current short-circuit resistance (R_{DCN}) is independent of the frequencies and RMS values of the harmonic currents.

The effective short-circuit resistance of each phase ($R_{cc,z}$) is a physical parameter, since its values accurately determine the load losses originating in each phase of the transformer ($P_{cc,z}$), which can be interchangeably calculated with Equations (12) and (13). These short-circuit resistances make it possible to calculate the total load losses of the three-phase transformers according to the following expression derived from (11).

$$P_{cc} = \sum_{z=A,B,C} P_{cc,z} = \sum_{z=A,B,C} \left(R_{cc,z} \cdot \sum_{h_z=1}^{h_{z,max}} I_{hz}^2 \right) \tag{19}$$

Figure 2 shows the actual operating model of the transformer based on the effective short-circuit resistance of each phase. The elements included in this equivalent model of the three-phase transformer have the following physical meanings.

- $R_{cc,A}$, $R_{cc,B}$ and $R_{cc,C}$ are the effective short-circuit resistances referring to each phase of the secondary, which represents the load losses in each phase of the three-phase transformers ($P_{cc,A}$, $P_{cc,B}$, $P_{cc,C}$) according to IEEE Standard C57.110-2018 [10], caused by the circulation of currents (I_{sA}, I_{sB}, I_{sC}) through each phase of the secondary winding. These resistances usually have different values in each phase when currents are distorted.
- X_{cc} is the short-circuit reactance referred to as the secondary, representing the transformer's scattered magnetic fluxes. This reactance has the same value in the three phases, because it depends only on the harmonic frequencies, not the current RMS values.
- R_a is the resistance that represents the transformer's core losses. This resistance has the same value in each phase because it is not practically affected by the voltage harmonic frequencies.
- X_μ is the magnetic reactance, which represents the main magnetic flux of the transformers and drives its electromotive forces. This reactance usually has the same values in each phase because of the same reasons indicated for X_{cc}.

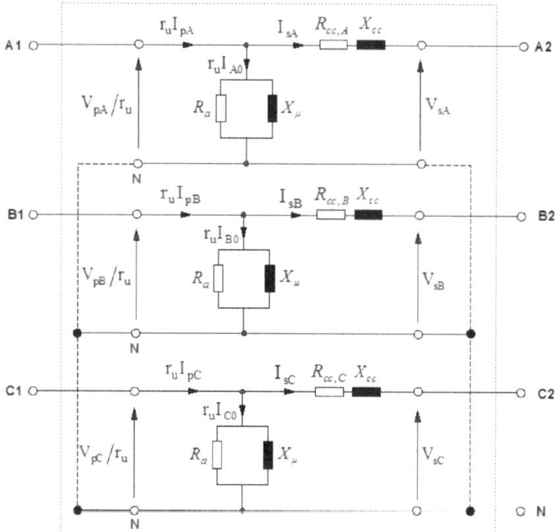

Figure 2. Physical equivalent model of three-phase transformers feeding non-linear loads, being ABC1 the primary and ABC2 the secondary phases.

2.2.2. Effective Short-Circuit Resistance of the Transformer

Similarly to the short-circuit resistances of each phase ($R_{cc,z}$), an effective short-circuit resistance of the transformer ($R_{cc,ef}$) can be defined. The short-circuit resistance $R_{cc,ef}$ determines the total load losses of the transformer according to the following expression:

$$P_{cc} = R_{cc,ef} \cdot \sum_{h=1}^{h_{max}} I_h^2 \tag{20}$$

I_h being the combined RMS value of the harmonic currents of order h_z:

$$I_h = \sqrt{\sum_{z=A,B,C} I_{hz}^2} \tag{21}$$

Matching expressions (11) and (20) and considering that there are usually harmonics of the same frequencies in the three phases of the transformer, the expression for the effective short-circuit resistance of the transformer ($R_{cc,ef}$) is obtained, as follows:

$$R_{cc,ef} = R_{DCN} + R_{ECN} \frac{\sum_{h=1}^{h_{max}} h^2 \cdot I_h^2}{\sum_{h=1}^{h_{max}} I_h^2} + R_{OSN} \frac{\sum_{h=1}^{h_{max}} h^{0.8} \cdot I_h^2}{\sum_{h=1}^{h_{max}} I_h^2} = R_{DCN} + R_{ECN} \cdot F_{HL} + R_{OSN} \cdot F_{HL-STR} \tag{22}$$

where

$$F_{HL} = \frac{\sum_{h=1}^{h_{max}} h^2 \cdot I_h^2}{\sum_{h=1}^{h_{max}} I_h^2} \qquad F_{HL-STR} = \frac{\sum_{h=1}^{h_{max}} h^{0.8} \cdot I_h^2}{\sum_{h=1}^{h_{max}} I_h^2} \tag{23}$$

are the transformer loss factors included in IEEE Standard C57.110-2018.

According to Equation (22), the effective short-circuit resistance of the transformer ($R_{cc,ef}$) usually has the same values in the three phases (Figure 3), because this parameter has been established using the values of the combined currents (I_h). Thus, the circulation of the phase currents through these resistances gives values of the losses in each phase that are different from those calculated with Equations (12) and (13). This result shows that $R_{cc,ef}$ is a merely mathematical parameter and that the three-phase transformer model based on these resistances (Figure 3) is not related to the actual operation of these machines.

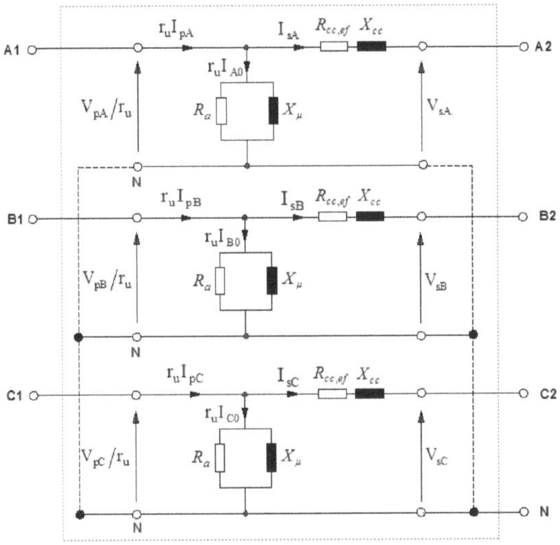

Figure 3. Mathematical equivalent model of three-phase transformers feeding non-linear loads based on $R_{cc,ef}$, being ABC1 the primary and ABC2 the secondary phases.

For the above reasons, the resistances $R_{cc,ef}$ can only be used to calculate the total load losses of the transformers (P_{cc}), according to expression (20).

The effective short-circuit resistance of the transformer ($R_{cc,ef}$) coincides with the short-circuit resistance of L. Sima et al. (R_k) [36], referred to as the secondary (R_k'), i.e.,

$$R_{cc,ef} = \frac{R_k}{r_u^2} = R_k' \tag{24}$$

Therefore, R_k has the same drawbacks found for $R_{cc,ef}$.

3. Results

The expressions for the losses and short-circuit resistances developed in Section 2.2 are used in this section to calculate the values of the load losses and short-circuit resistances of a 630 kVA distribution transformer when feeding to the low-voltage (LV) installations of residential consumers (homes) in a town near the city of Valencia (Spain). The RMS values and harmonic content of the currents absorbed by these installations were recorded by a Fluke 435 Series II analyzer, which was connected to the secondary of the transformer. The measurements were carried out at one-hour intervals over a week between 7 November and 13 November 2022.

The distribution transformer is oil-immersed with Dyn11 connections from the manufacturer Ormazabal, with nominal features summarized in Table 1.

Table 1. Ratings of the distribution transformer from the manufacturer Ormazabal.

POWER (kVA)	P_{DCN} (W)	P_{ECN} (W)	P_{OSN} (W)	Secondary Rated Current (A)	Transformation Ratio (r_u)
630	5900	200	400	866	24,000/420 V

The currents measured by the Fluke analyzer in the secondary phases of the transformer confirm the presence of unbalanced and non-linear loads in the residential installations. The imbalances are deduced from the RMS values of the fundamental-frequency currents recorded in each phase ($z = A, B, C$) throughout the 24 h of 11 November 2022 (Figure 4a). The non-linear loads are denoted by the total harmonic distortion (Figure 4b) of the currents measured in each secondary phase ($THDi\%_z$) by the Fluke analyzer as the quotient between the RMS value of all harmonics without the fundamental and the total RMS value of that phase current.

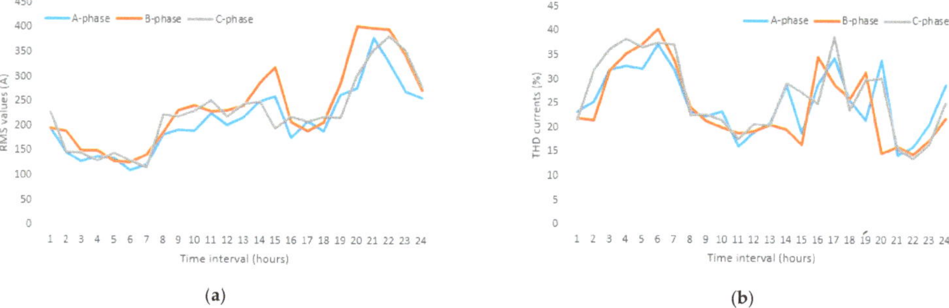

Figure 4. Transformer secondary currents registered on 11 November 2022: (**a**) fundamental-frequency RMS values and (**b**) total harmonic distortion ($THDi\%$).

Tables 2 and 3 summarize the RMS values of the first 25 harmonics, including the fundamental, of the currents recorded in the three secondary phases of the transformer at 6:55 p.m. and 0:55 a.m. on 11 November 2022. Based on the RMS values of the fundamental-frequency currents in the three phases of the secondary (Figure 4a) on 11 November 2022, it is observed that the transformer operates with more significant imbalances at 6:55 p.m. (Table 2) than at 0:55 a.m. (Table 3).

Table 2. RMS values of the harmonic currents at 6:55 PM on 11 November 2022.

Frequency (Hz)	Harmonic Order (h)	Secondary Currents (A)			
		A-Phase	B-Phase	C-Phase	Combined (I_h)
50	1	278.345	403.233	305.503	577.412
100	2	14.32	9.571	13.518	21.895
150	3	4.093	10.877	2.129	11.815
200	4	14.884	3.562	13.036	20.103
250	5	33.541	16.762	6.983	38.141
300	6	16.088	7.711	16.159	24.070
350	7	19.980	18.350	10.392	29.050
400	8	22.621	15.445	23.481	36.078
450	9	20.366	19.781	22.514	36.234
500	10	14.689	11.486	12.990	22.725
550	11	10.203	2.713	5.895	12.091
600	12	10.452	4.591	9.289	14.717
650	13	4.393	7.096	13.158	15.581
700	14	11.194	2.962	12.144	16.780
750	15	14.002	4.085	15.535	21.309

Table 2. Cont.

Frequency (Hz)	Harmonic Order (h)	Secondary Currents (A)			
		A-Phase	B-Phase	C-Phase	Combined (I_h)
800	16	14.919	5.887	12.204	20.153
850	17	33.999	22.709	38.182	55.942
900	18	21.875	11.785	20.646	32.305
950	19	10.553	7.124	10.106	16.255
1000	20	7.195	2.759	8.437	11.426
1050	21	7.697	3.225	8.727	12.075
1100	22	7.760	1.113	9.491	12.310
1150	23	9.949	4.723	9.599	14.609
1200	24	13.225	3.168	12.903	18.746
1250	25	11.977	9.367	9.618	17.991
	TOTAL	289.75	406.50	314.23	589.86

Table 3. RMS values of the harmonic currents at 0:55 AM on 11 November 2022.

Frequency (Hz)	Harmonic Order (h)	Secondary Currents (A)			
		A-Phase	B-Phase	C-Phase	Combined (I_h)
50	1	195.165	195.462	226.837	357.4210
100	2	13.411	16.062	15.774	26.2042
150	3	17.034	12.235	29.030	35.8133
200	4	16.350	17.177	20.707	31.4825
250	5	29.179	21.472	14.281	38.9411
300	6	6.029	14.787	8.218	17.9593
350	7	12.464	14.051	20.427	27.7497
400	8	7.344	6.755	6.903	12.1332
450	9	7.639	7.147	10.946	15.1409
500	10	3.483	6.534	2.550	7.8311
550	11	13.709	5.117	4.123	15.2026
600	12	3.437	4.745	3.060	6.6099
650	13	3.948	4.927	6.909	9.3593
700	14	1.423	2.793	0.139	3.1377
750	15	4.454	4.767	3.878	7.5895
800	16	1.929	1.979	2.086	3.4625
850	17	1.667	1.766	2.033	3.1671
900	18	1.998	1.404	1.454	2.8420
950	19	1.317	1.087	0.936	1.9473
1000	20	1.118	0.910	0.902	1.7004
1050	21	0.937	0.721	0.619	1.3345
1100	22	0.882	0.949	0.541	1.4040
1150	23	1.002	0.819	0.883	1.5666
1200	24	1.134	0.924	0.991	1.7668
1250	25	0.772	0.902	0.633	1.3454
	TOTAL	200.622	200.107	232.255	366.38

In addition, these residential installations were similarly distorted in both analyzed cases, with values of $THDi\%_z$ in each phase ($z = A, B, C$), as follows (Figure 4b): 24.17%, 17.63%, and 25.36%, respectively, at 6:55 p.m. and 27.81%, 24.32%, and 24.18%, respectively, at 0:55 a.m. (Tables 2 and 3).

Table 4 summarizes the values of the loss factors corresponding to the two cases analyzed: (1) consumption at 6:55 PM, with significant current imbalances (Table 2), and (2) consumption at 0:55 AM, with slight current imbalances (Table 3). The loss factors of each phase and the total of the transformer have been calculated according to Equations (15) and (23), respectively.

Table 4. Loss factors of each phase and total corresponding to the cases analyzed.

	A-Phase		B-Phase		C-Phase		Transformer	
	F_{HL}	F_{HL-STR}	F_{HL}	F_{HL-STR}	F_{HL}	F_{HL-STR}	F_{HL}	F_{HL-STR}
Case 1 (Table 2)	14.6768	1.4645	3.4485	1.09	12.9388	1.3815	8.8511	1.2631
Case 2 (Table 3)	3.07619	1.15411	2.70159	1.12849	2.37866	1.11319	2.68413	1.13002

Table 5 shows the values of the short-circuit resistances $R_{cc,z}$ and $R_{cc,ef}$, calculated with Equations (17) and (??), respectively, with the values of the nominal resistances $R_{DCN} = 2.262$ mΩ, $R_{ECN} = 0.088$ mΩ, and $R_{OSLN} = 0.177$ mΩ, obtained from (16), being the nominal secondary current $I_{sN} = 866$ A (Table 1).

Table 5. Transformer short-circuit resistances corresponding to the cases analyzed.

	$R_{cc,A}$ (mΩ)	$R_{cc,B}$ (mΩ)	$R_{cc,C}$ (mΩ)	$R_{cc,ef}$ (mΩ)
Case 1 (Table 2)	4.1874	3.1227	4.1802	3.6337
Case 2 (Table 3)	3.1010	3.0672	3.0317	3.0619

From Tables 4 and 5, obtained for similarly distorted loads as those indicated in Tables 2 and 3, it is verified that:

(1) the values of loss factors and short-circuit resistances increase with the imbalances of the current harmonics, and
(2) the effective short-circuit resistances ($R_{cc,z}$) have values different in each phase ($z = A, B, C$) and are different from the effective short-circuit resistance of the transformer ($R_{cc,ef}$); these differences increase with the imbalances in the harmonics, as noted in case 1.

4. Discussion

In this section, our short-circuit resistances of three-phase transformers have been compared with those developed by L. Sima et al. [36], which is the only known work in the technical literature that develops an expression for the short-circuit resistance of transformers based on IEEE Standard C57.110-2018.

Table 6 summarizes the RMS values of the currents of each secondary phase (I_{sA}, I_{sB}, I_{sC}), as well as the combined RMS value of the currents of the three phases of the secondary (I_s) and the primary (I_p). The RMS values of the currents of each secondary phase (I_{sA}, I_{sB}, I_{sC}) were obtained by the Fluke 435 Series II analyzer and are summarized in Tables 2 and 3 for each of the analyzed cases. The combined RMS value of the three secondary phases (I_s) is also indicated in Tables 2 and 3, and was obtained as follows:

$$I_s = \sqrt{I_{sA}^2 + I_{sB}^2 + I_{sC}^2} \tag{25}$$

Table 6. RMS values, in amperes (A), of transformer secondary and primary currents corresponding to the cases analyzed.

	I_{sA}	I_{sB}	I_{sC}	I_s	I_p
Case 1 (Table 2)	289.75	406.50	314.23	589.86	10.322
Case 2 (Table 3)	200.622	200.107	232.255	366.38	6.411

The combined RMS values of the currents of the primary phases (I_p) indicated in Table 6 were calculated as:

$$I_p \approx \frac{I_s}{r_u} \tag{26}$$

where $r_u \approx V_p / V_{s0}$ is the transformation ratio of the transformer.

It is observed in Table 7 that the total load losses have the same values in each of the analyzed cases (1264.321 W, in case 1, and 411.011, in case 2), either using IEEE Standard C57.110-2018 with Equation (11), or applying Equation (19) with the effective short-circuit resistances of each phase ($R_{cc,A}$, $R_{cc,B}$, $R_{cc,C}$), or using Equation (20), with the effective short-circuit resistance of the transformer ($R_{cc,ef}$).

Table 7. Transformer load losses, in watts (W), using IEEE Standard C57.110 and the effective short-circuit resistances in the two analyzed cases.

	Total Losses Using IEEE Std.C57.110	Total Losses Using $R_{cc,ef}$	Losses Using $R_{cc,z}$			
			A-Phase	B-Phase	C-Phase	Total
Case 1 (Table 2)	1264.321	1264.321	351.558	516.005	396.758	1264.321
Case 2 (Table 3)	411.011	411.011	124.814	122.658	163.539	411.011

The short-circuit resistance referred to the primary R_k developed by L. Sima et al. relates the total load losses defined by IEEE Standard C57.110 with the primary current (I_p), as indicated in Equation (2). Substituting the calculated values of the total load losses according to IEEE Standard C57.110, included in Table 7, and the RMS value of the combined primary currents, indicated in Table 6, the short-circuit resistances of L. Sima et al. referred to the primary (R_k) and the secondary (R'_k) have the values indicated in Table 8.

Table 8. Short-circuit resistances of L. Sima in the cases analyzed.

	Referred to Primary (R_k, Ω)	Referred to Secondary (R'_k, mΩ)
Case 1 (Table 2)	11.865	3.6337
Case 2 (Table 3)	9.998	3.0619

Comparing the values of the short-circuit resistance of L. Sima et al., referred to as the secondary (R'_k), indicated in Table 8, with those of our effective short-circuit resistance of the transformer ($R_{cc,ef}$), summarized in Table 5, it is noted that they are identical. This result shows that our $R_{cc,ef}$ is the L. Sima short-circuit resistance referred to as the secondary of the transformer, as had been advanced in Equation (24).

Table 9 shows the values of the load losses that would be obtained in each phase of the transformer using the effective short-circuit resistance $R_{cc,ef}$ (or the short-circuit resistance of L. Sima—R'_k) instead of the effective short-circuit resistances of each phase ($R_{cc,A}$, $R_{cc,B}$, $R_{cc,C}$).

Table 9. Load losses in watts (W), and relative loss errors calculated in each phase of the transformer with $R_{cc,ef}$ instead of with $R_{cc,A}$, $R_{cc,B}$, $R_{cc,C}$.

	Load Losses (W)				Relative Loss Errors (%)		
	A-Phase	B-Phase	C-Phase	Total	A-Phase	B-Phase	C-Phase
Case 1 (Table 2)	305.073	600.448	358.800	1264.321	13.222	−16.365	9.567
Case 2 (Table 3)	123.239	122.607	165.165	411.011	1.262	0.042	−0.994

Comparing the values of the losses in each phase indicated in Table 7 with those shown in Table 9, essential differences are observed at 6:55 PM of greater than 16%. These high errors in the calculation of the load losses of each phase confirm that the resistances $R_{cc,ef}$ and R_k are parameters not related to the energy phenomena that occur in the phases of the transformer, and thus these resistances should not be used to calculate the load losses of each phase of the transformers.

Specifically, it has been verified in this section that:

(1) The load losses calculated with our short-circuit resistances, referred to as the secondary of the three-phase transformers, developed in Section 2.2, are equal to those resulting from applying the IEEE Standard C57.110 2018.

(2) The short-circuit resistance of L. Sima, referred to as secondary (R_k'), coincides with our effective short-circuit resistance ($R_{cc,ef}$), referred to as the primary of the transformer.

(3) In general, the effective short-circuit resistance of the transformer ($R_{cc,ef}$) and therefore the resistance of L. Sima et al. cannot be used to calculate the load losses of each phase.

5. Conclusions

Efficiency, warm-up, and power transmission capacity, among other quantities that determine the proper steady-state operation of electrical transformers, depend on the values of load losses. In three-phase transformers, load losses can be calculated either by applying IEEE Standard C57.110 or by using short-circuit resistances derived from that standard. The last procedure has been used by L. Sima et al. with the development of their short-circuit resistance, referred to as the primary (R_k).

In this article, two types of short-circuit resistance referring to the secondary have been developed, deduced from IEEE Standard C57.110: (1) the effective short-circuit resistance of each phase ($R_{cc,z}$) and (2) the effective short-circuit resistance of the transformer ($R_{cc,ef}$). The total load losses of three-phase transformers (P_{cc}) can be calculated by any of these resistances, which have their own properties and applications.

In our opinion:

- The effective short-circuit resistances of the transformer ($R_{cc,ef}$) and therefore the short-circuit resistances of L. Sima et al. (R_k) are mathematical parameters unrelated to the energy phenomena of the transformer. The use of these resistances gives rise to errors in the calculation of the load losses in the transformer phases ($P_{cc,z}$), which increase with the harmonic imbalances. This fact has been verified in the operation of the transformer of an actual residential distribution network feeding two very differently unbalanced loads, both with the same $THDi\% \approx 25\%$. We have verified that if the loads are slightly unbalanced, the errors in the calculation of $P_{cc,z}$ barely exceed 1% in some phases, while with moderately unbalanced loads, the errors exceed 16% (Table 9).

- Based on the above, the effective short-circuit resistances of the transformer ($R_{cc,ef}$) can only be used to calculate the total load losses of three-phase transformers according to IEEE Standard C57.110, but their use is not suitable for monitoring the operation of three-phase transformers.

- The effective short-circuit resistances of each phase ($R_{cc,z}$) can be used to monitor the operating status of three-phase transformers. Both resistances are related to the energy phenomena that manifest in the transformer, since with them, the load losses of each phase ($P_{cc,z}$) and total (P_{cc}) of the transformer can be accurately calculated.
- The effective short-circuit resistances of each phase ($R_{cc,z}$) define the accurate operating model of three-phase transformers, represented in Figure 2.

6. Patents

This article has been based on our patent P202330968—"Use of short-circuit resistors, procedure and device for monitoring the operating state of a three-phase transformer in service"—lodged with the Spanish Patent and Trademark Agency.

Author Contributions: Conceptualization, V.L.-M., E.P.-L. and J.Á.S.-J.; methodology, V.L.-M. and E.P.-L.; software, V.L.-M. and C.A.-M.; validation, V.L.-M., E.P.-L. and J.Á.S.-J.; formal analysis, V.L.-M., E.P.-L. and C.A.-M.; investigation, V.L.-M., E.P.-L., C.A.-M. and J.Á.S.-J.; resources, V.L.-M., E.P.-L. and C.A.-M.; data curation, V.L.-M.; writing—original draft preparation, V.L.-M., E.P.-L., and J.Á.S.-J.; writing—review and editing, V.L.-M., E.P.-L. and C.A.-M.; visualization, V.L.-M., C.A.-M. and J.Á.S.-J.; supervision, V.L.-M. and J.Á.S.-J.; funding acquisition, E.P.-L. All authors have read and agreed to the published version of the manuscript.

Funding: This research and APC was funded by the Generalitat Valenciana within the ValREM Project (CIAICO/2022/007).

Data Availability Statement: Data are contained within the article.

Conflicts of Interest: The authors declare no conflict of interest. The funders had no role in the design of the study; in the collection, analyses, or interpretation of data; in the writing of the manuscript; or in the decision to publish the results.

References

1. Chorshanbiev, S.R.; Shvedov, G.V.; Sultan, H.M.; Nazirov, K.B.; Ismoilov, F.O. Structural Analysis of Power Losses in (6–10/0.4 kV) Urban Distribution Electric Networks of the City of Dushanbe, the Republic of Tajikistan. In Proceedings of the IEEE Conference of Russian Young Researchers in Electrical and Electronic Engineering (EIConRus), Saint Petersburg and Moscow, Russia, 28–31 January 2019. [CrossRef]
2. U4E. *Accelerating the Global Adoption of Energy-Efficient Transformers*; UN Environment, Economy Division, Energy & Climate Branch: Paris, France, 2017.
3. León-Martínez, V.; Andrada-Monrós, C.; Molina-Cañamero, L.; Cano-Martínez, J.; Peñalvo-López, E. Decarbonization of Distribution Transformers Based on Current Reduction: Economic and Environmental Impacts. *Energies* **2021**, *21*, 7207. [CrossRef]
4. León-Martínez, V.; Peñalvo-López, E.; Andrada-Monrós, C.; Cano-Martínez, J.; León-Vinet, A.; Molina-Cañamero, L. Procedure for Improving the Energy, Environmental and Economic Sustainability of Transformation Houses. *Appl. Sci.* **2022**, *9*, 4204. [CrossRef]
5. León-Martínez, V.; Peñalvo-López, E.; Montañana-Romeu, J.; Andrada-Monrós, C.; Molina-Cañamero, L. Assessment of Load Losses Caused by Harmonic Currents in Distribution Transformers Using the Transformer Loss Calculator Software. *Environments* **2023**, *10*, 177. [CrossRef]
6. Mikhak-Beyranvand, M.; Faiz, J.; Rezaeealam, B. Thermal analysis and derating of a power transformer with harmonic loads. *IET Gener. Transm. Distrib.* **2020**, *14*, 1233–1241. [CrossRef]
7. Gouda, O.E.; Amer, G.M.; Salem, W.A.A. Predicting transformer temperature rise and loss of life in the presence of harmonic load currents. *Ain Shams Eng. J.* **2012**, *3*, 113–121. [CrossRef]
8. Borodin, M.; Psarev, A.; Kudinova, T.; Mukhametzhanov, R. Improving power quality by calculating voltage losses. Energy Systems and Complexes. In Proceedings of the International Scientific and Technical Conference Smart Energy Systems 2019 (SES-2019), Kazan, Russia, 18–20 September 2019. [CrossRef]
9. Biryulin, V.I.; Gorlov, A.N.; Larin, O.M.; Kudelina, D.V. Calculation of power losses in the transformer substation. In Proceedings of the 2016 13th International Scientific-Technical Conference on Actual Problems of Electronics Instrument Engineering (APEIE), Novosibirsk, Russia, 3–6 October 2016. [CrossRef]
10. *C57.110-2018*; IEEE Recommended Practice for Establishing Liquid Immersed and Dry-Type Power and Distribution. Transformer Capability When Supplying Nonsinusoidal Load Currents. Transformers Committee of the IEEE Power and Energy Society: New York, NY, USA, 2018. [CrossRef]
11. Taher, M.A.; Kamel, S.; Ali, Z.M. K-Factor and transformer losses calculations under harmonics. In Proceedings of the 2016 Eighteenth International Middle East Power Systems Conference (MEPCON), Cairo, Egypt, 27–29 December 2016. [CrossRef]

12. Contreras-Ramírez, D.; Lata-García, J. K-Factor Analysis to Increase the Actual Capacity of Electrical Distribution Transformers. In *Communication, Smart Technologies and Innovation for Society*; Springer: Berlin/Heidelberg, Germany, 2021; pp. 367–379. [CrossRef]
13. Megahed, T.F.; Kotb, M.F. Improved design of LED lamp circuit to enhance distribution transformer capability based on a comparative study of various standards. *Elsevier Energy Rep.* **2022**, *8*, 445–465. [CrossRef]
14. *BSI-BS 7821-4*; Three Phase Oil-Three Phase Oil-Immersed Distribution Transformers. 50 Hz from 50 to 2500 kVA with Highest Voltage for Equipment Not Exceeding 36 kV Part 4: Determination of the Power Rating of a Transformer Loaded with Non-Sinusoidal Currents. BSI: London, UK, 1995. Available online: https://standards.globalspec.com/std/33779/BS%207821-4 (accessed on 30 October 2023).
15. *AS 2374.1.-2003*; Power Transformers Part 1: General. Standards Australia: Sydney, Australia, 2003. Available online: https://www.saiglobal.com/PDFTemp/Previews/OSH/as/as2000/2300/2374.1.2-2003(+A1).pdf (accessed on 30 October 2023).
16. *AS 2374.2*; Dry-Type Power Transformers. Standards Australia: Sydney, Australia, 2003. Available online: https://infostore.saiglobal.com/en-gb/standards (accessed on 30 October 2023).
17. *IEC Standard 60076–18, Ed. 1.0*; Measurement of Frequency Response. IEC: Geneva, Switzerland, 2012. Available online: http://atecco.ir/fa/wp-content/uploads/2021/07/IEC60076-part18-edition2012-FRA-measurements.pdf (accessed on 30 October 2023).
18. *IEEE Std C57.149–2012*; IEEE Guide for the Application and Interpretation of Frequency Response Analysis for Oil-Immersed Transformers. IEEE: New York, NY, USA, 2013; pp. 1–72. Available online: https://pdfcoffee.com/ieee-std-c57-149-ieee-guide-for-the-application-and-interpretation-of-frequency-response-analysis-for-oil-immersed-transformerspdf-pdf-free.html (accessed on 30 October 2023).
19. *ANSI/UL Std. 1561 Ed. 4*; Standard for Dry-Type General Purpose and Power Transformers. UL Standard: Northbrook, IL, USA, 2011; ANSI Approved 13 May 2019. Available online: https://webstore.ansi.org/standards/ul/ul1561ed2011 (accessed on 30 October 2023).
20. *ANSI/UL Std. 1562 Ed. 4*; Transformers, Distribution, Dry-Type, Over 600 V. UL Standard: Northbrook, IL, USA, 25 January 2013; Revised 11 August 2020. Available online: https://webstore.ansi.org/standards/ul/ul1562ed2013 (accessed on 30 October 2023).
21. Chen, H.; Huang, Y.; Hu, H.; Wang, J.; Wang, W. Analysis of the influence of voltage harmonics on the maximum load capacity of the power supply transformer for the LHCD system. In Proceedings of the IEEE 6th International Electrical and Energy Conference (CIEEC), Hefei, China, 12–14 May 2023. [CrossRef]
22. Hao, M.; Xin, Z.; Chi, Z.; Xiaoguang, M.; Jufang, W.; Minghui, D. A 110kV transformer breakdown resulting from insufficient short circuit resistance. In Proceedings of the 2021 China International Conference on Electricity Distribution (CICED), Shanghai, China, 7–9 April 2021. [CrossRef]
23. Chen, Q.; Wang, S.; Lin, D.; Wang, S.; Wang, S.; Yuan, D.; Li, H. Analysis of mechanical characteristics of transformer windings under short circuit fault. In Proceedings of the 2018 12th International Conference on the Properties and Applications of Dielectric Materials (ICPADM), Xi'an, China, 20–24 May 2018. [CrossRef]
24. Gao, S.; Sun, L.; Tian, Y.; Liu, H. Research on the Distribution Characteristics of Transformer Axial Vibration under Short-Circuit Conditions Considering Damping Parameters. *Appl. Sci.* **2022**, *12*, 8443. [CrossRef]
25. Ding, H.; Zhao, W.; Diao, C.; Li, M. Electromagnetic Vibration Characteristics of Inter-Turn Short Circuits in High Frequency Transformer. *Electronics* **2023**, *12*, 1884. [CrossRef]
26. Xian, R.; Wang, L.; Zhang, B.; Li, J.; Xian, R.; Li, J. Identification Method of Interturn Short Circuit Fault for Distribution Transformer Based on Power Loss Variation. *IEEE Trans. Ind. Inform.* **2023**, 1–11. [CrossRef]
27. Bu, L.; Han, S.; Feng, J. Short-Circuit Fault Analysis of the Sen Transformer Using Phase Coordinate Model. *Energies* **2021**, *14*, 5638. [CrossRef]
28. Cheema, M.A.M.; Fletcher, J.E.; Dorrell, D.; Junaid, M. A Novel Approach to Investigate the Quantitative Impact of Harmonic Currents on Winding Losses and Short Circuit Forces in a Furnace Transformer. *IEEE Trans. Magn.* **2013**, *49*, 2025–2028. [CrossRef]
29. Dawood, K.; Alboyaci, B.; Cinar, M.C. The impact of short-circuit electromagnetic forces in a 12-pulse converter transformer. In Proceedings of the 2017 10th International Conference on Electrical and Electronics Engineering (ELECO), Bursa, Turkey, 30 November–2 December 2017; Available online: https://ieeexplore.ieee.org/abstract/document/8266210 (accessed on 30 October 2023).
30. Wei, J.; Zhang, C.; Yao, C.; Li, S.; Duan, M.; Li, L.; Ma, X.; Ma, H. The cause analysis and preventive measures of transformer short circuit fault. In Proceedings of the 9th International Forum on Electrical Engineering and Automation (IFEEA), Zhuhai, China, 4–6 November 2022. [CrossRef]
31. Ye, Z.; Yu, W.; Gou, J.; Tan, K.; Zeng, W.; An, B.; Li, Y. A Calculation Method to Adjust the Short-Circuit Impedance of a Transformer. *IEEE Access* **2020**, *8*, 223848–223858. [CrossRef]
32. Zhihua, P.; Hongfa, Z.; Mingjian, T. Three-dimensional Leakage Magnetic Field Simulation and Short-circuit Impedance Calculation of Large Yoke Transformer. In Proceedings of the International Conference on Intelligent Computing, Automation and Systems (ICICAS), Chongqing, China, 29–31 December 2021. [CrossRef]
33. Song, H.; Gu, K.; Zheng, X. Simulation test of transformer short-circuit impedance based on equivalent model at different frequencies. *J. Vibroengineering* **2022**, *24*, 1174–1187. [CrossRef]
34. Zhao, R.; Gu, J.; Wang, C.; Wang, Y. Online Identification of High-Frequency Transformer Short-Circuit Parameters Based on Instantaneous Phasor Method. *IEEE J. Emerg. Sel. Top. Power Electron.* **2022**, *10*, 3677–3684. [CrossRef]

35. Han, M.; Zhao, R.; Zhang, Y.; Zhang, Y. On-line identification of power transformer short-circuit parameters based on instantaneous phasor. *IET Electr. Power Appl.* **2023**, *17*, 1101–1110. [CrossRef]
36. Sima, L.; Miteva, N.; Dagan, K.J. A novel approach to power loss calculation for power transformers supplying nonlinear loads. *Elsevier Electr. Power Syst. Res.* **2023**, *223*, 109582. [CrossRef]
37. Pires-Corrêa, H.; Teles-Vieira, F.H. An Approach to Steady-State Power Transformer Modeling Considering Direct Current Resistance Test Measurements. *Sensors* **2021**, *21*, 6284. [CrossRef] [PubMed]
38. *C57.12.90™-2006*; IEEE Standards Interpretation for IEEE Std. C57.12.90™-2006 IEEE Standard Test Code for Liquid-Immersed Distribution, Power, and Regulating Transformers. The Institute of Electrical and Electronics Engineers: New York, NY, USA, 2009. Available online: https://standards.ieee.org/wp-content/uploads/import/documents/interpretations/C57.12.90-2006 _interp.pdf (accessed on 30 October 2023).

Article

Optimal Dispatch Strategy for a Distribution Network Containing High-Density Photovoltaic Power Generation and Energy Storage under Multiple Scenarios

Langbo Hou [1], Heng Chen [1,*], Jinjun Wang [1], Shichao Qiao [1], Gang Xu [1], Honggang Chen [1] and Tao Liu [2]

[1] School of Energy Power and Mechanical Engineering, North China Electric Power University, Beijing 102206, China; houlangbo1@163.com (L.H.); wangjj160116@163.com (J.W.); gregory_qiao@outlook.com (S.Q.); xgncepu@163.com (G.X.); hgchen@ncepu.edu.cn (H.C.)

[2] Beijing Guo Dian Tong Network Technology Co., Ltd., Beijing 100086, China; 15210926258@163.com

* Correspondence: heng@ncepu.edu.cn; Tel.: +86-183-4017-0409

Abstract: To better consume high-density photovoltaics, in this article, the application of energy storage devices in the distribution network not only realizes the peak shaving and valley filling of the electricity load but also relieves the pressure on the grid voltage generated by the distributed photovoltaic access. At the same time, photovoltaic power generation and energy storage cooperate and have an impact on the tidal distribution of the distribution network. Since photovoltaic output has uncertainty, the maximum photovoltaic output in each scenario is determined by the clustering algorithm, while the storage scheduling strategy is reasonably selected so the distribution network operates efficiently and stably. The tidal optimization of the distribution network is carried out with the objectives of minimizing network losses and voltage deviations, two objectives that are assigned comprehensive weights, and the optimization model is constructed by using a particle swarm algorithm to derive the optimal dispatching strategy of the distribution network with the cooperation of photovoltaic and energy storage. Finally, a model with 30 buses is simulated and the system is optimally dispatched under multiple scenarios to demonstrate the necessity of conducting coordinated optimal dispatch of photovoltaics and energy storage.

Keywords: distribution network optimization; high-density photovoltaic; energy storage; multi-target; multi-scenario; improved particle swarm algorithm

Citation: Hou, L.; Chen, H.; Wang, J.; Qiao, S.; Xu, G.; Chen, H.; Liu, T. Optimal Dispatch Strategy for a Distribution Network Containing High-Density Photovoltaic Power Generation and Energy Storage under Multiple Scenarios. *Inventions* **2023**, *8*, 130. https://doi.org/10.3390/inventions8050130

Academic Editor: Om P. Malik

Received: 14 September 2023
Revised: 15 October 2023
Accepted: 17 October 2023
Published: 19 October 2023

1. Introduction

The rapid development of photovoltaic (PV) power generation provides a clean and efficient solution for the use of energy, and its use as a renewable energy source has great significance for the sustainable development of the energy industry. In terms of environmental pollution, the use of photovoltaic resources can reduce the burning of fossil fuels and alleviate some of the pollution caused by fossil fuel power generation. To build a better photovoltaic power grid system, the use of distributed PV consumption is an effective way to utilize high-density PVs. However, as the number of distributed power sources increases, the control scheme needs to change significantly, and more complex coordination and interaction between controllers is required. Recently, new challenges and opportunities for voltage control in transmission and distribution grids were reviewed, and layered voltage control was performed for high-penetration distributed power sources [1]. The PV consumption problem was solved in [2], in which the authors constructed a grid-connected PV storage system and proposed a coordinated control strategy. Now that the use of renewable energy is becoming more and more popular, in research on energy use in transportation and electricity, the proportion of new energy used in various energy use situations and the efficiency of energy use have become key research focuses [3–6].

For a grid operation strategy containing PVs and energy storage, it is necessary to determine the output characteristics of PVs and the charging/discharging characteristics of energy storage. By modeling the distribution network structure and circuit configuration, and controlling and managing the power side, the grid can avoid large transient voltage fluctuations and load collapse. At the same time, this approach maximizes the use of PV power generation in the grid-connected state, coordinates the source storage in the system, and realizes the distributed integrated control of PVs and energy storage systems under the microgrid [7]. Energy storage in solar-containing distribution grids has also demonstrated a unique economic value, while research has progressed further in both siting and sizing and dispatch optimization [8–10]. This paper differs from these studies by focusing more on mobilizing the grid as a whole, providing good conditions for the use of distributed power sources and energy storage, and realizing the reactive power optimization of the system according to the change in power output on the power side and the adjustment of the transformer.

Based on the demand for active power control in grid operation, a decentralized active power control system containing distributed energy storage and distributed PVs with hierarchical control is proposed [11]. The system decentralizes the regulation of power and performance among components to keep the total system frequency as a continuous and smooth signal, which avoids a sudden change in grid frequency caused by the access to renewable energy sources. A grid with access to a large number of renewable energy sources can also be operated alone during peak hours and blackout periods, and the PV arrays are controlled hierarchically to provide power, which sets the grid supply power constraints and improves the penetration of PVs into the grid [12]. By using the alternating direction method of multipliers (ADMM) within a model predictive control (MPC) framework, a shift from centralized to decentralized control is achieved [13].

For the wide application of distributed power within the distribution network, it is necessary to solve the problems of supply demand balance and peak loads therein, as well as to optimize the distribution of tidal currents within the distribution network, and to form a corresponding demand response scheme. Considering both unbalanced network constraints and reactive power limitations, an optimal tidal optimization of distribution networks with reactive power control is proposed [14]. Then, from the perspective of a hybrid distribution transformer, the remote coordinated control of PV arrays is carried out through reactive power optimization to ensure voltage regulation and network loss reduction [15]. Research on optimization of inverter-based PV integration has also attracted a lot of attention, and topology optimization through the PV location capacity and the design of the inverter also helps to improve the reliability and stability of the system [16–18]. The application of energy storage systems can alleviate some of the scheduling challenges brought about by renewable energy access and contribute to improving the power quality of the distribution network, among others. In some studies, the advantages of energy storage systems in the optimized scheduling of distribution grid operation and the corresponding scheduling methods have been investigated [19–22]. There has also been a significant breakthrough in the progress of research on the application methods of energy storage technology in the power grid, which presents the characteristics of large-scale integration and multi-objective co-regulation [23–26]. Although a lot of research work has been conducted on energy storage technology and power system trend optimization, there is still a lot of in-depth research that needs to be carried out to coordinate and control the active output of each distributed power source and reduce the energy loss under the premise of guaranteeing the quality of power system operation.

Due to the volatility and uncertainty of renewable energy output, the output characteristics of renewable energy cannot be accurately reflected in the practical application of optimal scheduling methods, and typical scenarios need to be divided to determine its maximum output. Coordinated optimal scheduling of energy systems under PV uncertainty can improve the economy and security of grid operation and realize real-time energy utilization [27]. A large number of studies have optimized the PV output scheduling of systems with uncertainty by dividing different scenarios to achieve the coordination of multiple distributed power sources, which makes the energy supply model more efficient and improves the economy and stability of the system [28–30]. The method of PV clustering is also reflected in many studies, and an equivalent computational model can be obtained by analyzing the clustering of high-density distributed PVs connected to the distribution grid [31]. Studies have also presented the clustering method in detail and have demonstrated how it works in simulation models [32–36]. According to different control strategies, different system models have been developed for full utilization of energy in different scenarios [37]. The scenario division in this paper, on the other hand, is based on the local climate and environment, and the light radiation intensity is clustered to reflect the characteristics of PV output under different light conditions.

This paper starts from the status quo of accessing high-density distributed photovoltaics in the power grid, seeking to solve the related problems created by PV uncertainty, dividing different light radiation intensity scenarios, accessing suitable-capacity energy storage devices in the system, and inducing charging and discharging of energy storage devices according to the fluctuation in electricity prices. Then, we establish an objective function of network loss and voltage deviation, and we carry out rational scheduling for the photovoltaic storage system in the distribution grid under different scenarios, so that each bus accesses distributed photovoltaics as much as possible under the premise of stabilizing the voltage, as well as reduces the network loss.

The results of this study show that the optimally dispatched system containing a high density of PV power generation and energy storage devices can effectively reduce energy losses, and we demonstrate that the system maintains good power quality even after a large amount of PV power is connected. Section 1 of the manuscript describes the need to develop a new type of power system with multiple distributed power sources, and Section 2 presents a model for connecting PV power generation and energy storage devices to the grid, as well as a methodology for clustering and optimizing their data. Section 3 presents the results obtained for the optimization of the system under the two optimization objectives of network losses and voltage deviation values, and finally, Section 4 describes the main conclusions and future work for this research.

2. Materials and Methods

Enabling high-density distributed PV access to the distribution network requires considering not only the PV consumption brought about by the voltage limit but also considering its coordination with the distributed power supply scheduling, to reduce the network losses during system operation. At the same time, the high proportion of distributed PV power output is characterized by strong randomness and fluctuation, which challenge the safety of grid operation, and the coordinated and optimal scheduling of PVs and energy storage in the distribution grid needs to be achieved in different scenarios. Distributed photovoltaic access to the grid requires a series of conversion processes. Photovoltaic power-generation devices need to convert sunlight into electrical energy, which is controlled by the inverter to form the power that can be used for the network, with the ability to output external voltage power. An energy storage device also has the characteristics of charging and discharging and is connected to the distribution network as a distributed power source together with PV power generation. In this way, a distribution grid power system with high-density photovoltaics and energy storage devices is formed (Figure 1).

To solve the optimization problem of a distribution network with high-density photovoltaics and energy storage, the following methods are applied in this paper, among which the application of a clustering algorithm solves the problem of inaccurate operational data and improves the solving efficiency and data accuracy of the algorithm; the application of a comprehensive evaluation method solves the assignment problem in the multi-objective decision process and makes a scientific selection and reasonable decision in the face of multiple conflicting objectives; and the improved particle swarm optimization algorithm searches for optimization according to the optimization objectives, while having a faster convergence speed and avoiding falling into the local optimal solution too early.

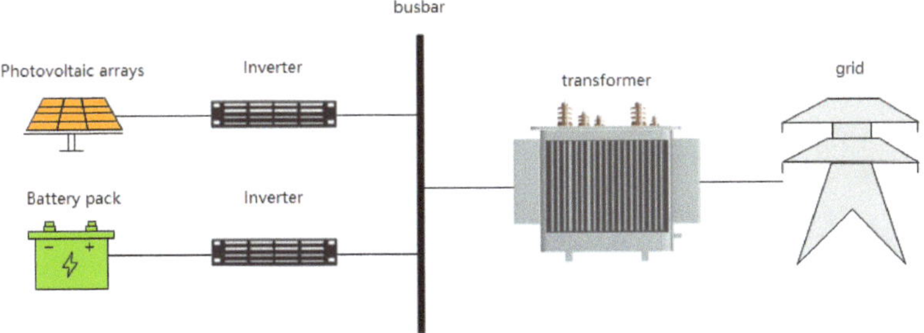

Figure 1. Distribution grid power supply model with high-density photovoltaic and energy storage batteries.

2.1. PV Output Modeling

High-density photovoltaic access to the distribution network requires the solution of two problems in terms of utilization, namely how to convert a large amount of solar energy into electrical energy and how to make this converted electrical energy available to the distribution network. For photovoltaic power generation, the photovoltaic model used to generate electricity is generally in the form of photovoltaic cells. When sunlight hits the surface of the cells, the carriers are subjected to the action of the P-N junction in the interior, and a closed circuit is formed in the exterior to generate a current. When the relevant parameters and inputs of the PV cell are given, the output characteristics of the PV cell can be obtained. However, this output method is not sufficient for the grid. When several PV cells are connected in series and parallel to form a PV array, they can output voltage and power. This electrical energy online needs to be controlled by an inverter, which can effectively control the transient current in the process of grid connection and regulate the voltage and current to guarantee the stability of the system operation.

After coordinated control of all aspects, PV power generation is equivalent to a distributed power supply for the load, while the structure of the distribution network is a radial power supply. The power supply mode is changed from the original single power supply to multiple distributed power supplies, which increases the reliability (Figure 2).

2.2. Energy Storage Modeling

The ESS energy storage system has both charging and discharging characteristics, charging when the electricity supply is sufficient and discharging when the electricity supply is insufficient and the price is high, to realize the peak regulation of the power grid.

It takes 24 h a day as an operation cycle to ensure that the energy storage battery is in the lowest charge state at the initial moment and can recover this state after one operation cycle. The 0/1 constraint is added to the control charging and discharging strategy to ensure that the charging and discharging of the energy storage battery will not be carried out simultaneously. Meanwhile, the self-discharge rate of ESS represents the coefficient of ESS power loss after a period of time. After each period of charging and discharging the battery, the power of ESS changes accordingly, which is expressed as SOC_{ESS}, and the battery state is expressed in periods as

$$SOC_{ESS,t} = (1 - \sigma_{ESS,t-1}) + (P_{ESS,t}^{charge}\,\eta_{ESS}^{charge}\,\Delta t - P_{ESS,t}^{discharge}\,\eta_{ESS}^{discharge}\,\Delta t)/C_{ESS} \tag{1}$$

where $P_{ESS,t}^{charge}$ and $P_{ESS,t}^{discharge}$ are the charging and discharging power of *ESS* at time t, η_{ESS}^{charge} and $\eta_{ESS}^{discharge}$ are the charging and discharging efficiency of *ESS*, and C_{ESS} is the battery capacity of the energy storage system.

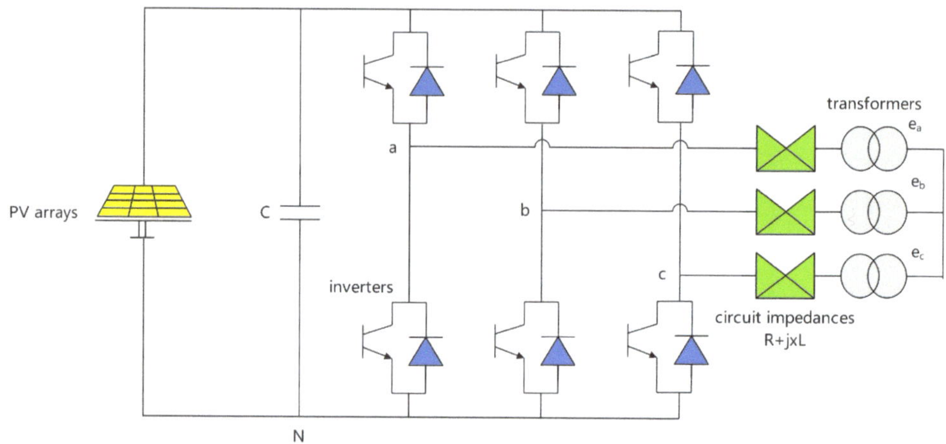

Figure 2. Distributed PV array on-grid control model.

2.3. Distributed PVs and Energy Storage Connected to the Distribution Network Modeling

An example analysis was carried out using the IEEE 30-bus test distribution system, which has a base capacity of SB = 100 MVA and a base voltage of VB = 135 KV. In the IEEE 30-bus distribution grid, distributed PVs are accessed at buses 2, 5, 8, 11, and 13, and energy storage ESS devices are accessed at buses 22 and 27. The bus types can be categorized into three: PQ buses, PV buses, and balanced buses. The voltage of balanced buses is 1.0 pu. The buses are analyzed where the buses accessing distributed PVs with energy storage devices can be classified as PQ buses. Bus 1, which is the balancing bus, is selected as the reference bus for trend calculation. There is one and only one balancing bus in the system, and in this system, bus 1 is connected to the higher grid for interaction. The rest of the buses are set as PQ buses and PV buses depending on the conditions under which the data measurements are obtained. The bus network of the 30-bus system is shown (Figure 3).

The distributed PV access to the distribution network for trend calculation, for the distribution network over a day of PVs and energy storage system coordination and optimization, can allow the PV output and storage charging and discharging to adapt to the time-sharing tariff step change, thus supporting different operating strategies. Influenced by the peak and valley periods of electricity prices, the energy storage system starts charging to accumulate power in a valley, and then it discharges to release power at a peak. At the same time, the application of high-density photovoltaics eases the problem of tight power supply during peak hours, and the application of multiple distributed power sources makes the supply of electric loads more secure.

In this system, transformer control is also an important factor in achieving reactive power optimization and voltage regulation. By controlling the gear changes of the tap's five gears, the system can be made to absorb energy, to avoid over-voltage during peak PV output, and to output energy, to avoid under-voltage at night. Therefore, when coordinating the power output of each distributed power source of the system, the action of the transformer taps should also be within the range of optimized dispatch, to achieve voltage regulation. The action of transformer taps in the system is determined according to the distribution of power tides, and when the load and distributed power generation change, guided by the optimization goal, the transformer taps will choose the appropriate gear to coordinate the control of photovoltaic power generation and energy storage charging and discharging.

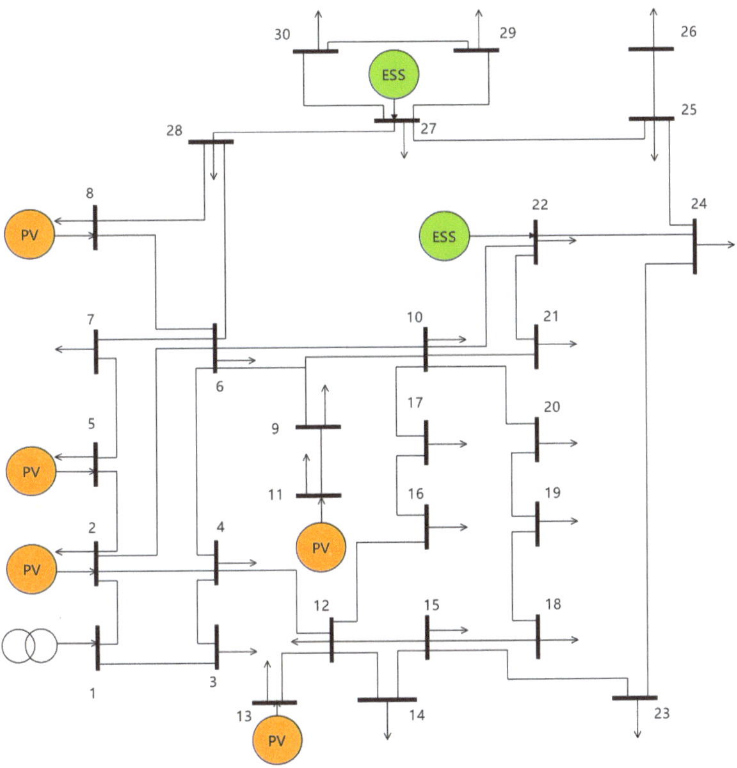

Figure 3. Arithmetic simulation for 30 buses.

2.4. Clustering Algorithm

The difficulty of PV output prediction lies in the uncertainty and uncontrollability of the power it emits. By organizing the historical data, using the clustering algorithm to filter more reliable data can provide help for system scheduling and real-time operation, and it can reduce the impact of distributed PV access on the grid. Therefore, the K-means clustering algorithm is used to categorize the historical data and gather them into K clusters according to their similarity. The process of using the K-means clustering algorithm to process the data is as follows:

Step 1: Select appropriate sample eigenvalues and normalize five statistical indicators, namely standard deviation, skewness coefficient, coefficient of variation, peaking coefficient, and total power, as the eigenvalues of the system. The formulas for the five indicators are as follows:

$$\sigma = \sqrt{\frac{\sum\limits_{i=1}^{N}(P_i - P_{avg})^2}{N}} \tag{2}$$

$$s = \frac{N \sum\limits_{i=1}^{N} (P_i - P_{avg})^2}{\sigma(N-1)(N-2)} \tag{3}$$

$$c = \frac{\sigma}{P_{avg}} \tag{4}$$

$$k_u = \frac{\sum\limits_{i=1}^{N} (P_i - P_{avg})^4}{\sigma(N-1)} \tag{5}$$

$$P_{sum} = \sum\limits_{i=1}^{N} P_i \tag{6}$$

where N is the number of sampling points, P_{avg} is the unit average PV power, and P_{sum} is the instantaneous power.

Step 2: Normalization of the indicators is calculated as

$$x_1 = \frac{x - x_{min}}{x_{max} - x_{min}} \tag{7}$$

Step 3: A sample is chosen at random as the first center of mass, denoted C_1.

Step 4: The shortest distance $D(x)$ between each sample and the center of mass C1 is computed, and the next center of mass is selected based on the result obtained from the probability $p(x)$ that each sample is selected as the center of mass. The probability of being selected as a center of mass is calculated as

$$p(x) = \frac{D^2(x)}{\sum\limits_{x \in X} D^2(x)} \tag{8}$$

Step 5: Repeat the previous step until K clustering centers are selected.

Step 6: Based on the distance of each sample from each center of mass, assign each sample to its nearest center of mass to form the corresponding cluster.

Step 7: Update the center of gravity of each cluster.

$$C_i = \frac{\sum\limits_{x \in C_i} x}{|C_i|} \tag{9}$$

Step 8: Repeatedly update each cluster with the center of mass until no change occurs.

Figure 4 shows the flow of the K-means clustering-based optimal scheduling method for high-density PV resources studied in this paper.

Take the data of light radiation intensity in Beijing, for example. Through K-means clustering, all the data are categorized into five typical scenarios. The final clustering center is determined through continuous iteration, and the distribution of light radiation intensity is obtained under different scenarios. We followed these steps, and the results are shown (Figure 5). All areas of similar climate can be included.

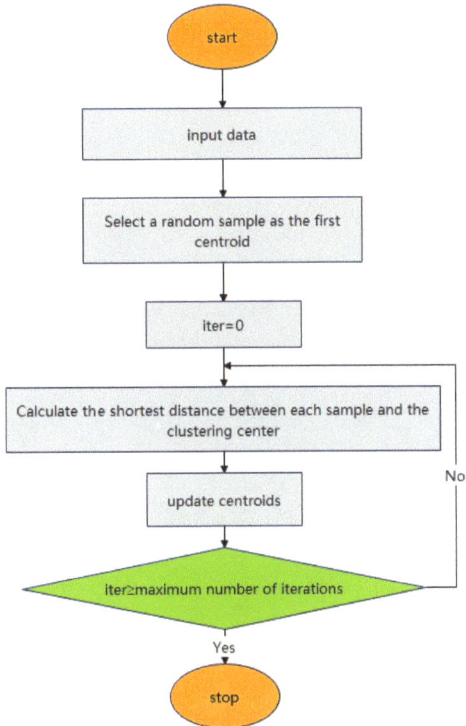

Figure 4. Flowchart of the K-means clustering.

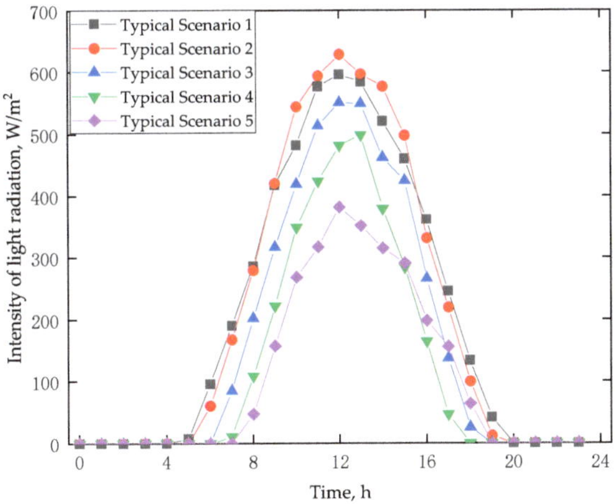

Figure 5. Typical scenario of light radiation intensity in a region.

2.5. Comprehensive Evaluation Methodology

To better evaluate the indicators of the distribution network, the hierarchical analysis method–entropy weight method is invoked to assign weights to the indicators, and the distributed power output situation is reasonably dispatched through the weight indicators. The comprehensive evaluation system of the distribution network is established with system network loss and voltage deviation as the optimization target and optimal scheduling of multiple objects in the system.

2.5.1. Hierarchical Analysis Method

When applying the hierarchical analysis method to the comprehensive evaluation system of distribution network optimization and dispatching, the optimization objectives are empowered by combining qualitative and quantitative methods, so that the optimization problem of the system has a hierarchy. The basic calculation steps are as follows:

Step 1: Establish the hierarchical structure of the system

The optimization problem is organized into a hierarchical architecture, and its elements are divided into the highest, middle, and lowest levels according to the goal, criterion, and object of decision-making. The highest level is the problem solved by the decision, the middle level comprises the criteria to be considered, and the lowest level covers the alternatives to achieve the goal.

Step 2: Construction of judgment matrix

According to the hierarchical structure, the factors are compared with each other two by two, and the relative importance of each indicator is compared using the 1–9 scale. Then, the comparison results are used as the elements of the judgment matrix in order to obtain the judgment matrix A.

$$A = \left(a_{ij}\right)_{m \times n} \tag{10}$$

where a_{ij} is the expert's empirical weighting, m denotes the number of matrix rows, and n denotes the number of matrix columns.

Step 3: Consistency test

First, the consistency index CI is defined, and the size of the CI is calibrated by using the corresponding random consistency index RI. Then, the consistency ratio CR is calculated with the formula:

$$CI = \frac{\lambda_{max} - n}{n - 1} \tag{11}$$

$$CR = \frac{CI}{RI} \tag{12}$$

where λ_{max} is the maximum eigenvalue of the judgment matrix. When $CR \leq 0.1$, the matrix satisfies the consistency test. Otherwise, the judgment matrix needs to be adjusted so that it meets the condition of the consistency test.

Step 4: Determine the integrated weights

Find the maximum eigenvector of the judgment matrix, which is used as the objective weight of the index.

2.5.2. Entropy Weight Method

The entropy weight method determines the weights of the indicators according to the magnitudes of their variability, obtains the respective entropy weights through the information entropies of the indicators, and utilizes the entropy weights to correct the magnitudes of the respective weights.

According to the proposed program data, establish the original information matrix X.

$$\mathbf{X} = \left(x_{ij}\right)_{m \times n} \tag{13}$$

Indicators are normalized to obtain a normalization matrix E_{ij}.

$$E_{ij} = \frac{x_{ij}}{\sum\limits_{j=1}^{m} x_{ij}} \tag{14}$$

When $E_{ij} = 0$, let $E_{ij} \ln E_{ij} = 0$.
Calculate the information entropy of the indicator E_j.

$$E_j = -\frac{1}{\ln m} \sum\limits_{i=1}^{m} E_{ij} \ln E_{ij} \tag{15}$$

Calculate the weight ω_j.

$$\omega_j = \frac{1 - e_j}{n - \sum\limits_{j=1}^{n} e_j} \tag{16}$$

2.5.3. Comprehensive Weight Calculation

According to the hierarchical analysis method and entropy weight method (used to obtain the weight matrices for $\lambda = [\lambda_1, \lambda_2, \cdots, \lambda_m]^T$ and $\omega = [\omega_1, \omega_2, \cdots, \omega_m]^T$, respectively, in order to form the substrate $[\lambda, \omega]$), the two methods' combined weight W is

$$W = \frac{\lambda \times \omega}{\sum\limits_{j=1}^{m} \lambda \times \omega} \tag{17}$$

2.6. Improved Particle Swarm Optimization Algorithm

The particle swarm optimization algorithm is an evolutionary computing technique that seeks the optimal solution of an objective through the iteration of a group of particles. However, the traditional particle swarm algorithm relies on inertia weights and learning factors, which make it easy to fall into the situation of local optimization. A single traditional intelligent algorithm may not be able to solve some problems effectively, while the use of intelligent algorithm fusion can improve traditional algorithms and obtain a better performance [38,39]. Accordingly, research in this area has been widely used in the field of power systems, seeking the optimal solution based on reactive power optimization [40,41].

Similarly, this paper improves the traditional particle swarm algorithm by searching for the optimum in an improved initialized population and adopting a chaotic search to improve the convergence and convergence speed of the algorithm (Figure 6).

The improved particle swarm algorithm containing power system tidal current calculation needs to constantly seek the optimal solution value of the objective, i.e., it constantly seeks the optimal value for the system network loss. The flow is shown in Figure 7.

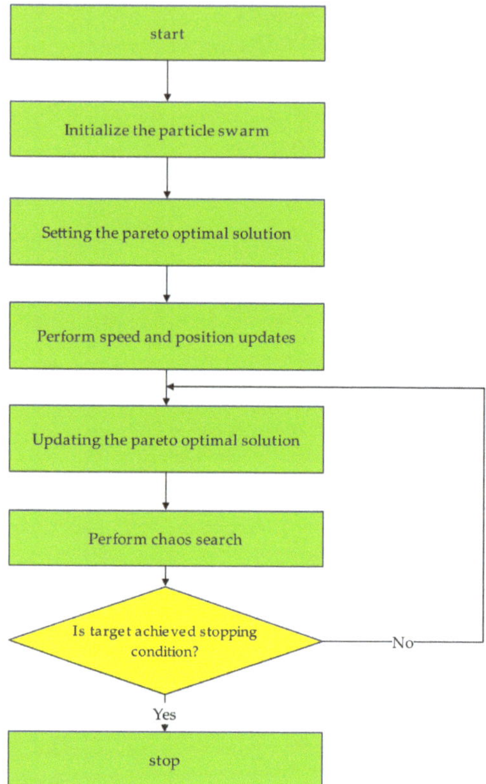

Figure 6. Improved particle swarm algorithm calculation process.

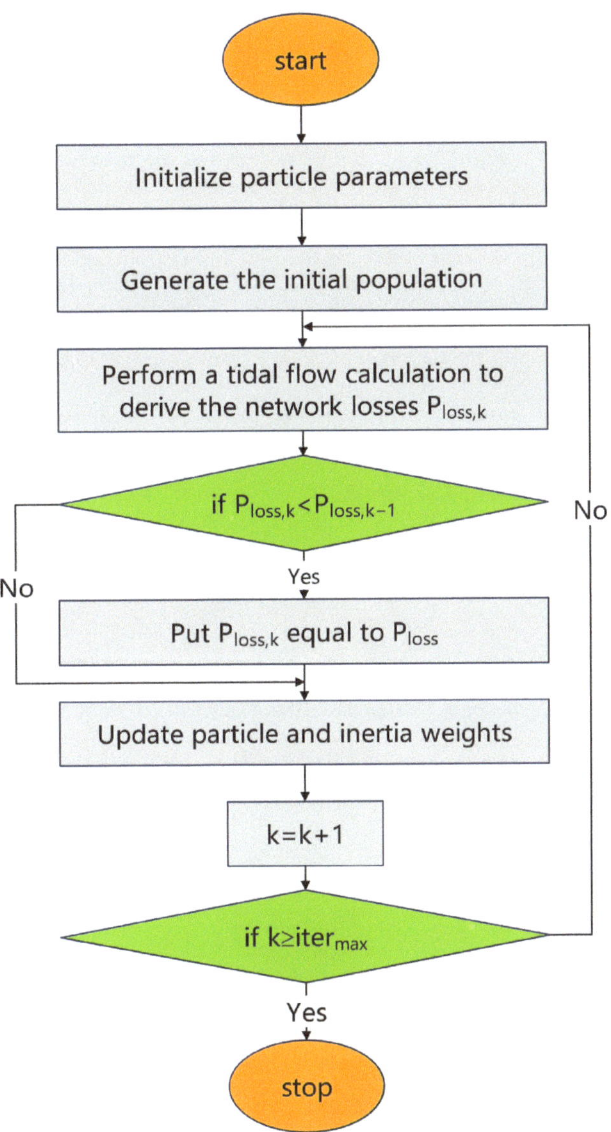

Figure 7. Flowchart of the particle swarm optimization algorithm involving power system current calculation.

3. Results

3.1. Integrated Evaluation Decision-Making for Distribution Network Optimization

The optimization objective of the distribution network includes both network loss and voltage offset. First, the active network loss of the system is used as the optimization objective, which is achieved by rationally allocating the size of the PV output. The active network loss objective function is defined as

$$\min P_{loss} = \sum \frac{P_i^2 + Q_i^2}{U_i^2} R_i \tag{18}$$

where P_i and Q_i denote the active and reactive power at bus i, respectively, U_i denotes the voltage at bus i, and R_i denotes the resistance at bus i.

Second, access to distributed energy sources will lead to changes in the voltage level; to ensure the stability of the voltage, it is necessary to set the voltage stability level as an objective function as well and adjust the voltage magnitude of each bus by adjusting the output of the distributed power supply in the system and the taps of the transformer. The voltage stability objective function is defined as

$$\min \delta U = \sum \frac{U_i - U_N}{U_N} \tag{19}$$

where U_i is the actual voltage at bus i and U_N is the rated voltage at bus i.

Setting the objective function in these two aspects at the same time is equivalent to making requirements for the economy and security of the system, respectively, and optimizing the distribution network dispatch for economical and safe operation.

The two optimization objectives of minimal network loss and voltage deviation are assigned, and the objective function expression is as follows:

$$\min F = \beta_1 \sum \frac{P_i^2 + Q_i^2}{U_i^2} R_i + \beta_2 \sum \frac{U_i - U_N}{U_N} \tag{20}$$

where the values of β_1 and β_2 can be calculated from the weights above.

The following constraints are also required for this distribution system, where the capacity constraints are as follows:

$$0 \leq P_i \leq P_{PV} \tag{21}$$

$$S_i \leq S_{imax} \tag{22}$$

where P_{PV} and S_{imax} denote the maximum PV output and the maximum capacity of branch transmission, respectively.

Set the voltage constraints as follows:

$$U_{imin} \leq U_i \leq U_{imax} \tag{23}$$

where U_{imin} and U_{imax} are the minimum and maximum values of voltage at bus i. The lower and upper limit values are set at 0.95 pu and 1.05 pu, respectively.

The current distribution constraints of the system itself are

$$P_i + P_{Gi} = P_{Li} + U_{Li} \sum_{i=1}^{N} U_i Y \tag{24}$$

$$Q_i + Q_{Gi} = Q_{Li} + U_{Li} \sum_{i=1}^{N} U_i Y \tag{25}$$

where P_i and Q_i are the active and reactive power at bus i; P_{Gi} and Q_{Gi} are the active and reactive power injected at the bus; P_{Li} and Q_{Li} are the active and reactive power of the load; U_i is the voltage at bus i; and Y is the branch conductance.

Set the energy storage charge/discharge constraint as follows:

$$0 \leq P_{ESS,t}^{charge} \leq P_{ESS,\max}^{charge} \tag{26}$$

$$0 \leq P_{ESS,t}^{discharge} \leq P_{ESS,\max}^{discharge} \tag{27}$$

$$SOC_{ESS,\min} \leq SOC_{ESS,t} \leq SOC_{ESS,\max} \tag{28}$$

where $P_{ESS,\max}^{charge}$ and $P_{ESS,\max}^{discharge}$ are the maximum charging and discharging power of the storage battery, and $SOC_{ESS,\min}$ and $SOC_{ESS,\max}$ are the minimum and maximum power of the storage battery, respectively.

Apply the improved particle swarm algorithm to the IEEE 30-bus system for multi-objective optimization in terms of network loss and voltage deviation, where the calculation steps are as follows:

Step 1: Determine the initial data matrix according to the relevant operation data of the distribution network;
Step 2: Initialize the particle swarm, and set the number of particles and the maximum number of iterations;
Step 3: Use the forward and backward generation method to calculate the current, and analyze the particle adaptation value to select the optimal solution;
Step 4: Update the individual optimal value and the group optimal value;
Step 5: Update the particle velocity and position, and iteratively carry out the last two calculations until the iteration stop condition is satisfied.

3.2. Application Example Optimization Results

The charging and discharging strategy of energy storage in the scheduling process of the distribution grid containing PVs and energy storage should ensure the consumption of PVs as much as possible and alleviate the pressure brought to the grid by high-density PV access. For the cooperative control of PVs and storage in different scenarios, the scheduling strategy is the same. Time-of-day regulation in the system can fully utilize the role of energy storage in distribution grid scheduling, where the charging and discharging of energy storage and the output of PVs are divided into 24 time periods of the day for coordinated control. According to the optimization objectives and constraints related to storage charging and discharging and PV output, the integration of distributed power sources can be obtained for storage charging and discharging and PV output under different periods on a given day under a large power grid (Figure 8).

Figure 9 shows the magnitudes of network losses in the five cases for the power system without storage compared to that with storage and with dispatch optimization. If reactive power compensation devices can also be added at each bus, the system network loss is further optimized. For the five scenarios of distributed PV access in this system, the network loss of the system after reasonable deployment is smaller than that of the original system, which proves that distributed power can reduce the network loss of the system. Meanwhile, according to the structure of the distribution network, the optimization of the capacity and location of distributed PVs can also reduce the network loss of the system and improve the power quality of the system operation.

After optimizing the distribution system containing high-density distributed photovoltaics and energy storage in five typical scenarios, the system still maintains a stable voltage level and ensures good power quality, which proves that the optimized grid structure tends to be reasonable and the power supply modes are more diverse (Figure 10).

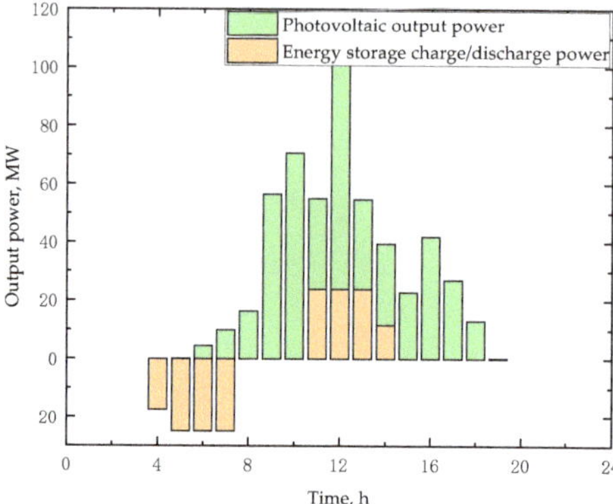

Figure 8. Photovoltaic output power and storage charging/discharging power over time.

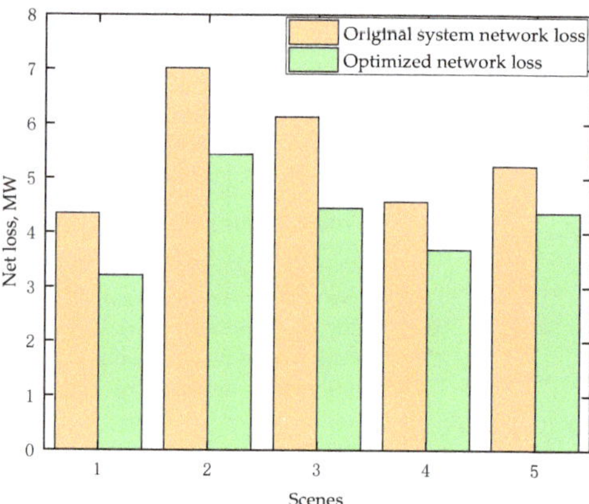

Figure 9. Comparison of network loss before and after optimization in different scenarios.

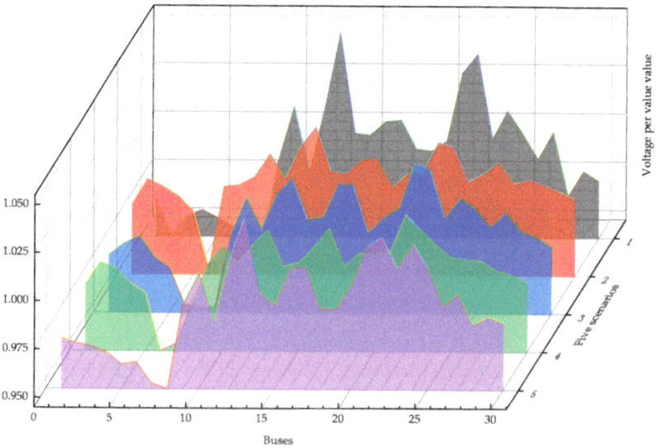

Figure 10. Bus voltage distribution under different scenario-optimized operation strategies.

The operation of the system was observed over a long period, under the control of five different operation strategies. The voltage level of each bus of the system could still be maintained at a relatively stable level, and the scheduling of the control strategies under each scenario was effective (Figure 11).

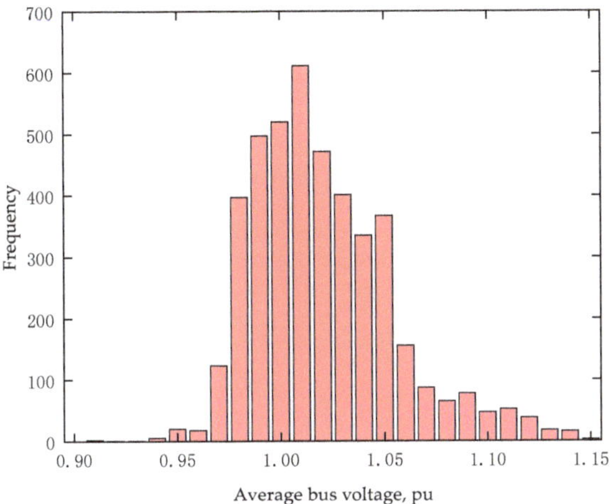

Figure 11. Bus voltage frequency distribution.

4. Conclusions

In this paper, based on the increasing high-density photovoltaic access to the distribution network and the rapid development of energy storage, the problems and solutions that may arise from the coordinated control of high PV access to the distribution network and energy storage were discussed, and different typical scenarios were delineated for the uncertainty of distributed PV generation. We tested out the coordinated scheduling of a distribution network that contained high-density PVs and energy storage, and multi-objective optimization was carried out, based on which the following can be concluded:

1. The original distribution network with high-density PVs, energy storage, and other distributed power supply modes was changed, and the coordinated optimization of PVs and energy storage could reduce the uncertainty brought about by distributed PV access. Through the protection of bus voltage stability at the same time, and distribution network loss optimization for multi-bus access to distributed PVs, energy use was more reasonable.
2. Due to the uncertainty of PV output, grid scheduling is difficult, but different typical scenarios can be divided and then optimized, which is close to the actual operation. The division of scenarios has guiding significance for the subsequent optimization, and the use of big data generation and analysis can improve the accuracy of the calculations continuously.
3. The IEEE 30-bus model simulation was carried out after considering the cooperative optimal scheduling of photovoltaic storage. We found that the deviations of each bus's voltage and the system's network loss were within a reasonable range, which proves the reasonableness of the algorithm's calculations. At the same time, this system can be further studied for optimization in dynamic operating situations.
4. Distributed photovoltaic access to the distribution network will have different impacts. The variety of distributed power supply modes will make the power supply more secure, but at the same time, the uncertainty of PV output will negatively impact on the grid scheduling and power quality. Reasonable use of an energy storage system to

configure the corresponding PV output can cut down the adverse effects, while the application of an energy storage system realizes the peak shaving and valley filling of the electricity load, and the coupling of multiple distributed power sources can also allow those play to each other's advantage.

This study examined the problems of and solutions to grid scheduling arising from high-density photovoltaic access to the grid. The maximum PV power is obtained by clustering the light intensity in the region, which, in turn, leads to the rational use of energy storage devices and reactive power optimization of the system for optimal scheduling of distributed power sources. The network losses and voltage offsets are optimized using an improved particle swarm algorithm for the actual model, and the optimization objective can be derived from the trend calculation. Since the network losses in the model are significantly reduced and the power quality is still maintained at a high level, it can be concluded that the proposed algorithm can be practically applied to compute the operating conditions of a power system with a high density of photovoltaics and energy storage devices. The model takes less account of aspects such as reverse power flow and load variations, and it is planned to complete the research by studying those aspects in the future. Future-focused modeling methodologies and theoretical studies of energy storage and distribution-grid-optimization models will also be taken into account [42–45].

Author Contributions: Conceptualization, L.H. and H.C. (Heng Chen); methodology, S.Q.; software, J.W.; validation, L.H., G.X. and T.L.; formal analysis, L.H.; investigation, S.Q.; resources, J.W.; data curation, H.C. (Heng Chen); writing—original draft preparation, L.H.; writing—review and editing, L.H.; visualization, H.C. (Heng Chen); supervision, H.C. (Honggang Chen); project administration, G.X.; funding acquisition, T.L. All authors have read and agreed to the published version of the manuscript.

Funding: This work was supported by the Science and Technology Project of the State Grid Corporation of China (Grant No. 5108-202218280A-2-142-XG, Research and application of the loss reduction technology for medium and lower voltage distribution networks considering high-proportion distributed power generation and multi-load interaction).

Data Availability Statement: The datasets used and/or analyzed during the current study are available from the corresponding author upon reasonable request.

Acknowledgments: We thank all individuals who participated in this study.

Conflicts of Interest: The authors declare no conflict of interest.

References

1. Lamnatou, C.; Chemisana, D.; Cristofari, C. Smart grids and smart technologies in relation to photovoltaics storage systems, buildings and the environment. *Renew. Energy* **2022**, *185*, 1376–1391. [CrossRef]
2. Meghraj, M.; Nitin, G.; Man, M.-G.; Ajay, K. A comprehensive review of grid-connected solar photovoltaic system: Architecture, control, and ancillary services. *Renew. Energy Focus* **2023**, *45*, 307–330.
3. Nedelcu, L.-I.; Tanase, V.-M.; Rusu, E. An Evaluation of the Wind Energy along the Romanian Black Sea Coast. *Inventions* **2023**, *8*, 48. [CrossRef]
4. Stokowiec, K.; Wciślik, S.; Kotrys-Działak, D. Innovative Modernization of Building Heating Systems: The Economy and Ecology of a Hybrid District-Heating Substation. *Inventions* **2023**, *8*, 43. [CrossRef]
5. Kulikov, A.; Ilyushin, P.; Suslov, K.; Filippov, S. Organization of Control of the Generalized Power Quality Parameter Using Wald's Sequential Analysis Procedure. *Inventions* **2023**, *8*, 17. [CrossRef]
6. Alam, F.; Haider Zaidi, S.S.; Rehmat, A.; Mutarraf, M.U.; Nasir, M.; Guerrero, J.M. Robust Hierarchical Control Design for the Power Sharing in Hybrid Shipboard Microgrids. *Inventions* **2023**, *8*, 7. [CrossRef]
7. Eghtedarpour, N.; Farjah, E. Control strategy for distributed integration of photovoltaic and energy storage systems in DC micro-grids. *Renew. Energy* **2012**, *45*, 96–110. [CrossRef]
8. Rodríguez-Gallegos, C.-D.; Gandhi, O.; Yang, D.; Alvarez-Alvarado, M.S.; Zhang, W.; Reindl, T.; Panda, S.K. A Siting and Sizing Optimization Approach for PV–Battery–Diesel Hybrid Systems. *IEEE Trans. Ind. Appl.* **2018**, *54*, 2637–2645. [CrossRef]
9. Giannelos, S.; Konstantelos, I.; Strbac, G. Investment Model for Cost-effective Integration of Solar PV Capacity under Uncertainty using a Portfolio of Energy Storage and Soft Open Points. In Proceedings of the 2019 IEEE Milan PowerTech, Milan, Italy, 23–27 June 2019; pp. 1–6.

10. Konstantelos, I.; Giannelos, S.; Strbac, G. Strategic Valuation of Smart Grid Technology Options in Distribution Networks. *IEEE Trans. Power Syst.* **2017**, *32*, 1293–1303.

11. Wu, D.; Tang, F.; Dragicevic, T.; Vasquez, J.-C.; Guerrero, J.-M. Autonomous Active Power Control for Islanded AC Microgrids With Photovoltaic Generation and Energy Storage System. *IEEE Trans. Energy Convers.* **2014**, *29*, 882–892. [CrossRef]

12. Sechilariu, M.; Wang, B.; Locment, F. Building Integrated Photovoltaic System with Energy Storage and Smart Grid Communication. *IEEE Trans. Ind. Electron.* **2013**, *60*, 1607–1618. [CrossRef]

13. Wang, T.; O'Neill, D.; Kamath, H. Dynamic Control and Optimization of Distributed Energy Resources in a Microgrid. *IEEE Trans. Smart Grid* **2015**, *6*, 2884–2894. [CrossRef]

14. Verma, R.; Padhy, N.-P. Optimal Power Flow Based DR in Active Distribution Network with Reactive Power Control. *IEEE Syst. J.* **2022**, *16*, 3522–3530. [CrossRef]

15. Zhang, L. Local and Remote Cooperative Control of Hybrid Distribution Transformers Integrating Photovoltaics in Active Distribution Networks. *IEEE Trans. Sustain. Energy* **2022**, *13*, 2012–2026. [CrossRef]

16. Abdel-Mawgoud, H.; Ali, A.; Kamel, S.; Rahmann, C.; Abdel-Moamen, M.-A. A Modified Manta Ray Foraging Optimizer for Planning Inverter-Based Photovoltaic with Battery Energy Storage System and Wind Turbine in Distribution Networks. *IEEE Access* **2021**, *9*, 91062–91079. [CrossRef]

17. Dall'Anese, E.; Dhople, S.-V.; Johnson, B.-B.; Giannakis, G.-B. Decentralized Optimal Dispatch of Photovoltaic Inverters in Residential Distribution Systems. *IEEE Trans. Energy Convers.* **2014**, *29*, 957–967. [CrossRef]

18. Dall'Anese, E.; Dhople, S.-V.; Johnson, B.-B.; Giannakis, G.-B. Optimal Dispatch of Residential Photovoltaic Inverters Under Forecasting Uncertainties. *IEEE J. Photovolt.* **2015**, *5*, 350–359. [CrossRef]

19. Li, X.; Wang, L.; Yan, N.; Ma, R. Cooperative Dispatch of Distributed Energy Storage in Distribution Network with PV Generation Systems. *IEEE Trans. Appl. Supercond.* **2021**, *31*, 1–4. [CrossRef]

20. Azizivahed, A. Energy Management Strategy in Dynamic Distribution Network Reconfiguration Considering Renewable Energy Resources and Storage. *IEEE Trans. Sustain. Energy* **2020**, *11*, 662–673. [CrossRef]

21. Li, X.; Ma, R.; Gan, W.; Yan, S. Optimal Dispatch for Battery Energy Storage Station in Distribution Network Considering Voltage Distribution Improvement and Peak Load Shifting. *J. Mod. Power Syst. Clean Energy* **2022**, *10*, 131–139. [CrossRef]

22. Hinov, N.; Gilev, B. Intelligent Design of ZVS Single-Ended DC/AC Converter Based on Neural Network. *Inventions* **2023**, *8*, 41. [CrossRef]

23. Li, X.; Wang, S. Energy management and operational control methods for grid battery energy storage systems. *CSEE J. Power Energy Syst.* **2021**, *7*, 1026–1040.

24. Shen, X.; Shahidehpour, M.; Han, Y.; Zhu, S.; Zheng, J. Expansion Planning of Active Distribution Networks with Centralized and Distributed Energy Storage Systems. *IEEE Trans. Sustain. Energy* **2017**, *8*, 126–134. [CrossRef]

25. Sugihara, H.; Yokoyama, K.; Saeki, O.; Tsuji, K.; Funaki, T. Economic and Efficient Voltage Management Using Customer-Owned Energy Storage Systems in a Distribution Network with High Penetration of Photovoltaic Systems. *IEEE Trans. Power Syst.* **2013**, *28*, 102–111. [CrossRef]

26. Stecca, M.; Elizondo, L.-R.; Soeiro, T.-B.; Bauer, P.; Palensky, P. A Comprehensive Review of the Integration of Battery Energy Storage Systems into Distribution Networks. *IEEE Open J. Ind. Electron. Soc.* **2020**, *1*, 46–65. [CrossRef]

27. Zhai, J.; Wu, X.; Zhu, S.; Yang, B.; Liu, H. Optimization of Integrated Energy System Considering Photovoltaic Uncertainty and Multi-Energy Network. *IEEE Access.* **2020**, *8*, 141558–141568. [CrossRef]

28. Cai, H.; Xiang, J.; Wei, W. Decentralized Coordination Control of Multiple Photovoltaic Sources for DC Bus Voltage Regulating and Power Sharing. *IEEE Trans. Ind. Electron.* **2018**, *65*, 5601–5610. [CrossRef]

29. Confrey, J.; Etemadi, A.-H.; Stuban, S.-M.; Eveleigh, T.-J. Energy Storage Systems Architecture Optimization for Grid Resilience with High Penetration of Distributed Photovoltaic Generation. *IEEE Syst. J.* **2020**, *14*, 1135–1146. [CrossRef]

30. Niknam, T.; Zare, M.; Aghaei, J. Scenario-Based Multiobjective Volt/Var Control in Distribution Networks Including Renewable Energy Sources. *IEEE Trans. Power Deliv.* **2012**, *27*, 2004–2019. [CrossRef]

31. Wu, H.; Liu, Z.; Chen, Y.; Xu, B.; Qi, X. Equivalent modeling method for regional decentralized photovoltaic clusters based on cluster analysis. *CPSS Trans. Power Electron. Appl.* **2018**, *3*, 146–153. [CrossRef]

32. Li, P.-X.; Gu, W.; Wang, L.-F.; Xu, B.; Wu, M.; Shen, W.-X. Dynamic equivalent modeling of two-staged photovoltaic power station clusters based on dynamic affinity propagation clustering algorithm. *Int. J. Electr. Power Energy Syst.* **2018**, *95*, 463–475. [CrossRef]

33. Chen, X.-H.; Li, S.-F.; Wang, F.-S.; Li, J.-P.; Tang, C.-H. Power estimation method of low-voltage distributed photovoltaic generation based on similarity aggregation. *Energy Rep.* **2021**, *7*, 1344–1351. [CrossRef]

34. Bai, R.-X.; Shi, Y.-T.; Yue, M.; Du, X.-N. Hybrid model based on K-means++ algorithm, optimal similar day approach, and long short-term memory neural network for short-term photovoltaic power prediction. *Glob. Energy Interconnect.* **2023**, *6*, 184–196. [CrossRef]

35. Fu, Y.; Li, W.; Xiong, N.; Fang, Y.; He, X.-M.; Wang, Z.-F. Influence of off-grid/grid-connected operation on stability of large-scale photovoltaic system. *Energy Rep.* **2023**, *9*, 904–910. [CrossRef]

36. Eftekharnejad, S.; Heydt, G.-T.; Vittal, V. Optimal Generation Dispatch with High Penetration of Photovoltaic Generation. *IEEE Trans. Sustain. Energy* **2015**, *6*, 1013–1020. [CrossRef]

37. Park, Y.-W.; Kim, T.-W.; Kim, C.-J. Compromised Vibration Isolator of Electric Power Generator Considering Self-Excitation and Basement Input. *Inventions* **2023**, *8*, 40. [CrossRef]

38. Gao, Y. An Improved Hybrid Group Intelligent Algorithm Based on Artificial Bee Colony and Particle Swarm Optimization. In Proceedings of the 2018 International Conference on Virtual Reality and Intelligent Systems (ICVRIS), Hunan, China, 10–11 August 2018; pp. 160–163.
39. Li, T.; Xiong, Q.; Ge, Q. Genetic and particle swarm hybrid QoS anycast routing algorithm. In Proceedings of the 2009 IEEE International Conference on Intelligent Computing and Intelligent Systems, Shanghai, China, 20–22 November 2009; pp. 313–317.
40. Liu, H.; Ge, S.-Y. Reactive power optimization based on improved particle swarm optimization algorithm with boundary restriction. In Proceedings of the 2008 Third International Conference on Electric Utility Deregulation and Restructuring and Power Technologies, Nanjing, China, 6–9 April 2008; pp. 1365–1370.
41. Dabhi, D.; Pandya, K. Uncertain Scenario Based MicroGrid Optimization via Hybrid Levy Particle Swarm Variable Neighborhood Search Optimization (HL_PS_VNSO). *IEEE Access* **2020**, *8*, 108782–108797. [CrossRef]
42. Giannelos, S.; Borozan, S.; Strbac, G. A Backwards Induction Framework for Quantifying the Option Value of Smart Charging of Electric Vehicles and the Risk of Stranded Assets under Uncertainty. *Energies* **2022**, *15*, 3334. [CrossRef]
43. Giannelos, S.; Borozan, S.; Aunedi, M.; Zhang, X.; Ameli, H.; Pudjianto, D.; Konstantelos, I.; Strbac, G. Modelling Smart Grid Technologies in Optimisation Problems for Electricity Grids. *Energies* **2023**, *16*, 5088. [CrossRef]
44. Rostmnezhad, Z.; Dessaint, L. Power Management in Smart Buildings Using Reinforcement Learning. In Proceedings of the 2023 IEEE Power & Energy Society Innovative Smart Grid Technologies Conference (ISGT), Washington, DC, USA, 16–19 January 2023; pp. 1–5.
45. Nema, S.; Mathur, A.; Prakash, V.; Pandžić, H. Optimal BESS Compensator Design for Fast Frequency Response. In Proceedings of the 2023 International Conference on Power, Instrumentation, Energy and Control (PIECON), Aligarh, India, 10–12 February 2023; pp. 1–5.

Article

Fault Location Method for Overhead Power Line Based on a Multi-Hypothetical Sequential Analysis Using the Armitage Algorithm

Aleksandr Kulikov [1], Pavel Ilyushin [2,*], Anton Loskutov [1] and Sergey Filippov [2]

[1] Department of Electroenergetics, Power Supply and Power Electronics, Nizhny Novgorod State Technical University n.a. R.E. Alekseev, 603950 Nizhny Novgorod, Russia; inventor61@mail.ru (A.K.); loskutov.nnov@gmail.com (A.L.)

[2] Department of Research on the Relationship between Energy and the Economy, Energy Research Institute of the Russian Academy of Sciences, 117186 Moscow, Russia; fil_sp@mail.ru

* Correspondence: ilyushin.pv@mail.ru

Abstract: The use of modern methods for determining the fault location (FL) on overhead power lines (OHPLs), which have high accuracy and speed, contributes to the reliable operation of power systems. Various physical principles are used in FL devices for OHPLs, as well as various algorithms for calculating the distance to the FL. Some algorithms for FL on OHPLs use emergency mode parameters (EMP); other algorithms use measurement results based on wave methods. Many random factors that determine the magnitude of the error in calculating the distance to the FL affect the operation of FL devices by EMP. Methods based on deterministic procedures used in well-known FL devices for OHPLs do not take into account the influence of random factors, which significantly increases the time to search for the fault. The authors have developed a method of FL on OHPLs based on a multi-hypothetical sequential analysis using the Armitage algorithm. The task of recognizing a faulted section of an OHPL is formulated as a statistical problem. To do this, the inspection area of the OHPL is divided into many sections, followed by the implementation of the procedure for FL. The developed method makes it possible to adapt the distortions of currents and voltages on the emergency mode oscillograms to the conditions for estimating their parameters. The results of the calculations proved that the implementation of the developed method has practically no effect on the speed of the FL algorithm for the OHPL by EMP. This ensures the uniqueness of determining the faulted section of the OHPL under the influence of random factors, which leads to a significant reduction in the inspection area of the OHPL. The application of the developed method in FL devices for OHPLs will ensure the required reliability of power supply to consumers and reduce losses from power outages by minimizing the time to search for a fault.

Keywords: overhead power line; fault location; emergency mode parameters; sequential analysis; Armitage algorithm

Citation: Kulikov, A.; Ilyushin, P.; Loskutov, A.; Filippov, S. Fault Location Method for Overhead Power Line Based on a Multi-Hypothetical Sequential Analysis Using the Armitage Algorithm. *Inventions* **2023**, *8*, 123. https://doi.org/10.3390/inventions8050123

Academic Editor: Om P. Malik

Received: 27 August 2023
Revised: 21 September 2023
Accepted: 22 September 2023
Published: 27 September 2023

1. Introduction

Overhead power lines (OHPLs) of various voltage classes are the main elements of power systems, both in terms of quantity and length. The output of power from all types of power plants and the transfer of power between regional power systems are carried out, as a rule, via high- and extra-high-voltage OHPLs [1,2]. This is due to the fact that the cost of construction and operation of cable power lines is much more expensive. At the same time, a number of countries have long-term programs to convert overhead power lines into cable ones [3,4]. This is due to the susceptibility of OHPLs to the influence of many natural and technogenic factors [5–7]. As a result, OHPLs have low reliability indicators compared to cable transmission lines [8,9].

Short circuits (SCs) on OHPLs that occur for various reasons are accompanied by voltage dips of various depths and durations. It depends on the type of SC, the magnitude of the transient resistance at the place of the SC, the operation algorithms and settings of the relay protection (RP) devices, as well as the intrinsic time of switching off the high-voltage circuit breakers [10,11]. Voltage dips adversely affect the operation of electrical receivers, especially electric motors, which are slowed down during an SC, and after the SC is eliminated, their self-starting begins. Self-starting occurs only if the electric motors have not been turned off by electrical or technological protection [12–14]. According to statistics, on OHPLs with a voltage of 110–500 kV, single-phase SCs are the main ones, accounting for up to 70% of the total. At the same time, two-phase and three-phase SCs account for 20% and 10%, respectively.

The negative impact of faults on OHPLs is mainly associated with either damage to electricity consumers, depending on their power supply scheme, or with violations of the stability of electric power systems.

In some countries, OHPLs with a voltage of 110–500 kV have a length of hundreds of kilometers, pass through mountainous, forested, and swampy areas, with a large number of crossings through water barriers (streams, rivers, lakes, and artificial reservoirs), etc. In addition, OHPLs are operated in difficult climatic conditions, for example, high wind pressures that cause vibration and dancing of wires, the formation of ice-frost deposits on wires, ground wires, and supports of overhead lines, as well as critically high or low air temperatures [15–17]. Under these conditions, determining the fault location (FL) on OHPLs by inspection requires considerable time and labor.

The reliability of the functioning of power systems as well as the ability to provide a reliable power supply to consumers depend on the availability of reserve OHPLs, as well as the time required to search for and eliminate a fault on OHPLs [18–20]. To eliminate the fault on OHPL, it is necessary to implement organizational and technical measures, which consist of the departure of the repair team to the place of fault and the implementation of emergency recovery work. The volume of repair work, the category of its complexity, as well as the need for materials and devices, are determined on the spot and depend on the scale of the fault and the reasons for its occurrence. The time for elimination of a fault on OHPLs largely depends on the accuracy of determining the fault location by FL devices, i.e., where the repair team should go to search for a fault on the overhead line.

For consumers powered by two OHPLs, one of which was taken out for repair and the second turned off in an emergency, the power supply will be completely disrupted. This will lead to damage from the disconnection of OHPLs, as well as losses from the undersupply of products during the restoration of power supply to the technological process [21–23]. To minimize damages and losses to consumers, the accuracy of determining the FL on OHPLs should be increased, and the time for performing emergency recovery work should be minimized [24,25].

For more than 70 years, power grid companies have been using FL devices installed at substations on one or two sides of OHPLs [26–29]. However, their accuracy remained unsatisfactory for many years, so the staff of the electric grid companies did not trust their testimony. In addition, according to the readings of the FL devices, it was necessary to carry out calculations in order to determine the distance from the substation to the place of fault. To accurately determine the fault location, it is necessary to use modern FL devices, which have sufficient speed and high accuracy and also do not require additional calculations. This will ensure the required reliability of power supply to consumers.

The FL methods by EMP have specific features when implemented on OHPLs of various designs and under different circuit conditions. For example, ref. [30] provides detailed studies of the use of measurements of currents and voltages of zero and negative sequences for the two-way FL of parallel (double-circuit) OHPLs. Features of the use of distributed generation sources in modern power supply systems lead to the advisability of using nonparametric methods in relay protection algorithms and FL on OHPLs [31], such as genetic algorithms, particle swarm methods, differential evolution, and others. It should be noted that achieving high accuracy and speed of FL methods on OHPLs is important in the conditions of reconfiguring the electrical network [32] to ensure system stability and reliability.

It is important to note that various random factors influence the magnitude of errors in calculating the distance to the fault in FL devices:

- Relative and angular errors of measuring current and voltage transformers;
- Harmonic components in currents and voltages recorded in an emergency mode [33];
- Current waveform distortions associated with saturation of electromagnetic measuring current transformers;
- Distortion of the sinusoidality of currents and voltages due to the influence of the load and devices based on power electronics elements [34,35];
- The presence of transient resistance at the site of fault on OHPLs;
- Uneven distribution of resistivity along the OHPL [36];
- Change in the resistance of the ground loops of OHPLs at different times of the year [37,38];
- Not taking into account the capacitive component of the OHPL relative to the ground in the FL algorithm;
- Neglect of mutual induction in the corridors of joint passage of OHPLs [39];
- Errors in the initial data on the resistivity of sections of OHPLs;
- Not taking into account the resistance of bypass connections, etc.

Under these conditions, it is required to use FL methods that allow calculating the distance to the fault location under the influence of random factors with high accuracy [40]. In the existing FL devices for OHPLs and various FL methods considered in the scientific literature, it was not previously proposed to apply a multi-hypothetical sequential analysis using the Armitage algorithm to determine the fault location on overhead lines.

The purpose of the study is to develop a new FL method for OHPLs according to the emergency mode parameters (EMP), based on a multi-hypothetical sequential analysis using the Armitage algorithm. The task of recognizing a faulted section of an OHPL is formulated as a statistical problem. The use of the developed FL method for OHPLs practically does not affect its performance but ensures the unambiguous identification of the faulted section of the OHPL under the influence of random factors.

2. Materials and Methods

In FL devices for OHPLs, they can use various physical principles as well as all kinds of algorithms to calculate the distance to the fault [41–44]. In some FL algorithms, emergency mode parameters are used, and in other algorithms, the results of the measurements are based on wave methods. In wave methods, either active probing of OHPLs is used or passive registration of wave processes at the ends of OHPLs (at substations) is used [45–47].

The relative error of wave FL methods, including those based on active probing of OHPLs, is much less than the error of FL devices by EMP [48–50]. However, the high cost of wave FL devices for OHPLs limits the possibility of their mass application in power grid companies. As experience shows, wave FL devices are used only on especially critical OHPLs.

FL devices for OHPLs by EMP are simple since the calculation algorithms are based on measurements of the components of currents and voltages of industrial frequency. One-sided [51,52], two-sided, and multi-sided measurements of currents and voltages are used in FL devices for OHPLs by EMP [53–55]. When implementing FL algorithms for OHPLs by EMP, it is not required to use analog-to-digital converters with a high sampling rate as well as high-performance processors. One-way FL algorithms for OHPLs do not require communication channels for information exchange since current and voltage measurements are made from one side of the OHPL. However, one-sided FL algorithms for OHPL have a large error compared to two-sided algorithms [56–62].

It is possible to implement FL algorithms for OHPLs by EMP in the form of specialized software in relay protection devices, emergency event recorders, automated process control systems for substations, devices for phasor measurement units (PMUs), etc. [63–65]. This makes it possible to not install separate FL devices for OHPLs, as has been the case for several decades, but implement this function in devices that are already installed at substations, which is more cost-effective.

In the regulatory and technical documents of large electric grid companies, the OHPL inspection zone means the estimated section of the OHPL (in km), determined on the basis of data from the FL devices (RP or others), which were obtained after an emergency shutdown of the OHPL. This information is the basis for planning the departure of the repair team to the OHPL in order to establish the actual location of the fault to the OHPL, identify the causes of fault, and carry out emergency recovery work. The permissible value of the OHPL inspection zone to search for the actual location of fault depends on the length of the OHPL, and it can reach up to $\pm 10\%$ of its length [66]. If the length of the OHPL is 50 km, then the permissible value of the inspection zone is ± 5 km, which is a lot, especially when the OHPL passes over rough terrain.

To reduce the inspection area of OHPLs, in [67,68], two FL methods for OHPLs were proposed and studied, which are used in electrical networks and the contact network of railways. In the first case, based on the use of the interval method [67], the problem of reducing the size of the OHPL section, including the fault site, is solved. The second method [68] offers provisions for splitting the OHPL into sections to solve the problem of determining the faulted section of the OHPL. However, the authors do not take into account the influence of random factors that determine the magnitude of the error when calculating the distance to the fault site, and the FL methods for OHPLs by EMP are implemented on the basis of deterministic procedures.

Let us formulate the problem of recognizing a faulted section of an OHPL as a classification problem, which consists of establishing whether the fault belongs to one of the sections of an OHPL within the zone of its inspection. Due to the influence of random factors, the decision-making process during recognition has a stochastic character, since it is based on the processing of emergency oscillograms of currents and voltages recorded over a limited time interval. The duration of the time interval is determined by the time of SC elimination.

To implement a sequential analysis when choosing a faulted area within the inspection zone of an OHPL, it is proposed to carry out k experiments with sample data at each step of the procedure. According to the results of each of the experiments, one of the $(M + 1)$ decisions is made:

- Complete the experiment by accepting the hypothesis H_1 (fault in section No. 1).
- Complete the experiment by accepting the H_2 hypothesis (fault in section No. 2).
- ....
- Complete the experiment by accepting the H_M hypothesis (fault in section No. M).
- Continue the experiment by making additional observations.

Thus, the procedure is implemented sequentially: based on the first observation, one of the $(M + 1)$ decisions is made, and when one of the first M decisions is chosen, the analysis process ends. If the solution numbered $(M + 1)$ is chosen, then the next (second) observation is made. Then, based on the first two sample data, one of the $(M + 1)$ decisions is made again. If the choice corresponds to the last $(M + 1)$th decision, then the third experiment is performed, and so on. The process continues until one of the first M solutions is chosen.

In the general case, the decision regarding the faulted section of the OHPL is made on the basis of the vector of parameters of currents and voltages x corresponding to the faulted section numbered m ($m = 1, \ldots, M$). In this case, the vector $x = \{x_1, x_2, \ldots\}$ of current and voltage parameters is generally random since it may include distorting components; for example, it is associated with deviations of power quality indicators from standard values [69,70].

Since the hypotheses $H_1, \ldots, H_M$ mutually exclude each other, exhausting all possible cases for the chosen values of the vector x, then one (and only one) of the hypotheses, $H_1, \ldots, H_M$, is consistent with a specific set of values of the vector x.

In order to form a rational decision rule in case of FL on OHPLs, it is necessary to introduce indicators of the effectiveness of a sequential analysis in recognizing a faulted section of an OHPL.

The most common probabilistic indicators of the effectiveness of the sequential analysis procedure should include a matrix of conditional probabilities for M—hypotheses, each of which corresponds to its faulted section within the inspection area of the OHPL:

$$\|P(k|i)\| = \|P_i(k)\| = \|P_{ik}\|, \tag{1}$$

where $i, k = 1, \ldots, M$; $P(k \mid i) = P_i(k) = P_{ik}$—the conditional probability of making a decision about the number k of the faulted section, provided that the fault belongs to section i.

The probabilistic indicators of recognition of a faulted section of an OHPL are directly related to the concept of resolution used in a number of physical problems [71]. In this case, the resolution of the FL algorithm for OHPLs should be understood as the minimum length of the OHPL section for which the problem of fault recognition with given performance indicators is implemented (Expression (1)). It is assumed that the resolution of the FL for OHPLs is the potentially achievable minimum length section of the OHPL, while the influence of undesirable random factors is minimized.

It is important to note that the actual resolution refers not only to the FL algorithm (device) for OHPLs, but also to a specific OHPL, which has design and operational features. For each fault on the OHPL, the accuracy of calculating the distance to the fault site will be determined by the resolution of the FL algorithm and the values of random factors affecting the OHPL.

Let us consider the option of using a multi-hypothetical sequential analysis with the use of the Armitage algorithm [72,73] when determining the fault location. In the algorithm developed by the authors, at each step of the analysis, $M \cdot (M - 1)$ paired likelihood ratios are calculated:

$$\lambda_m(x \mid H_{k,l}) = p_m(x \mid H_k) / p_m(x \mid H_l) \tag{2}$$

where $k,l = 1, \ldots, M$; $k \neq l$; $p_m(x \mid H_k)$ $p_m(x \mid H_l)$—multi-dimensional probability density of the vector x observed at the m-th step.

At each observation step m, the calculated likelihood ratios (Expression (2)) are compared with the thresholds and the condition is checked: if $\lambda_m(x \mid H_{k,l}) > \lambda_m^{\text{threshold}}(x \mid H_{k,l})$, $l = 1, \ldots, M$, $k \neq l$, then a decision is made in favor of the hypothesis H_k. Otherwise, a decision is made to continue observations. The sequential analysis procedure is implemented until the condition is met when all $(M - 1)$ likelihood ratios $\lambda_m(x \mid H_{k,l}) > \lambda_m^{\text{threshold}}(x \mid H_{k,l})$, $l = 1, \ldots, M$, $k \neq l$, characteristic of the hypothesis H_k will simultaneously exceed the corresponding thresholds $\lambda_m^{\text{threshold}}(x \mid H_{k,l})$.

The thresholds $\lambda_m^{\text{threshold}}(x \mid H_{k,l})$ for each of the tested hypotheses regarding the faulted section of the OHPL are formed on the basis of probabilistic indicators of the quality of the probability, combined into a matrix of conditional probabilities (Expression (1)). It was shown in [72] that the probability of making a correct decision about a faulted section of an OHPL increases and approaches unity as the number of observations m increases.

The values of the probabilities of correct recognition of the faulted section P_{kk} and the threshold values $\lambda_m^{\text{threshold}}(x \mid H_{k,l})$ in the multi-hypothetical sequential analysis using the Armitage algorithm are interconnected:

$$P_{kk} > 1 - \sum_{k \neq l} \left[1 / \lambda_m^{\text{threshold}}(x \mid H_{k,l}) \right] \tag{3}$$

$$\lambda_m^{\text{threshold}}(x \mid H_{k,l}) = (1 / P_{kl}) \cdot [1 - \sum_{k \neq l} P_{kl}]. \tag{4}$$

For $M = 2$, the threshold values are identical to those for sequential Wald analysis [74]. Thus, the Armitage algorithm can be considered a combination of $M \cdot (M - 1)$ binary consecutive Wald analyses, where the threshold ratios are determined by Expression (4).

The use of a multi-hypothesis sequential analysis using the Armitage algorithm for FL on OHPLs helps reduce the number of observations while maintaining the simplicity of the approach based on a comparison of pairwise likelihood ratios. For this, threshold values should be set, which depend on the number of observations m:

$$\lambda_m^{\text{threshold}*}(x \mid H_{k,l}) = \lambda_m^{\text{threshold}}(x \mid H_{k,l}) / \left(m^{(r)} \right), \quad k,l = 1, \ldots, M; k \neq l, \tag{5}$$

where $\lambda_m^{\text{threshold}}(x \mid H_{k,l})$—threshold determined by Expression (4); r—positive constant [75].

The developed FL method for OHPLs does not impose restrictions on the maximum required number of observations, but new threshold values reduce the likelihood of conducting a large number of them. It is noted in [75] that for $r = 1$, the new thresholds significantly reduce the average required number of observations, having an insignificant effect on the classifier error probabilities.

Figure 1 shows a structural diagram of the FL device that implements a multi-hypothetical sequential analysis using the Armitage algorithm. The FL device has a multi-channel structure, including $M \cdot (M - 1)$ channels, where M characterizes the number of sections into which the OHPL inspection zone is divided. For example, for $M = 3$, the number of channels will be $3 \times (3 - 1) = 6$, where paired hypothesis testing will be implemented: $H_{1,2}$, $H_{1,3}$; $H_{2,1}$, $H_{2,3}$; $H_{3,1}$, and $H_{3,2}$. There can be any number of sections of OHPLs, and they are determined by the personnel of electric grid companies based on their operating experience.

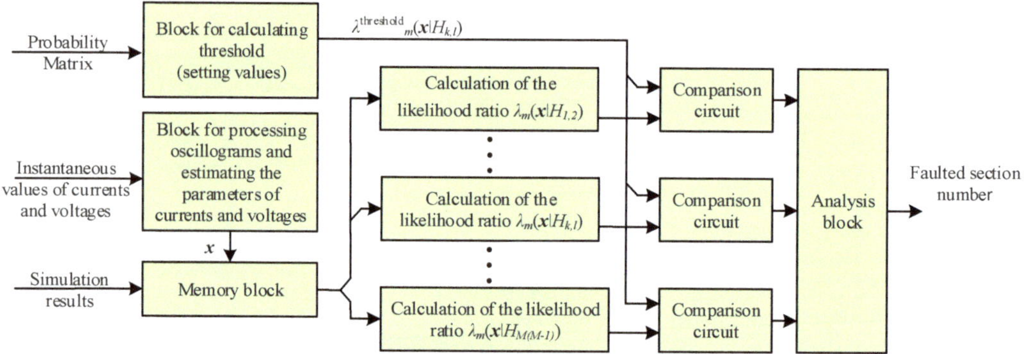

Figure 1. Structural diagram of the FL device for OHPLs that implements the sequential analysis procedure using the Armitage algorithm.

The instantaneous (complex) values of currents and voltages obtained from the oscillograms of the emergency mode are received at the input of the FL device (Figure 1). Based on this information, the components of the vector x are calculated in the block for processing oscillograms and estimating the parameters of currents and voltages. The composition of the vector x includes quantities characterizing the faulted section of the OHPL (active resistance, reactance, reactive power value, current distribution coefficient value, etc.). In addition, it includes values calculated according to various FL algorithms for OHPLs by EMP, which have various systematic and random errors.

Further, in each of the blocks for calculating the likelihood ratio, based on the vector x, the likelihood ratios are calculated by the Expression (2) $\lambda_m(x \mid H_{k,l}) = p_m(x \mid H_k) / p_m(x \mid H_l)$. When calculating each $\lambda_m(x \mid H_{k,l})$ from the memory block, for the oscillograms received in the block for processing oscillograms and estimating the parameters of currents and voltages, the values of x are received by the corresponding values $p_m(x \mid H_k)$ and $p_m(x \mid H_l)$.

Values $p_m(x \mid H_k)$ and $p_m(x \mid H_l)$ are formed for different combinations of hypotheses $H_{k,l}$, $k \neq l$ based on simulation results. Simulation modeling is performed in advance, until the moment of implementation of the FL algorithm and its results are recorded in the memory block. Probability distributions $p_m(x \mid H_k)$ and $p_m(x \mid H_l)$ can be obtained on the basis of statistical data, taking into account errors detected by repair teams during inspections of OHPLs after emergency shutdowns. They can also be obtained on the basis of the normalized value of the OHPL inspection zone (averaged values for OHPLs of various lengths and voltages), which is less accurate.

The corresponding threshold values $\lambda_m^{\text{threshold}}(x \mid H_{k,l})$ are fed to the first inputs of the comparison circuits from the block for calculating threshold (setting) values to implement the comparison $\lambda_m(x \mid H_{k,l}) > \lambda_m^{\text{threshold}}(x \mid H_{k,l})$. The calculated values of the likelihood ratio for each of the hypotheses $\lambda_m(x \mid H_{k,l})$ are fed to the second inputs of the comparison circuits, starting from the first and up to $M \cdot (M - 1)$. When the step m of the sequential procedure reaches the value $\lambda_m(x \mid H_{k,l}) > \lambda_m^{\text{threshold}}(x \mid H_{k,l})$ from the output of the comparison circuit, a logical signal is sent to the analysis block [76].

As noted above, the sequential analysis procedure is implemented until the condition is met when all $(M - 1)$ likelihood ratios $\lambda_m(x \mid H_{k,l}) > \lambda_m^{\text{threshold}}(x \mid H_{k,l})$, $l = 1, \ldots, M$, $k \neq l$, characteristic of the hypothesis H_k, do not simultaneously exceed the corresponding thresholds $\lambda_m^{\text{threshold}}(x \mid H_{k,l})$. In this case, the sequential analysis procedure stops and a decision is made that the fault on the OHPL is located in the section numbered k. From the output of the analysis block of the FL device (Figure 1), information is provided to the repair personnel about the faulted section (in the form of its number) within the inspection area of the OHPL.

3. Results and Discussion

Full-scale experiments to determine the location of fault on OHPLs are expensive and require the development of special organizational measures. Therefore, the advantages of the proposed method of FL on OHPLs are illustrated by a calculated example.

Let us consider the implementation of the FL method for OHPLs based on a multi-hypothetical sequential analysis using the Armitage algorithm using the example of a 110 kV OHPL with a length of $l = 50$ km and two-sided power supply (Figure 2) [77].

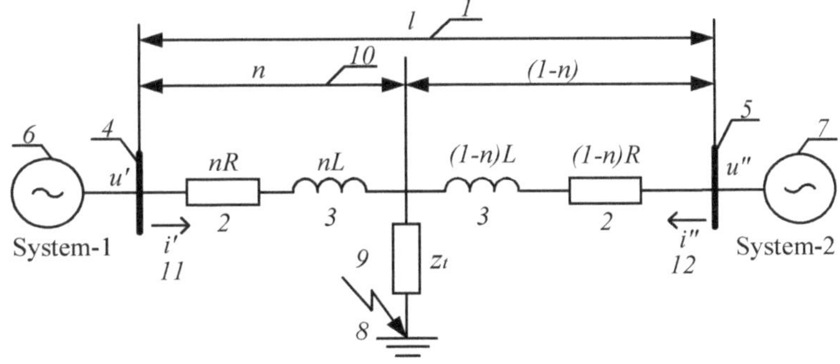

Figure 2. Single-line equivalent circuit for 110 kV overhead lines.

Figure 2 shows the equivalent circuit of a 110 kV OHPL with a length l (1) in relation to the calculation example, phase active resistance R (2), and inductance L (3) connecting buses (4) and (5) of two power systems (6) and (7). On the OHPL, a short circuit (8) is shown behind the transition resistance Zt (9) at a distance $x = n \cdot l$ (10) from one of the ends of the OHPL. In the event of an SC, current i' flows through the OHPL from the busbars (4) and current i'' from the busbars (5). At the time of a short circuit, the instantaneous values of phase currents (i'_A, i'_B, i'_C), (i''_A, i''_B, i''_C) and voltages (u'_A, u'_B, u'_C), (u''_A, u''_B, u''_C) are measured from both ends of the overhead line, which are not synchronized in time.

The relative distance to the fault location n is determined in accordance with the following well-known Expression (6):

$$n = [(u'(m) - u''(m)) + R \cdot I''(m) + L \cdot I''(m)/dt_m]/[R \cdot (I'(m) + I''(m)) + L \cdot (I'(m)/dt_m + I''(m)/dt_m)]. \tag{6}$$

This FL method for OHPLs has small errors in calculating the distance to the fault location under short-circuit conditions with undistorted (sinusoidal) currents and voltages in emergency mode oscillograms [77]. Let us assume that from the power System-1 side (Figure 2), discrete instantaneous values of current $i'(m)$ are distorted by flicker [78–80]. The distorted current signal $i'(m)$ is shown in Figure 3a.

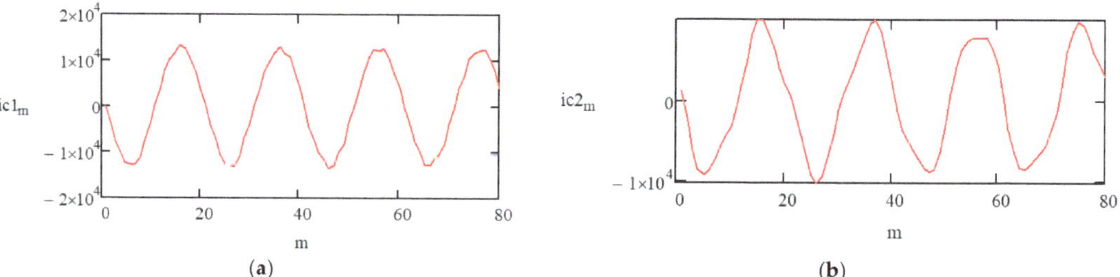

Figure 3. Oscillograms of overhead line currents distorted by the following: (**a**) flicker from power System-1; (**b**) interharmonics with frequency $f_i = 135$ Hz from power System-2.

Let us assume that there is a non-linear load on the side of power System-2 (Figure 2), which outputs interharmonics into the electrical network [81,82]. In the example, the instantaneous values of current $i''(m)$ are distorted by interharmonics with frequency $f_i = 135$ Hz, amplitude $I_i = 0.15 \cdot I''$, and a zero initial phase, as shown in Figure 3b.

The calculation expression for determining the fault location in the presence of flicker and frequency interharmonics $f_i = 135$ Hz will correspond to the equation:

$$\begin{aligned}
n_i(m) &= \{(u'(m) - u_i''(m)) + I''[R \cdot sin(2\pi f(t_d + m \cdot t_s)) + L \cdot cos(2\pi f(t_d + m \cdot t_s))] \\
&\quad + 0.15I''[R \cdot sin(2\pi f_i(t_d + m \cdot t_s)) + L \cdot cos(2\pi f_i(t_d + m \cdot t_s))]\}/ \\
&\quad \{(I'(1 - k \cdot rnd(m)) + I'') \cdot [R \cdot sin(2\pi f(t_d + m \cdot t_s)) + L \cdot cos(2\pi f(t_d + m \cdot t_s))] \\
&\quad + 0.15I''[R \cdot sin(2\pi f_i(t_d + m \cdot t_s)) + L \cdot cos(2\pi f_i(t_d + m \cdot t_s))]\}; \\
u'(m) &= U + I' \cdot (1 - k \cdot rnd(m)) \cdot [nR \cdot sin(2\pi f(t_d + m \cdot t_s)) + nL \cdot cos(2\pi f(t_d + m \cdot t_s))], \\
u_i''(m) &= U + (1 - n) \cdot R[I'' \cdot sin(2\pi f(t_d + m \cdot t_s)) + 0.15I'' \cdot sin(2\pi f_i(t_d + m \cdot t_s))] \\
&\quad + (1 - n)L \cdot [I'' \cdot sin(2\pi f(t_d + m \cdot t_s)) + 0.15I'' \cdot sin(2\pi f_i(t_d + m \cdot t_s))],
\end{aligned} \tag{7}$$

where k—number (constant coefficient) characterizing the "depth of distortion" by flicker; $rnd(m)$—random number (for example, distributed according to a uniform law in the interval [0; 1], formed at each discrete time value m); U—voltage at the fault location; t_d—delay time; t_s—sampling interval.

The technical characteristics of the OHPL and the measured parameters of the emergency mode from the two ends of the OHPL are given in Table 1. The measurement data of the emergency mode parameters were obtained from the emergency event recorder during a real short circuit on the 110 kV OHPL considered in the example.

Table 1. Technical characteristics of overhead lines and measured parameters of the emergency mode.

Parameter	I' (A)	I'' (A)	f (Hz)	t_s (s)	L (H)	R (Ohm)	f_i (Hz)	U (V)	n	k	t_d (s)
Meaning	13,908.15	9030.13	50	0.0025	0.0643	12.5	135	29,323.83	0.5	0.15	0.003

Substitution of numerical values from Table 1 into Expression (7) allowed us to obtain the following results:

- At $m = 20$; n_i and $(20) = 0.486$; $\Delta x = l \cdot (n - n_i) = 50 \times (0.5 - 0.486) = 0.7$ (km);
- At $m = 60$; n_i and $(60) = 0.526$; $\Delta x = l \cdot (n - n_i) = 50 \times (0.5 - 0.526) = -1.30$ (km).

The analysis of the obtained calculation results shows that the FL errors can have both positive and negative signs. In addition, they are distributed unevenly with respect to different points in time [83].

Since the length of the OHPL is $l = 50$ km, the inspection area by the repair team to search for the actual fault location, in accordance with the requirements of regulatory and technical documents, should not exceed $\pm 10\%$ of the length of the OHPL [66]. Therefore, $\Delta l = \pm 50 \times 0.1 = \pm 5$ km relative to the fault location.

Taking into account the normal law of FL error distribution for OHPLs by EMP and the three-sigma rule [84], we assume that the standard deviation (root-mean-square deviation) of the normal distribution law of errors of the FL device is $\sigma \approx (2 \cdot \Delta l)/6 = 10/6 = 1.67$ km.

Let us consider the process of implementing a sequential analysis when determining the FL on OHPL with dividing the inspection area of an OHPL into three sections relative to the place of its fault (Figure 4), corresponding to three hypotheses: H_1: $\mu = -\sigma$; H_2: $\mu = 0$; and H_3: $\mu = \sigma$. Each of the hypotheses corresponds to making a decision about the compliance of the fault location with the value of mathematical expectations μ.

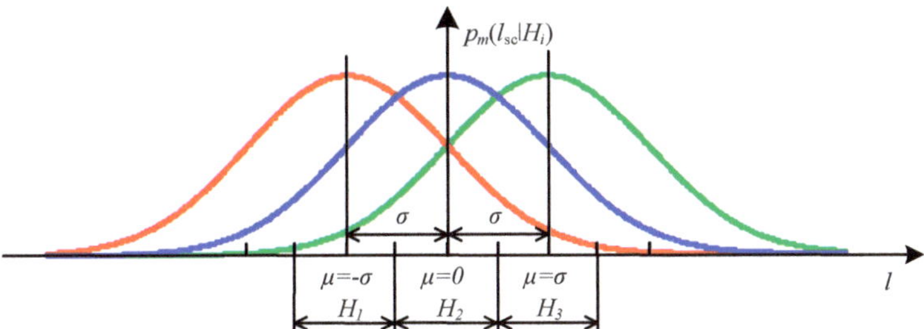

Figure 4. Probability distribution of hypotheses $p_m(l_{sc} | H_i)$.

As a result of calculating the distance to the fault location (Expression (7)), based on the instantaneous values of the current (Figure 3a,b) and voltage oscillograms, ten consecutive sample values l_{sc} were obtained, which are given in Table 2.

Table 2. Sample values of the distance to the fault location obtained by measuring the instantaneous values of the oscillograms of currents and voltages of the emergency mode.

m	1	2	3	4	5	6	7	8	9	10
l_{sc} (km)	25.85	24.9	23.7	26.35	24.6	25.6	25.7	27.36	23.75	25.05

Due to the scatter of the sample values of l_{sc}, it is impossible to make an unambiguous decision regarding the validity of the hypotheses H_1, H_2, H_3. The mathematical expectation of the sample values l_{sc} (Table 2) is $M[l_{sc}] = 25.185$ (km).

To implement a sequential analysis, we introduce a matrix of conditional probabilities (Expression (1)) for making a decision regarding the faulted section of the OHPL:

$$||P(k|i)|| = ||P_i(k)|| = ||P_{ik}|| = \begin{Vmatrix} 0.70 & 0.15 & 0.15 \\ 0.15 & 0.70 & 0.15 \\ 0.15 & 0.15 & 0.70 \end{Vmatrix}. \tag{8}$$

The choice of matrix elements should take into account the operational features of the OHPL, as well as the economic consequences of making an incorrect decision when determining the fault location.

Let us calculate the threshold values necessary for the implementation of a multi-hypothesis sequential analysis using the Armitage algorithm, taking into account the components of the matrix $||P(k \mid i)||$ (Expression (1)):

$$\lambda_m^{\text{threshold}}(x \mid H_{1,2}) = \lambda_m^{\text{threshold}}(x \mid H_{1,3}) = \lambda_m^{\text{threshold}}(x \mid H_{2,1})$$
$$= \lambda_m^{\text{threshold}}(x \mid H_{2,3}) = \lambda_m^{\text{threshold}}(x \mid H_{3,1}) = \lambda_m^{\text{threshold}}(x \mid H_{3,2}) \tag{9}$$
$$= (1/P_{12}) \cdot [1 - \sum_{k \neq l} P_{12}] = (1/0.15) \times [1 - 0.15] = 6.667 \times 0.85 = 5.667$$

It is advisable to calculate the likelihood ratios $\lambda_m(x \mid H_{k,l}) = p_m(x \mid H_k)/p_m(x \mid H_l)$ for each ratio $k \neq l$ using the standard Gaussian function, the tables of which are given in the following [84]:

$$f(x) = \left(1/\sqrt{2\pi}\right) \cdot exp\left\{-x^2/2\right\}, \tag{10}$$

At the first step of sequential analysis in relative units $I^{\text{st}}_{\text{sc}}(1) = 0.398$, then we achieve the following:

$$\lambda_1(0.398 \mid H_{1,2}) = p_1(0.398H_1)/p_2(0.398 \mid H_2) = 0.0485/0.369 = 0.131;$$
$$\lambda_1(0.398 \mid H_{1,3}) = p_1(0.398H_1)/p_2(0.398 \mid H_3) = 0.0485/0.18 = 0.269;$$
$$\lambda_1(0.398 \mid H_{2,1}) = p_2(0.398H_2)/p_1(0.398 \mid H_1) = 0.369/0.0485 = 7.608;$$
$$\lambda_1(0.398 \mid H_{2,3}) = p_2(0.398H_2)/p_3(0.398 \mid H_3) = 0.369/0.18 = 2.05;$$
$$\lambda_1(0.398 \mid H_{3,1}) = p_3(0.398H_3)/p_1(0.398 \mid H_1) = 0.18/0.0485 = 3.711;$$
$$\lambda_1(0.398 \mid H_{3,2}) = p_3(0.398H_3)/p_2(0.398 \mid H_2) = 0.18/0.369 = 0.488.$$

The calculation results show that at the first step of the sequential analysis, which is only one likelihood ratio $\lambda_1(0.398 \mid H_{2,1}) = 7.608$, exceeds the specified threshold value. This is due to the introduced condition that all $(M - 1)$ likelihood ratios $\lambda_m(x \mid H_{k,l}) > \lambda_m^{\text{threshold}}(x \mid H_{k,l})$, $l = 1, \ldots, M$, $k \neq l$, characteristic of the hypothesis H_k, must simultaneously exceed the corresponding thresholds $\lambda_m^{\text{threshold}}(x \mid H_{k,l})$. If this condition is not met, then the sequential analysis continues.

Likewise, the likelihood ratios are calculated for the next steps m of sequential analysis using the Armitage algorithm, as shown in Table 3.

Table 3. Likelihood ratio calculation results required to implement sequential analysis using the Armitage algorithm.

m	1	2	3	4	5	6	7	8	9	10
$\lambda_m(x \mid H_{1,2})$	0.131	0.043	0.047	0.004	0.002	–	–	–	–	–
$\lambda_m(x \mid H_{1,3})$	0.269	0.475	9.362	0.917	2.953	–	–	–	–	–
$\lambda_m(x \mid H_{2,1})$	7.608	23.098	23.236	298.12	669.88	–	–	–	–	–
$\lambda_m(x \mid H_{2,3})$	2.05	10.98	196.25	247.66	1790	–	–	–	–	–
$\lambda_m(x \mid H_{3,1})$	3.711	2.10	1.05	1.07	0.332	–	–	–	–	–
$\lambda_m(x \mid H_{3,2})$	0.488	0.091	0.005	0.004	0.0006	–	–	–	–	–

Visually, the process of implementing sequential analysis using the Armitage algorithm is shown in Figure 5.

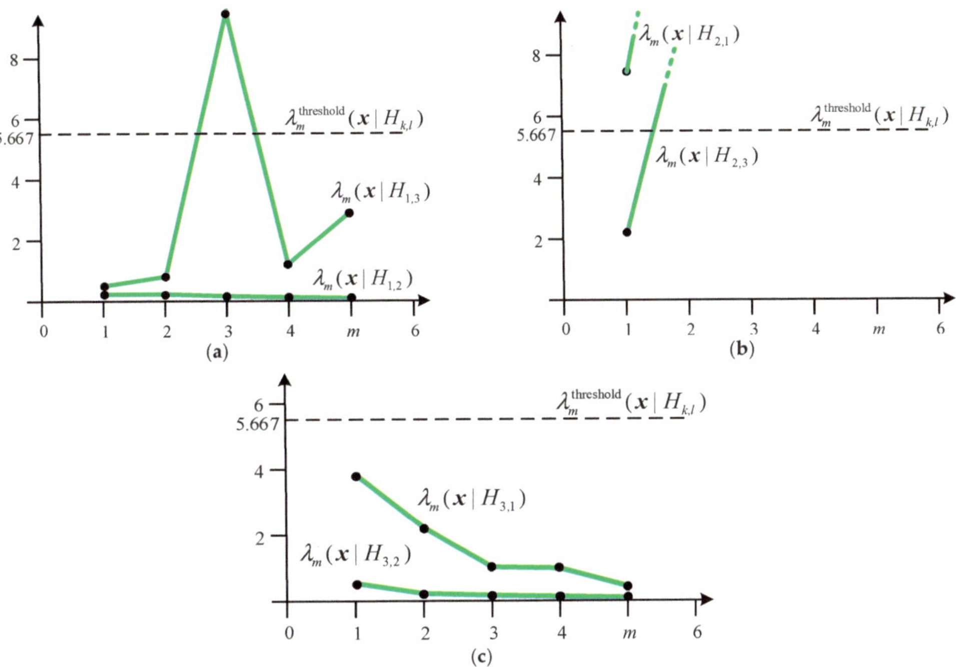

Figure 5. The process of implementing sequential analysis using the Armitage algorithm for a 110 kV overhead line with a length of l = 50 km: (**a**) likelihood ratios $\lambda_m(x \mid H_{1,2})$, $\lambda_m(x \mid H_{1,3})$; (**b**) likelihood ratios $\lambda_m(x \mid H_{2,1})$, $\lambda_m(x \mid H_{2,3})$; (**c**) likelihood ratios $\lambda_m(x \mid H_{3,1})$, $\lambda_m(x \mid H_{3,2})$.

Analysis of Figure 5 allows us to draw the following conclusions:

- Multi-criteria sequential analysis using the Armitage algorithm as applied to FL for OHPLs by EMP leads to the selection of a faulted section in the interval $M[l_{\mathrm{sc}}] \pm \sigma/2 = 25.185 \pm 0.835$ (km);
- The sequential analysis procedure does not require significant time costs, allowing us to make a decision about the faulted section in two steps, practically without affecting the speed of the OHPL fault algorithm;
- There is no need to use special computational methods to increase the speed of FL for OHPLs;
- A comparison of the likelihood ratios shown in Figure 5 allows us to state that the H_3 hypothesis is the least probable; therefore, the inspection of the OHPL should be started from Section 25.185 + 0.835 (km) toward power System-1 (Figure 2), i.e., the most likely location of the fault;
- The speed of making a decision on the fault location on OHPLs when implementing sequential analysis depends on the degree of distortion of currents and voltages in emergency mode oscillograms, including deviations of power quality parameters from standard values [85].

The proposed method is applicable to determining the location of faults on OHPLs accompanied by one-, two-, and three-phase SCs. The reasons for such SCs may be as follows:

- Overlaps as a result of thunderstorms;
- Falling of trees onto wires without breaking the wire or overlapping onto tree branches;
- Overlap with the destruction of insulators, for example, due to unauthorized persons shooting at the garland from a hunting rifle;
- Overlap from the wire to the support body as a result of strong winds, ice, and frost deposits;
- Blocking the wire from passing large-sized machinery and agricultural machinery;
- Breakage of lightning protection cables followed by an SC of the phase wire(s) to the ground;
- A break with a wire falling to the ground;
- Uncoupling of the insulator string;
- Throwing metal objects onto overhead line wires by unauthorized persons;
- Other reasons.

It should be noted that the implementation of a multi-hypothesis sequential analysis slightly increases the time of the decision-making process regarding the faulted area when determining the FL. Based on statistical calculations, the number of stages of the sequential procedure does not exceed 8–10 steps [74]. Moreover, as shown above, the number of processing operations at each step of the sequential procedure is a small number. Thus, for modern FL devices, the practical implementation of the proposed algorithm does not require significant computational costs and time. On the other hand, information about the FL is necessary in network control centers (dispatch centers) to assign the line crew a zone to bypass the faulted OHPL. The time from the moment of fault detection on OHPLs (triggering of relay protection devices) until the departure of the line crew, as a rule, in Russian practice ranges from tens of minutes to an hour and a half. Therefore, the complication of calculations and the additional time spent on them for the proposed FL method do not have any effect on the overall speed of eliminating faults on OHPLs.

4. Conclusions

The influence of various random factors, including deviations of power quality indicators from standard values, leads to the need to use statistical procedures when determining the fault location on an overhead power line based on emergency mode parameters.

A method has been developed for determining the location of the fault based on emergency mode parameters, including dividing the overhead power line bypass zone into many sections, followed by the implementation of a statistical procedure for recognizing the faulted area based on the Armitage multi-hypothesis sequential analysis algorithm.

The use of the Armitage algorithm allows us to adapt the decision-making process regarding the faulted area to the distortion features of emergency oscillograms.

An analysis of the computational operations of the proposed fault location method using the EMP shows an insignificant dependence of its performance on the distortion of emergency oscillograms, but at the same time, unambiguous decision-making is ensured regarding the faulted section of the overhead power line.

Author Contributions: Conceptualization, A.K. and P.I.; methodology, A.K. and S.F.; software, A.L.; validation, P.I. and S.F.; formal analysis, A.K. and A.L.; investigation, P.I. and S.F.; resources, A.K.; data curation, P.I.; writing—original draft preparation, A.K. and P.I.; writing—review and editing, A.L. and S.F.; visualization, A.L.; supervision, A.K.; project administration, P.I.; funding acquisition, S.F. All authors have read and agreed to the published version of the manuscript.

Funding: This work is supported by the Russian Science Foundation under grant 21-79-30013 for the Energy Research Institute of the Russian Academy of Sciences.

Data Availability Statement: Not applicable.

Abbreviation

FL	fault location
OHPL	overhead power line
EMP	emergency mode parameters
RP	relay protection
SC	short circuit
PMU	phasor measurement unit

References

1. Voropai, N. Electric Power System Transformations: A Review of Main Prospects and Challenges. *Energies* **2020**, *13*, 5639. [CrossRef]
2. Suliman, A.S.; Ahmed, B.M.; Elareefi, M.B.; Elrahman, E.A.E.A.; Arbab, E.A.; Abdulwahab, M.M. Monitoring System for Overhead Power Transmission Lines in Smart Grid System Using Internet of Things. *Univ. Khartoum Eng. J.* **2022**, *12*, 1. [CrossRef]
3. Savina, N.V.; Varygina, A.O. Selection of an optimal cable brand for high-voltage overhead power lines based on criterion analysis. *Ipolytech J.* **2023**, *27*, 339–353. [CrossRef]
4. Korotkevich, M.A.; Podgaisky, S.I. On the Expediency of Laying Cable Power Lines with a Voltage of 6–35 kV Outside Settlements Instead of Overhead Power Lines. *ENERGETIKA Proc. CIS High. Educ. Inst. Power Eng. Assoc.* **2022**, *65*, 463–476. [CrossRef]
5. Nazemi M Dehghanian, P.; Darestani, Y.M.; Jinshun, S.J. Parameterized Wildfire Fragility Functions for Overhead Power Line Conductors. *Power Syst. IEEE Trans. Power Syst.* **2023**, *99*, 1–11. [CrossRef]
6. Gonçalves, A.C.R.; Marques, M.C.; Loureiro, S.; Nieto, R.; Liberato, M.L.R. Disruption risk analysis of the overhead power lines in Portugal. *Energy* **2022**, *263*, 125583. [CrossRef]
7. Li, P.; Qiu, R.; Wang, M.; Wang, X.; Jaffry, S.; Xu, M.; Huang, K.; Huang, Y. Online Monitoring of Overhead Power Lines Against Tree Intrusion via a Low-cost Camera and Mobile Edge Computing Approach. *J. Phys. Conf. Ser.* **2023**, *2422*, 012018. [CrossRef]
8. Farkhadzeh, E.M.; Muradaliev, A.Z.; Abdullaeva, S.A.; Nazarov, A.A. Quantitative Assessment of the Operational Reliability of Overhead Power Transmission Lines. *Power Technol. Eng.* **2022**, *55*, 790–796. [CrossRef]
9. Listyukhin, V.A.; Pecherskaya, E.A.; Artamonov, D.; Zinchenko, T.O.; Anisimova, A.A. Improving the reliability of overhead power transmission lines through the introduction of information and measurement systems for monitoring their parameters. *Meas. Monit. Manag. Control* **2022**, *3*, 62–68. (In Russian) [CrossRef]
10. Ilyushin, P.; Volnyi, V.; Suslov, K.; Filippov, S. Review of Methods for Addressing Challenging Issues in the Operation of Protection Devices in Microgrids with Voltages of up to 1 kV that Integrates Distributed Energy Resources. *Energies* **2022**, *15*, 9186. [CrossRef]
11. Listyukhin, V.; Pecherskaya, E.; Gurin, S.; Anisimova, A.; Shepeleva, A. Analysis and classification of cause and effect factors affecting the operating modes of overhead power lines. *AIP Conf. Proc.* **2022**, *2767*, 020019. [CrossRef]
12. Ilyushin, P.; Filippov, S.; Kulikov, A.; Suslov, K.; Karamov, D. Specific Features of Operation of Distributed Generation Facilities Based on Gas Reciprocating Units in Internal Power Systems of Industrial Entities. *Machines* **2022**, *10*, 693. [CrossRef]
13. Gurevich, Y.E.; Kabikov, K.V. *Peculiarities of Power Supply, Aimed at Failure-Free Operation of Industrial Consumers*; ELEKS-KM: Moscow, Russia, 2005. (In Russian)
14. Ilyushin, P.V. Emergency and post-emergency control in the formation of micro-grids. *E3S Web Conf.* **2017**, *25*, 02002. [CrossRef]
15. Maly, A.S.; Shalyt, G.M.; Eisenfeld, A.I. *Determination of the Places of Damage to Power Transmission Lines According to the Parameters of the Emergency Mode*; Energiya: Moscow, Russia, 1972; 215p. (In Russian)
16. Kezunovic, M.; Knezev, M. Selection of optimal fault location algorithm. In Proceedings of the 2008 IEEE Power and Energy Society General Meeting—Conversion and Delivery of Electrical Energy in the 21st Century, Pittsburgh, PA, USA, 20–24 July 2008. [CrossRef]
17. Panahi, H.; Zamani, R.; Sanaye-Pasand, M.; Mehrjerdi, H. Advances in Transmission Network Fault Location in Modern Power Systems: Review, Outlook and Future Works. *IEEE Access* **2021**, *9*, 158599–158615. [CrossRef]
18. Schwan, M.; Ettinger, A.; Gunaltay, S. Probabilistic reliability assessment in distribution network master plan development and in distribution automation implementation. In Proceedings of the CIGRE, 2012 Session, Paris, France, 26–30 August 2012, Rep. C4-203.
19. Shor, Y.B. *Statistical Methods of Analysis and Quality Control and Reliability*; Soviet Radio: Moscow, Russia, 1962. (In Russian)
20. Bollen, M.H.J.; Dirix, P.M.E. Simple model for post-fault motor behaviours for reliability/power quality assessment of industrial power systems. *IEE Proc.—Gener. Transm. Distrib.* **1996**, *143*, 56–60. [CrossRef]
21. Powanga, L.; Kwakwa, P.A. Determinants of Electricity Transmission and Distribution Losses in South Africa. *J. Renew. Energy* **2023**, *2023*, 2376449. [CrossRef]
22. Thomas, D.; Fung, J. Measuring downstream supply chain losses due to power disturbances. *Energy Econ.* **2022**, *114*, 106314. [CrossRef]
23. Koks, E.E.; Pant, R.; Thacker, S.; Hall, J. Understanding Business Disruption and Economic Losses Due to Electricity Failures and Flooding. *Int. J. Disaster Risk Sci.* **2019**, *10*, 421–438. [CrossRef]
24. Lacommare, K.; Eto, J. Cost of power interruptions to electricity consumers in the United States (US). *Energy* **2006**, *31*, 1845–1855. [CrossRef]

25. Edomah, N. Effects of voltage sags, swell and other disturbances on electrical equipment and their economic implications. In Proceedings of the CIRED 2009—20th International Conference and Exhibition on Electricity Distribution—Part 1, Prague, Czech Republic, 8–11 June 2009; pp. 1–4.
26. Shalyt, G.M. *Determination of the Place of Damage in Electrical Networks*; Energoatomizdat: Moscow, Russia, 1982; 312p. (In Russian)
27. Krzysztof, G.; Kowalik, R.; Rasolomampionona, D.D.; Anwar, S. Traveling wave fault location in power transmission systems: An overview. *J. Electr. Syst.* **2011**, *7*, 287–296.
28. Lachugin, V.F. Wave methods for determining the location of damage on overhead power lines. *Relay Prot. Autom.* **2023**, *1*, 58–61. (In Russian)
29. Obalin, M.D.; Kulikov, A.L. Application of adaptive procedures in algorithms for determining the location of damage to power lines. *Ind. Power Eng.* **2013**, *12*, 35–39. (In Russian)
30. Abbasi, F.; Abdoos, A.; Hosseini, S.M.; Sanaye-Pasand, M. New ground fault location approach for partially coupled transmission lines. *Electr. Power Syst. Res.* **2023**, *216*, 109054. [CrossRef]
31. Sampaio, F.; Tofoli, F.L.; Melo, L.S.; Barroso, G.C.; Sampaio, R.F.; Leao, R. Adaptive fuzzy directional bat algorithm for the optimal coordination of protection systems based on directional overcurrent relays. *Electr. Power Syst. Res.* **2022**, *211*, 108619. [CrossRef]
32. Aziz, T.; Lin, Z.; Waseem, M.; Liu, S. Review on optimization methodologies in transmission network reconfiguration of power systems for grid resilience. *Int. Trans. Electr. Energy Syst.* **2021**, *31*, e12704. [CrossRef]
33. Ilyushin, P.V.; Pazderin, A.V. Approaches to organization of emergency control at isolated operation of energy areas with distributed generation. In Proceedings of the International Ural Conference on Green Energy (URALCON), Chelyabinsk, Russia, 4–6 October 2018. [CrossRef]
34. Gurevich, Y.E.; Libova, L.E. *Application of Mathematical Models of Electrical Load in Calculation of the Power Systems Stability and Reliability of Power Supply to Industrial Enterprises*; ELEKS-KM: Moscow, Russia, 2008. (In Russian)
35. Ilyushin, P.V.; Filippov, S.P. Under-frequency load shedding strategies for power districts with distributed generation. In Proceedings of the 2019 International Conference on Industrial Engineering, Applications and Manufacturing (ICIEAM), Sochi, Russia, 25–29 March 2019. [CrossRef]
36. Voitovich, P.A.; Lavrov, Y.A.; Petrova, N.F. Innovative technical solutions for the construction of ultra-compact high-voltage overhead power transmission lines. *New Russ. Electr. Power Ind.* **2018**, *8*, 44–57. (In Russian)
37. Kosyakov, A.A.; Kuleshov, P.V.; Pogudin, A.L. The influence of the grounding device structure of a substation on the voltage of conducted interference of lightning currents. *Russ. Electr. Eng.* **2019**, *90*, 752–755. [CrossRef]
38. Sharipov, U.B.; Égamnazarov, G.A. Calculating currents in lightning protection cables and in optical cables built into them during asymmetric short circuits in overhead transmission lines. *Power Technol. Eng.* **2017**, *50*, 673–678. [CrossRef]
39. Shevchenko, N.Y.; Ugarov, G.G.; Kirillova, S.N.; Lebedeva, Y.V. Review and analysis of the design features of overhead power line wires with increased resistance to icy-wind loads. *Quest. Electr. Technol.* **2018**, *4*, 53–63. (In Russian)
40. Kulikov, A.; Ilyushin, P.; Loskutov, A.; Suslov, K.; Filippov, S. WSPRT Methods for Improving Power System Automation Devices in the Conditions of Distributed Generation Sources Operation. *Energies* **2022**, *15*, 8448. [CrossRef]
41. Saha, M.M.; Izykowski, J.; Rosolowski, E. *Fault Location on Power Networks*; Springer: London, UK, 2010; 437p.
42. Visyashchev, A.N. *Devices and Methods for Determining the Location of Damage on Power Transmission Lines: A textbook. At 2 h. h. 1*; Publishing House of IrSTU: Irkutsk, Russia, 2001; 188p. (In Russian)
43. Lebedev, V.; Filatova, G.; Timofeev, A. Increase of accuracy of the fault location methods for overhead electrical power lines. *Adv. Mater. Sci. Eng.* **2018**, *2018*, 3098107. [CrossRef]
44. Arzhannikov, E.A.; Lukoyanov, V.Y.; Misrikhanov, M.S. *Determining the Location of a Short Circuit on High−Voltage Power Transmission Lines*; Shuin, V.A., Ed.; Energoatomizdat: Moscow, Russia, 2003; 272p. (In Russian)
45. Ilyushin, P.V.; Shepovalova, O.V.; Filippov, S.P.; Nekrasov, A.A. The effect of complex load on the reliable operation of solar photovoltaic and wind power stations integrated into energy systems and into off-grid energy areas. *Energy Rep.* **2022**, *8*, 1515–1529. [CrossRef]
46. Kulikov, A.L. *Remote Determination of Power Line Damage Sites by Active Sensing Methods*; Energoatomizdat: Moscow, Russia, 2006; 148p. (In Russian)
47. Minullin, R.G. Detecting the faults of overhead electric-power lines by the location-probing method. *Russ. Electr. Eng.* **2017**, *88*, 61–70. [CrossRef]
48. Kulikov, A.L.; Lukicheva, I.A. Determination of the place of damage to the power transmission line by the instantaneous values of the oscillograms of emergency events. *Bull. Ivanovo State Power Eng. Univ.* **2016**, *5*, 16–21. (In Russian)
49. Ustinov, A.A.; Visyashchev, A.N. Iterative methods for determining the location of damage by the parameters of the emergency mode during one-way measurements on overhead power lines. *Bull. IrSTU* **2010**, *5*, 260–266. (In Russian)
50. Kulikov, A.; Ilyushin, P.; Suslov, K.; Filippov, S. Estimating the Error of Fault Location on Overhead Power Lines by Emergency State Parameters Using an Analytical Technique. *Energies* **2023**, *16*, 1552. [CrossRef]
51. Aboshady, F.M.; Thomas, D.W.P. A Wideband Single End Fault Location Scheme for Active Untransposed Distribution Systems. *IEEE Trans. Smart Grid* **2020**, *11*, 2115–2124. [CrossRef]
52. Takagi, T.; Yamakoshi, Y.; Yamaura, Y.; Kondow, R.; Matsushima, T. Development of a new type fault locator using the one-terminal voltage and current data. *IEEE Trans. Power Appl. Syst.* **1982**, *PAS-101*, 2892–2898. [CrossRef]

53. Simeon, O.; Faithpraise, F.O.; Ibanga, J. Iterative Newton-Raphson-Based Impedance Method for Fault Distance Detection on Transmission Line. *Int. Multiling. J. Sci. Technol.* **2020**, *5*, 2805–2810. [CrossRef]
54. Nagendra Reddy, P.L.V.; Mukunda, V.K.S.; Sushyanth, C.; Vanitha, V. Implementation of Novosel Simple Impedance Algorithm for fault location. *IJCTA* **2016**, *9*, 7589–7596.
55. Schweitzer, E.O. A review of impedance-based fault locating experience. In Proceedings of the 14th Annual Iowa–Nebraska System Protection Seminar, Omaha, NE, USA, 12 October 1990; pp. 1–31.
56. Suslov, K.V.; Solonina, N.N.; Solonina, Z.V.; Akhmetshin, A. Operational determination of the point of a short circuit in power lines. *Power Eng. Res. Equip. Technol.* **2023**, *25*, 71–83. [CrossRef]
57. Bahmanyar, A.; Jamali, S.; Estebsari, A.; Bompard, E. A comparison framework for distribution system outage and fault location methods. *Electr. Power Syst. Res.* **2017**, *145*, 19–34. [CrossRef]
58. Liao, Y. Transmission Line Fault Location Algorithms Without Requiring Line Parameters. *Electr. Power Compon. Syst.* **2008**, *36*, 1218–1225. [CrossRef]
59. Stringfield, T.W.; Marihart, D.J.; Stevens, R.F. Fault location methods for overhead lines. *IEEE Trans. Power Appar. Syst.* **1957**, *76*, 518–530. [CrossRef]
60. Born, E.; Jaeger, J. Device locates point of fault on transmission lines. *Electr. World* **1967**, *168*, 133–134.
61. Senderovich, G.A.; Gryb, O.G.; Karpaliuk, I.T.; Shvets, S.V.; Zaporozhets, A.O.; Samoilenko, I.A. Automation of determining the location of damage of overhead power lines. *Stud. Syst. Decis. Control* **2021**, *359*, 35–53.
62. Senderovich, G.A.; Gryb, O.G.; Karpaliuk, I.T.; Shvets, S.V.; Zaporozhets, A.O.; Samoilenko, I.A. Experimental studies of the method for determining location of damage of overhead power lines in the operation mode. *Stud. Syst. Decis. Control* **2021**, *359*, 55–77.
63. Lyamets, Y.Y.; Antonov, V.I.; Efremov, V.A. *Diagnostics of Power Transmission Lines. Electrotechnical Microprocessor Devices and Systems: Interuniversity Collection of Scientific Papers*; Publishing House of the Chuvash State University: Cheboksary, Russia, 1992. (In Russian)
64. Yu, C.S.; Liu, C.W.; Jiang, J.A. A new fault location algorithm for series compensated lines using synchronized phasor measurements. In Proceedings of the 2000 Power Engineering Society Summer Meeting, Seattle, WA, USA, 16–20 July 2000. [CrossRef]
65. Kezunovic, M.; Meliopoulos, S.; Venkatasubramanian, V.; Vittal, V. *Application of Time-Synchronized Measurements in Power System Transmission Networks*; Springer: New York, NY, USA, 2014.
66. *STO 56947007–29.240.55.159–2013*; Standard Instructions for the Organization of Work to Determine the Places of Damage to Overhead Power Lines with a Voltage of 110 kV and above. Standard of the Organization of PJSC FGC UES: Moscow, Russia, 2013.
67. Martynov, M.V. Method of interval determination of the place of damage to the power transmission line. Patent of the Russian Federation No. 2720949, 15 May 2020. (In Russian).
68. Bykadorov, A.L.; Zarutskaya, T.A.; Muratova-Milekhina, A.S. Application of pattern recognition theory in determining the location of a short circuit in AC traction networks. *Bull. RSUPS* **2021**, *2*, 119–128. [CrossRef]
69. Kulikov, A.L.; Shepovalova, O.V.; Ilyushin, P.V.; Filippov, S.P.; Chirkov, S.V. Control of electric power quality indicators in distribution networks comprising a high share of solar photovoltaic and wind power stations. *Energy Rep.* **2022**, *8*, 1501–1514. [CrossRef]
70. Won, D.J.; Chung, I.Y.; Kim, J.M.; Moon, S.I.; Seo, J.C.; Choe, J.W. A new algorithm to locate power-quality event source with improved realization of distributed monitoring scheme. *IEEE Trans. Power Deliv.* **2006**, *21*, 1641–1647. [CrossRef]
71. Shirman, Y.D. *Radio-Electronic Systems: Fundamentals of Construction and Theory*, 2nd ed.; Radio Engineering: Moscow, Russia, 2007; 510p. (In Russian)
72. Armitage, P. Sequential analysis with more than two alternative hypotheses, and its relation to discriminant function analysis. *J. R. Stat. Soc.* **1950**, *12*, 137–144. Available online: https://api.semanticscholar.org/CorpusID:126310625 (accessed on 1 September 2023). [CrossRef]
73. Ghosh, B.K.; Sen, P.K. *Handbook of Sequential Analysis (Statistics: A Series of Textbooks and Monographs)*, 1st ed.; CRC Press: Boca Raton, FL, USA, 1991; 664p.
74. Wald, A. *Sequential Analysis*; John Wiley and Sons: New York, NY, USA, 1947; 232p.
75. Jouny, I.; Garber, F.D. Mary sequential hypothesis tests for automatic target recognition. *IEEE Trans. Aerosp. Electron. Syst.* **1992**, *28*, 473–483. [CrossRef]
76. Kulikov, A.L.; Ilyushin, P.V.; Suslov, K.V.; Karamov, D.N. Coherence of digital processing of current and voltage signals at decimation for power systems with a large share of renewable power stations. *Energy Rep.* **2022**, *8*, 1464–1478. [CrossRef]
77. Visyashchev, A.N.; Plenkov, E.R.; Tiguntsev, S.G. A method for Determining the Location of a Short Circuit on an Overhead Power Transmission Line with Unsynchronized Measurements from Its Two Ends. Patent of the Russian Federation No. 2508556, 27 February 2014. (In Russian).
78. Varma, R.K.; Rahman, S.A.; Vanderheide, T.; Dang, M.D.N. Harmonic impact of a 20-MW PV solar farm on a utility distribution network. *IEEE Power Energy Technol. Syst. J.* **2016**, *3*, 89–98. [CrossRef]
79. Torquato, R.; Freitas, W.; Hax, G.R.T.; Donadon, A.R.; Moya, R. High frequency harmonic distortions measured in a Brazilian solar farm. In Proceedings of the 17th International Conference on Harmonics and Quality of Power (ICHQP), Belo Horizonte, Brazil, 16–19 October 2016; pp. 623–627.

80. *IEEE STD 519-2014*; Recommended Practices and Requirements for Harmonic Control in Electrical Power Systems/Transmission and Distribution Committee of the IEEE Power and Energy Society. The Institute of Electrical and Electronics Engineers: New York, NY, USA, 2014.
81. Adineh, B.; Keypour, R.; Sahoo, S.; Davari, P.; Blaabjerg, F. Robust optimization based harmonic mitigation method in islanded microgrids. *Int. J. Electr. Power Energy Syst.* **2022**, *137*, 107631. [CrossRef]
82. Saha, S.; Saleem, M.I.; Roy, T.K. Impact of high penetration of renewable energy sources on grid frequency behavior. *Int. J. Electr. Power Energy Syst.* **2022**, *145*, 108701. [CrossRef]
83. Ribeiro, P.F.; Duque, C.A.; Silveira, P.M.; Cerqueira, A.S. *Power Systems Signal Processing for Smart Grids*; John Wiley & Sons Ltd.: Hoboken, NJ, USA, 2014.
84. Wentzel, E.S. *Probability Theory: Textbook for University Students*, 10th ed.; Academia: Moscow, Russia, 2005; 571p. (In Russian)
85. Kulikov, A.; Ilyushin, P.; Suslov, K.; Filippov, S. Organization of Control of the Generalized Power Quality Parameter Using Wald's Sequential Analysis Procedure. *Inventions* **2023**, *8*, 17. [CrossRef]

 *inventions*

Article

A Rank Analysis and Ensemble Machine Learning Model for Load Forecasting in the Nodes of the Central Mongolian Power System

Tuvshin Osgonbaatar [1], Pavel Matrenin [1,2], Murodbek Safaraliev [2,*], Inga Zicmane [3], Anastasia Rusina [1] and Sergey Kokin [2]

[1] Faculty of Energy, Novosibirsk State Technical University, 20 K. Marx Ave., 630073 Novosibirsk, Russia; osgonbaatar@corp.nstu.ru (T.O.); p.v.matrenin@urfu.ru (P.M.); rusina@corp.nstu.ru (A.R.)
[2] Ural Power Engineering Institute, Ural Federal University, 19 Mira Str., 620002 Yekaterinburg, Russia; s.e.kokin@urfu.ru
[3] Faculty of Electrical and Environmental Engineering, Riga Technical University, 12/1 Azenes Str., 1048 Riga, Latvia; inga.zicmane@rtu.lv
* Correspondence: murodbek_03@mail.ru

Abstract: Forecasting electricity consumption is currently one of the most important scientific and practical tasks in the field of electric power industry. The early retrieval of data on expected load profiles makes it possible to choose the optimal operating mode of the system. The resultant forecast accuracy significantly affects the performance of the entire electrical complex and the operating conditions of the electricity market. This can be achieved through using a model of total electricity consumption designed with an acceptable margin of error. This paper proposes a new method for predicting power consumption in all nodes of the power system through the determination of rank coefficients calculated directly for the corresponding voltage level, including node substations, power supply zones, and other parts of the power system. The forecast of the daily load schedule and the construction of a power consumption model was based on the example of nodes in the central power system in Mongolia. An ensemble of decision trees was applied to construct a daily load schedule and rank coefficients were used to simulate consumption in the nodes. Initial data were obtained from daily load schedules, meteorological factors, and calendar features of the central power system, which accounts for the majority of energy consumption and generation in Mongolia. The study period was 2019–2021. The daily load schedules of the power system were constructed using machine learning with a probability of 1.25%. The proposed rank analysis for power system zones increases the forecasting accuracy for each zone and can improve the quality of management and create more favorable conditions for the development of distributed generation.

Keywords: forecasting; machine learning; rank models; daily load schedule; power supply zone; node substations; central power system of Mongolia

Citation: Osgonbaatar, T.; Matrenin, P.; Safaraliev, M.; Zicmane, I.; Rusina, A.; Kokin, S. A Rank Analysis and Ensemble Machine Learning Model for Load Forecasting in the Nodes of the Central Mongolian Power System. *Inventions* **2023**, *8*, 114. https://doi.org/10.3390/inventions8050114

Academic Editor: Om P. Malik

Received: 21 July 2023
Revised: 19 August 2023
Accepted: 23 August 2023
Published: 5 September 2023

1. Introduction

Modern electric power systems (EPSs) are complex and include a large number of structural elements that are connected hierarchically. They are characterized by a large share of generation from renewable energy sources as well as intellectualization. These factors complicate the functioning of EPSs, which must make their own adjustments to reliability assessment processes when planning and managing operation modes. In addition, the growing availability of renewable energy sources increases the instability of the power balance of the power system, as there is additional uncertainty in terms of electricity production. The combination of these factors makes the short-term forecasting of power consumption a critical aspect of ensuring the reliability and efficiency of the power system. The reliability of the power supply to individual consumer groups and the economic efficiency of the functioning of the power system as a whole depends on the accuracy of such short-term forecasting to a significant extent. Increasing the accuracy of forecasting saves energy resources and determines the efficiency of power supply management and the consequent increase in the profits of energy enterprises. This, in turn, is determined by the transition to market relations between the subjects of the wholesale market, as well as responsibility for the results of actions based on the forecast.

In the wholesale electricity market, the forecasting problem is solved on different time horizons: long term (for several years ahead), medium term (for a period from one month to one year ahead), and short term (for an hour, a day, or a week ahead, respectively, solved using hourly, daily, or weekly data). One of the biggest difficulties in the short-term forecasting of the electrical load is the unpredictable behavior of the observed objects, which are influenced by various external factors, including user actions [1].

The problem of forecasting power consumption is that it is necessary to simultaneously take into account a huge number of factors that have an impact on the change in energy consumption during the period under consideration. Experts in energy companies who forecast such dependencies acquire experience gradually, over months and years of work. At the same time, there is always the possibility of unforeseen load surges. Consequently, it is extremely important to use electrical load forecasting software that could minimize the number of such incidents through carefully analyzing historical trends.

It should also be noted that the solution to the above scientific and practical problems is of great interest to both manufacturers and consumers of electric energy. Thus, for electric power producers, load forecasting is significant from the point of view of optimizing the supply and reservation of electric energy, the convenience of carrying out preventive maintenance, and ensuring the safety of the operation of the EPS. For consumers, load forecasting is useful for minimizing costs associated with the payment of fines when exceeding capacity limits or overpayment for declared but unused capacity, as well as predicting downtime of technological equipment in case of a power shortage in the EPS.

Currently, the methodology used by the electric power industry has been extensively researched in numerous studies, for example [2–5]. There are many formalized methods for predicting power consumption (approximately 150 in total, but in practice only 20 to 30 are used), which can be conditionally divided into five main groups.

1. Regulatory methods (methods of "direct counting") are based on the use of energy consumption standards for the main types of products and sectors of the economy. The use of regulatory methods presupposes the prediction of specific power consumption rates per unit of production [6]. From the point of view of the proposed model, the advantages of this method include the fact that it is quite simple and does not require any complex calculations.

2. Technological methods take into account the policy of energy saving, efficient use of energy, justification of rational types of energy carriers, and modes of operation of electric receivers. The complexity of such accounting limits the scope of application of these methods by individual enterprises, while regulatory methods can be applied to relatively large territorial units (network nodes and energy districts). Difficulties in predicting specific indicators of electricity consumption constrain the use of both of the above methods [7].

3. Methods of processing consumer applications, for example, for connecting additional loads, are effective for individual substations but are much less effective for energy districts [8]. In other words, the comparative effectiveness of this method decreases with the enlargement of the territorial division, that is, with the increase in the number of consumers.

4. Forecasting methods based on mathematical models, including trend extrapolation methods (simple regression models) consist of establishing an analytical relationship between a certain modeled indicator (power consumption, load, balance indicators, etc.) and a set of parameters affecting it. The tasks of regression analysis are establishing the form of dependence, selecting a regression model, and evaluating model

parameters. Note that there is no minimally necessary data set that is required to prepare a reliable model [9]. However, the above listed methods rely on data obtained from consumers or on some standards obtained empirically, while others are based on statistical data processing using various mathematical methods or their combinations. Regression models and time series models should be noted as the most successful.

5. Economic–statistical and econometric methods have the main purpose of identifying future tendencies for predicting the load for the time period under consideration. The method studies and makes provisions for seasonal changes in energy consumption, the reduction of electricity consumption of large consumers due to the suspension of factories, equipment repairs, temperature factors, the shutdown of energy-intensive industries, and consumer withdrawal from the unified energy system due to high tariffs, as well as the reduction of electricity consumption by large enterprises, etc.

It is known that there are a large number of variables that affect the mode of power consumption and, accordingly, the accuracy of its forecast. These variables differ a lot and can be divided into explicit and implicit (latent), exogenous (originated outside the power system) and endogenous (conversely, born by the EPS itself) [7]. Power consumption, frequency, power losses, and overflows are clearly endogenous variables. Meteorological variables and the type of day are explicit exogenous variables [10,11]. Higher temperatures lead to an increase in power demand as people turn on air conditioning units to cool their homes and offices. Wind speed also has an impact on power consumption. For example, high wind speed at low temperatures leads to more intense heat removal from buildings. Cloudiness affects the cost of electricity for lighting.

To date, a number of different approaches to short-term forecasting have been proposed, starting from regression methods [12,13] and ending with machine learning approaches based on neural networks [8,14,15] and hybrid or analog forecasting methods [16,17]. A significant part of modern publications devoted to this problem are focused on the development and improvement of new information technologies for predicting time series, such as neural, fuzzy networks, genetic algorithms, etc. This is due to the ability of these methods to make a forecast in such conditions as the uncertainty of the initial data (the presence of telemetric distortions), the lack of a priori information, the complex non-stationary behavior of the predicted time series. A great part of the work is related to the development of forecasting algorithms for the entire power system or a separate node from which a large enterprise is supplied, while relatively little attention is paid to the problem of short-term forecasting in the nodes of the power system [18].

In [19], the authors proposed two approaches featuring the autoregression method (so called 'bottom–up' and 'top–down'). The first method separately predicted the daily load curve at each substation (or node). Then, the load profile of the system was formed as the sum of the load curves of the individual nodes in the second method, the daily load schedules of this power system were predicted using the autoregression algorithm, and through multiplying system consumption by the load distribution coefficient, the load profiles for each node were obtained.

Tan et al. [20,21] processed node and power system data using the deep learning method to obtain a consumption forecast for both the entire power system and for each node. The method implies that at the initial stage, the participation coefficient is determined in p.u. values; then, at the next stage, the daily load curves for each node are obtained with the use of these coefficients.

In [22–25], the neural network method was used to predict electricity consumption at the level of both the power system and its nodes. Data on the consumption of the power system and its nodes, as well as such exogenous variables as meteorological factors and calendar features, were used as input data. The prediction of the consumption of each node was implemented similarly to the total consumption, but the difference is in the calculation of the participation coefficient for the respective node.

Bruce Steven et al. [26,27] compared the results of load forecasting in 22 nodes performed using such methods as the proportional one, the linear regression method, the integrated moving average autoregression model, and the machine learning method. In these studies, the proportional method containing the coefficient of node participation showed sufficient accuracy in predicting the load in the nodes.

Wang et al. [28] proposed to use the support vector machine to predict electricity consumption in northern China. They considered, the analysis and processing of the initial data to be most important stage that leads to obtaining the correct result. Nonlinear initial data were transformed into linear ones using autoregression—a moving average. Based on this, the final result was formed taking into account seasonal factors according to the method of support vectors. Using fuzzy logic [29], load graphs were developed reflecting the influence of temperature, type of day, and time of year in Turkey. In that work, fuzzy logic played the role of an auxiliary tool for neural networks.

The above-mentioned works show the positive results of using a coefficient expressing the share of consumption of each node in total consumption. Moreover, the implementation of the methodology is quite simple and requires a small amount of information obtained on the basis of statistical data or calculation results, rather than big data in each node.

The advantages of classical methods are statistical significance, fast and easy implementation, prevalence, and that these methods are well investigated. They have some disadvantages, including low efficiency in predicting complex time series, a dependence on unreliable assumptions, a limited ability to use additional variables, and sensitivity to distortions [16]. In contrast, machine learning is more flexible than classical methods and has the possibility of using many different factors. However, the process of using them is quite complicated.

The purpose of this article is to present an approach to load forecasting for both power systems in general and for their individual nodes through the example of the central power system of Mongolia. For this purpose, a machine learning method based on an ensemble of decision trees was used to obtain a model of the total consumption of the power system [30–32]. The daily load curves of the entire power system, data on meteorological factors, and calendar features for 2021 were used as input data. Due to the lack of complete information on power consumption for each node of the system, these data were modeled via the calculation of the IEEE reliability test system (IEEE reliability test system), and the results were used further on for forecasting [33,34]. The establishment of the zone's participation coefficient was carried out via the method of rank models [35]. Rank models together with the machine learning method made it possible to predict consumption in each zone. Thus, calculating the various operating conditions of the EPS, a model was created.

The contributions of this study include the following:

- For the first time, a methodology was proposed to make load profile forecasts for the nodes of the EPS of Mongolia with hourly resolution. It can improve the accuracy of planning the EPS's operation.
- In contrast to existing studies on forecasting the power consumption of large energy systems, it was proposed to divide the power system into zones for predicting their power consumptions using rank analysis. This approach allows us to increase the forecasting accuracy for each zone, improve the quality of management, and create more favorable conditions for the development of distributed generation.
- It has been established that the accurate prediction of power consumption in Mongolia requires the use of temperature forecasting; other meteorological factors have little influence on consumption.
- It has been discovered that despite the cyclic nature of power consumption, statistical methods, such as ARIMA, are inferior to machine learning algorithms that are able to take into consideration additional factors, such as the type of day (weekends, holidays) and temperature.

The organization of this paper is as follows: Section 2 provides information about the research methods. Section 3 contains a description of the dataset and power system under consideration and the results of the research. Section 4 provides a discussion about the results. Finally, the conclusions are given in Section 5.

2. Research Methods

2.1. Autoregressive Integrated Moving Average (ARIMA) Model

The ARIMA model assumes that the forecast value is determined using a linear function of several previous values of the original time series and random errors. Modeling is implemented according to the following sequences [36].

The first stage is the calculation of the integrated component into which the data are integrated. It is achieved through subtracting each value from the previous retrospective value. The purpose of this stage is to create a time series without a trend. In other words, the time series is transformed to a stationary form from a non-stationary form to approximate modeling. There are three methods that are widely used to transform time series into stationary ones, including trend removal, seasonality, and differentiation [37]. In this paper, the differentiation method was used. The method is expressed using the following equation:

$$P_t' = P_t - P_{t-1},\tag{1}$$

where P_t'—converted value and P_t—actual value.

At the next stage, autoregression is performed. It calculates the forecast value based on the weighted sum of the previous values:

$$P_t = \sum_{k=1}^{p} \alpha_k P_{t-k},\tag{2}$$

where P_t is forecast value, α is the value of the weighting factor, and p is the order of the autoregression polyline.

The last step of the ARIMA is to calculate the moving average. The calculation of the moving average is performed similarly to autoregression, but errors are taken instead of actual previous values, as shown in the equation below:

$$u_t = \sum_{j=1}^{q} \beta_j u_{t-j},\tag{3}$$

where u_t is the value of the random error, β is the value of the weighting factor, and q is the order of the moving average polynomial.

As a result of the listed stages, the time series model was developed as follows.

$$P_t = \sum_{k=1}^{p} \alpha_k P_{t-k} + \sum_{j=1}^{q} \beta_j u_{t-j}.\tag{4}$$

It can be seen from the equation that the ARIMA model (p, d, q) is determined by the values of the polynomial degree p and q, which are calculated using the autocorrelation function (ACF) and the partial autocorrelation function (PACF) [38]. The value d reflects the number of steps required to bring the series to a stationary form. When constructing an ARIMA (p, d, q) time series model, it is necessary to strive to minimize the number of its parameters.

2.2. Ensemble Models

Ensemble models are machine learning algorithms that combine multiple individual models to improve the accuracy and robustness of predictions. Three popular types of ensemble models are Random Forest, AdaBoost, and XGBoost.

Random Forest is a decision tree-based ensemble model that creates multiple decision trees and aggregates their predictions to make a final prediction [39]. It is a powerful algorithm for both classification and regression tasks and is known for its ability to handle high-dimensional data with many features.

AdaBoost (Adaptive Boosting) is another ensemble model that combines weak learners to create a strong learner [40]. It works through iteratively training weak models on the same dataset and adjusting the weights of misclassified samples in each iteration to improve the overall accuracy of the model.

XGBoost (Extreme Gradient Boosting) is a gradient boosting algorithm that uses decision trees as base models [41]. It is known for its speed and scalability, making it a popular choice for large datasets. XGBoost also includes regularization techniques to prevent overfitting and improve generalization performance.

Consider a certain time series, expressed as follows:

$$S_n = \{(X_1, Y_1), (X_2, Y_2), \ldots, (X_n, Y_n)\},\tag{5}$$

where X are the source vectors containing the functions $f(X)$; Y are the output scalars or labels; S_n are the studied samples of the time series (X_n, Y_n) with observation n.

To develop an algorithm, it is necessary to divide the data into parts for training and testing. After training on the data, the algorithm should be built with a model that calculates the dependencies between the corresponding variables. In other words, at the end of the learning process, the algorithm outputs the function $\hat{h}(X, S_n)$ of the time series model.

The basic principle of the Random Forest method is that samples of size n from the training time series S_n are randomly selected and placed on the decision trees. Regression analysis and the classification of random samples are carried out on each tree, and their models are derived, which express the dependencies between random variables. The aggregation of results, which is performed via averaging the output of all decision trees, will become a Random Forest model. The main advantage of aggregation is that it is immune to outliers, since independent trees with different training samples are generated:

$$\hat{Y}' = \frac{1}{q}\sum_{i=1}^{q} \hat{h}\left(X, S_n^i\right),\tag{6}$$

where S_n^i is i-th random sample; q is the number of the decision tree.

A model developed using the Random Forest method has the ability to take into account several factors affecting electricity consumption simultaneously. The Random Forest method contains all the advantages of machine learning methods and is preferable to the method of support vectors and neural networks since it does not require a complex theory.

To build a model with sufficient probability using this algorithm, it is necessary to specify the number of decision trees and their depth. The more decision trees, the better the probability, but the time to build a Random Forest also increases proportionally. Also, the probability of the model depends on the depth of the decision trees. Despite the fact that increasing the depth improves the quality of both training and testing, the smaller the depth of the trees, the faster this algorithm is built and works. Hence the need for an optimal choice of the number of decision trees and their depth approach.

2.3. Rank Models

Rank models allow predicting the structural properties of objects. Such tasks are quite common when calculating operating conditions and optimizing them [34]. In particular, to predict consumption in the nodes of the power system, the method of rank models is used, whose basic idea is to establish the coefficient of participation of a node in total power consumption. If we imagine the power system as a hierarchical structure with several nodes that differ in the predicted value, then we can use these models. The rank can be determined in p.u. values for tie-substations, power supply areas, power plants, and other parts of the power system. The rank is determined using the following equation:

$$R_i = \frac{P_i}{P_{\Sigma}},\tag{7}$$

in which

$$P_{\Sigma} = \sum_{i=1}^{n} P_i,\tag{8}$$

where P_i is the forecast parameter for the ith part, P_Σ—is the total value for the object under study, R_i is the rank coefficient, and n is the number of parts.

If the value of the total load or consumption is known, it can be distributed among nodes using rank coefficients. In other words, to obtain a forecast load curve for the nodes, the values of the predicted daily load profiles of the entire power system must be multiplied by the rank coefficient:

$$P_i = P\left(\hat{Y}'\right) * R_i, \tag{9}$$

where $P\left(\hat{Y}'\right)$ is the value of the daily load curve of the power system, R_i is the rank coefficient, and P_i is the value of the daily load curve of the ith node.

If the rank remains stable at different time periods, using rank models together with other forecasting methods, it is possible to make a time series of consumption at each node. Otherwise, it is necessary to determine the rank coefficient for different time periods.

The prediction error should be measured using the average modulo error (MAE is main absolute error) and the average modulo error in percent (MAPE is main absolute percentage error). They are expressed using the following equations:

$$MAE = \frac{1}{N}\sum_{m=1}^{N}\left|P_m - P'_m\right|, \tag{10}$$

$$MAPE = \frac{1}{N}\sum_{m=1}^{N}\left|\frac{P_m - P'_m}{P_m}\right| * 100, \tag{11}$$

where N is number of hours in the data set, P'_m is the power forecast value in the mth hour, and P_m is actual value in the mth hour.

3. Results

The Mongolian power system contains five regions including the central power system, which accounts for 97% of the country's energy consumption. In 2021, the consumption of the central power system reached 9.8 million kWh, and generation—7484 million kWh. In 2021, combined heat and power (CHP) plants covered 92% of the total generation. Solar and wind power plants generated the other 8% [42–44]. The total installed capacity of renewable energy plants in the central power system is 268 MW, comprising 23 MW of hydropower plants, 67 MW of solar photovoltaic stations, and 155 MW of wind turbine plants. The main share of power consumption is occupied by the household sector, since industry is relatively underdeveloped. The central power system provides electricity to approximately 40 settlements. Although the mining industry has been on the rise recently, the household sector is expected to remain the leader in electricity consumption in the immediate future, according to the forecast.

3.1. The Result of the Autoregressive Integrated Moving Average Model

Any retrospective is non-stationarity by nature, since the shape of the daily load curves included in the time series is influenced by some factors. Therefore, according to the rules of analysis of the model, it is necessary to convert it to a stationary form. As an example, Figure 1 partially shows a time series of the source data. The data set is received from the system operator of the power system of Mongolia and depicts hourly load consumption.

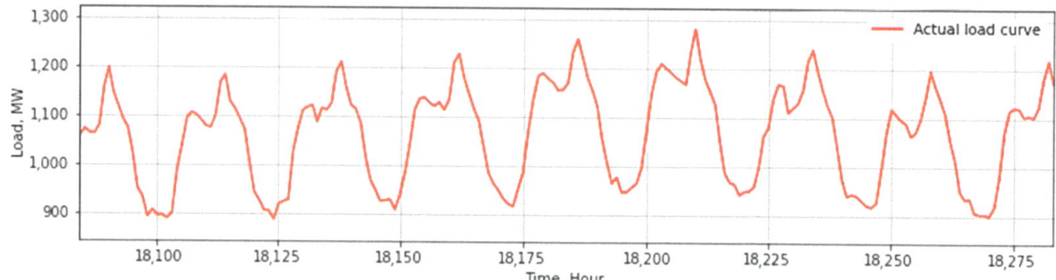

Figure 1. Example of a time series of source data.

After removing the trend and seasonality via differentiation, the time series is turned into the following pattern, as shown in Figure 2.

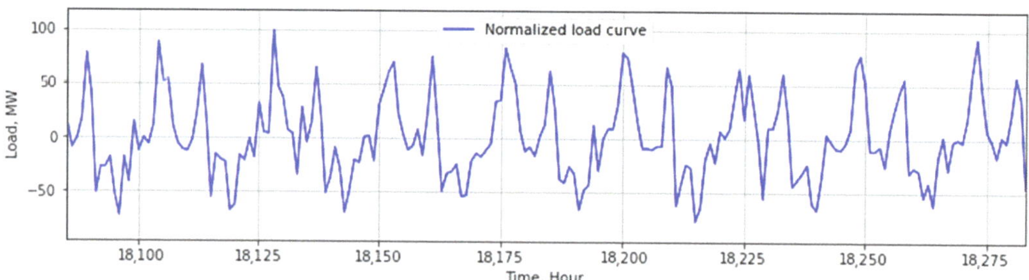

Figure 2. Transformed time series.

The initial data were divided into training and testing parts. As a training set, hourly consumption data were selected for the last 30 days before the forecast day. The analysis of autoregression and moving average was carried out on a training set, and a time series model was built. Figure 3 shows the modelled and the actual time series.

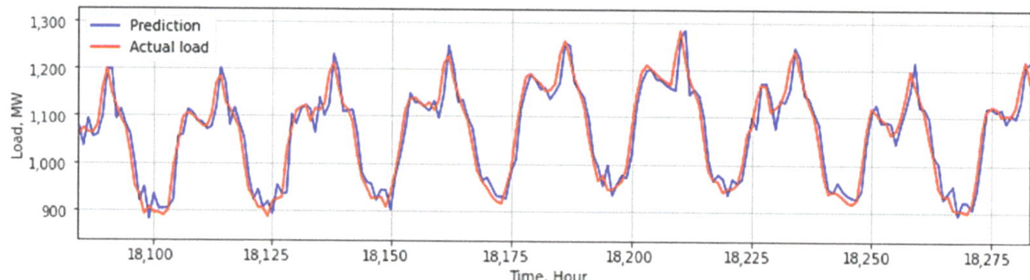

Figure 3. Time series modeling.

According to the developed model, the daily load curves of forecast days were predicted, which were randomly selected from each month. Since regression analysis considers the relationship between only two variables, ARIMA does not have the ability to take into account additional variables, including weather factors and other factors that affect electricity consumption.

3.2. The Result of Ensemble Models

As input data, the daily load profiles of the central power system of Mongolia were taken. These load profiles account for most of Mongolia's electricity consumption and generation. The observation period is 1 January 2019 to 31 December 2021. Meteorological factors, including wind speed, humidity, and outdoor air temperature, were used as input data for model construction. In addition to these data, calendar features were calculated, including the type of day (weekdays and weekends) and the day of the week (*Monday, Tuesday, . . ., Sunday*) in order to reflect the difference in daily consumption.

Table 1 shows a part of the of initial data, which include the number (*time*), day in the week (*wd*), type of day (*wh*), outdoor temperature (*temp*), outdoor humidity (*hum*), wind speed (*wind*), electricity consumption (*load*), and electricity consumption for the *i*–th day ahead (*load–i*).

Table 1. Fragment of initial data.

Year	Month	Day	Hour	Wd	Wh	Temp	Hum	Wind	Load-7	Load-6	...	Load-1
2019	1	8	0	2	1	−33	67	3	947	894	...	917
2019	1	8	1	3	1	−31	68	5	888	838	...	850
2019	1	8	2	3	1	−29	68	4	825	825	...	819
2019	1	8	3	3	1	−33	67	4	795	811	...	813
2019	1	8	4	3	1	−33	67	5	773	808	...	804

In many works regarding load forecast, the influence of individual variables listed above is investigated. In this paper, the effects of these variables are determined via correlation analysis, and the results are shown in Table 2. It can be seen that the most significant influence on electricity consumption is exerted by the outdoor air temperature, and the correlation coefficient ranges from −0.56 to −0.59, while other exogenous variables do not significantly affect consumption. In addition, it is obvious from Figure 4 that the predicted consumption significantly depends on the previous days. Thus, these factors must be considered in the development of a multifactorial model of the daily load.

Table 2. Correlation values between initial and forecasting data.

	Wd	Wh	Wind	Hum	Temp	Load-7	Load-6	Load-5	Load-4	Load-3	Load-2	Load-1
load	0.004	0.05	0.15	−0.03	−0.59	0.96	0.96	0.96	0.97	0.97	0.97	0.98

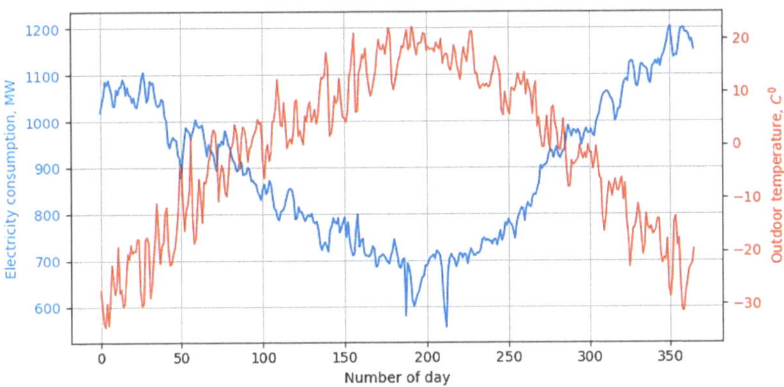

Figure 4. Graph of average daily consumption and average daily temperature for 2021.

To develop ensemble models, the data set is randomly divided into training and test sets in a ratio of 70 to 30. In the process of training, a list of the most influential variables is established, and a regression analysis is carried out between the predicted and actual values. Figure 5 shows the importance of the features as a measure of the relative importance of each feature in a machine learning model. The higher the feature importance score, the more important the feature is in predicting the target variable.

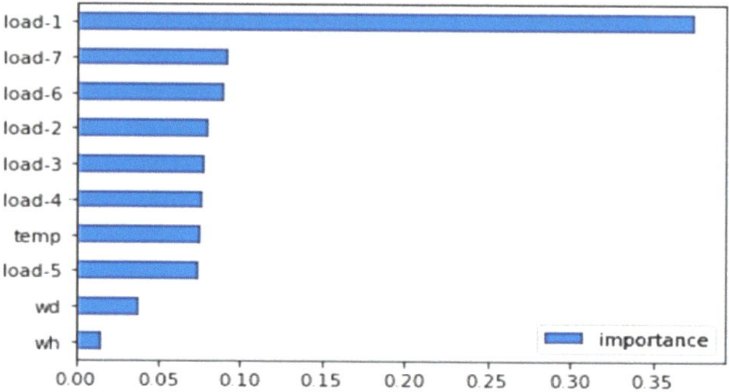

Figure 5. The role of initial variables in the learning process.

Figure 5 shows that the parameters of the daily load profile for forecast days depend on the previous day and prior week. As for exogenous variables, the outdoor temperature strongly influences the creation of the model. Despite the calendar features, including the type of day and the day of the week, they play their role in developing the model, but no more than the other variables considered. It is worth noting that the same behavior was observed when predicting the daily load for working days using statistical analysis in [45].

When developing algorithms, the most important task is to determine the number ($n\_estimators$) and depth (max_depth) of decision trees included in the ensemble. The goal of the task is to set the parameters so that the quality of the model is the best, reducing the volume of the algorithm. The choice of these parameters has a significant impact on the volume of the algorithm and the probability of the model. The dependence of the model estimate on these parameters is shown in Figure 6.

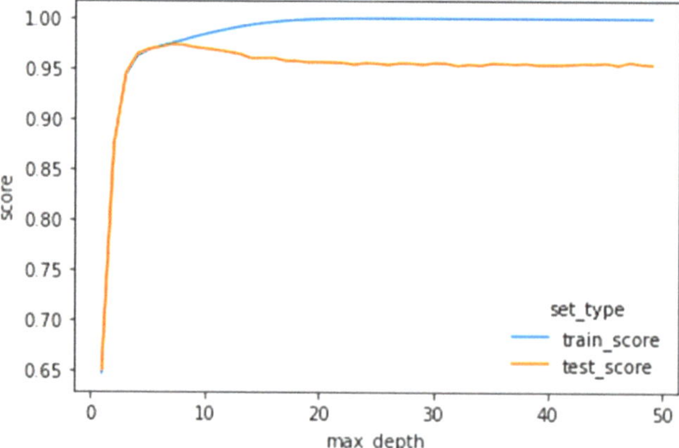

Figure 6. *Cont.*

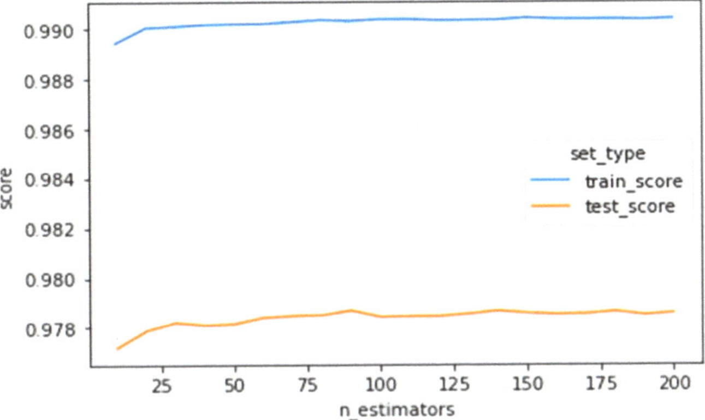

Figure 6. Dependence of the model probability on the depth and number of trees.

From Figure 6, it is obvious that the optimal tree depth value coincides with the point at which the training score reached its maximum value, and from that moment, the test score stabilizes. In terms of the number of trees, it does not greatly affect the quality of the model, given that the model contains more than 100 trees. Thus, in order to minimize the volume of the algorithm, it is necessary to set the depth and number of trees to these values. Setting the depth and number of trees is implemented using the GridSearchCV function (sklearn library), selecting the best model parameters (Table 3).

Table 3. Hyperparameters of models.

	Random Forest	**AdaBoost**	**XGBoost**
Depth of trees	12	12	12
Number of trees	100	100	100
MAPE [%]	2.44	2.38	2.35

The quality of the model was estimated using the MAE and MAPE. For the test set, MAE was 18.8 MW, and MAPE was 2.44%. Figure 7 shows the segment of the test set of the model.

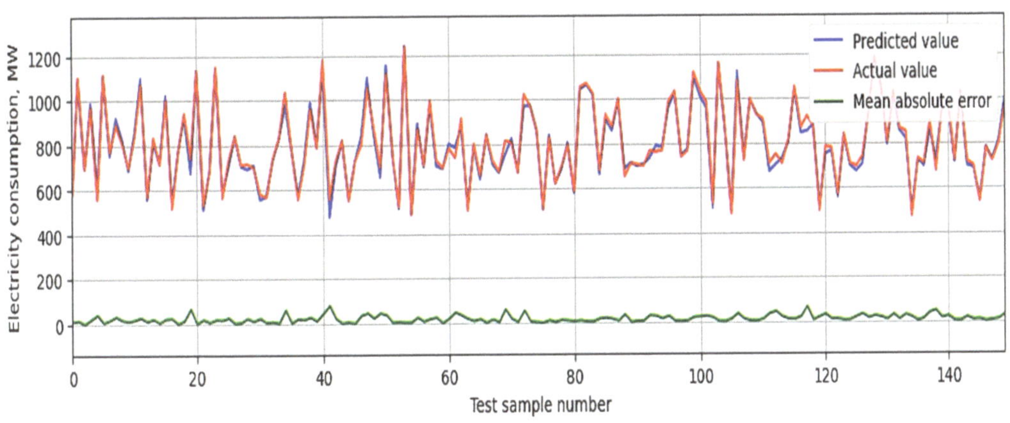

Figure 7. The segment of the model testing process.

The results of Random Forest, Adaptive boosting (AdaBoost), and Extreme Gradient Boosting (XGBoost) algorithms were analyzed. The results of the analysis confirm that the ensemble models have high accuracy, as shown in Table 4.

Table 4. Forecasting accuracy.

Month	Naive		AR		ARIMA		Random Forest		AdaBoost		XG Boost	
	MAE [MW]	MAPE [%]	MAE [MW]	MAPE [%]	MAE [MW]	MAPE [%]	MAE [MW]	MAPE [%]	MAE [MW]	MAPE [%]	MAE [MW]	MAPE [%]
January	20.90	1.97	26.11	2.45	19.90	1.90	10.24	0.92	10.82	1.04	5.84	0.55
February	26.54	2.70	23.12	2.21	33.70	3.05	16.67	1.63	13.73	1.33	11.38	1.11
March	20.53	2.23	31.29	3.26	25.86	2.64	25.04	2.66	11.70	1.23	9.38	0.97
April	20.22	2.45	26.42	3.17	20.06	2.22	9.80	1.32	10.45	1.26	8.75	1.02
May	22.76	2.98	29.51	3.98	28.26	3.88	16.62	2.26	11.08	1.47	9.43	1.26
June	27.04	3.67	11.68	1.69	28.25	4.06	7.69	1.17	12.65	1.80	12.03	1.70
July	30.19	5.54	11.73	1.78	19.91	3.05	11.20	1.66	8.64	1.31	6.89	1.06
August	22.69	3.21	33.18	4.52	15.06	2.08	8.87	1.21	19.36	2.64	9.25	1.24
September	24.09	2.98	23.78	3.03	17.21	2.14	8.91	1.08	16.64	2.17	15.54	2.09
October	19.64	2.07	43.28	4.61	21.81	2.46	11.93	1.27	18.39	1.93	17.48	1.79
November	22.51	2.14	21.76	2.18	18.61	1.93	14.31	1.37	15.52	1.58	14.83	1.48
December	20.29	1.78	30.53	2.83	17.39	1.66	9.20	0.87	10.69	0.96	8.33	0.76
Result	23.14	2.81	26.03	2.96	22.17	2.59	12.54	1.45	13.26	1.56	10.76	1.25

According to the constructed model, daily load profiles from each month were predicted and compared against the statistical analysis method developed in the previous work [45]. Table 4 shows the results of the ensemble models, ARIMA, and the simplest linear autoregression (AR) as a simple benchmarking method. In addition, a naïve forecast algorithm was applied (the values recorded on a previous day were used as the next-day forecast; for example, 1 p.m. is taken as the forecast for 1 p.m. for the next day).

As shown in Table 4, the machine learning method has the highest prediction accuracy with an average error of 1.25% or 10.76 MW. Therefore, this machine learning algorithm has the ability to predict power consumption with sufficient accuracy and to take into account several variables that affect load profile.

3.3. Consumption Forecasting in the Nodes of the Energy System

The central energy system of Mongolia consists of five energy supply zones, including Ulaanbaatar (energy supply zone "U"), Erdenet–Bulgan (energy supply zone "H"), Darkhan–Selenge (energy supply zone "T"), Baganuur–Choir (energy supply zone "B") and Gov (energy supply zone "G"). Figure 8 shows the simulation network of the power system.

The daily load curves for each substation were modeled on the basis of information from the IEEE reliability testing system, where the required data were gathered via instrumentation, control, and automation equipment. Through summing up the data for the respective substations, the daily load curves of each power supply zone were calculated. Since the power system is characterized by a hierarchical structure with nodes differing in the amount of electricity consumed, the coefficients of individual node participation in the total power consumption can be determined using rank models in per-unit values. Table 5 presents the results of the rank models' application from the point of view of the energy supply zone. The rank model parameters are visualized in Figures 9 and 10.

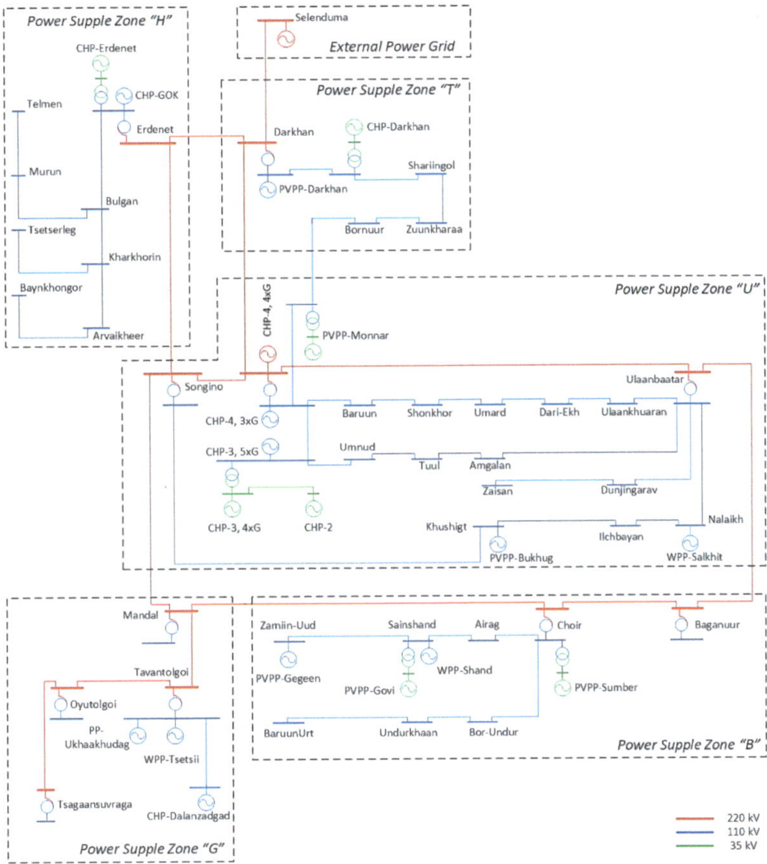

Figure 8. Simulation network representing the central power system of Mongolia.

Table 5. Results of rank models.

Name of the Energy Supply Zone	Name of the Calculation	Rank Number	Percentage of the Total Power Load Participation Rate, %
Ulaanbaatar	'U'	I	54.34
Erdenet-Bulgan	'H'	II	25.73
Darkhan-Selenge	'T'	III	8.75
Frog	'B'	IV	8.63
Gobi	'G'	V	2.55

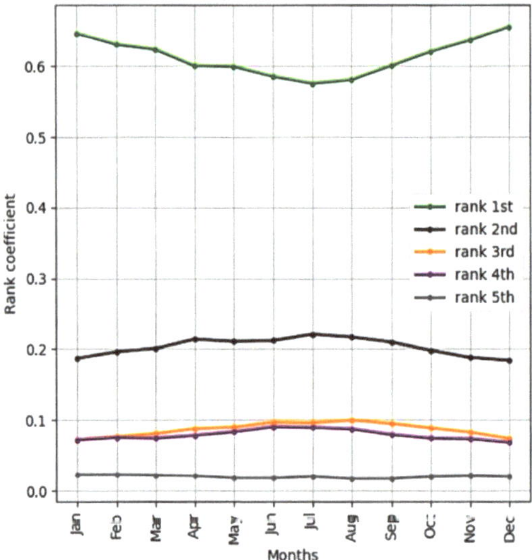

Figure 9. Rank model for the power supply zone.

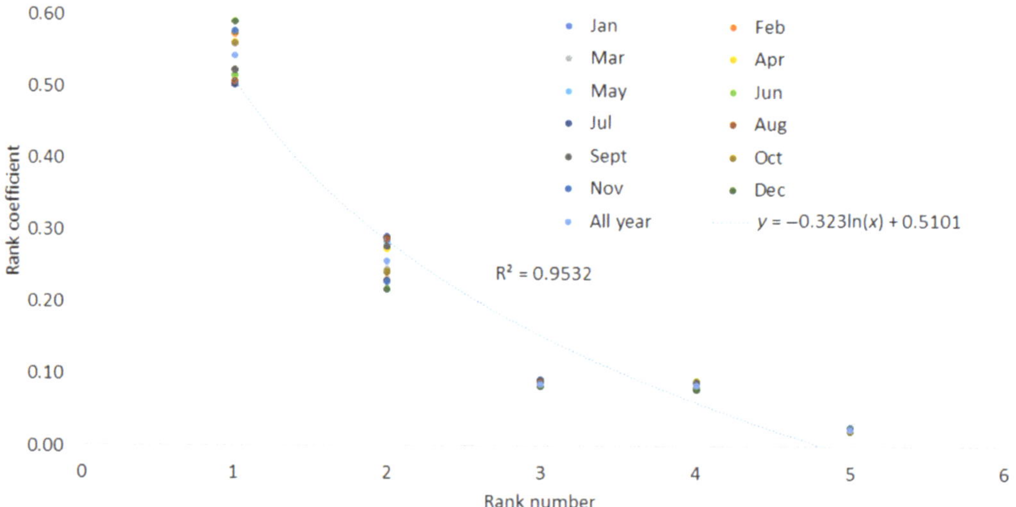

Figure 10. Dependence between rank coefficient and rank number.

It can be seen from Figure 10 that fairly good models have been obtained for the power supply zone according to the R^2 criteria. It is worth noting that during the observations, the rank orders of rank in the studied time series did not change. Moreover, for the entire energy supply zone, fluctuations in the coefficients of participation are of small range. Hence, it can be concluded that the use of rank models allows us to precisely predict the electricity consumption in each energy supply zone.

Using machine learning methods, electricity consumption was predicted at one-hour intervals with an average error of 1.25%, and daily load curves for each energy supply zone were obtained through applying rank coefficients. As an example, Figure 11 shows the load curve for overall power consumption in the power system and load profiles per every supply zone.

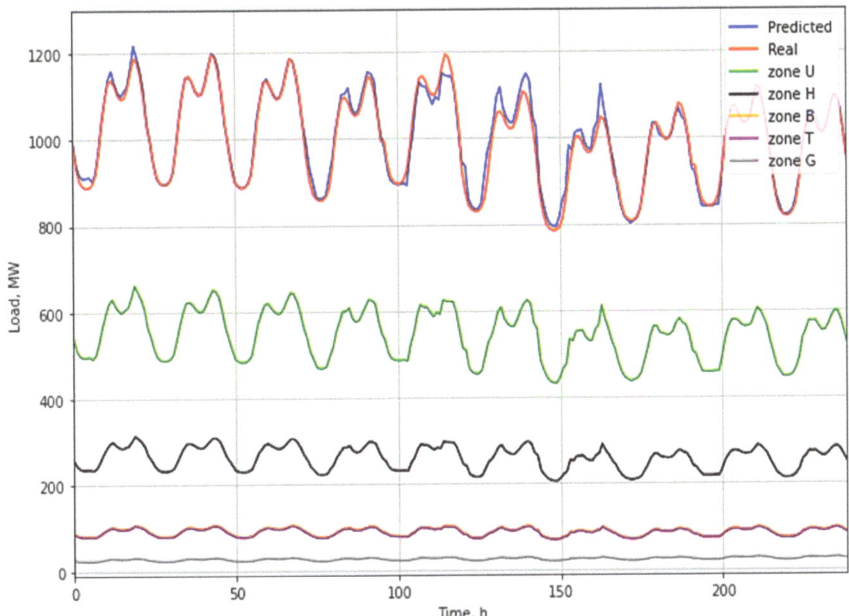

Figure 11. Daily load curves of the power system and power supply zones.

It can be seen that the graphs of each energy supply zone have a different shape, since the values of the rank coefficients are constantly changing over time. In other words, the graphs show what kind of load affects the form of the total consumption curve. The final results of forecasting electricity consumption in power supply zones are shown in Table 6. Also, Figures 12 and 13 depict the accuracy of the consumption model for each power supply zone. From this, it can be concluded that the proposed models can be used to calculate the operating conditions of the power system, since the average error of the models was not more than 2.0%. Also, from the point of view of power supply zones, the method can determine the loads of respective substations.

Table 6. Final results of consumption forecasting in energy supply zones.

Rank Number	Zone U		Zone H		Zone B		Zone T		Zone G	
	MAE [MW]	MAPE [%]	MAE [MW]	MAPE [%]	MAE [MW]	MAPE [%]	MAE [MW]	MAPE [%]	MAE [MW]	MAPE [%]
January	2.81	0.37	0.83	0.37	0.41	0.48	0.38	0.49	0.29	1.38
February	7.70	1.24	2.52	1.27	1.04	1.21	0.86	1.18	0.31	1.46
March	3.69	0.58	1.16	0.56	0.49	0.56	0.54	0.70	0.23	1.29
April	8.05	1.66	3.36	1.65	1.66	1.83	1.15	1.68	0.24	1.86
May	1.38	0.32	0.60	0.34	0.30	0.40	0.25	0.41	0.15	1.43
June	2.55	0.66	1.18	0.70	0.57	0.78	0.48	0.69	0.16	1.46
July	5.05	1.43	2.41	1.43	0.90	1.34	0.94	1.37	0.18	1.48
August	6.73	1.58	2.46	1.59	1.14	1.58	1.11	1.60	0.30	2.12
September	1.37	0.32	0.62	0.37	0.29	0.41	0.24	0.35	0.22	1.78
October	3.56	0.80	1.44	0.80	0.58	0.84	0.55	0.86	0.22	1.41
November	6.68	1.39	2.65	1.41	1.06	1.37	0.95	1.40	0.33	1.79
December	1.77	0.31	0.72	0.36	0.35	0.46	0.37	0.49	0.23	1.10
Result	4.2	0.88	1.66	0.90	0.73	0.93	0.65	0.93	0.23	1.54

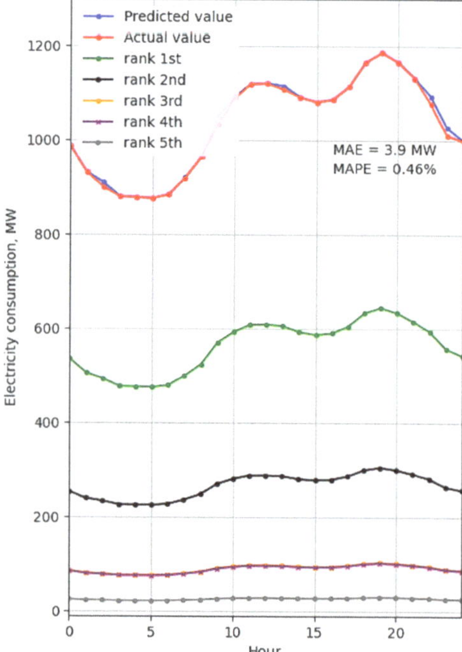

Figure 12. Rank model for the power supply zone.

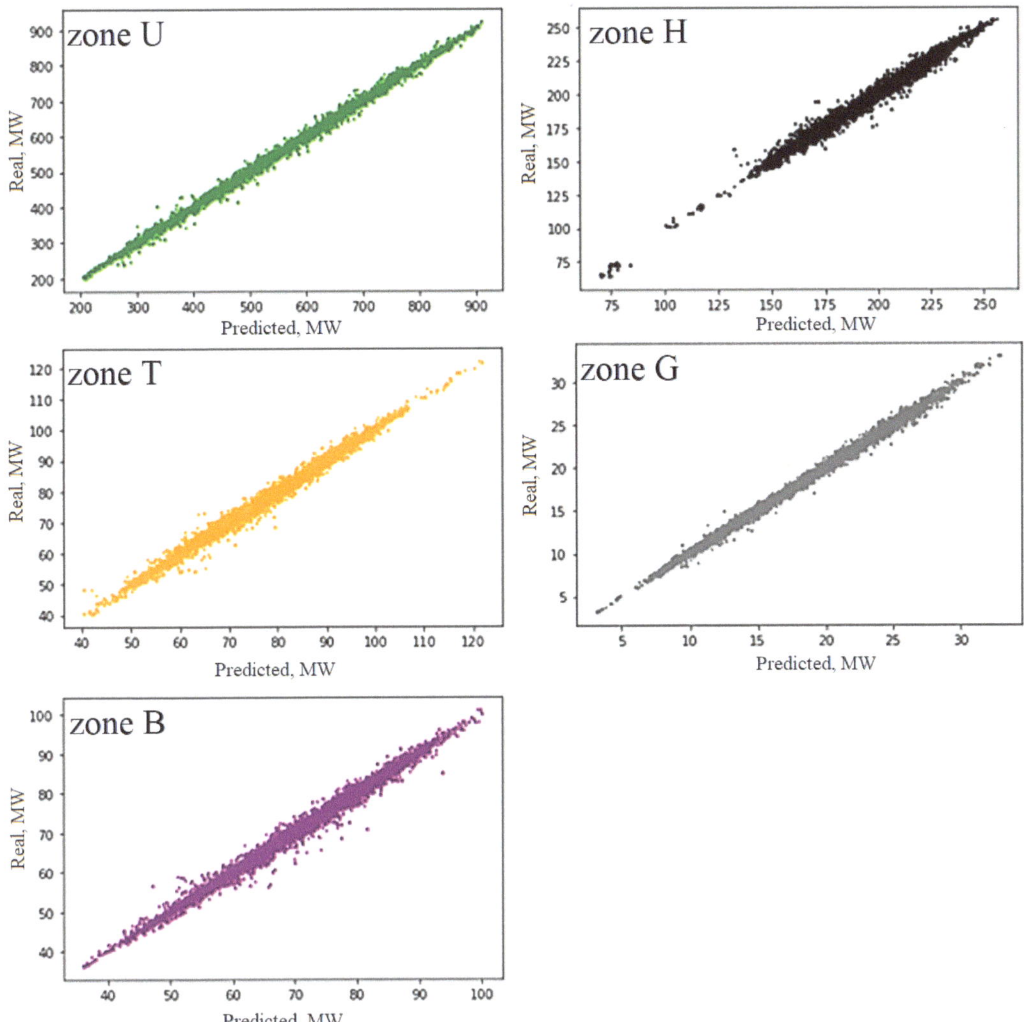

Figure 13. Accuracy of consumption models of power supply zones.

4. Discussion

The task of EPS modeling and forecasting of the processes appearing in the whole system and in its individual elements is a key objective of power system management. The solution of this task would make it possible to plan electricity generation more economically and reliably, as well as to optimize possible EPS running arrangements. However, the adopted short-term forecasting methods do not fulfill completely the needs of EPS planning and control.

The main disadvantage of the existing methods is the need to develop a load model and constantly refine the obtained model. Another disadvantage of these methods is the inaccurate determination of the relationships between input and output variables, as the dependencies between them are nonlinear. Hence, to ensure the high-quality short-term forecasting of electrical consumption in enterprises, a specific forecasting system is required. It ensures the efficient acquisition and use of the necessary data, carries out all stages of forecasting, and is controlled through a graphical user interface. The forecasting system should be adaptive, use modern methods of data analysis, and make full use of the computing power of modern computers.

This study considers the problem of modeling and forecasting the daily load profile in the nodes of a power system. A forecasting method using the coefficients of mode individual participation in the total consumption of the EPS is proposed in this work.

In the paper, the electricity consumption of the entire power system was predicted with an average error of 1.25%. The use of the machine learning method reduced the error of the ARIMA model, which was 2.58%, to 1.25%. Thus, the error reduction was 1.33 percentage points or 52%. Despite the fact that the results obtained using the ARIMA statistical method show the possibility of implementation in practice, the method is not designed to consider some important additional variables, including meteorological factors and calendar features.

For Mongolia, the proposed method of power consumption forecasting, which makes a provision for meteorological factors and is aided by machine learning algorithms and rank analysis techniques, was performed for the first time. Therefore, the results obtained are unique and cannot be found in other studies. However, the resulting accuracy is in line with state-of-the-art research in this field. The day-ahead forecasting error for large power systems is usually 1–4% [46].

The received power consumption forecast does not result in reduced electricity production since there are always standby generating capacities. But the more accurate the forecast, the more efficiently the problem of load sharing among utilities can be solved.

5. Conclusions

This paper presents techniques, such as the autoregressive integrated moving average model and a machine learning method based on the ensemble of decision trees, and demonstrates their effective use for the central power system of Mongolia for the first time in history.

This study considers the problem of modeling and forecasting the daily load profile in the nodes of the power system. The daily load curves of energy supply zones are modeled using the coefficients indicating the participation of individual nodes in total power consumption, which are established using the method of rank models. It can be concluded that the proposed models of power consumption in power supply zones can be practically applied to calculate the operating conditions of this power system, since the average error of the models was no more than 1.5%. With such accurate models, the consumption of any node of the power system can be determined using rank coefficients calculated at the appropriate level of the power system, such as tie-substations, power supply zones, and other parts of the system.

The disadvantage of the Id forecasting system is the dependence on the accuracy of meteorological forecasts, as well as the need for regular revision of power consumption coefficients for each node of the power system. The proposed approach should be generalized for other electric power systems, which is planned to be accomplished in the future.

Author Contributions: All authors have made valuable contributions to this paper. Conceptualization, T.O., P.M., M.S., I.Z. and A.R.; methodology, T.O., P.M., S.K., M.S., I.Z. and A.R.; software, P.M., T.O. and M.S.; validation, T.O., P.M., S.K. and I.Z.; formal analysis, T.O., P.M., S.K., M.S., I.Z. and A.R.; investigation, T.O., P.M., M.S. and A.R.; writing—original draft preparation, T.O., P.M., S.K., M.S., I.Z. and A.R.; writing—review and editing, T.O., S.K. and I.Z.; supervision, A.R. All authors have read and agreed to the published version of the manuscript.

Funding: The research funding from the Ministry of Science and Higher Education of the Russian Federation (Ural Federal University Program of Development within the Priority-2030 Program) is gratefully acknowledged.

Institutional Review Board Statement: Not applicable.

Informed Consent Statement: Not applicable.

Data Availability Statement: Not applicable.

Conflicts of Interest: The authors declare no conflict of interest.

References

1. Kychkin, A.V.; Chasparis, G.C. Feature and model selection for day-ahead electricity-load forecasting in residential buildings. *Energy Build.* **2021**, *249*, 111200. [CrossRef]
2. Alfares, H.K.; Nazeeruddin, M. Electric load forecasting: Literature survey and classification of methods. *Int. J. Syst. Sci.* **2002**, *33*, 23–34. [CrossRef]
3. Ghalehkhondabi, I.; Ardjmand, E.; Weckman, G.R.; Young, W.A. An overview of energy demand forecasting methods published in 2005–2015. *Energy Syst.* **2017**, *2*, 411–447. [CrossRef]
4. Abdurahmanov, A.; Volodin, M.; Zybin, E.; Ryabchenko, V. Forecasting methods in electricity distribution networks (review). *Russ. Internet J. Electr. Eng.* **2016**, *3*, 3–23. [CrossRef]
5. Patel, H.; Shah, M. Energy Consumption and Price Forecasting Through Data-Driven Analysis Methods: A Review. *SN Comput. Sci.* **2021**, *2*, 315. [CrossRef]
6. Makoklyuev, B.I. *Analysis and Planning of Electricity Consumption*; Energoatomizdat: Moscow, Russia, 2008.
7. Matrenin, P.; Antonenkov, D.; Arestova, A. Energy Efficiency Improvement of Industrial Enterprise Based on Machine Learning Electricity Tariff Forecasting. In Proceedings of the 2021 15th International Scientific-Technical Conference on Actual Problems of Electronic Instrument Engineering, APEIE 2021, Novosibirsk, Russia, 19–21 November 2021. [CrossRef]
8. Matrenin, P.V.; Manusov, V.Z.; Khalyasmaa, A.I.; Antonenkov, D.V.; Eroshenko, S.A.; Butusov, D.N. Improving Accuracy and Generalization Performance of Small-Size Recurrent Neural Networks Applied to Short-Term Load Forecasting. *Mathematics* **2020**, *8*, 2169. [CrossRef]
9. Kamalov, F.; Smail, L.; Gurrib, I. Stock price forecast with deep learning. In Proceedings of the 2020 International Conference on Decision Aid Sciences and Application (DASA), Sakheer, Bahrain, 8–9 November 2020; pp. 1098–1102.
10. Chen, Y.; Tang, Y.; Zhang, S.; Liu, G.; Liu, T. Weather Sensitive Residential Load Forecasting Using Neural Networks. In Proceedings of the 2023 IEEE 6th International Electrical and Energy Conference (CIEEC), Hefei, China, 12–14 May 2023. [CrossRef]
11. Xu, F.; Xu, W.; Qiu, Y.; Wu, M.; Wang, R.; Li, Y.; Fan, P.; Yang, J. A Short-term Load Forecasting Model Based on Neural Network Considering Weather Features. In Proceedings of the 2021 IEEE 4th International Conference on Automation, Electronics and Electrical Engineering (AUTEEE), Shenyang, China, 19–21 November 2021. [CrossRef]
12. Chodakowska, E.; Nazarko, J.; Nazarko, Ł. ARIMA Models in Electrical Load Forecasting and Their Robustness to Noise. *Energies* **2021**, *14*, 7952. [CrossRef]
13. López, J.C.; Rider, M.J.; Wu, Q. Parsimonious Short-Term Load Forecasting for Optimal Operation Planning of Electrical Distribution Systems. *IEEE Trans. Power Syst.* **2019**, *34*, 1427–1437. [CrossRef]
14. Sun, X.; Luh, P.B.; Cheung, K.W.; Guan, W.; Michel, L.D.; Venkata, S.S.; Miller, M.T. An efficient approach to short-term load forecasting at the distribution level. *IEEE Trans. Power Syst.* **2015**, *31*, 2526–2537. [CrossRef]
15. Fernandes, K.C.; Sardinha, R.; Rebelo, S.; Singh, R. Electric load analysis and forecasting using artificial neural networks. In Proceedings of the 2019 3rd International Conference on Trends in Electronics and Informatics (ICOEI), Tirunelveli, India, 23–25 April 2019; pp. 1274–1278. [CrossRef]
16. Alobaidi, M.H.; Chebana, F.; Meguid, M.A. Robust ensemble learning framework for day—Ahead forecasting of household based energy consumption. *Appl. Energy* **2018**, *212*, 997–1012. [CrossRef]
17. Shi, J.; Wang, Z. A Hybrid Forecast Model for Household Electric Power by Fusing Landmark-Based Spectral Clustering and Deep Learning. *Sustainability* **2022**, *14*, 9255. [CrossRef]
18. Huyghues-Beaufond, N.; Tindemans, S.; Falugi, P.; Sun, M.; Strbac, G. Robust and automatic data cleansing method for short-term load forecasting of distribution feeders. *Appl. Energy* **2020**, *261*, 114405. [CrossRef]
19. Hayes, B.P.; Gruber, J.K.; Prodanovic, M. Multi-nodal short-term energy forecasting using smart meter data. *IET Gener. Transm. Distrib.* **2018**, *12*, 2988–2994. [CrossRef]
20. Tan, M.; Hu, C.; Chen, J.; Wang, L.; Li, Z. Multi-node load forecasting based on multi-task learning with modal feature extraction. *Eng. Appl. Artif. Intell.* **2022**, *112*, 104856. [CrossRef]
21. Tan, M.; Liu, Y.; Meng, B.; Su, Y. Multinodal forecasting of industrial power load using participation factor and ensemble learning. In Proceedings of the 2020 IEEE 4th Conference on Energy Internet and Energy System Integration (EI2), Wuhan, China, 30 October–1 November 2020; pp. 745–750. [CrossRef]

22. Abreu, T.; Amorim, A.J.; Santos-Junior, C.R.; Lotufo, A.D.; Minussi, C.R. Multinodal load forecasting for distribution systems using a fuzzy-artmap neural network. *Appl. Soft Comput.* **2018**, *71*, 307–316. [CrossRef]
23. Rai, S.; De, M. Effect of Load Contribution Factor on Multinodal Load Forecasting. In Proceedings of the IEEE EUROCON 2021—19th International Conference on Smart Technologies, Lviv, Ukraine, 6–8 July 2021; pp. 455–459. [CrossRef]
24. Amorim, A.J.; Abreu, T.A.; Tonelli-Neto, M.S.; Minussi, C.R. A new formulation of multinodal short-term load forecasting based on adaptive resonance theory with reverse training. *Electr. Power Syst. Res.* **2020**, *179*, 106096. [CrossRef]
25. Ferreira, A.B.A.; Minussi, C.R.; Lotufo, A.D.P.; Lopes, M.L.M.; Chavarette, F.R.; Abreu, T.A. Multinodal load forecast using euclidean ARTMAP Neural network. In Proceedings of the 2019 IEEE PES Innovative Smart Grid Technologies Conference-Latin America (ISGT Latin America), Gramado, Brazil, 15–18 September 2019; pp. 1–6. [CrossRef]
26. Stephen, B.; Telford, R.; Galloway, S. Non-Gaussian residual based short term load forecast adjustment for distribution feeders. *IEEE Access* **2020**, *8*, 10731–10741. [CrossRef]
27. Stephen, B.; Tang, X.; Harvey, P.R.; Galloway, S.; Jennett, K.I. Incorporating practice theory in sub-profile models for short term aggregated residential load forecasting. *IEEE Trans. Smart Grid* **2015**, *8*, 1591–1598. [CrossRef]
28. Wang, J.; Zhu, W.; Zhang, W.; Sun, D. A trend fixed on firstly and seasonal adjustment model combined with the ε-SVR for short-term forecasting of electricity demand. *Energy Policy* **2009**, *37*, 4901–4909. [CrossRef]
29. Çevik, H.H.; Çunkaş, M. Short-term load forecasting using fuzzy logic and ANFIS. *Neural Comput. Appl.* **2015**, *26*, 1355–1367. [CrossRef]
30. Lahouar, A.; Slama, J.B.H. Day-ahead load forecast using random forest and expert input selection. *Energy Convers. Manag.* **2015**, *103*, 1040–1051. [CrossRef]
31. Moon, J.; Kim, Y.; Son, M.; Hwang, E. Hybrid short-term load forecasting scheme using random forest and multilayer perceptron. *Energies* **2018**, *11*, 3283. [CrossRef]
32. Lahouar, A.; Slama, J.B.H. Hour-ahead wind power forecast based on random forest. *Renew. Energy* **2017**, *109*, 529–541. [CrossRef]
33. Barrows, C.; Bloom, A.; Ehlen, A.; Ikaheimo, J.; Jorgenson, J.; Krishnamurthy, D.; Lau, J.; McBennett, B.; O'Connell, M.; Preston, E. The IEEE reliability test system: A proposed 2019 update. *IEEE Trans. Power Syst.* **2019**, *35*, 119–127. [CrossRef]
34. Saranchimeg, S.; Nair, N.K.C. A novel framework for integration analysis of large-scale photovoltaic plants into weak grids. *Appl. Energy* **2021**, *282*, 116141. [CrossRef]
35. Rusina, A.G.; Sidorkin, Y.M.; Kalinin, A.E. Application of rank models for structural forecasting. In Proceedings of the 2016 11th International Forum on Strategic Technology (IFOST 2016), Novosibirsk, Russia, 1–3 June 2016; pp. 271–275. [CrossRef]
36. Velasco, L.C.P.; Polestico, D.L.L.; Macasieb, G.P.O.; Reyes, M.B.V.; Vasquez, F.B., Jr. Load forecasting using autoregressive integrated moving average and artificial neural network. *Int. J. Adv. Comput. Sci. Appl.* **2018**, *9*, 23–29. [CrossRef]
37. Kamalov, F. A note on time series differencing. *Gulf J. Math.* **2021**, *10*, 50–56. [CrossRef]
38. Kamalov, F. A note on the autocovariance of p-series linear process. *Gulf J. Math.* **2020**, *9*, 40–45. [CrossRef]
39. Breiman, L. Random Forests. *Mach. Learn.* **2001**, *4*, 5–32. [CrossRef]
40. Chen, T.; Guestrin, C. XGBoost: A Scalable Tree Boosting System. Available online: https://arxiv.org/abs/1603.02754 (accessed on 22 May 2023).
41. Drucker, H. Improving Regressors Using Boosting Techniques. Available online: http://citeseerx.ist.psu.edu/viewdoc/download?doi=10.1.1.31.314&rep=rep1&type=pdf (accessed on 22 May 2023).
42. Matrenin, P.V.; Osgonbaatar, T.; Sergeev, N.N. Overview of Renewable Energy Sources in Mongolia. In Proceedings of the 2022 IEEE International Multi-Conference on Engineering, Computer and Information Sciences (SIBIRCON), Yekaterinburg, Russia, 11–13 November 2022. [CrossRef]
43. Bumtsend, U.; Safaraliev, M.; Ghulomzoda, A.; Ghoziev, B.; Ahyoev, J.; Ghulomabdolov, G. The Unbalanced Modes Analyze of Traction Loads Network. In Proceedings of the 2020 Ural Symposium on Biomedical Engineering, Radioelectronics and Information Technology (USBEREIT), Yekaterinburg, Russia, 14–15 May 2020; pp. 0456–0459. [CrossRef]
44. Manusov, V.Z.; Bumtsend, U.; Demin, Y.V. Analysis of the power quality impact in power supply system of Urban railway passenger transportation—The city of Ulaanbaatar. *IOP Conf. Ser. Earth Environ. Sci.* **2018**, *177*, 012024.
45. Vivas, E.; Allende-Cid, H.; Salas, R. A Systematic Review of Statistical and Machine Learning Methods for Electrical Power Forecasting with Reported MAPE Score. *Entropy* **2020**, *22*, 1412. [CrossRef]
46. Rusina, A.G.; Tuvshin, O.; Matrenin, P.V. Forecasting the daily energy load schedule of working days using meteofactors for the central power system of Mongolia. *Power Eng. Res. Equip. Technol.* **2022**, *24*, 98–107. [CrossRef]

inventions

Article

Inductive Compensation of an Open-Loop IPT Circuit: Analysis and Design

Mario Ponce-Silva *, Alan R. García-García, Jaime Arau, Josué Lara-Reyes and Claudia Cortés-García

Tecnológico Nacional de México—CENIDET, Cuernavaca 62490, Mexico;
alan.garcia18ee@cenidet.edu.mx (A.R.G.-G.)
* Correspondence: mario.ps@cenidet.tecnm.mx

Abstract: The main contribution of this paper is the inductive compensation of a wireless inductive power transmission circuit (IPT) with resonant open-loop inductive coupling. The variations in the coupling coefficient k due to the misalignment of the transmitter and receiver are compensated with only one auxiliary inductance in the primary of the inductive coupling. A low-power prototype was implemented with the following specifications: input voltage $Vin = 27.5$ V, output power $Po = 10$ W, switching frequency $f = 500$ kHz, output voltage $Vo = 12$ V, transmission distance $d = 1.5$ mm. Experimental results varying the distance "d" with several values of the compensation inductor demonstrate the feasibility of the proposal. An efficiency of 75.10% under nominal conditions was achieved. This proposal is a simple compensation topology for wireless chargers of cellular phones presenting small distances between the transmitter and receiver.

Keywords: cellular phones; inductive power transfer; resonant converter; DC-DC converter

Citation: Ponce-Silva, M.;
García-García, A.R.; Arau, J.;
Lara-Reyes, J.; Cortés-García, C.
Inductive Compensation of an
Open-Loop IPT Circuit: Analysis and
Design. *Inventions* **2023**, *8*, 104.
https://doi.org/10.3390/
inventions8040104

Academic Editor: Francesco
Della Corte

Received: 19 July 2023
Revised: 7 August 2023
Accepted: 14 August 2023
Published: 17 August 2023

1. Introduction

Nowadays, wireless power transmission (WPT) has become a trending research topic due to many applications involving no wired connections such as charging of electric vehicles, cell phones, hand tools, biomedical implants and portable electronics [1,2].

WPT has been used mainly in two applications—in the sending of information and electrical energy, or only for transmitting electrical energy. The second case shall be addressed in this work, with the aim of transmitting electrical power from a power supply to an electrical load, using a resonant inductive coupling [3–5].

WPT can be classified into two categories—far-field and near-field WPT. The far-field category refers to the transmission of a large amount of energy between two positions, e.g., sending energy by electromagnetic waves. Near-field WPT refers to the distance of transfer energy within the wavelength (λ) of the transmitter antenna [6–8]. This paper will focus on near-field WPT.

As shown in Figure 1, inductive power transmission (IPT) circuits consist of multiple stages of energy conversion to achieve the desired operation, usually as follows: low-frequency rectifier, high-frequency (HF) inverter with a resonant tank, inductive coupling with a coupling factor k, high-frequency rectifier, and load [9–11]. This work is focused on the stages from the HF inverter only.

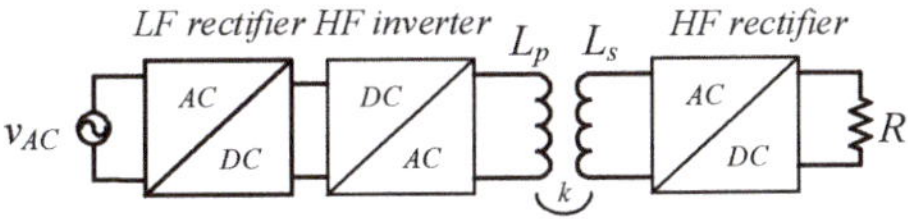

Figure 1. Typical structure of inductive power transmission (IPT) circuit.

An indispensable stage in IPT circuits is the resonant tank, also known as resonant network, which is an electrical circuit designed with capacitors and inductors which is capable of storing electrical energy and oscillating when it is operating at the resonant frequency. The circuit has zero reactance during the resonance condition, allowing the maximum transfer of power from a power supply to an electrical load with high efficiency [12,13].

In low-power resonant IPT applications, such as wireless chargers for smartphones and portable electronic devices (less than 20 W), the literature reports efficiencies of approximately 50–70% in open-loop operation and 70–80% in closed-loop operation [14,15].

Usually, the IPT topologies require compensation networks to operate the circuit in resonance. The basic compensation networks are presented in Figure 2; as can be seen, the compensation networks include a capacitor in series or parallel with the transmitter and the receiver to compensate for their inductance and to operate at resonance [16,17].

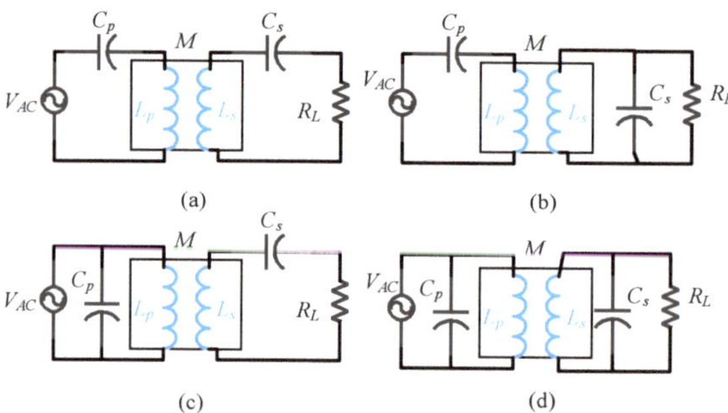

Figure 2. Basic compensation networks. (**a**) series-series, (**b**) series-parallel, (**c**) parallel-series, (**d**) parallel-parallel.

The full circuit needs many components to achieve optimal operation; these components cause additional losses, thus diminishing the efficiency. One solution is to consider the parasitic components (leakage inductances) and use them as part of the design to achieve high efficiency, fewer components, and better power transfer in inductive coupling [18,19].

Many compensation topologies have been proposed to compensate for the leakage inductances. The most used compensation topologies consist of adding one capacitor in series with each leakage inductance C_p and C_s (Figure 3). Both capacitors are chosen to be in resonance with each leakage inductance [20–22].

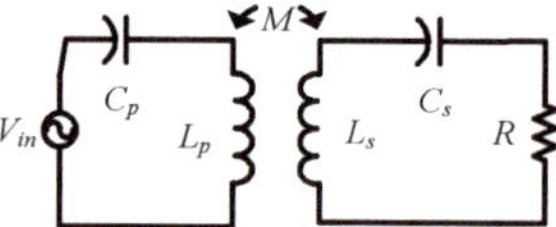

Figure 3. Resonance conditions used in the literature for IPT circuits.

Table 1 shows some reported circuits with their most important operating characteristics. This table shows that a higher operating power results in better overall efficiency of the circuit. In addition, Table 1 allows comparing the results obtained in this paper with those reported in the literature, where even though in this paper a low power of 10 W was used, an efficiency of up to 75% was achieved, which is higher than some of the circuits shown in Table 1 with similar power levels.

The main contribution of this paper is to eliminate the compensation capacitor C_s in the receiver, compensate for the misalignment of the transmitter and the receiver with only one series capacitor in the transmitter, and add an inductor in series with this capacitor to compensate for the changes in the mutual inductance M.

Table 1. Comparison of reported resonant circuits with applications in wireless power transmission.

Frequency	Output Power	Coupling Factor	Efficiency	Ref.
140 kHz	80 W	0.24	60.70%	[23]
40 kHz	45 W	0.25	70.60%	[24]
82.3 kHz	10 W	0.059	50%	[25]
1 MHz	100 W	0.25	92%	[26]
800 kHz	12 W	0.1	56%	[27]
450 kHz	300 W	0.32	75%	[28]
500 kHz	2 W	0.05	65%	[29]
120 kHz	10 W	0.3	69%	[30]

The proposal simplifies the secondary circuit and reduces the complexity of the circuit, with a higher efficiency than the works reported in Table 1. This contribution is particularly relevant in low-power applications where low size and cost are important, such as biomedical implants, portable electronics, electro mobility, and Internet of Things applications.

2. Analysis of the Circuit

The IPT circuits are usually modelled as two coupled inductors as shown in Figure 4a, with a coupling factor k defined as:

$$k = \frac{M}{\sqrt{L_1 L_2}} \tag{1}$$

where M is the mutual inductance, L_1 is the self-inductance of the primary (transmitter) and L_2 is the self-inductance of the secondary (receiver). This model is commonly employed and useful in the literature.

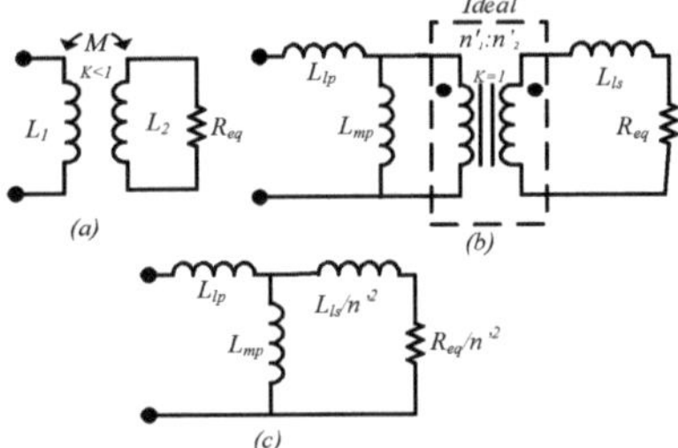

Figure 4. Models of IPT circuits. (**a**) Typical model using coupling inductors. (**b**) Model T of the coupled inductors using an ideal transformer, (**c**) model T with secondary reflected to the primary.

The main disadvantage is that the model separates the analysis of the circuit into two parts—primary and secondary. To merge both circuits in only one, model T is the usual selection [31,32].

Model T could be rearranged as the circuit in Figure 4b. This circuit models the mutual inductance M as an ideal transformer with coupling factor $k = 1$ and turn relationship $n' = n'_2/n'_1$; turns n'_1 and n'_2 represent the cross section of coils L_1 and L_2 that share the same common flux φ_M and are different from the total turn number of each inductor L_1 (n_1) and L_2 (n_2). The components of this model are related to the model in Figure 3b as shown in Table 2.

Figure 4c shows the elements of the secondary reflected to the primary of the ideal transformer; this last model was the model used in this analysis. The proposed circuit to be analyzed in this paper is presented in Figure 5. The HF inverter consists of a half bridge inverter with a compensation topology that is only made up of the compensation inductor L_C and the capacitor C_p. The IPT transformer is modelled like the model in Figure 4b, where the HF AC–DC conversion consists of a conventional full bridge rectifier.

Table 2. Relationships among the coupling inductances models.

Parameter	Description	Value
L_{lp}	Primary leakage inductance	$L_{lp} = L_1 - M$
L_{ls}	Secondary leakage inductance	$L_{ls} = L_2 - M$
L_{mp}	Primary magnetizing inductance	$L_{mp} = \frac{n_1}{n_2} M = \frac{M}{n_l}$

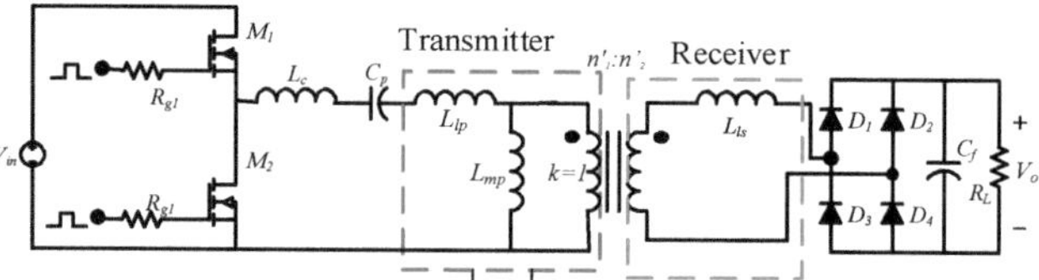

Figure 5. IPT circuit analyzed.

To compensate for the variations due to the misalignment of the transmitter and receiver, as well as variations caused by the increase in the distance between the transmitter and receiver, the compensation inductor L_c will have a value greater than 90% of the inductive reactance at resonant conditions. The remaining percentage of the inductive reactance belongs to the primary dispersion inductance L_{lp} and the equivalent inductance L_{eq1} obtained from the inductive coupling.

To analyze the resonant network, the resonant converter shown in Figure 5 is simplified until an equivalent circuit is obtained in series, in which the resonance condition will be established. Figure 6 shows the simplification process.

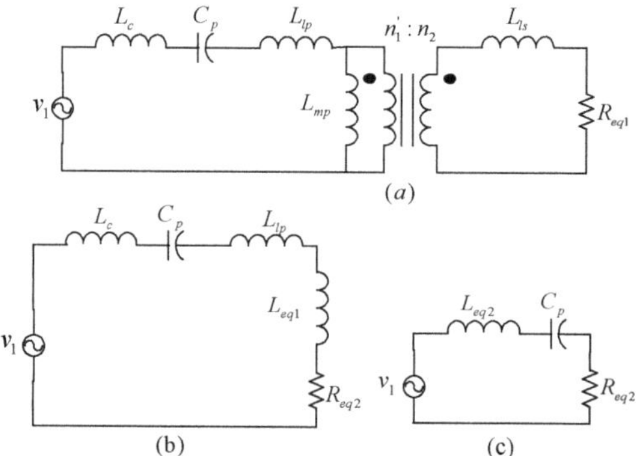

Figure 6. Analysis of the IPT circuit to obtain the resonance condition, (**a**) simplified IPT circuit with the equivalent resistance of the rectifier, (**b**) circuit obtained by reflecting the circuit secondary to the primary circuit, (**c**) total series equivalent circuit where the resonance condition is established.

Figure 6a shows that the half bridge inverter is modelled as the fundamental component of the square waveform V_{in} and the full bridge rectifier as a resistance R_{eq1} that consumes the same average power. In Figure 6b, the leakage inductance of the receiver L_{ls} and the equivalent resistance R_{eq1} are reflected to the transmitter side, the equivalent impedance, in parallel with the primary magnetizing inductance L_{mp}, is reduced into an equivalent inductance and resistance L_{eq1} and R_{eq2}, and finally, all the inductances are reduced in only one inductor L_{eq2} (see Figure 6c).

The first step of the analysis is to obtain the equivalent resistance of the full bridge rectifier R_{eq1}. A simplified analysis is carried out in [33], which assumes that the efficiency of the rectifier is 100%. To improve accuracy, it is possible to consider a different value for the efficiency of the rectifier (η_r) as follows:

$$\eta_r = \frac{P_o}{P_R} = \frac{P_o}{4P_D + P_o} \tag{2}$$

where P_o is the power delivered to the load (R_L) and P_R is the power delivered to the rectifier stage.

With this consideration, the equivalent resistant R_{eq1} is expressed as:

$$R_{eq_1} = \frac{8\eta_r R_L}{\pi^2} \tag{3}$$

where R_L is the load resistance.

As indicated in the introduction, the HF rectifier is one of the stages with higher losses. So, to minimize these losses, the efficiency of the rectifier is evaluated. To evaluate this efficiency, the following simplifications will be assumed:

1. The voltage at the input of the full bridge rectifier (V_o) is a square waveform in phase with the current, similar to the waveform in Figure 7, where V_o is the output voltage in the load resistance.

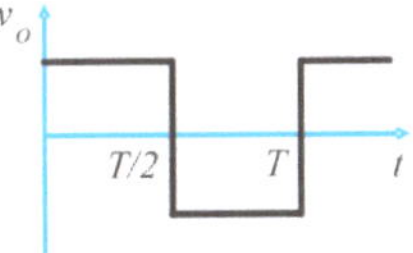

Figure 7. Assumed waveform at the input of the rectifier.

2. The input current in the rectifier is a sinusoidal waveform. Therefore, the losses in one diode can be evaluated with Equation (4).

$$P_D = I_D V_f = \frac{I V_f}{\pi} \tag{4}$$

where "I" is the maximum value of the input sinusoidal current waveform of the rectifier. This current could be expressed as $I = V_{o1}/R_{eq1}$, where V_{o1} is the fundamental of the square voltage applied to the input of the rectifier and can be expressed as: $V_{o1} = 4V_o/\pi$. Substituting these expressions into (4):

$$P_D = \frac{V_o V_f}{2\eta_r R_L} \tag{5}$$

Substituting Equation (5) into (2), the efficiency of the rectifier could be evaluated as:

$$\eta_r = 1 - \frac{2V_f}{V_o} \tag{6}$$

This equation is plotted in Figure 8 assuming a forward voltage in the diodes $V_f = 0.7$ V. As can be seen in this figure, for an output voltage below 12 V, the efficiency abruptly decreases, so, to maintain a high efficiency, it is recommended that the output voltage verifies $V_o > 12$ V.

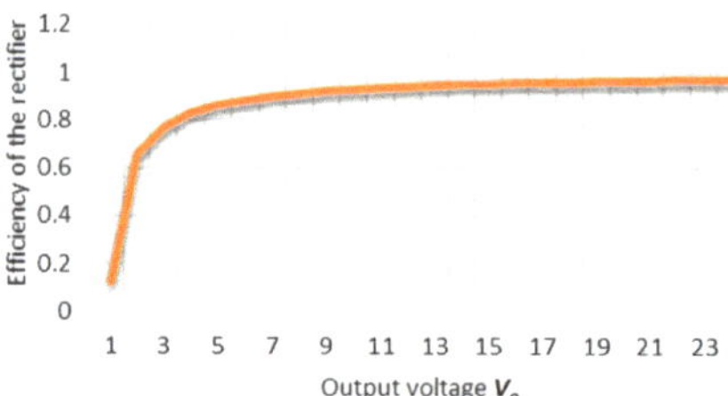

Figure 8. Efficiency of the rectifier η_r vs. the output voltage V_o ($V_f = 0.7$ V).

The rectifier efficiency graph in Figure 8 is presented just to show and clarify that it would work with a certain voltage level due to the efficiency of the rectifier. The graph is obtained from a simplified analysis in [33], which assumes that the efficiency of the rectifier is 100%. For this reason, the verification of this graph in the experimental tests is not necessary.

The next step is to reflect the secondary circuit of the inductive coupling to the primary circuit obtaining the circuit shown in Figure 6b. Afterwards, to obtain the equivalent impedance formed by L_{ls} and R_{eq1} (Z_{eq}), the parallel circuit of both elements is evaluated:

$$Z_{eq} = \frac{jX_{Lmp}\left(\frac{R_{eq1}+jX_{Lls}}{n^2}\right)}{jX_{Lmp} + \frac{R_{eq1}+jX_{Lls}}{n^2}} \tag{7}$$

Solving and simplifying Equation (7), the equivalent resistance R_{eq2} (real part of Z_{eq}) and equivalent inductance L_{eq1} (imaginary part of Z_{eq}) are obtained:

$$R_{eq2} = \frac{\frac{R_{eq1}X_{Lmp}^2}{n^2}}{\left(\frac{R_{eq1}}{n^2}\right)^2 + \left(X_{Lmp} + \frac{X_{Lls}}{n^2}\right)^2} \tag{8}$$

$$L_{eq1} = \frac{X_{Lmp}\left[\frac{X_{Lls}}{n^2}\left(X_{Lmp} + \frac{X_{Lls}}{n^2}\right) + \left(\frac{R_{eq1}}{n^2}\right)^2\right]}{\omega_o\left[\left(\frac{R_{eq1}}{n^2}\right)^2 + \left(X_{Lmp} + \frac{X_{Lls}}{n^2}\right)^2\right]} \tag{9}$$

As can be seen in Figure 6c, the resulting equivalent circuit is a serial circuit composed of three inductances, a capacitor, and an equivalent resistance. Finally, the equivalent inductance L_{eq2} "seen" by the square input voltage V_{in} is:

$$L_{eq2} = L_c + L_{lp} + L_{eq1} \tag{10}$$

The final equivalent series circuit is shown in Figure 6c. This circuit will be in resonance when the following condition is satisfied:

$$X_{C_p} = X_{Leq2} \tag{11}$$

where X_{cp} is the reactance of C_p and X_{Leq2} is the reactance of the inductance L_{eq2}. Equation (13) can be used to calculate the necessary capacitance value C_p in order to operate the circuit at resonance. The resonance condition is important so the switching losses in the MOSFET M_1 and M_2 are as small as possible. Capacitance C_p can be calculated with the following expression:

$$C_p = \frac{1}{\omega_o^2 L_{eq2}} \tag{12}$$

Finally, to obtain the value of the compensation inductor, the following equation can be used:

$$L_c = a\left(L_{lp} + L_{eq_1}\right) \tag{13}$$

where a is a proposed value, and the higher the value of a, the lower the variation of L_{eq2}. The worst case is when the distance d is large, then $L_{lp} + L_{eq1} \approx 0$, and the relationship m between the worst case of L_{eq2} and the nominal value of L_{eq2} is:

$$m = \frac{L_{eq2w}}{L_{eq2n}} = \frac{L_c}{L_c + L_{lp} + L_{eq1}} = \frac{a}{a+1} \tag{14}$$

$$\%error = (1 - m)100 \tag{15}$$

Equation (16) was plotted as shown in Figure 9.

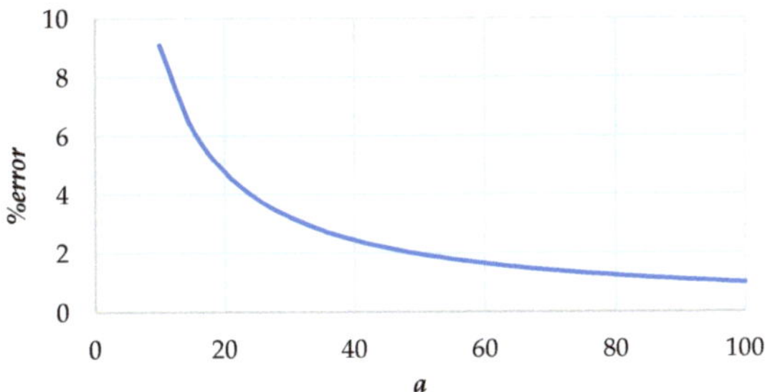

Figure 9. Percentage variations of L_{eq2} (%error) in function the factor a.

According to Figure 9, the higher the value of a, the lower the percentage of L_{eq2} variations (%error), so the compensation of the distance d variation in the transmitter and receiver misalignments will be more efficient. For example, for $a = 100$, the percentage variations of L_{eq2} (%error) are almost 1, which means no changes in L_{eq2} for large variations of d.

In this way, it is possible to compensate for the variation in the inductive coupling inductances with respect to the transmission distance and their misalignment. The circuit is expected to remain operating close to the resonance condition so that the maximum energy transfer can be achieved.

To evaluate the performance of the compensation inductor in relation to the resonance and efficiency of the circuit, three values of a will be proposed. The values will be $a = 10$, $a = 20$, and $a = 30$.

The last stage to analyze is the HF inverter, which consists of a half bridge inverter. The RMS current I_{RMS} of the resonant network (when it is operating at resonance) is obtained by applying Ohm's law using the values of the input voltage V_{in} and R_{eq2}:

$$I_{RMS} = \frac{V_{in}}{\pi R_{eq2}} \tag{16}$$

Another important characteristic to obtain is the value of average power P_M dissipated in each MOSFET of the inverter because this power represents the losses present in the inverter. In this analysis, it is recommended to select a desired dissipated power in the order of mW so that it does not affect the total efficiency of the circuit. Using the I_{RMS} and P_M values, the value of Drain-Source ON Resistance $R_{DS(on)}$ that each MOSFET should have is calculated to obtain the estimated losses in the inverter:

$$R_{DS(on)} = \frac{P_M}{I_{RMS}^2} \tag{17}$$

Substituting Equation (16) into (17):

$$R_{DS(on)} \leq \frac{P_M R_{eq2}{}^2 \pi^2}{V_{in}{}^2} \tag{18}$$

Finally, to calculate the total efficiency η value in the proposed IPT circuit, the output power P_o (present in the load resistance) and the input power P_{in} (applied to the HF inverter) are used, which are substituted into the property of $P = VI$. Then:

$$\eta = \frac{P_o}{P_{in}}(100\%) = \frac{V_o I_o}{V_{in} I_{in}}(100\%) \tag{19}$$

3. Design Methodology

To validate the above equations, a step-by-step design methodology for the resonance condition is proposed in this work. The considered application is to charge smartphones, so the nominal distance between the transmitter and the receiver was $d = 1.5$ mm.

The design methodology to be shown is when the compensating inductor has a value of 10 times the sum of L_{lp} y L_{eq1} ($a = 10$). The proposed methodology is similar in cases where $a = 20$ or $a = 30$, and only the values of L_c, C_p and Q change, respectively, in each case.

Table 3 shows the design specifications, and a power of 10 W was considered for wireless chargers for smartphones and the output voltage of 12 V was selected according to Figure 8. The input voltage is considered from a small photovoltaic panel.

Table 3. Design parameters.

Parameter	Description	Value
P_o	Average output power	10 W
V_o	Output voltage	12 V
V_{in}	Input voltage	27.5 V
f_{sw}	Switching frequency	500 kHz
f_o	Resonance frequency	500 kHz
Δ_{vo}	Output voltage ripple	$5\% V_o$
a	Relationship between L_c and $(L_{lp} + L_{eq1})$	10

The design obtains the properties of the inductive coupling using the following procedure:

i.   Coupling the inductor L_p (transmitter) and L_s (receiver), correctly aligned using an acrylic plastic as the core, the thickness of the acrylic is 1.5 mm; this distance will be the transmission distance d. The inductors should not be connected to any power source or any electrical load during this measurement process.

ii.   When the inductors are coupled, an LCR tester will be used. A Hioki model 3532-50 was used in this work which is set to measure inductance at a frequency of 500 kHz; this value will be the resonance frequency at which the circuit will operate. The inductance of the primary inductor L_p (primary self-inductance) is measured by the LCR tester. Then, the inductance of the secondary inductor L_s (secondary self-inductance) is measured.

iii.   Maintaining the same conditions in the inductive coupling, the inductors L_p and L_s are connected in series, and then the measurement of the total series inductance L_T is performed with the LCR tester.

iv.   Using the values obtained in the measurements, the value of the mutual inductance M is calculated:

$$M = \frac{L_T - L_p - L_s}{2} \tag{20}$$

v.   The value of the coupling coefficient k is calculated by substituting the values of L_p, L_s and M into Equation (1).

vi. The primary magnetizing inductance L_{mp}, primary leakage inductance L_{lp} and secondary leakage inductance L_{ls} are calculated, and the values of these inductances are obtained using the equations presented in Table 2.

Using the method described above, the value of the inductive coupling properties shown in Table 4 can be calculated, and then the value of the IPT circuit components can be calculated using the design methodology described in Table 5.

Table 4. Characteristics of the wireless transmitter and receiver.

Parameter	Description	Value
d	Transmitter–receiver distance	1.5 mm
	Transmission medium thickness (inductive coupling core): acrylic plastic	1.5 mm
L_p	Primary self-inductance	8.625 µH
L_s	Secondary self-inductance	8.700 µH
L_T	Total series inductance	31.727 µH
L_{lp}	Primary leakage inductance. Measured short-circuiting the receiver and measuring the inductance of the primary with the nominal distance d.	1.42 µH

Table 5. Design methodology.

Parameter	Description	Equation	Value
M	Mutual inductance.	$\frac{L_T - L_p - L_s}{2}$	7.201 µH
k	Coupling coefficient.	$\frac{M}{\sqrt{L_p L_s}}$	0.831
L_{mp}	Primary Magnetizing inductance	$L_p - L_{lp}$	7.201 µH
n'	Turns relationship of the ideal transformer $n' = n'_2/n'_1$, the turns n'_1 and n'_2 represent the cross section of coils L_1 and L_2 that share the same common flux φ_M.	$\frac{M}{L_{mp}}$	1
L_{ms}	Secondary magnetizing inductance.	Mn'	7.201 µH
L_{ls}	Secondary leakage inductance.	$L_s - L_{ms}$	1.50 µH
I_o	Output current.	$\frac{P_o}{V_o}$	0.833 A
R_L	Load resistance.	$\frac{V_o^2}{P_o}$	14.4 Ω
C_f	Output filter capacitor.	$\frac{V_o}{2R_L f_o \Delta_{vR}}$	3.49 µF
n_{Rect}	Full bridge rectifier efficiency.	$1 - \frac{2V_f}{V_o}$	0.89
I_D	Diode average current	$\frac{I_o}{2}$	0.417 A
L_c	Compensation inductor.	$a\left(L_{lp} + L_{eq_1}\right)$	34.31 µH
C_p	Primary capacitor.	$\frac{1}{(2\pi f_o)^2 (1.1 L_c)}$	2.711 nF
Q	Quality factor	$\frac{X_{C_p}}{R_{eq_2}}$	18.88
$R_{DS(on)}$	Drain-Source ON Resistance	$\leq \frac{P_M R_{eq_2}^2 \pi^2}{V_{in}^2}$	≤20 mΩ
η	Total efficiency of the IPT circuit	$\frac{P_o}{P_{in}} (100\%)$	75%

Table 4 presents the characteristics of the wireless transmitter and receiver, the IPT inductors were model WE760308111 from Würth Electronics.

Table 5 presents the design methodology, where the nominal distance d = 1.5 mm, which is the thickness of the acrylic between the transmitter and the receiver.

For the implementation of the full bridge rectifier located in the receiver circuit, the MUR840 diodes were selected. This diode model supports an average current of 8A.

The MOSFET IRFZ46N was selected for the high-frequency inverter located in the transmitter circuit. This MOSFET model has an $R_{DS(on)}$ of 16.5 mΩ

3.1. Implementation of Compensation Inductor Lc

The compensation inductor used for the TPI circuit was manufactured with Litz wire, which is composed of a wire number N_w of 160 wires and each copper wire is 44 AWG. The length of the conductor l_c is 3.56 m. The compensation inductor is made up of 63 turns, using an ETD29 model that is shown in Figure 10.

Figure 10. Picture of the compensation inductor L_C picture used in the IPT circuit.

The compensation inductor designed for the TPI circuit has an air core, so the only losses that occur in the component are conduction losses in the copper of the inductor. The copper presents a parasitic resistance that will oppose the flow of electric current through the inductor. Equation (21) presents the formula to obtain the parasitic resistance of the inductor R_{Lc}.

$$R_{Lc} = \frac{\rho_{cu} l_c}{N_w A_c} \tag{21}$$

where ρ_{cu} is the copper resistivity (17.2×10^{-9} $\Omega \cdot$m), l_c is the length of the conductor (3.56 m), N_w is the wire number (160 wires) and A_c is the wire cross sectional area (2.463×10^{-9} m^2). Substituting these values into Equation (21):

$$R_{Lc} = \frac{\left(17.2 \times 10^{-9}\ \Omega \cdot m\right)\left(3.56\ m\right)}{(160)\left(2.463 \times 10^{-9}\ m^2\right)} = 155.37\ m\Omega \tag{22}$$

3.2. Determination of the Properties of the Inductive Coupling with Different Distances d

In the design methodology, the value of the IPT circuit components was obtained to work correctly at distance d = 1.5 mm. As the circuit is oriented to wireless charging applications of smartphones, it is important to consider that during wireless charging there may be changes in the transmission distance with different causes; these changes directly affect the properties of the inductive coupling, modifying their value.

For that reason, before starting with the implementation of the designed circuit, the measurement and calculation of the inductive coupling properties at different distances d must be performed using the procedure shown at the beginning of Section 3. In the procedure, it is established that the transmission inductors during the measurements should not be connected to any power supply or any electrical load; for this reason, the measurements of the inductive coupling should be performed before the assembly of the IPT circuit on the PCB.

Ten transmission distances were selected as test points to perform measurements of the inductive coupling, with the objective of watching the changes in the coupling with respect to the distance d. This circuit being of the near-field WTP type, the inductor L_s will be moving away from L_p and the plastic acrylic (coupling core) from a distance of 1.5 mm to 4 mm; therefore, there will be 11 measured values for each property of the inductive coupling. The measured and calculated values at each transmission distance are L_p, L_s, L_T, M and k.

The LCR tester was used to measure the inductances, while a digital Vernier was used for the transmission distances d. The measured and calculated results are presented in Table 6.

Table 6. Obtained values of inductive coupling properties at different transmission distances d.

d (mm)	L_p (μH)	L_s (μH)	L_T (μH)	M (μH)	k
1.5	8.625	8.700	31.727	7.201	0.831
1.8	8.558	8.624	31.288	7.053	0.821
2	8.392	8.467	30.547	6.844	0.812
2.2	8.268	8.343	29.931	6.660	0.802
2.5	8.146	8.208	29.306	6.476	0.792
2.8	8.053	8.097	28.794	6.322	0.783
3	7.934	7.989	28.233	6.155	0.773
3.2	7.819	7.864	27.649	5.983	0.763
3.5	7.708	7.746	27.090	5.818	0.753
3.8	7.582	7.623	26.517	5.656	0.744
4	7.496	7.531	26.055	5.514	0.734

4. Experimental Results

A prototype was implemented to experimentally test the circuit in Figure 5 to validate the design methodology shown in Table 5. Figures 11 and 12 show the implementation of the inductive IPT circuit.

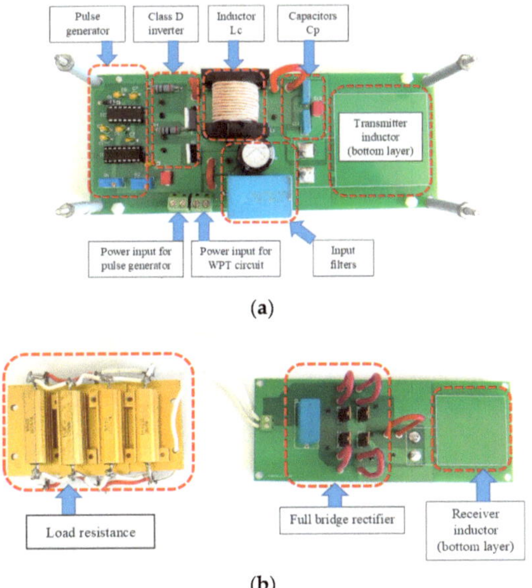

Figure 11. Implementation of the inductive IPT circuit, (**a**) transmitter device, (**b**) receiver device.

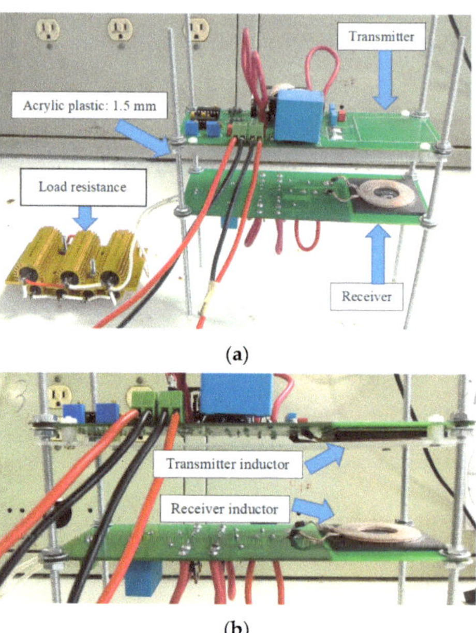

Figure 12. Prototype experimental assembly, (**a**) prototype test bench, (**b**) view of transmission distance setting (adjustable range from 1.5 mm to 4 mm).

To validate the design conditions, the prototype was adjusted with an initial transmission distance of 1.5 mm. Experimental results are shown in Figure 13. Figure 13a shows the output voltage, output current, and output power. Figure 13b shows the input voltage, input current and input power. The measured values are shown in Table 7, which comply with the design parameters established in Table 3.

Table 7. Experimental results obtained at nominal conditions.

Parameter	Description	Measured Value
P_o	Output power	10.59 W
V_o	Output voltage	12 V
I_o	Output current	0.881 A
V_{in}	Input voltage	27.5 V
I_{in}	Input current	0.5129 A
P_{in}	Input power	14.1 W
η	Total circuit efficiency	75.10%
f_o	Resonance frequency	497.4 kHz
ϕ	Phase offset angle	0.1074°

The input voltage was adjusted to deliver the nominal voltage to the load. Figure 13c shows the square voltage applied to the resonant tank and the current in the compensation inductor. As can be seen in this figure, the current is in phase with the voltage; therefore, the circuit is in resonance. These results validate the proposed design methodology since they are very similar to the theoretical ones.

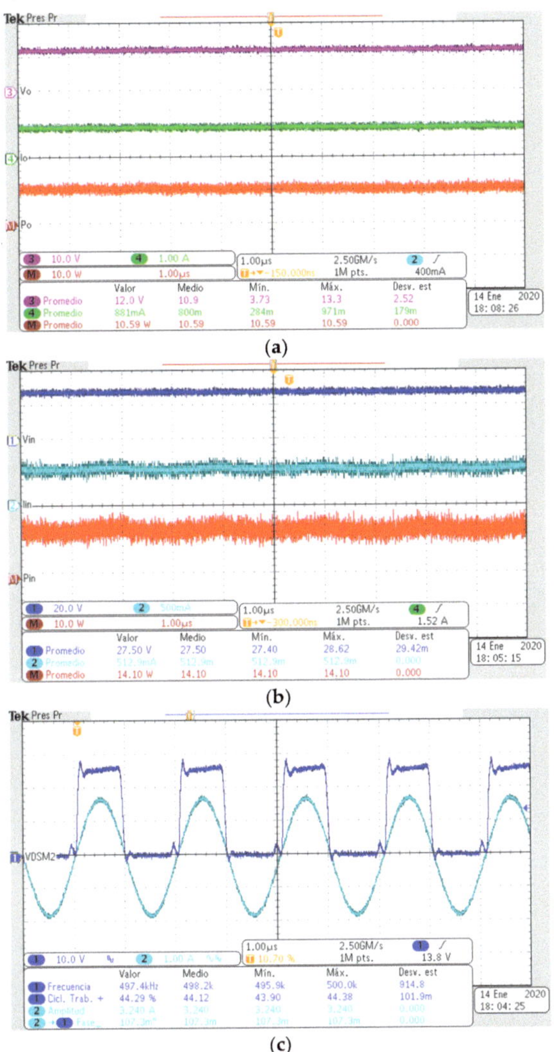

Figure 13. (**a**) Channel 3: output voltage V_o, 10 V/div (purple). Channel 4: output current I_o, 1 A/div (green). Math Channel: output power P_o, 10 W/div (red). (**b**) Channel 1: input voltage, 20 V/div (navy blue). Channel 2: input current, 500 mA/div (light blue). Math Channel: input power, 10 W/div (red). (**c**) Channel 1: class D inverter output voltage, 10 V/div (navy blue). Channel 2: resonant network current, 1 A/div (light blue).

The total efficiency η of the IPT circuit is calculated using Equation (19) and the values obtained in the measurements of P_{in} and P_o, which are presented in Figure 13a,b.

$$\eta = \frac{10.59\ \text{W}}{14.10\ \text{W}}(100\%) = 75.10\%$$ (23)

Experimental Results with Different Distances "d"

To test the robustness of the circuit for changes in the transmission distance. Ten different test points were performed from 1.5 mm to 4 mm. The following variables were measured: output voltage (V_o), output power (P_o), input power (P_{in}), total efficiency (η) and phase offset (ϕ). The experimental results obtained were compared with the k values obtained in Section 3.2, and the values together are shown in Table 8.

Table 8. Experimental results obtained by modifying the transmission distance from 1.5 mm to 4 mm (a = 10).

d (mm)	k	V_o (V)	P_o (W)	P_{in} (W)	η (%)	ϕ (°)
1.5	0.831	12.00	10.59	14.10	75.10	0.10
1.8	0.821	11.71	10.07	13.51	74.55	−7.74
2	0.812	11.70	10.05	13.60	73.93	−14.40
2.2	0.802	11.36	9.46	12.84	73.70	−16.74
2.5	0.792	11.11	9.07	12.35	73.41	−17.50
2.8	0.783	10.83	8.61	11.85	72.65	−18.68
3	0.773	10.57	8.21	11.58	70.84	−19.44
3.2	0.763	10.34	7.85	11.61	67.61	−21.61
3.5	0.753	10.14	7.54	11.84	63.72	−22.15
3.8	0.744	9.97	7.30	12.31	59.31	−37.46
4	0.734	8.56	5.62	10.37	54.29	−39.18

Using the same design methodology already shown, the value of the compensation inductor L_c is calculated. The value of L_c is obtained using Equation (14) when it is 20 (a = 20) and 30 (a = 30) times the sum of L_{lp} y L_{eq1}. The calculation of L_c and C_p when a = 20 is shown in Table 9, while the a = 30 calculation is shown in Table 10.

Table 9. Calculation of L_c and C_p when a = 20.

Parameter	Description	Equation	Value
L_c	Compensation inductor.	$a\left(L_{lp} + L_{eq_1}\right)$	68.62 µH
C_p	Primary capacitor.	$\dfrac{1}{(2\pi f_o)^2(1.03L_c)}$	1.406 nF

Table 10. Calculation of L_c and C_p when a = 30.

Parameter	Description	Equation	Value
L_c	Compensation inductor.	$a\left(L_{lp} + L_{eq_1}\right)$	102.93 µH
C_p	Primary capacitor.	$\dfrac{1}{(2\pi f_o)^2(1.03L_c)}$	0.952 nF

These new values of L_c and C_p were implemented in the IPT circuit for each case, and experimental tests were performed at different transmission distances using the same method as the tests shown in Table 8. The results are shown in Tables 11 and 12.

Analyzing the results obtained in Table 7, we can see that the coupling coefficient has a value of 0.734 for a transmission distance of 4 mm. This table shows the output voltage, which falls to 8.56 V when the transmission distance is 4 mm — more than double the nominal value. The output power in this table falls to 5.62 W for a transmission distance of 4 mm. The input power decreases with the transmission distance until a transmission distance of 4 mm. For d = 4 mm, the phase offset angle (ϕ) increases from zero to 39°. The efficiency falls to 54.29% for d = 4 mm.

Comparing the results obtained in Table 8 with the results of Tables 11 and 12, it can be observed that the coupling coefficient, the output voltage, the output power, and the phase angle present the same behaviour, slightly changing their values when the transmission distance increases up to 4 mm.

Table 11. Experimental results obtained when (a = 20).

d (mm)	k	V_o (V)	P_o (W)	P_{in} (W)	η (%)	ϕ (°)
1.5	0.831	12.00	10.57	14.49	72.95	0.15
1.8	0.821	11.86	10.33	15.26	67.67	−3.24
2	0.812	11.75	10.14	15.21	66.65	−7.02
2.2	0.802	11.54	9.78	15.14	64.59	−10.44
2.5	0.792	11.38	9.51	15.08	63.04	−11.98
2.8	0.783	11.16	9.14	14.95	61.16	−12.96
3	0.773	11.11	9.06	14.93	60.70	−14.40
3.2	0.763	10.86	8.66	14.88	58.19	−15.30
3.5	0.753	10.34	7.85	14.05	55.89	−18.00
3.8	0.744	9.70	6.91	13.44	51.43	−18.36
4	0.734	9.09	6.07	12.22	49.68	−21.78

Table 12. Experimental results obtained when (a = 30).

d (mm)	k	V_o (V)	P_o (W)	P_{in} (W)	η (%)	ϕ (°)
1.5	0.831	12.00	10.68	15.18	70.40	0.13
1.8	0.821	11.98	10.64	15.71	67.74	−1.31
2	0.812	11.95	10.60	16.13	65.74	−3.29
2.2	0.802	11.93	10.55	16.16	65.28	−8.87
2.5	0.792	11.86	10.54	16.75	62.90	−9.05
2.8	0.783	11.84	10.39	17.23	60.30	−10.25
3	0.773	11.72	10.38	17.36	59.81	−10.89
3.2	0.763	11.57	10.27	17.44	58.87	−11.58
3.5	0.753	11.38	10.17	17.41	58.41	−12.53
3.8	0.744	11.19	10.09	17.28	58.40	−13.42
4	0.734	10.97	9.90	17.03	58.15	−15.35

The results obtained in the tables show the following:

- The output voltage value decreases to 10.97 V, although it stays closer to 12 V, as the value of a increases.
- The output power value decreases to 9.9 W, staying closer to 10 W, as the value of a increases.
- The total efficiency at a transmission distance of 1.5 mm decreases minimally as the value of a increases. This is because the value of a increases and the value of the inductor wire resistance increases with it.
- The 10-phase angle (ϕ) stays closer to resonance as the value of a increases. As the inductor value a increases, it will better compensate for the variations caused by misalignment and transmission distance.

From these results, it is observed that the compensation inductor can compensate for changes at the transmission distance of twice the nominal distance of 1.5 mm, achieving a total efficiency of 75.10%. It is important to clarify that the prototype was designed to operate over a wide range of powers and distances between the transmitter and receiver. A specific design for a fixed distance "d", optimizing the design of the inductor L_c as well as the selection of the semiconductors, and changing them for SiC or GaN semiconductors, can substantially increase the efficiency.

There is a tradeoff between the size of the inductor L_c and the distance "d"; the greater the distance "d", the larger the inductor L_c required. However, with a higher switching frequency, an excessively large inductor L_c will not be required.

The main disadvantage of the proposal is that the proposed solution is passive. To compensate better for the misalignment and greater distances between the transmitter and receiver, it is recommended to combine this proposal with others such as closed-loop operation and solutions for better coupling factors.

5. Conclusions

This work introduced a very simple compensation topology that consists of only one inductor useful for resonant IPT circuits operating in an open loop for low-power applications and small transmission distance applications (near-field applications), such as wireless chargers of cell phones. To compensate for changes in the transmission distance and misalignment of the transmitter and receiver, only one compensation inductor is added to the resonant tank.

A sufficiently bigger inductor is chosen to maintain the resonant tank operating near resonance for changes until the nominal transmission distance is doubled. The analysis and design methodology for the inductive coupling was presented to establish the resonance condition. The experimental results validate the proposed design methodology. In addition, they confirm the correct functioning of the circuit.

The experimental results obtained under the design conditions validate the proposed design methodology with a maximum error of 5.9% in the output power. A total efficiency of 75% was achieved at a fixed transmission distance of 1.5 mm, using acrylic as a transmission medium.

The implemented circuit operates very near to the resonance condition; comparing the theoretical design values with the experimental results, errors of 0.059% in the phase shift angle and 0.02% in the resonance frequency are obtained.

The results obtained by modifying the transmission distance demonstrate the presence of inductance variations in the inductive coupling as the transmitter inductor moves away from the receiver inductor. However, these variations could be compensated for with only one inductor used instead of more complex compensation topologies. For small variations of the transmission distances (twice or three times), this inductor compensated for the variations in the mutual inductance and self-inductances of the transmitter due to the changes in the coupling factor.

According to the obtained results, the proposed circuit is a good option as a wireless charger for cell phones in which the application transmits power over very short distances.

The circuit proposed in this work is limited to applications where the transmission distance is larger because these applications are of the far-field WPT type. In this type of application, where the distance is large, it would affect the design of the circuit presented in several ways, such as the following:

- The size of the inductors L_p and L_s would increase considerably in order to realize the wireless power transmission through inductive coupling.
- The input voltage V_{in} would need to be increased considerably to compensate for the losses in the inductive coupling, because over long distances, the inductive coupling coefficient k is usually small and the power transmission efficiency from L_p to L_s is low.
- The size and the value of the compensation inductor L_c would increase because it must compensate for the misalignment of the inductors L_p and L_s in addition to trying to keep the circuit in resonance at a larger transmission distance. As the size of the inductor L_c increases, the series parasitic resistance R_{Lc} in the inductor also increases due to the length of the copper conductor, presenting a higher amount of losses in the IPT circuit.
- As the value of the inductor L_c increases, as a consequence, the value of the primary capacitor C_p will decrease considerably, achieving a value that is difficult to implement because it will be valued in the order of pF or less.
- The total efficiency will be low due to the losses present in the inductive coupling. In this type of application, it is recommended to combine the circuit proposed in this work with a closed-loop method that allows to adjust the resonance and bring the maximum power transfer from L_p to L_s.

The IPT circuit proposed in this work was validated and works adequately inside the limits previously described; therefore, it is recommended to be used in near-field WPT-type applications and for low-power electronic devices.

6. Discussion

This section will present an in-depth analysis of the results obtained, some of which were not mentioned in the previous sections, as well as some general characteristics, recommendations, contributions of this work, and possible future works to give continuity to the presented topic.

6.1. Relevant Conclusions

Once the work developed in this paper is completed, the following conclusions can be drawn:

- The resonant network suitable for low-power inductive IPT applications is the series–series topology with inductive coupling; due to this resonant network, operating in resonance has a unity voltage gain. For this reason, the series–series topology was selected for the IPT circuit designed.
- An analysis and design methodology was developed for the proposed resonant network, which will allow the proposed IPT circuit to operate according to the desired design specifications.
- The experimental results obtained validate the analysis and design methodology proposed in this work.
- The implemented IPT circuit enables wireless power transmission from a DC voltage source to an electrical load using resonant inductive coupling to perform the transmission.

6.2. General Characteristics of the Implemented IPT Circuit

The IPT circuit implemented in this work has the following characteristics:

- The implemented IPT circuit works according to the desired design specifications: 12 volts at the output, 10 watts at the output, the resonant network is in resonance, and the circuit performs the wireless power transmission.
- The IPT circuit achieves a total efficiency of 75.10% under the design conditions (stationary distance, transmission medium: plastic acrylic of 1.5 mm thickness).
- It is an easy-to-implement circuit because it operates in an open loop.
- The prototype is suitable for inductive wireless charging applications, as long as low-power applications are no greater than 10 W.
- The IPT circuit allows a horizon of wireless battery-charging applications, such as cell phones, electric toothbrushes, electric shavers, portable hearing aids, and hand lamps, among others.
- The transmitting inductors do not present a temperature rise because they only provide 10% of the inductive reactance of the resonant network.

6.3. Recommendations

To achieve a successful implementation of the proposed IPT circuit, it is recommended to consider the following:

- It is important to consider that this circuit is for near-field WPT-type applications, especially for wireless charging of low-power wireless devices.
- For the selection of the diodes of the full bridge rectifier in the secondary circuit, it is recommended to use ultra-fast diodes that are capable of operating correctly at high switching frequencies.
- For the implementation of the compensation inductor Lc, it is recommended to use Litz wire because it will reduce the series parasitic resistance R_{Lc} of this inductor.
- The transmission inductors L_p and L_s must be identical in structure, inductance, design and size to achieve an efficient coupling coefficient k. It is recommended to use two inductors of the same model designed for IPT applications.
- For the selection of the MOSFETs of the high-frequency inverter, it is recommended to use a device with the lowest possible $R_{DS(on)}$ to avoid the highest possible conduction losses in the device.

6.4. Contributions of This Work

The work developed has the following contributions:

- An analysis methodology is contributed for the resonant network of an inductive IPT circuit, in which all the inductances of the inductive coupling are considered as part of a single resonant tank. The literature uses two independent resonant tanks for the analysis—one in the transmitter circuit and one in the receiver circuit.
- A resonant network topology used for inductive IPT is presented, where a single capacitor is used, which is different from the literature, where two are used (one in the transmitter and one in the receiver); in addition, a compensation inductor is used that contributes 90% of the inductive reactance in the resonant condition [34].
- A prototype for inductive IPT that is capable of performing wireless power transmission is provided. Such a device is useful in wireless battery-charging applications for low power.

6.5. Possible Future Work

Considering the importance of the resonance and coupling coefficient of the IPT circuit together with the experience obtained in this work, the following recommendations are presented for possible future work:

- To apply a PLL control loop that allows maintaining the resonance of the resonant network under variations in the inductive coupling. Such variations can be caused by the temperature, the misalignment of the transmission inductors or the transmission distance between both inductors.
- To use a control technique with active diodes in the rectification stage, which is located in the receiver device, in order to improve the efficiency of the rectifier and consequently the total efficiency of the IPT circuit.
- Investigate and apply techniques that improve the inductive coupling coefficient in order to improve the efficiency of wireless power transmission. The techniques can be related to the geometry of the transmitting inductors and the use of multiple transmitting inductors.
- In order to improve the inductive coupling coefficient, the resonant network can be used as an analogue frequency multiplier. The resonant network receives at the input the switching frequency of the inverter, and then a sinusoidal signal with a frequency multiplied "n" times the switching frequency can be delivered to the output. The signal delivered at the output will be the resonant frequency of the resonant network. This method is presented in [4].

These proposals are suggested to give continuity to this work.

Author Contributions: Conceptualization, M.P.-S., A.R.G.-G. and J.A.; Data curation, J.A., J.L.-R. and C.C.-G.; Formal analysis, M.P.-S., A.R.G.-G., J.L.-R. and C.C.-G.; Funding acquisition, M.P.-S., A.R.G.-G. and J.L.-R.; Investigation, M.P.-S., A.R.G.-G., J.A. and C.C.-G.; Methodology, M.P.-S., A.R.G.-G., J.A. and J.L.-R.; Project administration, M.P.-S. and C.C.-G.; Resources, J.A., J.L.-R. and C.C.-G.; Software, A.R.G.-G., J.L.-R. and C.C.-G.; Supervision, M.P.-S., J.A. and C.C.-G.; Validation, J.A., J.L.-R. and C.C.-G.; Visualization, J.A. and C.C.-G.; Writing—original draft, M.P.-S.; Writing—review and editing, M.P.-S., A.R.G.-G., J.A. and J.L.-R. All authors have read and agreed to the published version of the manuscript.

Funding: This research received no external funding.

Data Availability Statement: Not applicable.

Conflicts of Interest: The authors declare no conflict of interest.

References

1. Tran, M.T.; Thekkan, S.; Polat, H.; Tran, D.-D.; El Baghdadi, M.; Hegazy, O. Inductive Wireless Power Transfer Systems for Low-Voltage and High-Current Electric Mobility Applications: Review and Design Example. *Energies* **2023**, *16*, 2953. [CrossRef]
2. Laha, A.; Kalathy, A.; Pahlevani, M.; Jain, P. A Comprehensive Review on Wireless Power Transfer Systems for Charging Portable Electronics. *Eng* **2023**, *4*, 1023–1057. [CrossRef]

3. Zhang, Z.; Pang, H. *The Era of Wireless Power Transfer*; John Wiley & Sons, Inc.: Hoboken, NJ, USA, 2023.
4. Lara-Reyes, J.; Ponce-Silva, M.; Hernández-González, L.; DeLeón-Aldaco, S.E.; Cortés-García, C.; Ramirez-Hernandez, J. Series RLC Resonant Circuit Used as Frequency Multiplier. *Energies* **2022**, *15*, 9334. [CrossRef]
5. Manivannan, B.; Kathirvelu, P.; Balasubramanian, R. A review on wireless charging methods—The prospects for future charging of EV. *Renew. Energy Focus* **2023**, *46*, 68–87. [CrossRef]
6. Yoo, J.-S.; Gil, Y.-M.; Ahn, T.-Y. High-Power-Density DC–DC Converter Using a Fixed-Type Wireless Power Transmission Transformer with Ceramic Insulation Layer. *Energies* **2022**, *15*, 9006. [CrossRef]
7. Ibrahim, H.H.; Singh, M.J.; Al-Bawri, S.S.; Ibrahim, S.K.; Islam, M.T.; Alzamil, A.; Islam, M.S. Radio frequency energy harvesting technologies: A comprehensive review on designing, methodologies, and potential applications. *Sensors* **2022**, *22*, 4144. [CrossRef]
8. Mahesh, M.; Kumar, K.V.; Abebe, M.; Udayakumar, L.; Mathankumar, M. A review on enabling technologies for high power density power electronic applications. *Mater. Today Proc.* **2021**, *46*, 3888–3892. [CrossRef]
9. Shakil, S.; Rashid, M.H. The potential impacts of wireless power transfer on the global economy, society, and environment. In Proceedings of the IEEE 14th Power Electronics, Drive Systems, and Technologies Conference (PEDSTC), Babol, Iran, 31 January–2 February 2023; pp. 1–5.
10. Iam, I.-W.; Choi, C.-K.; Lam, C.-S.; Mak, P.-I.; Martins, R.P. A constant-power and optimal-transfer-efficiency wireless inductive power transfer converter for battery charger. *IEEE Trans. Ind. Electron.* **2023**, *71*, 450–461. [CrossRef]
11. Okasili, I.; Elkhateb, A.; Littler, T. A Review of Wireless Power Transfer Systems for Electric Vehicle Battery Charging with a Focus on Inductive Coupling. *Electronics* **2022**, *11*, 1355. [CrossRef]
12. Wu, M.; Su, L.; Chen, J.; Duan, X.; Wu, D.; Cheng, Y.; Jiang, Y. Development and Prospect of Wireless Power Transfer Technology Used to Power Unmanned Aerial Vehicle. *Electronics* **2022**, *11*, 2297. [CrossRef]
13. Lara-Reyes, J.; Ponce-Silva, M.; Cortés-García, C.; Lozoya-Ponce, R.E.; Parrilla-Rubio, S.M.; García-García, A.R. High-Power-Factor LC Series Resonant Converter Operating Off-Resonance with Inductors Elaborated with a Composed Material of Resin/Iron Powder. *Electronics* **2022**, *11*, 3761. [CrossRef]
14. Benalia, N.; Laroussi, K.; Benlaloui, I. Improvement of the Magnetic Coupler design for Wireless Inductive Power Transfer. In Proceedings of the IEEE 5th International Conference on Power Electronics and their Applications (ICPEA), Hail, Saudi Arabia, 29–31 March 2022; Volume 1, pp. 1–6.
15. Oshimoto, N.; Sakuma, K.; Sekiya, N. Improvement in Power Transmission Efficiency of Wireless Power Transfer System Using Superconducting Intermediate Coil. *IEEE Trans. Appl. Supercond.* **2023**, *33*, 1501004. [CrossRef]
16. Rayan, B.A.; Subramaniam, U.; Balamurugan, S. Wireless Power Transfer in Electric Vehicles: A Review on Compensation Topologies, Coil Structures, and Safety Aspects. *Energies* **2023**, *16*, 3084. [CrossRef]
17. Chowdhury, S.A.; Kim, S.; Kim, S.; Moon, J.; Cho, I.; Ahn, D. Automatic Tuning Resonant Capacitor to Fix the Bidirectional Detuning with ZVS in Wireless Power Transfer. *IEEE Trans. Ind. Electron.* **2023**, 1–10. [CrossRef]
18. Brovont, A.D.; Aliprantis, D.; Pekarek, S.D.; Vickers, C.J.; Mehar, V. Magnetic Design for Three-Phase Dynamic Wireless Power Transfer with Constant Output Power. *IEEE Trans. Energy Convers.* **2023**, *38*, 1481–1484. [CrossRef]
19. Zhou, Y.; Qiu, M.; Wang, Q.; Ma, Z.; Sun, H.; Zhang, X. Research on Double LCC Compensation Network for Multi-Resonant Point Switching in Underwater Wireless Power Transfer System. *Electronics* **2023**, *12*, 2798. [CrossRef]
20. Yoo, J.-S.; Gil, Y.-M.; Ahn, T.-Y. Steady-State Analysis and Optimal Design of an LLC Resonant Converter Considering Internal Loss Resistance. *Energies* **2022**, *15*, 8144. [CrossRef]
21. Li, Q.; Duan, S.; Fu, H. Analysis and Design of Single-Ended Resonant Converter for Wireless Power Transfer Systems. *Sensors* **2022**, *22*, 5617. [CrossRef]
22. Kamali, R.; Gholamian, S.A.; Abdollahi, S.R. An Investigation on Selecting of Resonant Inverter Topologies for Efficient Wireless Power Transmission. In Proceedings of the IEEE 14th Power Electronics, Drive Systems, and Technologies Conference (PEDSTC), Babol, Iran, 31 January–2 February 2023; pp. 1–6.
23. Feng, H.; Cai, T.; Duan, S.; Zhao, J.; Zhang, X.; Chen, C. An LCC-Compensated Resonant Converter Optimized for Robust Reaction to Large Coupling Variation in Dynamic Wireless Power Transfer. *IEEE Trans. Ind. Electron.* **2016**, *63*, 6591–6601. [CrossRef]
24. Dai, X.; Qiu, N. Insensitive design for wireless power transfer system using LLC resonant converter. In Proceedings of the 2016 Progress in Electromagnetic Research Symposium (PIERS), Shanghai, China, 8–11 August 2016; pp. 5193–5198.
25. Jang, Y.; Han, J.-K.; Baek, J.-I.; Moon, G.-W.; Kim, J.-M.; Sohn, H. Novel multi-coil resonator design for wireless power transfer through reinforced concrete structure with rebar array. In Proceedings of the IEEE 3rd International Future Energy Electronics Conference and ECCE Asia (IFEEC 2017-ECCE Asia), Kaohsiung, Taiwan, 3–7 June 2017; pp. 2238–2243.
26. Sun, K.; Wang, J.; Burgos, R.; Boroyevich, D. A series-series-CL resonant converter for wireless power transfer in auxiliary power network. In Proceedings of the IEEE Applied Power Electronics Conference and Exposition (APEC), New Orleans, LA, USA, 15–19 March 2020; pp. 813–818.
27. Aldhaher, S.; Luk, P.C.-K.; Whidborne, J.F. Tuning Class E Inverters Applied in Inductive Links Using Saturable Reactors. *IEEE Trans. Power Electron.* **2013**, *29*, 2969–2978. [CrossRef]
28. Iannuzzi, D.; Rubino, L.; Di Noia, L.P.; Rubino, G.; Marino, P. Resonant inductive power transfer for an E-bike charging station. *Electr. Power Syst. Res.* **2016**, *140*, 631–642. [CrossRef]

29. Shi, L.; Alou, P.; Oliver, J.A.; Cobos, J.A. A Novel Self-adaptive Wireless Power Transfer System to Cancel the Reactance of the Series Resonant Tank and Deliver More Power. In Proceedings of the IEEE Energy Conversion Congress and Exposition, Baltimore, MD, USA, 29 September–3 October 2019; pp. 4207–4211.
30. Berger, A.; Agostinelli, M.; Vesti, S.; Oliver, J.A.; Cobos, J.A.; Huemer, M. A Wireless Charging System Applying Phase-Shift and Amplitude Control to Maximize Efficiency and Extractable Power. *IEEE Trans. Power Electron.* **2015**, *30*, 6338–6348. [CrossRef]
31. Ondin, U.; Balikci, A. A Transformer Design for High-Voltage Application Using LLC Resonant Converter. *Energies* **2023**, *16*, 1377. [CrossRef]
32. Yao, W.; Lu, J.; Taghizadeh, F.; Bai, F.; Seagar, A. Integration of SiC Devices and High-Frequency Transformer for High-Power Renewable Energy Applications. *Energies* **2023**, *16*, 1538. [CrossRef]
33. Erickson, R.W.; Maksimovic, D. *Fundamentals of Power Electronics*; Springer Science & Business Media: Berlin/Heidelberg, Germany, 2007; pp. 712–713.
34. García-García, A.R.; Ponce-Silva, M.; Cortés-García, C. Dispositivo para Transmisión de Potencia Inalámbrica—Device for Wireless Power Transmission. Mexico, Patent Application: MX/a/2020/011304, Mexican Institute of Industrial Property. 2020. Available online: https://vidoc.impi.gob.mx/visor?usr=SIGA&texp=SI&tdoc=E&id=MX/a/2020/011304 (accessed on 2 July 2023).

Performance Analysis of Harmonic-Reduced Modified PUC Multi-Level Inverter Based on an MPC Algorithm

Umapathi Krishnamoorthy [1], **Ushaa Pitchaikani** [2], **Eugen Rusu** [3,*] **and Hady H. Fayek** [4]

[1] Department of Biomedical Engineering, KIT-Kalaignarkarunanidhi Institute of Technology, Coimbatore 641402, India; umapathi.uit@gmail.com

[2] Department of Electrical and Electronics Engineering, M. Kumararsamy College of Engineering, Karur 639113, India; ushap25061992@gmail.com

[3] Department of Mechanical Engineering, Faculty of Engineering, 'Dunarea de Jos' University of Galati, 47 Domneasca Street, 800008 Galati, Romania

[4] Electromechanics Engineering Department, Heliopolis University, Salam, 11785, Egypt; hadyhabib@hotmail.com or hady.habib@hu.edu.eg

* Correspondence: eugen.rusu@ugal.ro

Abstract: Renewable and distributed energy generation includes wind turbines, fuel cells, solar cells, and batteries. These distributed energy sources need special power converters in order to connect them to the grid and make the generated power available for public use. Solar energy is the most readily available energy source; hence, if utilized properly, it can power up both domestic and industrial loads. Solar cells produce DC power, and this should be converted to an AC source with the help of inverters. A multi-level inverter for an application is selected based on a trade-off between cost, complexity, losses, and total harmonic distortion (THD). A packed U-cell (PUC) topology is composed of power switches and voltage sources connected in a series-parallel fashion. This basic unit can be extended to a greater number of output voltage levels. The significance of this design is the reduced use of power switches, gate drivers, protection circuits, and capacitors. The converter presented in this paper is a 31-level topology switched by a variable switching frequency-based model predictive controller that helps in achieving optimal output with reduced harmonics to a great extent. The gate driver circuit is also optimized in terms of power consumption and size complexity. A comparison of the 9-level and the 31-level PUC inverters is carried out to study the impact of the number of levels on the total harmonic distortion. The simulation results depict that the total harmonic distortion (THD) for a nominal modulation index of 0.8 is 11.54% and 3.27% for the 9-level multi-level inverter (MLI) and the 31-level modified packed U-cell multi-level inverter (MPUC-MLI), respectively. The reduction in THD is attributed to the increased number of steps in the output when using the model predictive controller.

Keywords: packed U-cell (PUC); inverter; multi-level inverter (MLI); model predictive control (MPC); total harmonic distortion (THD); harmonic reduction

Citation: Krishnamoorthy, U.; Pitchaikani, U.; Rusu, E.; Fayek, H.H. Performance Analysis of Harmonic-Reduced Modified PUC Multi-Level Inverter Based on an MPC Algorithm. *Inventions* **2023**, *8*, 90. https://doi.org/10.3390/inventions8040090

Academic Editor: Om P. Malik

Received: 12 June 2023
Revised: 30 June 2023
Accepted: 10 July 2023
Published: 13 July 2023

1. Introduction

Clean power and renewable energy generation are always evergreen topics of research, and many studies have shown that such energy generation is made possible by distributed small generation units. Some sources of distributed generation include wind turbines, fuel cells, solar cells, and batteries. These distributed energy sources need special power converters in order to connect them to the grid and make the generated power available for public use. Solar energy is the most readily available energy source; hence, if utilized properly, it can power up both domestic and industrial loads. Solar cells produce DC power, and this should be converted to an AC source with the help of inverters. Traditional inverters have some demerits, such as distorted output voltage rich in harmonics, high stress, losses, etc., and this induces the necessity for developing efficient inverters. The designed modified packed U-cell (MPUC) inverter is tested by designing a hardware prototype, and results were produced.

As multi-level power converter topologies have been proven to improve the power quality, reduce harmonic distortion, and, hence, act as efficient power converters, a comprehensive review of the different multi-level converters along with their merits and demerits is detailed in [1]. Contrary to power electronic converters, a simple active filter is proven to be active in reducing harmonics in [2]. A medium voltage multi-level inverter (MLI) [3] is designed for reducing harmonic distortion by reducing the switching frequency. Different industrial-grade multi-level inverters are the topics of discussion in [4], along with the various modulation and control techniques. In [5], a packed-U-cell-based multi-level inverter along with a sinusoidal modulation is proposed. A variety of application areas of power electronic converters, such as transportation, renewable energy production, and industrial drives, are the topics covered in [6]. A cascaded multi-level inverter design was proposed in order to reduce losses due to switching and distortions caused by harmonics [7]. A detailed review of the different PUC-based inverter designs, their operation, outputs, and modeling, along with a comparison with other multi-level converters, is presented in [8].

An H-Bridge inverter of 17 levels designed using flying and floating capacitors is proposed in [9]. A difference in modulation technique is introduced by a model predictive controller for inverter switching proposed in [10]. Similarly, a five-level PUC inverter that uses modulation switching based on a sensor-less voltage control is presented in [11]. An application-specific solution for the case of a solar powered water pump is given in [12]. It details the use of a common (VSI) voltage source inverter controlled by switches for a brushless DC (BLDC) motor load. An efficient controller design is proposed in [13] for industrial multi-level converters. A single phase five-level inverter designed from two T-type bridges is detailed in [14]. In [15], a single DC source with a smaller number of solid-state switches and gate drivers is used in the design of a three-level boost cascaded H-Bridge inverter, whereas a controller with fault tolerance for a cascaded H-Bridge is detailed in [16]. A cascaded H-bridge MLI design for a photo-voltaic (PV) grid with inverted decoupling is proposed in [17]. Detailed reviews of various predictive control techniques for the grid and converter level are presented in [18].

A hybrid PUC architecture with a reduced number of switches that is capable of producing 7- and 15-level outputs is proposed in [19] for high power, low voltage applications. An experimental review of different control configurations of a PUC converter that uses a PI controller and a model predictive controller for five- and seven-level inverters is provided in [20]. A PWM-based switching for a PUC converter and its effectiveness in distributing losses and reducing harmonics, which, in turn, increases the reliability of the converter, is explained in [21]. In [22], a new topology of a seven-level PUC converter with a reduced device count and cost factor is explained, along with relevant results. Switching techniques that use a multiple-carrier-based PWM method are used for a seven-level PUC converter, and its resulting enhancements are explained in [23]. A seven-level PUC converter that uses asymmetrical DC links for producing a reduced harmonic output signal is proposed in [24], along with the results. A frequency-dependent level shifted sinusoidal PWM switched PI controller used for a 15-level PUC is explained in [25].

A comparison of a cascaded hybrid bridge converter and a PUC converter with a varied height modulation is depicted in [26]. A 31-level H-cascaded asymmetrical inverter is designed and verified in [27] with the aim of reducing harmonics. A PI controller switched seven-level PUC converter is studied in [28] and is found to reduce harmonics. Fault analysis of multi-level inverters is the topic of interest in [29], and a machine learning algorithm is used for classification and diagnosis of the fault detected from signal processing. A CNN-LSTM-based switching control of an LC filter converter for a three phase multi-level inverter is the topic of discussion in [30], and it is found that CNN can produce effective switching, whereas LC filters help in reducing harmonics. A hybrid MLI controlled by a damped second order integral controller as applied to solar PV cells is studied in [31], and its efficiency is found. A review of recent optimization techniques following the design and control of MLI is studied in [32]. A sliding mode control along with a finite control-set-based control algorithm for a transformer-less voltage restorer is the topic discussed in [33]. A suitable weight function is used to enhance balancing of the inverter. A zeta converter to regulate output from PV array and a level-shifted switching-based MLI inverter is discussed in [34], and it is found to give promising results for PV array. In [35], a unified power quality controller is proposed, and this model uses an MLI based on ANN and soft computing techniques for mitigating various power quality issues. A switched capacitor MLI to achieve 19-level output is explained in [36], along with its results.

Identification of a faulty switch in an inverter circuit by metering the near field using wavelet packet transform is proposed in [37]. A PUC five-level inverter, which is gated by a finite control set MPC controller, is used for testing the fault identification algorithm. An algorithmic method that uses voltages of capacitors in addition to the cost function, while determining the switching states at each sampling time, is used for controlling various PUC architectures in [38]. An artificial-intelligence-based auto-adjusting cost function optimization for controlling a seven-level PUC inverter is proposed and experimentally verified in [39]. A seven-level PUC inverter for a PV system with maximum power point tracking and a finite set model predictive algorithm is experimentally studied in [40]. A modified five-level PUC inverter is studied in [41], where a finite control-set-based prediction control is used for determining the switching states. PUC capacitor voltage estimates based on a switched observer are used for determining the switching states of an MPC-based seven-level PUC inverter in [42].

As distributed energy generation has started flourishing, the need for efficient power conversion is very rapidly growing; thus, development in the power electronics field is inevitable. Multi-level inverters are a promising solution for effective high-power exchange with low harmonic content and reduced switching losses. Conventional MLIs have many demerits, which include: i) implementation cost; ii) complexity of design with an increase in voltage levels; and iii) the number of solid-state devices increases with voltage levels. As there are difficulties, there are also opportunities for improvement. A lot of research has been carried out in designing efficient multi-level inverters, and one among them is the PUC design, which uses power switches and capacitors to form the basic U-cell unit. Such a design has both the advantages of a cascaded H-Bridge (CHB) and flying capacitors (FC). Figure 1a reveals the components that make up the basic U-cell. It consists of one capacitor that connects two solid-state devices on either side.

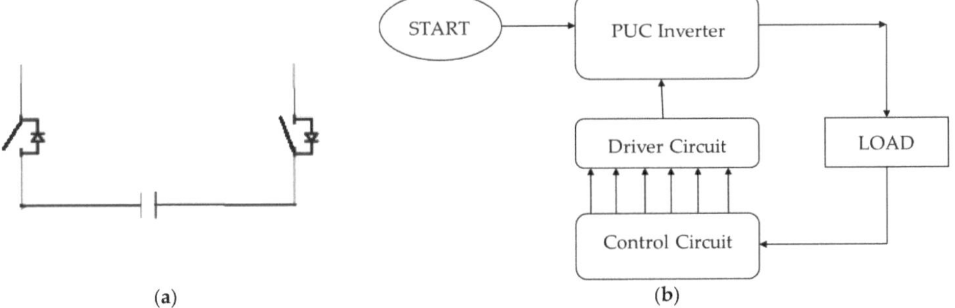

Figure 1. (**a**) A basic U-cell; (**b**) Block diagram of MPUC multi-level inverter.

As there are a lot of merits to the multi-level inverters, this paper proposes a 31-level MPUC inverter, which is gated in a dynamic manner by a variable switching frequency model predictive controller, which determines the switching states based on a cost function optimization for optimal grid current and capacitor voltages. This switching algorithm is implemented in a PIC microcontroller. The prediction efficiency of the switching algorithm is verified by simulation for 9-level and 31-level PUC inverters. There are multiple algorithmic implementations available in the literature for smaller-level MLIs where the number of switching states is smaller, whereas, in this paper, the cost function optimization has been used for switching a 31-level inverter efficiently. A 31-level design is prototyped, and the hardware results are verified.

2. Materials and Methods

The primary objective of using the multi-level inverters (MLI) is to increase the operating voltage and the current of the converters used. And, MLIs come with the advantages of higher operational voltage than the rated voltage of the components, stepped outputs with lower harmonic content, and reduced dv/dt changes. However, these MLIs come with a huge number of power switches that need to be reduced for optimizing cost and efficiency.

Packed U-cell (PUC) topologies are preferred over the other MLIs because of the reduced number of components used. For example, while considering a 9-level MLI, the components used by different inverter topologies are given in Table 1.

Table 1. Components used by different inverter topologies.

Inverter topology	Number of power switches	Number of capacitors	Number of dc sources
Manufacturer	Formax Electronics pvt ltd	Asoka Electronics	Generic Electronics
Component Description	15 V, 1 A MOSFET switches	1000uF/100V Electrolytic Capacitor	KBPC1510 15 A 1000 V Bridge Rectifier
Packed U Cell	8	2	1
Cascaded H Bridge	16	0	4
Flying Capacitor	16	36	1

A varied number of stage-based MLIs were studied in the literature, ranging from 5-level PUCs in [11,14,37,41], 7-level PUCs in [19,22–24,28,39,40,42], 15-level PUCs in [19,25], 19-level PUCs in [36], and a 31-level PUC in [27], each being handled by different control mechanisms. As the number of steps in a PUC increases, the stress on the power switches decreases, as the handling voltage reduces. As the number of steps in the output increases, the total harmonic distortion (THD) tends to reduce. As the optimal function of the MLI is desired, a 31-level PUC inverter was considered for design, and the output quality was compared with a 9-level PUC in this work.

2.1. A 31-Level MPUC Inverter

A 31-level MPUC inverter circuit was designed, and its performance with respect to output voltage and the harmonics produced were analyzed. The proposed modified PUC inverter used five pairs of power electronic switches (S1–S6, S2–S7, S3–S8, S4–S9, and S5–S10) connected across the DC source connected to dc1 and three capacitors (C2, C3, and C4, respectively). Considering dc_1 as a reference, then the other voltage levels are given as:

$dc_2 = 7/15(dc_1)$;
$dc_3 = 3/15(dc_1)$;
$dc_4 = 1/15(dc_1)$.

Figure 2 depicts the MLI design, and Table 2 lists the 31-level voltages' steps that yield the desired output. The different voltage levels are:

Table 2. Voltage levels of the 31-level MPUC inverter.

$\pm(-dc_4)$	$\pm(-dc_3 + dc_4)$	$\pm(-dc_3)$	$\pm(-dc_2 + dc_3)$	$\pm(-dc_2 + dc_3\text{-}dc4)$
$\pm(-dc_2 + dc_4)$	$\pm(-dc_2)$	$\pm(-dc_1 + dc_2)$	$\pm(-dc_1 + dc_2\text{-}dc_4)$	$\pm(-dc_1 + dc_2\text{-}dc_3 + dc_4)$
$\pm(-dc_1 + dc_2\text{-}dc_3)$	$\pm(-dc_1 + dc_3)$	$\pm(-dc_1 + dc_3\text{-}dc_4)$	$\pm(-dc_1 + dc_4)$	$\pm(-dc_1)$ and zero

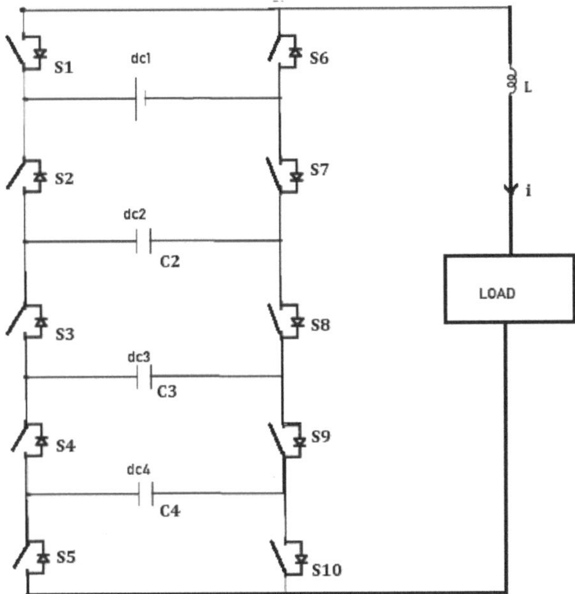

Figure 2. A 31-level MPUC inverter.

The DC power available from the solar cells has to be converted to a usable form by power converters before being fed to the load or connected to the power grid. Effective conversion of DC to AC power depends on the efficiency of the inverter. Thus, highly efficient and low THD inverter designs are required. Thus, this work detailed an efficient MPUC inverter design verified by simulation and then validated by hardware prototype implementation.

2.2. Modulation Technique

The process of gating the solid-state devices' ON and OFF in order to convert the input to the desired output form is called modulation. The modulation pulses are generated at different time intervals with different pulse widths based on some algorithm. The effective prediction of these pulses determines the nearness of the generated output to the desired output, and, thus, the modulation strategy plays an important role in the design of any power inverter/converter. In addition, the modulation algorithm should have good voltage quality and it should be simple, low-cost, and of modular design; at the same time, simultaneous switching of voltage levels is be avoided, and switching frequency should be reduced.

There are several modulation techniques used for switching the power devices in an MLI. These include Pulse Width Modulation (PWM) [21,23,25,34], Voltage balancing [11,33], Algorithmic-Controller-based PWM [37,38,40–42], and Artificial-Intelligence-based Controller [30,35,39]. In this work, a model predictive control algorithm was used to predict the width and instance of the switching pulses based on a reference sine wave. A model predictive control algorithm was used in this inverter, and its workflow is depicted below in Figure 3.

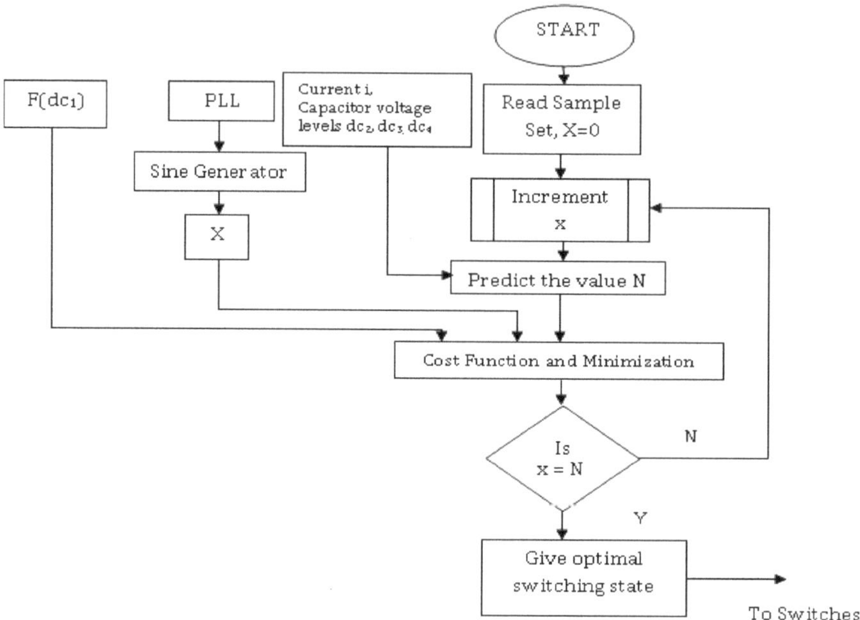

Figure 3. Flow chart depicting the workflow of the model predictive controller.

2.3. Model Predictive Controller

A model predictive controller gives the optimal switching state as the output, which gates the inverter switches. A prediction of the output is made from the present values of the load or the grid current I and the capacitor voltage levels dc_2, dc_3, and dc_4. The cost function is determined from the DC reference voltage input dc_1 and the sine wave generated from the PLL. If the error is minimal between the predicted and desired values, then the output is given, i.e., the switching pulses to trigger the power electronic switches are generated and given to the inverter.

2.4. Model Predictive Controller for 31-Level MPUC Inverter

The capacitor voltages dc_2, dc_3, and dc_4 were at $7/15(dc_1)$, $3/15(dc_1)$, and $1/15(dc_1)$, respectively. The switches (1,6), (2,7), (3,8), (4,9), and (5,10) were switched alternatively by the controller. The relationship between the load current i, the capacitor voltages dc_2, dc_3, and dc_4, and the switching states represented by SWx are given by Equations (1)–(4). A 31-level inverter requires 5 different variables to represent all its switching states (say, sw_1 to sw_5, each carrying a value of 0 or 1). When sw_1 is '0', the switch S6 conducts, and when sw_1 is '1', S1 conducts. The variables that constitute the equations are defined as dc_1, which is the voltage input from source; dc_2, dc_3, and dc_4, which are the capacitor voltages; sw_1 to sw_5, which represent the switching states whose value takes up {1,0} for ON and OFF states, respectively; L, which is the inductance value; i(t), which is the grid or load current; $v_g(t)$, which is the grid voltage; and C_2 to C_4, which are the capacitances:

$$C_2 \frac{ddc_2\,(t)}{dt} = (sw_3 - sw_2)i(t) \tag{1}$$

$$C_3 \frac{ddc_3\,(t)}{dt} = (sw_4 - sw_3)i(t) \tag{2}$$

$$C_4 \frac{ddc_4(t)}{dt} = (sw_5 - sw_4)i(t) \tag{3}$$

$$L\frac{di(t)}{dt} = (sw_1 - sw_2)dc_1(t) + (sw_2 - sw_3)dc_2(t) + (sw_3 - sw_4)dc_3(t) + (sw_4 - sw_5)dc_4(t) - v_g(t) \tag{4}$$

The objective of employing an MPC is to predict the capacitor voltages and the output current in order to switch the PUC inverter accordingly. Here, the term (k + 1) is the representation of the next state; thus, the next state of the voltage levels dc_2, dc_3, and dc_4 in terms of the current 'i' and switching states are found by applying the Euler Approximation; they are given by Equations (5)–(8)

$$dc_2(k+1) = dc_2(k) + \frac{T_s}{c_2}(sw_3 - sw_2)i(k) \tag{5}$$

$$dc_3(k+1) = dc_3(k) + \frac{T_s}{c_3}(sw_4 - sw_3)i(k) \tag{6}$$

$$dc_4(k+1) = dc_4(k) + \frac{T_s}{c_4}(sw_4 - sw_3)i(k) \tag{7}$$

$$i(k+1) = i(k) + \frac{T_s}{L}(sw_1 - sw_2)dc_1(k) + (sw_2 - sw_3)dc_2(k) + (sw_3 - sw_4)dc_3(k) + (sw_4 - sw_5)dc_4(k) - V_g(k) \tag{8}$$

The maximum allowable error between the actual measurement and the prediction values of different state variables for the model predictive controller is formulated. The maximum state variations are used to obtain the normalized state variables, as in Equations (9)–(12), and the cost function is given by Equation (13).

$$\Delta dc_{2max} = \frac{2i}{C_2}T_s \tag{9}$$

$$\Delta dc_{3max} = \frac{2i}{C_3}T_s \tag{10}$$

$$\Delta dc_{4max} = \frac{2i}{C_4}T_s \tag{11}$$

$$\Delta i_{max} = \frac{2dc_1}{L}T_s \tag{12}$$

$$g = \lambda_v \frac{dc_2^* - dc_2(k+1)}{\Delta dc_{2max}} + \lambda_v \frac{dc_3^* - dc_3(k+1)}{\Delta dc_{3max}} + \lambda_v \frac{dc_4^* - dc_4(k+1)}{\Delta dc_{4max}} + \lambda_i \frac{i^* - i(k+1)}{\Delta i_{max}} \tag{13}$$

Once the cost function is minimized, the prediction represented by (k + 1) and the reference values given by (*) will be very close to each other; furthermore, the output is given, i.e., the switching pulses to trigger the power devices are generated and given to the inverter.

3. Results

3.1. Simulation Results

3.1.1. Simulation of 9-Level MLI

The inputs used for the simulation of the nine-level MLI and the corresponding outputs are depicted below. Figure 4a–c represents the input current sources one, two, and three of the nine-level multi-level inverter. The ranges of the different input sources used are: (i) current sources one, two, and three, ranging up to 3.24 Amps; (ii) current source two, ranging up to 3.24 Amps; and (iii) input current source three, which is 3.24 Amps.

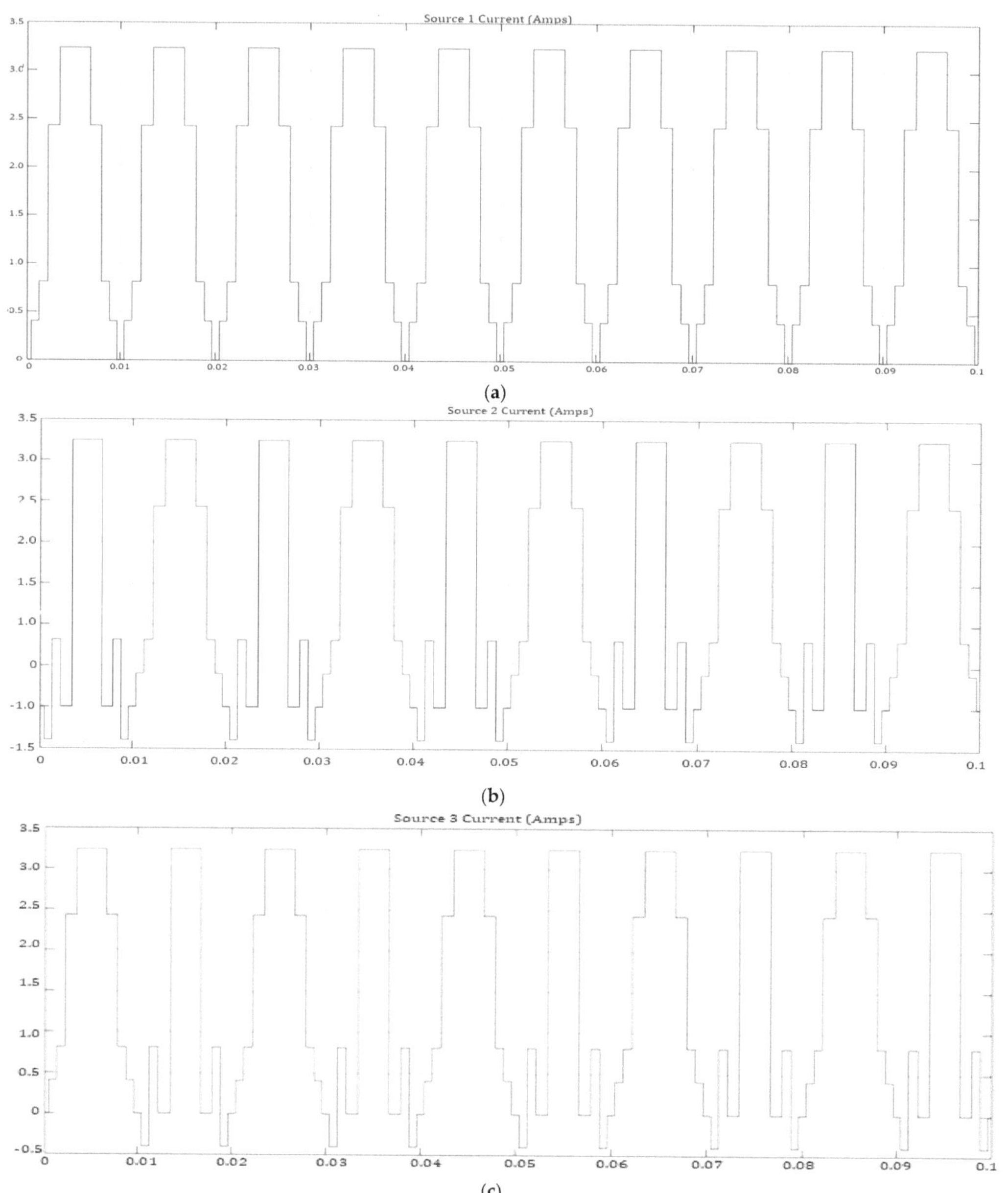

(a)

(b)

(c)

Figure 4. (**a**) Input source 1; (**b**) Input source 2; (**c**) Input source 3.

3.1.2. Output Voltage and Current

The simulation output of the nine-level MLI is as given below; Figure 5a represents the output voltage of the nine-level MLI, and it is approximately equal to 325 V. Figure 5b represents the output current whose value is 3.24 A. The Table 3 shows the Input and output values of 9-level MPUC inverter.

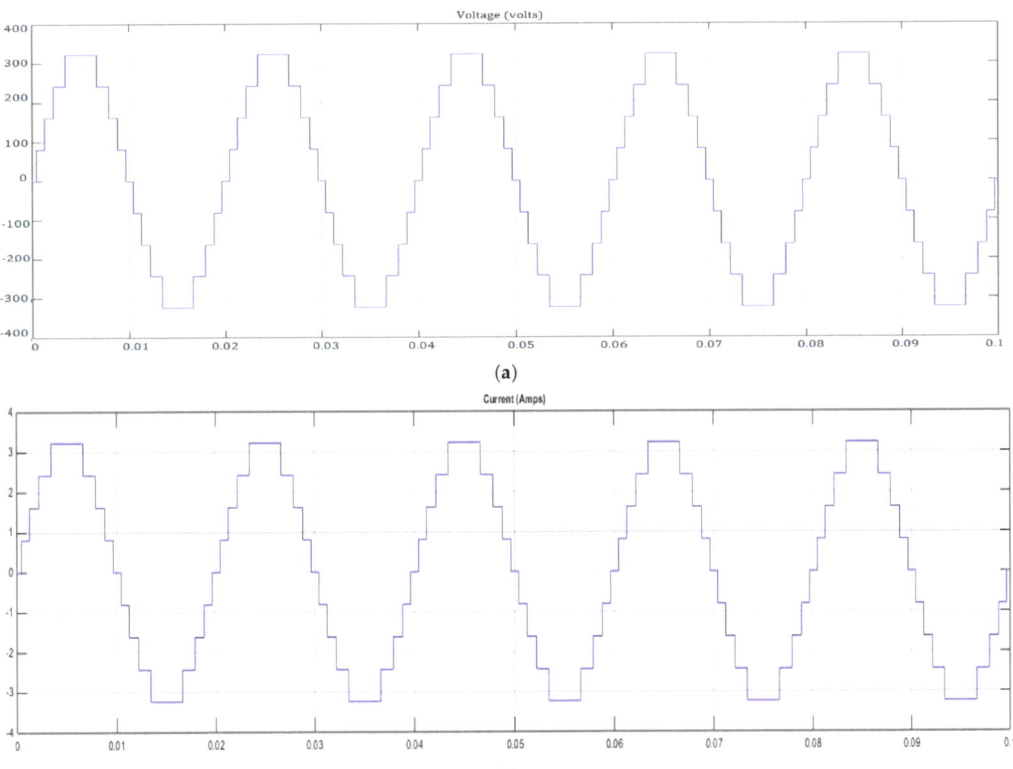

Figure 5. (**a**) Output voltage; (**b**) Output current.

Table 3. Input and output of 9-level MPUC inverter.

Input Current (Peak) in Amps			Output (Peak Values)	
Source 1	Source 2	Source 3	Voltage in V	Current in I
3.24	3.24	3.24	325	3.24

3.2. Simulation Result for the Proposed System

3.2.1. Input Sources

The inputs used for the simulation of the 31-level MPUC ML inverter are as given below. Figure 6a–d represents the input current sources one, two, three, and four of the 31-level MPUC MLI. The improvement in level shifting demands additional input sources in the circuit, and their corresponding values are: (i) current source one requires 3.24 A; (ii) input current source two ranges up to 1.5 A; (iii) current source three is approximately 2.3 A; and (iv) input current source four ranges up to 2.8 A. A varied range of input currents aids in providing performance improvement.

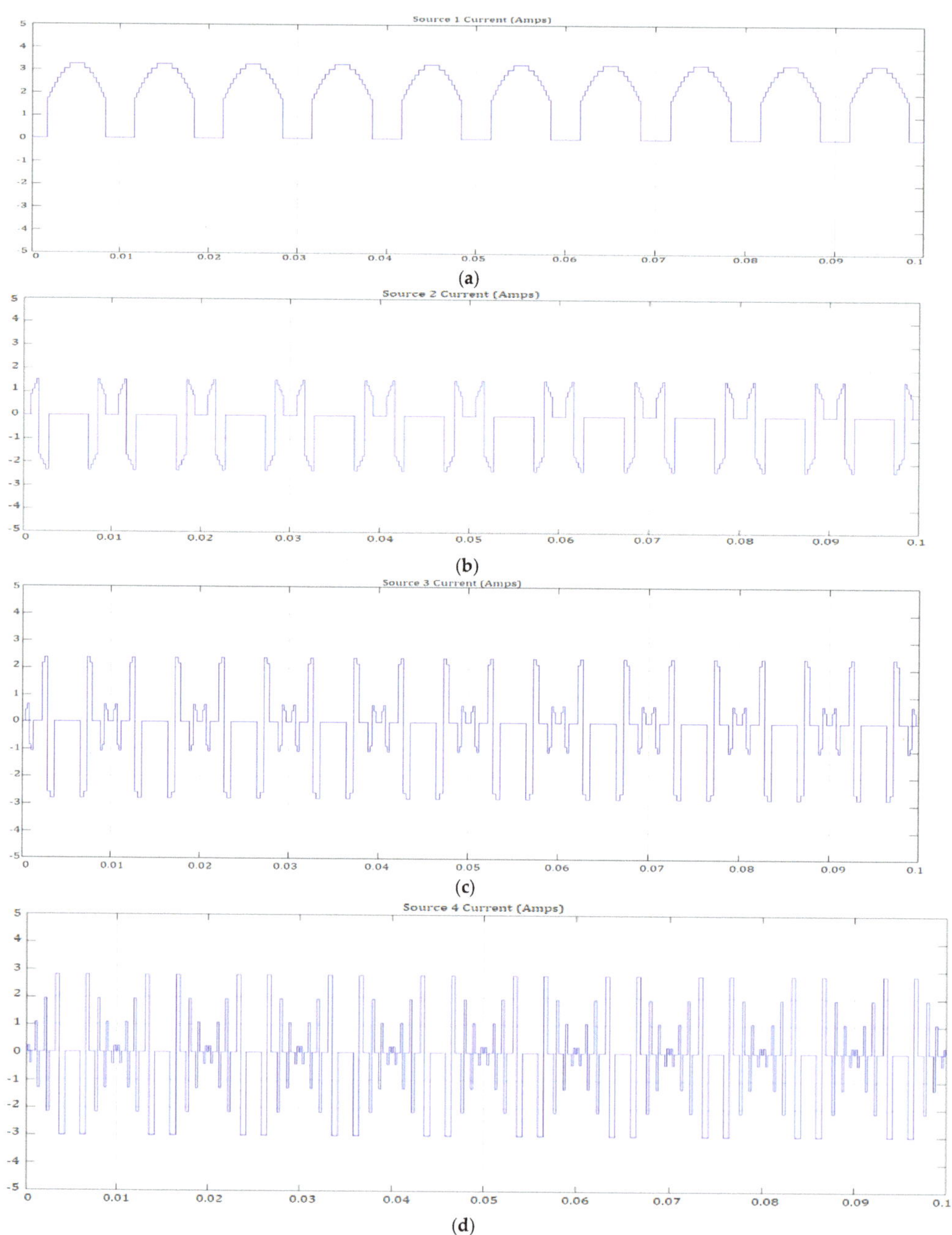

Figure 6. (**a**) Input source current 1 = 3.24 A; (**b**) Input source current 2 = 1.5 A; (**c**) Input source current 3 = 2.3 A; (**d**) Input source current 4 = 2.8 A.

3.2.2. Output Voltage and Current

The simulation output of the 31-level MPUC multi-level inverter is as given below in Figure 7a,b. Figure 7a represents the output voltage of the 31-level multi-level inverter, and Figure 7b represents the output current. The 31-level MLI has an output voltage of 325 V with reduced switching loss and harmonic distortion. The overall output current for the 31-level MLI is 3.24 A. The Table 4 shows the Input and output values of 31-level MPUC inverter.

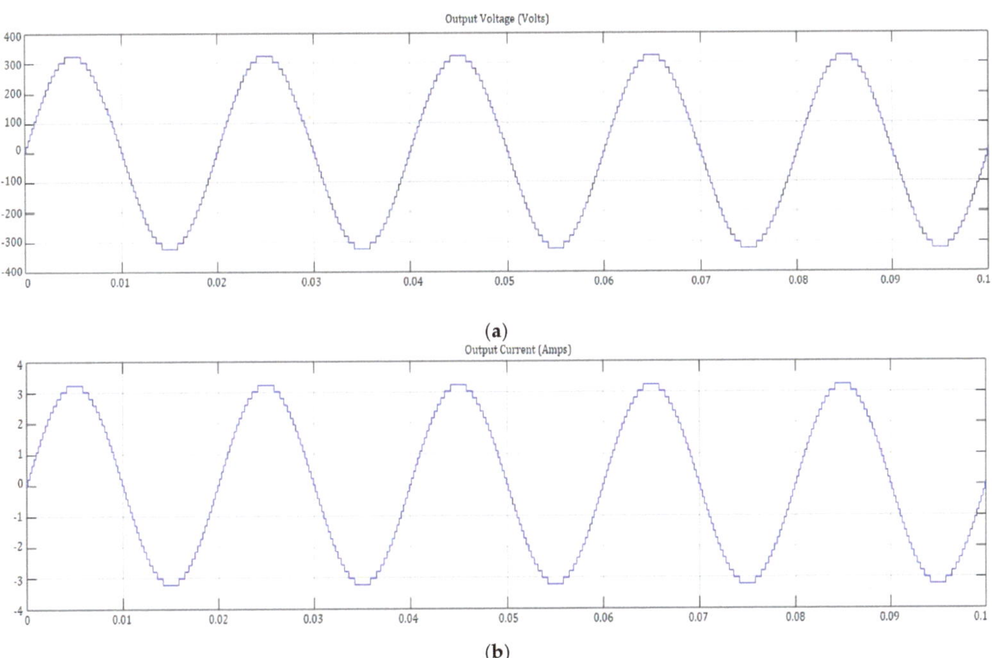

Figure 7. (**a**) Output voltage, 325 V; (**b**) Output current, 3.24 A.

Table 4. Input and output of 31-level MPUC inverter.

Input Current (Peak) in Amps				Output (Peak Values)	
Source 1	Source 2	Source 3	Source 4	Voltage in V	Current in I
3.24	1.5	2.3	2.8	325	3.24

3.3. Total Harmonic Distortion Study by FFT Analysis

The simulation results of the FFT Analysis of the 31-level packed U-cell multi-level inverter for different modulation indices (say, 1, 0.8, 0.6, and 0.4) are depicted by Figure 8a–d. Table 5 represents the comparison table of the THD Analysis of the MLI. And, Figure 8e represents the comparison of the THD Analysis chart.

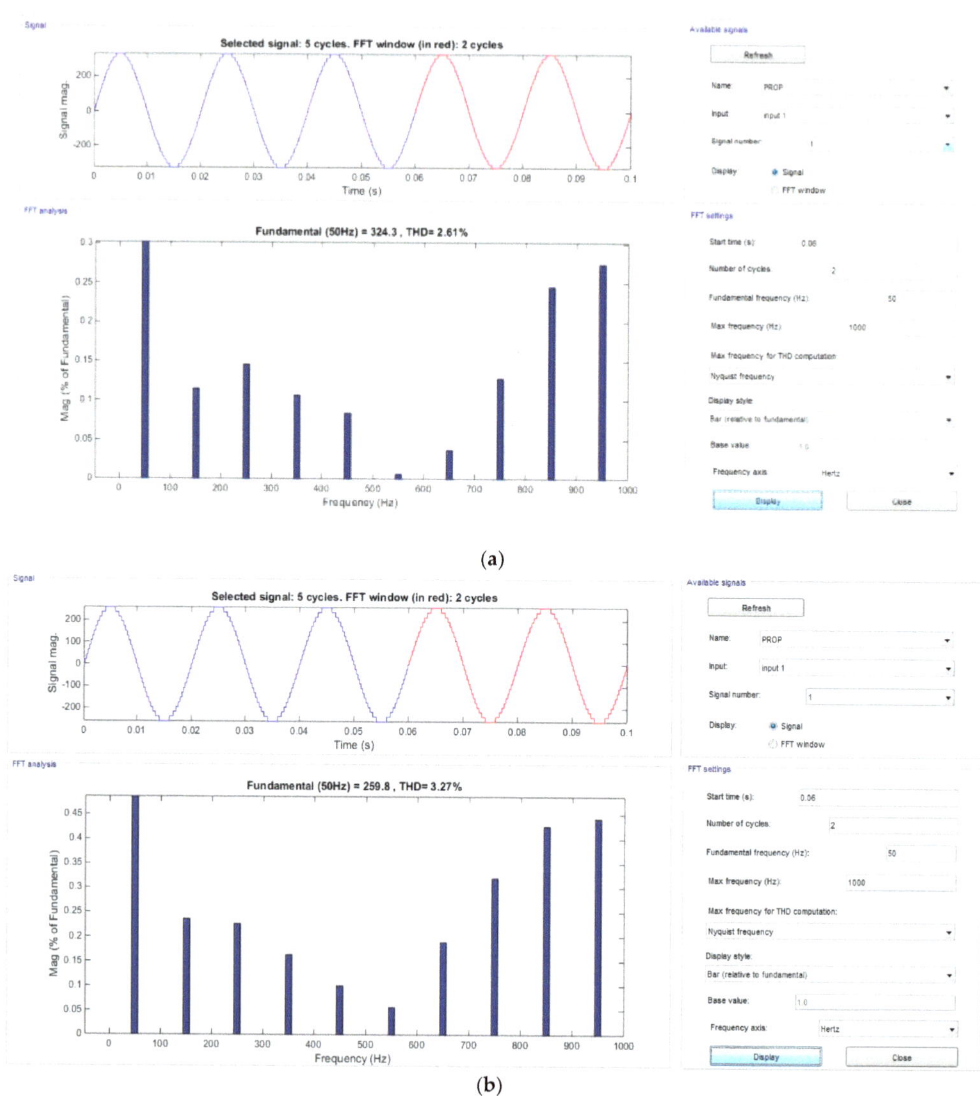

Figure 8. *Cont.*

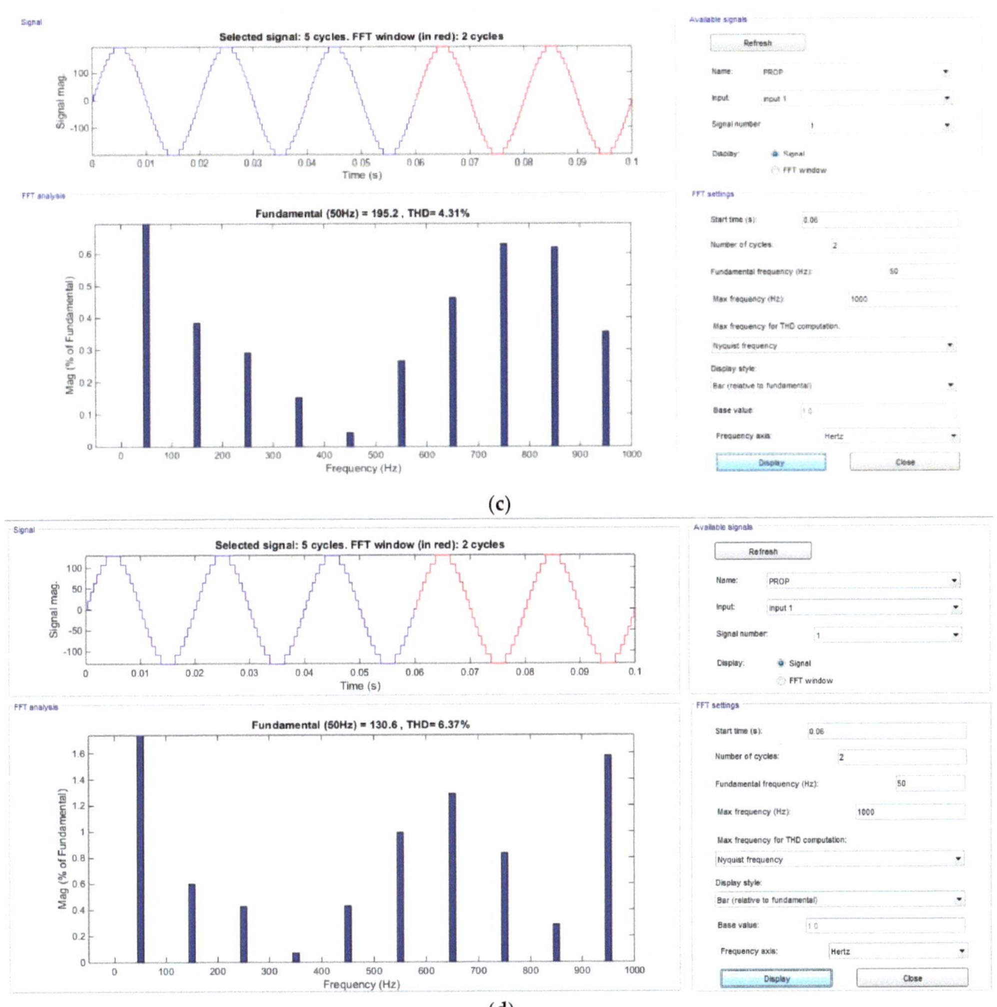

(c)

(d)

Figure 8. *Cont.*

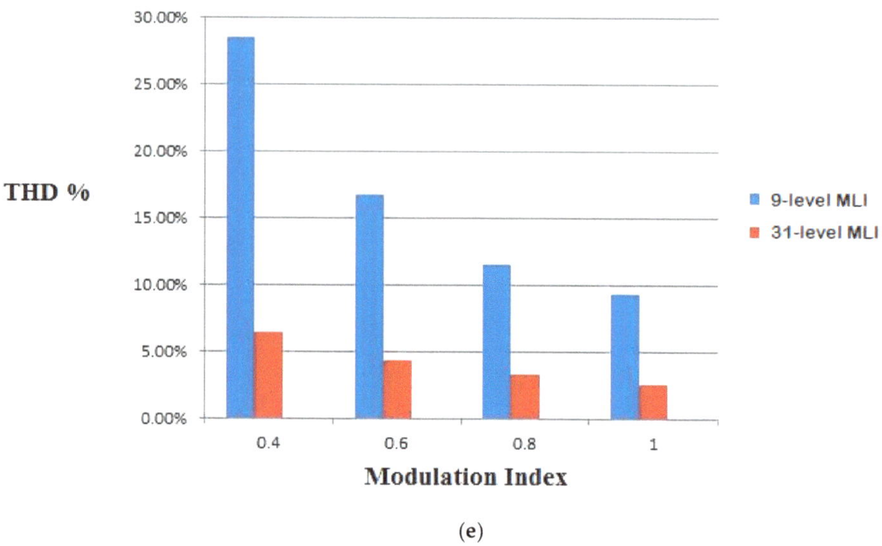

(**e**)

Figure 8. (**a**) FFT Analysis for the modulation index 1; (**b**) FFT Analysis for the modulation index 0.8; (**c**) FFT Analysis for the modulation index 0.6; (**d**) FFT Analysis for the modulation index 0.4; (**e**) THD Analysis of various MIs for the conventional and proposed multi-level inverters.

Table 5. MI and THD for 9-level and 31-level multi-level inverters.

Si. No.	Modulation Index	9-Level THD%	31-Level THD%
1	0.4	28.51%	6.37%
2	0.6	16.71%	4.31%
3	0.8	11.54%	3.27%
4	1.0	9.36%	2.61%

The simulation results reveal the output voltage current, which proves the efficiency of the 31-level MPUC MLI over its 9-level MLI counterpart. The FFT analysis proves that the harmonic distortions of the 31-level MPUC MLI for various modulation indices outperform its 9-level counterpart. The THD for the lowest modulation index of 0.4 for the 31-level MPUC MLI is 6.37%, whereas for the 9-level MLI, it is 28.51%. This proves the efficient harmonic reduction of the proposed design.

3.4. Hardware Results

Considering the simulation results and the FFT analysis, it is found that the designed 31-level MPUC multi-level inverter performs better than the 9-level MLI. As a next step, a hardware prototype for the proposed model is designed, and the various input and output results are obtained. The model predictive control algorithm is implemented on a PIC microcontroller, and the hardware prototype results are obtained. The complete hardware setup of the 31-level MPUC MLI is given in Figure 9. Figure 10a–e shows the switching pulses that trigger that inverter circuit. Figure 11a–d represents the four inputs at the inverter circuit. The output of the inverter is given in Figure 12.

Figure 9. Hardware prototype of the 31-level MPUC MLI.

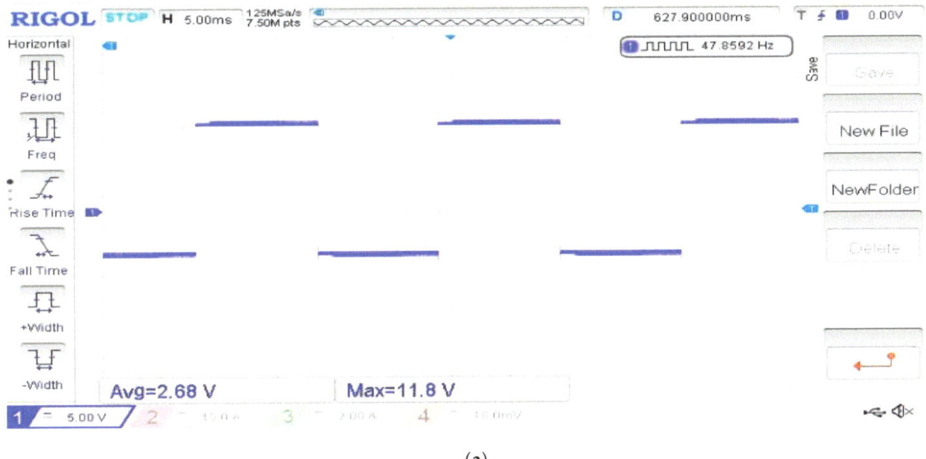

(**a**)

Figure 10. *Cont.*

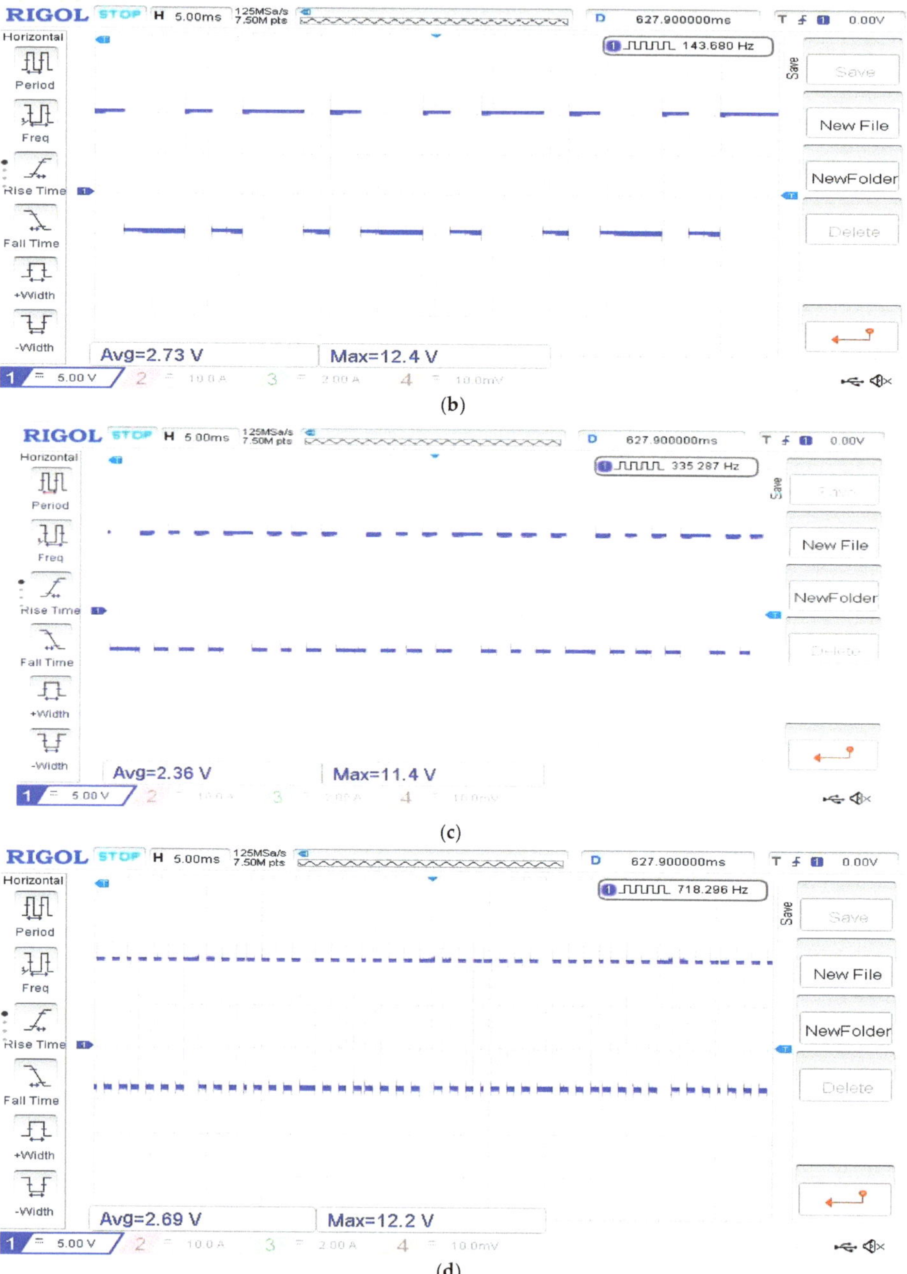

(**b**)

(**c**)

(**d**)

Figure 10. *Cont.*

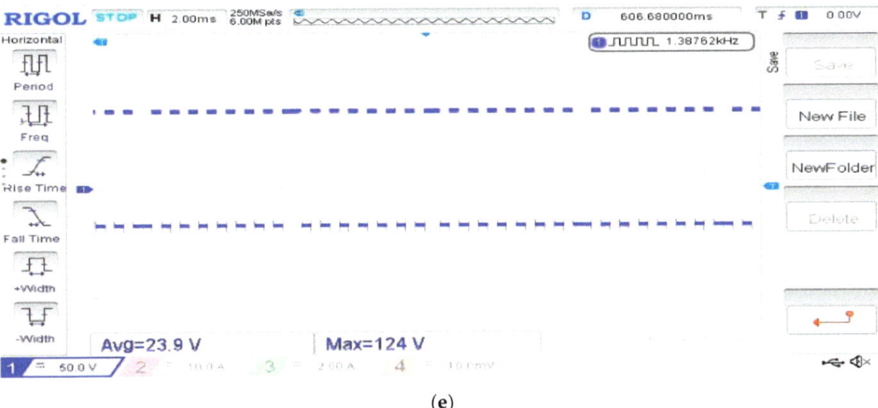

(**e**)

Figure 10. (**a**) Switch S1; (**b**) Switch S2; (**c**) Switch S3; (**d**) Switch S4; (**e**) Switch S5.

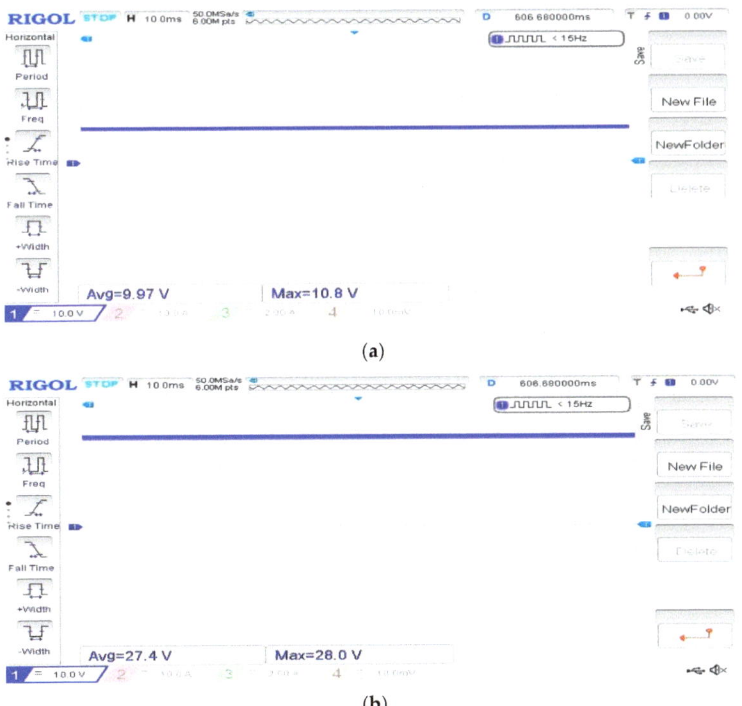

(**a**)

(**b**)

Figure 11. *Cont.*

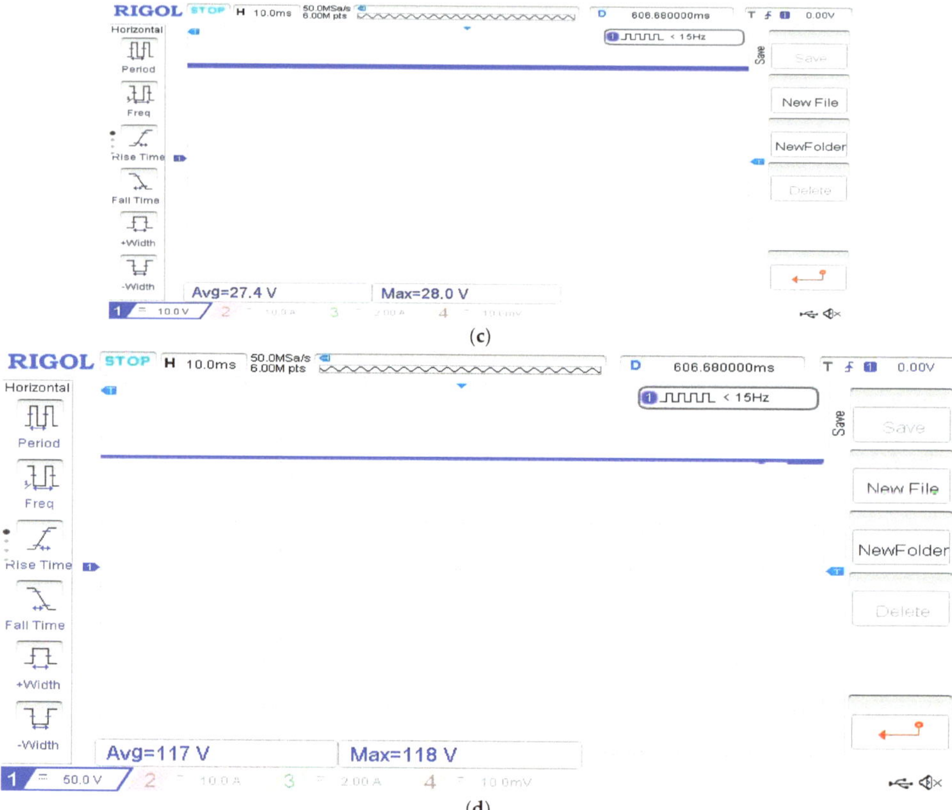

Figure 11. (**a**) Source 1; (**b**) Source 2; (**c**) Source 3; (**d**) Source 4.

3.4.1. Switching Pulses

The modulation signals given to switches, their switching carried out, and their switching output pulses are observed. There are five switches, and their switching pulses are given by Figure 10a–e.

3.4.2. Input Source Voltage

The DC source voltage is given to the inverter as input after the rectification, several inputs are taken, and the system is operated. Thus, the conversion of source to input voltage is taken with the help of the rectifier circuit.

3.4.3. Multi-Level Inverter Output

Inverters convert the DC input voltage to AC voltage for utilization. The 31-level MPUC multi-level inverter output voltage with level shifts and improved efficiency with a minimum amount of THD is taken out with the help of DSO (Digital storage Oscilloscope), and the output is as shown by Figure 12.

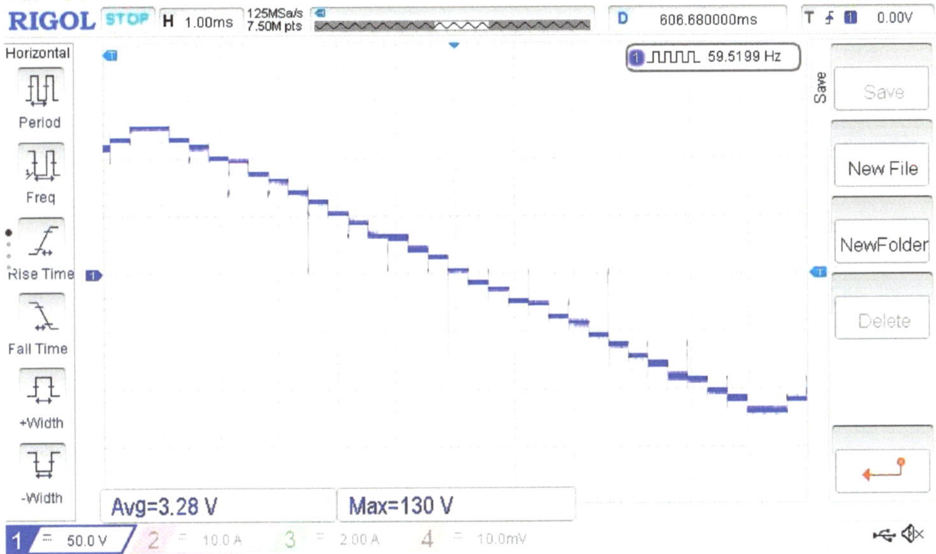

Figure 12. Output voltage from the hardware prototype.

4. Discussion

Solar energy is an abundant source for renewable energy generation. Still, the major limiting factor is the efficiency of the harvesting systems. In order to obtain maximum power from the solar generation units, almost every harvesting system tracks the maximum power point to achieve greater efficiencies. However, due to challenges in tracking, such as losses, oscillatory system behavior, sluggish system response, and low accuracy, the efficiency of the maximum power point tracking system is affected. Reducing the associated harmonics and minimizing oscillatory system responses are the two major areas of research that gain importance when it comes to optimizing distributed solar generation. The drawbacks are mainly related to the efficiency of the solid-state power converters used with the solar power generation unit; thus, efficient gating of these power devices plays a major role in determining the efficiency of the solar power generation system. Different control techniques have been used for this purpose, and the model predictive control method is one of the frequently used techniques due to its advantages. MPC-based switching techniques help improve power quality as they are devoid of filters, which are the major sources that affect the quality of power. Such MPC controllers, in addition to switching inverters, can be used to work along with maximum power point tracking for efficient and quick tracking. Thus, in this paper, a Variable Switching Model Predictive Controller implementation for a 31-level PUC multi-level inverter is presented with the objective of reducing the harmonics and improving power quality control. A model predictive controller is implemented using a PIC controller, and this is capable of providing a faster response under dynamic environmental conditions. Both the simulation and hardware results show a reduction in total harmonic distortion with increased step levels. The FFT analysis reveals that the total harmonic distortion for the lowest modulation index of 0.4 for the 31-level MPUC MLI is 6.37%, whereas for the 9-level MLI, it is 28.51%. Also, for a nominal modulation index of 0.8, the THD level is 11.54% and 3.27% for the 9-level MLI and the 31-level MPUC MLI, respectively. This depicts the reduction in total harmonic distortion achieved by the 31-level MPUC Inverter with model predictive control.

5. Conclusions

The modified multi-level PUC inverter (MPUC) can generate the output waveform in 31 steps, which helps reduce the low harmonic content in the output. Conventional MLIs have many demerits, which include: (i) implementation cost; (ii) complexity of design with increase in voltage levels; and (iii) the number of solid-state devices, which increases with voltage levels. Unlike the other available PUC topology, the proposed 31-level MPUC inverter uses a minimal number of power switches at the same time; therefore, it is capable of producing quality output voltage in terms of harmonic content. Also, the output power is greater, as it sums up the DC bus amplitudes. The efficiency of the system is attributed to the Variable Switching Model Predictive Control algorithm used for the generation of modulating pulses that switch the power devices in the inverter circuit efficiently. The significance of this design resides in the use of a reduced number of solid-state devices, capacitors, and filters; hence, this aids in compact and low-cost installations. The primary merit of using less switching and charging devices is to offer better power quality and reduced distortions. The FFT analysis results reveal that the total harmonic distortion THD is reduced for the 31-level design. This design can be further studied by (i) using different power switches, and (ii) using advanced prediction algorithms for switching inverters.

Author Contributions: Conceptualization, U.K. and U.P.; methodology U.K. and U.P.; software, U.K. and U.P.; validation, H.H.F. and E.R.; formal analysis, U.K. and U.P.; investigation, H.H.F. and E.R.; resources, H.H.F.; data curation, H.H.F. and E.R.; writing—original draft preparation, U.K. and U.P.; writing—review and editing, H.H.F. and E.R.; visualization, H.H.F. and E.R.; project administration, H.H.F. and E.R.; funding acquisition, E.R.; All authors have read and agreed to the published version of the manuscript.

Funding: This research received no external funding.

Data Availability Statement: No new data were created.

Acknowledgments: The work of Eugen Rusu was carried out in the framework of the research project DREAM (Dynamics of the REsources and Technological Advance in Harvesting Marine Renewable Energy), supported by the Romanian Executive Agency for Higher Education, Research, Development and Innovation Funding—UEFISCDI, grant number PN-III-P4-ID-PCE-2020-0008.

Conflicts of Interest: The authors declare no conflict of interest.

References

1. Franquelo, L.G.; Rodriguez, J.; Leon, J.I.; Kouro, S.; Portillo, R.; Prats, M.A. The age of multilevel converters arrives. *IEEE Ind. Electron. Mag.* **2008**, *2*, 28–39. [CrossRef]
2. Ribeiro, E.; Barbi, I. Harmonic Voltage Reduction Using a Series Active Filter Under Different Load Conditions. *IEEE Trans. Power Electron.* **2006**, *21*, 1394–1402. [CrossRef]
3. Abu-Rub, H.; Holtz, J.; Rodriguez, J.; Baoming, G. Medium-Voltage Multilevel Converters—State of the Art, Challenges, and Requirements in Industrial Applications. *IEEE Trans. Ind. Electron.* **2010**, *57*, 2581–2596. [CrossRef]
4. Kouro, S.; Malinowski, M.; Gopakumar, K.; Pou, J.; Franquelo, L.G.; Wu, B.; Rodriguez, J.; Perez, M.A.; Leon, J.I. Recent Advances and Industrial Applications of Multilevel Converters. *IEEE Trans. Ind. Electron.* **2010**, *57*, 2553–2580. [CrossRef]
5. Ounejjar, Y.; Al-Haddad, K.; Grégoire, L.-A. Packed U Cells Multilevel Converter Topology: Theoretical Study and Experimental Validation. *IEEE Trans. Ind. Electron.* **2010**, *58*, 1294–1306. [CrossRef]
6. Abu-Rub, H.; Mariusz, M.; Al-Haddad, K. *Power Electronics for Renewable Energy Systems, Transportation and Industrial Applications*; John Wiley & Sons: Hoboken, NJ, USA, 2014.
7. Mohamad, A.S.; Mariun, N.; Sulaiman, N.; Radzi, M.A.M. A new cascaded multilevel inverter topology with minimum number of conducting switches. In Proceedings of the 2014 IEEE Innovative Smart Grid Technologies—Asia (ISGT ASIA), Kuala Lumpur, Malaysia, 20–23 May 2014; pp. 164–169. [CrossRef]
8. Vahedi, H.; Kanaan, H.Y.; Al-Haddad, K. PUC converter review: Topology, control and applications. In Proceedings of the IECON 2015—41st Annual Conference of the IEEE Industrial Electronics Society, Yokohama, Japan, 9–12 November 2015; pp. 004334–004339. [CrossRef]
9. Kumar, P.R.; Kaarthik, R.S.; Gopakumar, K.; Leon, J.I.; Franquelo, L.G. Seventeen-Level Inverter Formed by Cascading Flying Capacitor and Floating Capacitor H-Bridges. *IEEE Trans. Power Electron.* **2015**, *30*, 3471–3478. [CrossRef]
10. Vazquez, S.; Rodriguez, J.; Rivera, M.; Franquelo, L.G.; Norambuena, M. Model Predictive Control for Power Converters and Drives: Advances and Trends. *IEEE Trans. Ind. Electron.* **2017**, *64*, 935–947. [CrossRef]

11. Vahedi, H.; Labbé, P.-A.; Al-Haddad, K. Sensor-Less Five-Level Packed U-Cell (PUC5) Inverter Operating in Stand-Alone and Grid-Connected Modes. *IEEE Trans. Ind. Inform.* **2016**, *12*, 361–370. [CrossRef]
12. Kumar, R.; Singh, B. Single Stage Solar PV Fed Brushless DC Motor Driven Water Pump. *IEEE J. Emerg. Sel. Top. Power Electron.* **2017**, *5*, 1377–1385. [CrossRef]
13. Vahedi, H.; Al-Haddad, K. Method and System for Operating a Multilevel Inverter. U.S. Patent US9923484B2, 20 March 2018.
14. Narendrababu, A.; Yalla, N.; Agarwal, P. A modified T-type single phase five-level inverter with reduced switch voltage stress. In Proceedings of the 2018 International Conference on Power, Instrumentation, Control and Computing (PICC), Thrissur, India, 18–20 January 2018; pp. 1–5. [CrossRef]
15. Sulake, N.R.; Venkata, A.K.D.; Choppavarapu, S.B. FPGA Implementation of a Three-Level Boost Converter-fed Seven-Level DC-Link Cascade H-Bridge inverter for Photovoltaic Applications. *Electronics* **2018**, *7*, 282. [CrossRef]
16. Kang, J.-W.; Hyun, S.-W.; Ha, J.-O.; Won, C.-Y. Improved Neutral-Point Voltage-Shifting Strategy for Power Balancing in Cascaded NPC/H-Bridge Inverter. *Electronics* **2018**, *7*, 167. [CrossRef]
17. Lee, S.; Kim, J. Optimized Modeling and Control Strategy of the Single-Phase Photovoltaic Grid-Connected Cascaded H-bridge Multilevel Inverter. *Electronics* **2018**, *7*, 207. [CrossRef]
18. Hu, J.; Shan, Y.; Guerrero, J.M.; Ioinovici, A.; Chan, K.W.; Rodriguez, J. Model predictive control of microgrids—An overview. *Renew. Sustain. Energy Rev.* **2021**, *136*, 110422. [CrossRef]
19. Shyam, D.; Premkumar, K.; Sivamani, D.; Nazarali, A.; Narendiran, S.; Ramkumar, R. PUC Optimal Switching Strategies for Renewable Applications in Single Phase Inverter. In Proceedings of the IEEE 9th Uttar Pradesh Section International Conference on Electrical, Electronics and Computer Engineering (UPCON), Prayagraj, India, 2–4 December 2022; pp. 1–5. [CrossRef]
20. Iqbal, H.; Tariq, M.; Sarfraz, M.; Anees, M.A.; Alhosaini, W.; Sarwar, A. Model predictive control of Packed U-Cell inverter for microgrid applications. *Energy Rep.* **2022**, *8*, 813–830. [CrossRef]
21. Abarzadeh, M.; Peyghami, S.; Al-Haddad, K.; Weise, N.; Chang, L.; Blaabjerg, F. Reliability and Performance Improvement of PUC Converter Using a New Single-Carrier Sensor less PWM Method with Pseudo Reference Functions. Electrical and Computer Engineering Faculty Research and Publications. (681) 2021. Available online: https://epublications.marquette.edu/electric_fac/681 (accessed on 1 March 2023).
22. Phukan, H.; Debela, T.; Singh, J. A modified seven-level cross-connected PUC boost multilevel inverter with reduced cost factor and device count. *Int. J. Emerg. Electr. Power Syst.* **2022**. [CrossRef]
23. Upreti, S.; Singh, B.; Kumar, N. Harmonics Minimization in PUC Type Solar Multilevel Converter with Multicarrier Switching Schemes. *IETE J. Res.* **2022**. [CrossRef]
24. Junior, S.C.S.; Jacobina, C.; Fabricio, E.L.L. A Single-Phase 35-levels Cascaded PUC Multilevel Inverter Fed by a Single DC-Source. In Proceedings of the IEEE Energy Conversion Congress and Exposition (ECCE), Vancouver, BC, Canada, 10–14 October 2021; pp. 2467–2474. [CrossRef]
25. Muthukuri, N.K.; Mopidevi, S. Optimal Controller Design of PUC-15 MLI Topology for Smart Grid Applications with Reduced Power Components. In Proceedings of the International Conference on Smart Technologies and Systems for Next Generation Computing (ICSTSN), Virtual, 25–26 March 2022; pp. 1–4. [CrossRef]
26. Owona, J.L.; Fendji, M.D.; Effa, J.Y.; Fankem, E.D.K. Control of Reduced Switches Count and Classical Multilevel Inverters: A Comparison. *E3S Web Conf.* **2022**, *354*, 02002. [CrossRef]
27. Venkedesh, R.; Kumar, A.; Renukadevi, G. Multilevel Inverter Design with Reduced Switches & THD Using Fuzzy Logic Controller. *Int. J. Electr. Electron. Eng.* **2022**, *9*, 1–21. [CrossRef]
28. Bellah, F.A.; Abouloifa, A.; Echalih, S.; Hekss, Z.; Naftahi, K.; Lachkar, I. Control Design of a Seven-Level Packed U Cell Inverter. *IFAC-PapersOnLine* **2022**, *55*, 677–682. [CrossRef]
29. Khan, F.A.; Shees, M.M.; Alsharekh, M.F.; Alyahya, S.; Saleem, F.; Baghel, V.; Sarwar, A.; Islam, M.; Khan, S. Open-Circuit Fault Detection in a Multilevel Inverter Using Sub-Band Wavelet Energy. *Electronics* **2021**, *11*, 123. [CrossRef]
30. Kumar, A.; Jain, S. Predictive Switching Control for Multilevel Inverter using CNN-LSTM for Voltage Regulation. *ADBU J. Eng. Technol. (AJET)* **2022**, *11*, 0110203197.
31. Mukundan, C.M.N.; Jayaprakash, P.; Subramaniam, U.; Almakhles, D.J. Binary Hybrid Multilevel Inverter-Based Grid Integrated Solar Energy Conversion System with Damped SOGI Control. *IEEE Access* **2020**, *8*, 37214–37228. [CrossRef]
32. Salem, M.; Richelli, A.; Yahya, K.; Hamidi, M.N.; Ang, T.-Z.; Alhamrouni, I. A Comprehensive Review on Multilevel Inverters for Grid-Tied System Applications. *Energies* **2022**, *15*, 6315. [CrossRef]
33. Komurcugil, H.; Bayhan, S.; Guler, N.; Abu-Rub, H.; Bagheri, F. A simplified sliding-mode control method for multi-level transformerless DVR. *IET Power Electron.* **2022**, *15*, 764–774. [CrossRef]
34. Azeem, A.; Ansari, M.K.; Tariq, M.; Sarwar, A.; Ashraf, I. Design and Modeling of Solar Photovoltaic System using Seven-Level Packed U-Cell (PUC) Multilevel Inverter and Zeta Converter for Off-Grid Application in India. *Electrica* **2019**, *19*, 101–112. [CrossRef]
35. Jeevanantham, Y.A.; Srinath, S. ANN Based Reduced Switch Multilevel Inverter in UPQC for Power Quality Improvement. *Intell. Autom. Soft Comput.* **2022**, *33*, 909–921. [CrossRef]
36. Fahad, M.; Tariq, M.; Faizan, M.; Ali, A.; Sarwar, A.; Tafti, H.D.; Ahmad, S.; Mohamed, A.S.N. A Dual Source Switched-Capacitor Multilevel Inverter with Reduced Device Count. *Electronics* **2021**, *11*, 67. [CrossRef]

37. Abari, I.; Hamouda, M.; Sleiman, M.; Slama, J.B.H.; Kanaan, H.Y.; Al-Haddad, K. Open-Circuit Fault Detection and Isolation Method for Five-Level PUC Inverter Based on the Wavelet Packet Transform of the Radiated Magnetic Field. *IEEE Trans. Instrum. Meas.* **2022**, *71*, 1–11. [CrossRef]
38. Kaymanesh, A.; Chandra, A. Computationally Efficient MPC Technique for PUC-Based Inverters Without Weighting Factors. In Proceedings of the 2021 IEEE Industry Applications Society Annual Meeting (IAS), Vancouver, BC, Canada, 10–14 October 2021; pp. 1–5. [CrossRef]
39. Mohamed-Seghir, M.; Krama, A.; Refaat, S.S.; Trabelsi, M.; Abu-Rub, H. Artificial Intelligence-Based Weighting Factor Autotuning for Model Predictive Control of Grid-Tied Packed U-Cell Inverter. *Energies* **2020**, *13*, 3107. [CrossRef]
40. Khawaja, R.; Sebaaly, F.; Kanaan, H.Y. Design of a 7-Level Single-Stage/Phase PUC Grid-Connected PV Inverter with FS-MPC Control. In Proceedings of the 2020 IEEE International Conference on Industrial Technology (ICIT), Buenos Aires, Argentina, 26–28 February 2020; pp. 751–756. [CrossRef]
41. Sahli, A.; Krim, F.; Laib, A.; Talbi, B. Energy management and power quality enhancement in grid-tied single-phase PV system using modified PUC converter. *IET Renew. Power Gener.* **2019**, *13*, 2512–2521. [CrossRef]
42. Trabelsi, M.; Ghanes, M.; Mansouri, M.; Bayhan, S.; Abu-Rub, H. An original observer design for reduced sensor control of Packed U Cells based renewable energy system. *Int. J. Hydrogen Energy* **2017**, *42*, 17910–17916. [CrossRef]

MDPI

Article

A Decentralized Blockchain-Based Energy Market for Citizen Energy Communities

Peyman Mousavi [1], Mohammad Sadegh Ghazizadeh [1,*] and Vahid Vahidinasab [2,*]

1   Faculty of Electrical Engineering, Shahid Beheshti University, Tehran 19839-69411, Iran; s_mousavimobarkeh@sbu.ac.ir
2   Department of Engineering, School of Science and Technology, Nottingham Trent University, Nottingham NG11 8NS, UK
*   Correspondence: m_ghazizadeh@sbu.ac.ir (M.S.G.); vahid.vahidinasab@ntu.ac.uk (V.V.)

Abstract: Despite the fact that power grids have been planned and utilized using centralized networks for many years, there are now significant changes occurring as a result of the growing number of distributed energy resources, the development of energy storage systems and devices, and the increased use of electric vehicles. In light of this development, it is pertinent to ask what an efficient approach would be to the operation and management of future distribution grids consisting of millions of distributed and even mobile energy elements. Parallel to this evolution in power grids, there has been rapid growth in decentralized management technology due to the development of relevant technologies such as blockchain networks. Blockchain is an advanced technology that enables us to answer the question raised above. This paper introduces a decentralized blockchain network based on the Hyperledger Fabric framework. The proposed framework enables the formation of local energy markets of future citizen energy communities (CECs) through peer-to-peer transactions. In addition, it is designed to ensure adequate load supply and observe the network's constraints while running an optimal operation point by consensus among all of the players in a CEC. An open-source tool in Python is used to verify the performance of the proposed framework and compare the results. Through its distributed and layered management structure, the proposed blockchain-based framework proves its superior flexibility and proper functioning. Moreover, the results show that the proposed model increases system performance, reduces costs, and reaches an operating point based on consensus among the microgrid elements.

Keywords: decentralized local energy market; blockchain; peer-to-peer trading; distributed energy resources; electric vehicles; citizen energy communities; distribution girds operation

Citation: Mousavi, P.; Ghazizadeh, M.S.; Vahidinasab, V. A Decentralized Blockchain-Based Energy Market for Citizen Energy Communities. *Inventions* 2023, 8, 86. https://doi.org/10.3390/inventions8040086

Academic Editor: Om P. Malik

Received: 19 May 2023
Revised: 22 June 2023
Accepted: 24 June 2023
Published: 30 June 2023

1. Introduction

Community-led energy systems, also called Citizen Energy Communities (CEC), enable citizens to take action that contributes to a clean energy transition while empowering them. As a result, they make it easier to attract private investment in clean energy projects and to increase public acceptance of renewable energy projects. They can assist citizens in reducing energy bills, create local jobs, and increase energy efficiency. Using energy communities, future energy systems can be reshaped by harnessing energy locally and enhancing quality of life while allowing citizens to participate actively in the transition to renewable energy by providing flexibility services to the electricity system through demand-side response and storage [1].

The CEC approach is part of the activities that are currently ongoing toward a more distributed energy system. Moving towards decentralized management and using blockchain networks reflects developed societies' inherent tendency to reduce the government's active involvement and increase flexibility in daily life. Today, microgrids consist of a wide range of distributed energy supply and consumption systems, making the electricity industry a primary field of interest for this technology. Blockchain technology affects this industry in various ways as well [2]. This paper aims to integrate blockchain technology into the operation of local energy networks. The aim is to present a decentralized model for electric microgrid management. Such a management system should be able to operate the microgrid economically while ensuring the adequacy and security of the system. Moreover, it should offer intrinsic flexibility and fast response compared to conventional models, mainly as a result of using smart contracts to push the operating point of the microgrid toward its global optimum point. Providing economic infrastructure and incentives to interest private investors expands the network, and decentralization of system operations leads to more optimal democratic use of the microgrids, making these networks desirable for conventional microgrid operation as well.

1.1. Aim and Scope

Blockchain networks stand out among other technologies due to their distributabilty and decentralized management capability. Low setup costs, diverse environments and setup tools, the ability to manage even small-scale elements of the microgrid, creating a self-sufficient economic cycle, protecting privacy, and emphasizing democratic decisions make this technology thrive and compete with the traditional centralized power grid.

New elements such as electric vehicles with varying locations, energy storage devices, and small renewable energy resources installed in any location have changed the electricity industry's outlook [3]. Localization of energy supplies and the tremendous increase in the amount of their information brings about new opportunities and challenges to the microgrid. Intel forecasts that each automatic vehicle will generate four hundred gigabytes of average daily data by 2025 [4]. These data will have various uses in intelligent vehicle control, urban traffic, routing, and finding parking places. It is possible that such data will directly or indirectly affect the time and place of the vehicle's charging and its amount [5]. These elements may be owned by individuals. The intentions and functions of these individuals could be greatly affected by circumstances and uncertainties of time and place, and their priorities might change at any moment [6]. Decentralized networks providing a distributing structure with layered management can be an appropriate solution to these changes. They emphasize the local energy supply while managing energy exchange at the microgrid level. This paper proposes a new decentralized framework for the operation and planning of the power grid of the future that contains millions of small distributed and mobile elements. The proposed framework is shown to have better performance than existing networks, and can represent a stepping stone toward the power grids of the future [7].

1.2. Literature Survey

Research on utilizing decentralized networks in the electricity industry can be categorized as follows.

The first category consists of designs for blockchain microgrids as communication infrastructure. Such papers mainly try to manage the issues around extensive networks with massive data [8]. Inspired by the consensus mechanism in blockchain networks, these papers look for a robust, scalable, and economic dispatch to optimize the significant number of small element transactions according to the computational structure of the microgrid [9]. These studies have proposed decentralized networks based on reputable microgrids such as Ethereum. They try to present an efficient model based on existing blockchain structures [10]. A decentralized microgrid model mainly depends on its consensus mechanism. Costly consensus mechanisms are not good choices for microgrids. The existing literature has tried to replace Proof of Work as a consensus mechanism, e.g., by Proof of Stake by Reputation (POWR), intended to reduce transaction confirmation and block creation delay [11]. Using the Internet of Things (IoT), smart power meters and small producers of renewable energies are of prime significance in microgrid structure design [12]. The transaction structures can be designed to be executable in a different blockchain infrastructure, such as Solidity, without any third-party intervention [13]. Due to the increase in renewable energies and islanding of enormous power grids in the form of small microgrids, several of these studies have addressed the issue of creating decentralized networks to interact between such islands. This increases their reliability along with mutual financial interests [14]. All these papers use the main chain of the blockchain as the pivotal part of their work. However, many processes and decision makings can be done in the layers before this chain. This will improve the microgrid response time substantially. This paper proposes a layered structure for a decentralized method to facilitate decentralized process management.

The second group involves management of the market's financial settlement tasks through a blockchain system. These financial systems include contracts as well as online and multi-stage settlement systems [15,16]. Papers have focused on P2P trades between electric vehicles or other two players in the microgrid [17]. Others emphasize smart contracts, which can be crucial in a decentralized microgrid's management [18]. Another group of these papers discusses managing request sending in various algorithms such as iceberg. This prevents price jumps in the microgrid. The microgrid uses algorithms and other settlement systems to avoid very short-term changes in the microgrid [19]. A significant point in this study is evaluating the effect of financial transactions in a decentralized microgrid which is considered a purely-economic market, without considering its technical utilization. This bears more significance in microgrids than in bank and insurance networks. The present paper fixes this issue using a two-stage consensus mechanism that simultaneously considers the economic and technical approaches in a decentralized microgrid. It computes the system's operating point by considering both mentioned issues simultaneously.

Finally, the third group consists of research that aims to improve the system's power quality [20] and control the voltage using blockchain infrastructure. These papers present a reward–punishment system based on voltage changes and emphasize the sensitivity matrix of each bus to control the voltage at the end of the radial distribution networks [21]. Most of the performed research tries to improve the performance of the microgrid for central planning and utilization of the system. Their objective function consists of increasing the power quality and a decentralized microgrid management which is more significant in large systems [8]. Others focus on mixing centralized and decentralized approaches to achieve better system performance. They try to dispatch the decentralized loads in the microgrid [22]. The main point is that all these studies try to improve the performance of the decentralized system for financial transactions in the microgrid. On the contrary, the present paper aims to design and present a decentralized system to replace the current centralized utilization microgrid. Such a microgrid improves performance by utilizing the microgrid at its global optimum. A comparison of the references discussed in the literature review is shown in Table 1.

1.3. Contributions

Considering the works reported in the literature and the highlighted research gaps, the main contribution of this paper can be summarized as follows:

- Proposing a decentralized framework for planning and operation of the future microgrids that takes care of both economical and technical constraints;
- Providing a two-stage consensus mechanism to maintain adequacy, security, and a global optimal operating point for the microgrid;
- Designing a layered structure based on contracts and P2P exchanges to increase the microgrid's response time.

1.4. Organization of the Paper

The rest of the paper is structured as follows. The second section discusses the design of a blockchain-based microgrid according to the Hyperledger model. It presents microgrid layers, definitions, duties, and the interaction of members in each layer. In section three, we offer a decentralized management mechanism consisting of the processes and microgrid rules. These rules lay the foundations for decentralized management and form the microgrid's consensus model. Section four demonstrates the proposed decentralized microgrid's performance in facing the challenges. This ensures maintaining adequacy, compliance with the system's technical requirements, and global optimal utilization of the microgrid. Section five compares the performance of the proposed decentralized microgrid with its centralized peer using an open-source program. The performance was challenged by introducing uncertain electrical vehicle and small renewable energy sources to prove the model's performance and show its advantages over its peers. Finally, the last section ends the paper by analyzing the performance result for the microgrid.

Table 1. Comparison of references in the literature.

Ref	Category	Innovation
[8]	Infrastructure	Manage the issues of extensive networks with massive data
[9]	Infrastructure	Look for a robust, scalable mechanism in V2G network
[10]	Infrastructure	A model based on existing blockchain structures
[11]	Infrastructure	A model mainly depends on its consensus mechanism
[12]	Infrastructure	Using the Internet of Things and small producers of renewable energies
[13]	Infrastructure	The transaction structures are designed to be executable such as Solidity
[14]	Infrastructure	Decentralized networks to interact between such islands
[15]	Financial	Blockchain multi settlement base Peer-to-peer trading framework
[16]	Financial	Financial systems include contracts, online, and multi stage settlement
[17]	Financial	Contract model for electric vehicle base peer to peer
[18]	Financial	Smart contracts managea decentralized microgrid
[19]	Financial	Managing request sending in various algorithms such as iceberg
[20]	Power Quality	Control the voltage using blockchain infrastructure
[21]	Power Quality	A reward-punishment system based on voltage changes
[22]	Power Quality	Focus on mixing centralized and decentralized approaches

2. Designing a Blockchain Network

Designing a decentralized microgrid requires an infrastructure where all authorized users can access the latest operating point of the system and basic network information. This enables members to work on the same operating point at any moment. We use a private blockchain [2] infrastructure for this purpose. A chain of confirmed blocks supports this distributed ledger. Such a network records all the information of the microgrid, the last operating point in the distributed ledger, and the last confirmed blocks. This enables all members to access the latest microgrid information and work on the same operating point. People with access to the microgrid information receive a license for their access from the representative of the network independent entity, and have layered access to the information [23]. Their real identity is not published in all layers to protect the customers' privacy. We used Hyperledger Fabric [24], one of IBM's most reputable blockchain platforms, to design our microgrid [25]. It is defined as a multi-layer microgrid where each layer has a specific duty and role. This increases the efficiency and performance of the microgrid and makes managing large systems possible. The platform has been used in the banking, insurance, and social services industries [26]. Here, we redefine its concepts according to the different nature of energy. This paper proposes a Hyperledger blockchain model based on the definitions in Figure 1 for decentralized management of microgrids.

- Client: Clients are the first layer in the microgrid, consisting of microgrid players [26]. The microgrid players might consist of a wide range of load or energy producers. Considering the progress in IoT, every kilowatt hour of energy shortly contains the identity and personal details of the producer and end consumer. However, major players such as electric vehicles and small energy sources such as renewable energy, for which performance is always highly uncertain, are the largest loads. Blockchain, a private microgrid, is not accessible to all. Any player needs an endorsement to access it. Endorsers are companies that provide identity certificates to the microgrid players and review violations of rules and complaints. They are licensed by the organization supervising the microgrid. These companies can assign identity tokens to approved players. The player can use that token to participate in the microgrid directly or by proxy. This microgrid defines gas fee concepts used in all layers. In the blockchain, the gas fee is the fee each player suggests for performing the intended process. Each layer receives part of this fee according to the definitions of the microgrid. The higher the gas of an order is, the quicker it is considered in the subsequent layers. Gas has a significant impact on the market direction.

- Committer : As in the case of trading stocks on financial markets, each actor must use a brokerage firm to connect to the network and use its services [26]. However, because most network actors are ordinary citizens who are not interested in complicated energy purchase systems, committers provide a set of different smart contracts to their clients. Smart vehicles can perform more optimally based on these smart contracts. A new concept is added to the network called "GAS" as the request-to-request fee, determined by contract type or the direct offer by the requestor.

- Anchor: Anchor's particular advantage of using blockchain in a microgrid is the P2P process. This advantage is crucial over the life of the microgrid. Anchors [26] are specialized companies that manage P2P in the microgrid. Committers' orders must be forwarded to them before being presented to the main chain of the blockchain. Each anchor has a unique area of activity in the microgrid. Each bus can be precisely in one anchor's scope of activity. Committers forwarded their request to the relevant anchor based on the microgrid bus it belongs to. This way, all the orders tradable as P2P are received and settled before being offered to the main chain, making for a fast response time and less working load on the main chain. An anchor must run its simple optimization program according to the objective function and constraints (Equation 1). The optimization result should maximize the charge of all units capable of P2P trade and the company's earnings in these exchanges.

$$Max\left(\sum_{CVh}\sum_{T}(Pc_{(V,Bus,T)} + Pd_{(V,Bus,T)}) \times Gas_{V,T}\right) \tag{1}$$

$$\sum_{T} Pc_{(V,Bus,T)} \leq Charge_{(V,Bus)} \tag{2}$$

$$\sum_{T} Pd_{(V,Bus,T)} \leq DisCharge_{(V,Bus)} \tag{3}$$

$$\sum_{Bus} Pc_{(V,Bus,T)} \leq \sum_{Bus} Pd_{(V,Bus,T)} \tag{4}$$

Here, $Pc_{(V,Bus,T)}$ is the energy ordering vehicle's charge, $Pd_{(V,Bus,T)}$ is the ordering vehicle's discharge, $Charge_{(V,Bus)}$ is the total charge ordered by each vehicle for that bus, and $DisCharge_{(V,Bus)}$ is the proposed discharge per bus per vehicle variable. All other orders not settled in the P2P mechanism are shared as a recommended package in a block pool.

- Orderer: In Hyperledger, orderers play the role of miners in a standard blockchain [26]. They are the most critical layer of the network. In addition to the market supervision entity, they are the only blockchain members with access to the last operating point and the main chain. These companies' function is the main difference between blockchains in the electricity industry and in other organizations. Orderers receive the orders by connecting to the block pool. To register a new order on the microgrid's main chain, these companies must provide a new operating point based on the last operating point while observing all system constraints. These companies must provide an economic dispatch according to the last operating point by adding new orders. The new operating point is confirmed by the consensus mechanism of the microgrid. If the operating point is confirmed by consensus, they receive a part of the gas for the recommended package. Receiving the other part of the gas depends on the confirmation of the optimality of the operating point. This optimality is determined in the second phase of the consensus mechanism. Orderers are advanced algorithms on powerful servers that perform a high number of computations in the least possible time. They earn money by registering other players' orders in the chain while observing the microgrid constraints.

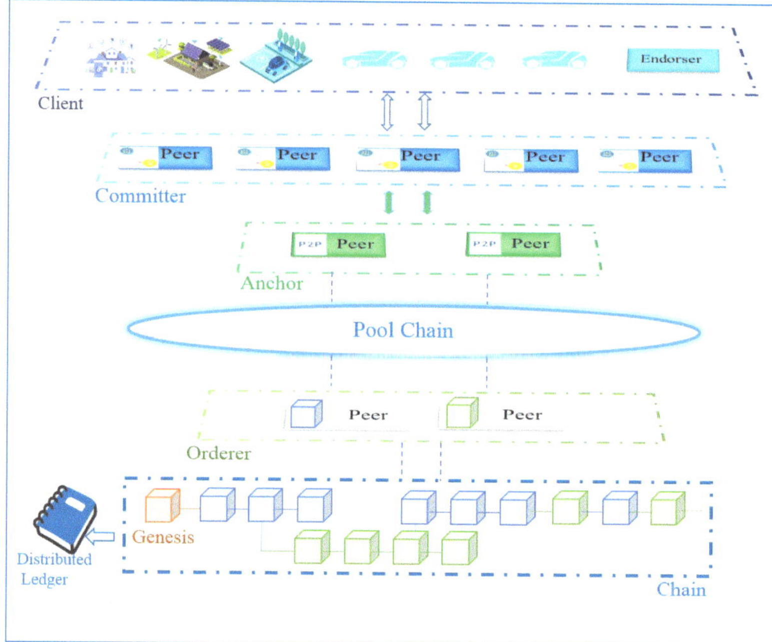

Figure 1. The microgrid graph for the proposed decentralized microgrid.

3. Designing a Decentralized Microgrid

A decentralized microgrid management system should maintain system adequacy and utilize the microgrid at its global optimal point while observing all system constraints. The rules and processes employed in this paper ensure fast system response, system adequacy, and optimal utilization. These processes are defined as follows:

- The time interval of the main chain is one day. The market supervision entity registers the chain's first block, the genesis block, which means the initiation of the chain. This block consists of the market's daily initial operating point and the microgrid's complete information. The decentralized microgrid can begin daytime activity based on this operating point.

- Each new block with a new microgrid operating point, microgrid consensus token, and the security code of hash of the last block in the main chain can be registered as a new block.

- If two blocks contain the main chain's last security code, the block with a lower utilization cost for the rest of the microgrid is considered the main chain block. Otherwise, if the utilization costs are the same, the block registered earlier is regarded as the main chain block.

One of the critical features of decentralized networks is information transparency, which enhances network efficiency. However, various methods consider the network participants' privacy concerns where it can be observed in the different cryptocurrencies in which they have used multiple approaches to preserve privacy of the users. These processes are defined as follows:

- The Hyperledger Fabric blockchain framework used in this paper provides a secure and private network for transactions. Using Hyperledger Fabric, access to data is limited to authorized parties, who participate in a permissioned network. In this way, the confidentiality of user information and the security of transaction data are ensured.
- In this network, private participants such as electric vehicles and smart homes can adopt anonymous identities to ensure that their information remains confidential. This feature applies to major cryptocurrencies such as Bitcoin as well, where although one can observe all participants' transactions in the distributed ledger, their identities cannot be uncovered.
- Private participation is arduous for major and legal participants because they occupy a specific point in the network and exist uniquely. Even if they possess anonymous identities, the possibility of identification remains. However, it is crucial to consider which participant information is made public and which is accessible to specifically authorized entities. In this model, only selected representatives, namely, the committers, have access to all the details of the participants' proposals as well as their smart and financial contracts. In practice, upon network approval orderers gain access solely to the limited information received from the committers. Consequently, all their information is be discernible to all network participants or to their competitors.

4. Consensus Algorithm

Consensus algorithms realize decentralized management of a microgrid. They are categorized according to their function, e.g., Proof of Stake, Delegated Proof of Stake, Proof of Work, and Proof of Vote [27]. The first three consensus types are better suited for public distributed networks. Due to their high block formation latency, they cannot be used in the energy industry.

4.1. Proof of Work Algorithm (POW)

The proof of work algorithm is regarded as one of the most prevalent blockchain network consensus algorithms. Bitcoin utilizes this algorithm to demonstrate the validity of its transactions. One of the most significant benefits of this algorithm is the remarkable scalability and decentralization it affords the network. This has resulted in the widespread adoption of Bitcoin globally. This algorithm has devised a competitive computation to solve an objective function whereby the fastest solver among all groups can generate a new block. Other participants in this computing network pause their computations and validate the resulting outcome [28]. If the outcome is correct, the block is stored in their network and they tackle a new one. Otherwise, they persist with their computation to acquire the correct solution. In Bitcoin mining, the objective function is used to find a hashed array with a particular structure from an input array. This array is generated by introducing a parameter named Nonce to the block input data and implementing an iterative algorithm. The initial group that uncovers this array adds the solution and the nonce value to authenticate its accuracy and incorporate all other groups in the network. Resolving this issue is considered time-intensive and takes over ten minutes. This delay is attributed to the network s enhanced resilience against attackers [29]. As the network's processing power increases, the objective function's constraints become more complex, not to solve the issue in a shorter time. This is referred to as an increase in network difficulty. In practice, this objective function can vary for each network. As in Bitcoin, solving a mathematical problem involves an array; in the power network, it can entail providing network operator constraints and quality control indicators as well as maintaining network security.

4.2. Proof of Stake Algorithm (POS)

Unlike the proof of work algorithm, there are no objective functions to solve in the proof of stake algorithm, and network members are not on the same level. Additionally, in this algorithm, each member is valued based on an index indicating their share in the network [30]. Those with greater power in the network play critical roles. The network also determines the power index. This index can be interpreted as network ownership percentage, processing power, or technical and probabilistic indicators. In a more common type, the proof of stake algorithm is delegated. In other words, influential network members select some individuals to generate a new block. Then, voting takes place, and votes are weighted to the extent of shared values. Ultimately, the selected members can generate a new block [31]. This process substantially reduces the algorithm's costs and enhances the efficiency of the proof-of-work algorithm. In power networks, members' valuation can be conducted using technical indicators. This fact empowers the leading members of the traditional network to play productive roles in guiding the new decentralized network. This permits the responsibility of generating a new block to be assigned to individuals who do not compromise the security and efficiency of the network.

4.3. Proof of Vote-Based Algorithm (POV)

The vote-based algorithm represents the third consensus algorithm employed in blockchain networks. Under this algorithm, all members must undergo identity verification before joining the network and contribute to validating new blocks. In this respect, safeguarding the network against attackers and sabotage is paramount. This algorithm is well-suited for private networks and is classified into Byzantine Fault Tolerance (BFT) and resilience against failure [32]. One of the powerful algorithms within the vote-based algorithm subset is the Ripple algorithm, which is utilized in Ripple. The algorithm's structure, which functions based on two-step verification, is open-source [33]. In this algorithm, each validated network member examines the information of the new block. Then based on the network constraints, each member votes for it and sends it to another validated member for feedback. When more than %50 of the network members validate the block, it chis confirmed in an open ledger. However, to achieve final confirmation and entry into a distributed public ledger, validation from %80 of the members is necessary [34]. A comparison of the performance of consensus algorithms is shown in Table 2.

Table 2. Comparison of consensus algorithms.

	POW	POS	POV
Energy consumption	High	Medium	Low
Scalability	High	High	Medium
Decentralization	High	High	Low
Resilience against attacks	%50	%50	%20
Average response time	10 T/s	10–1000 T/s	1000 T/s

This study uses the practical byzantine method, which has different variants itself. Networks based on Hyperledger usually use Proof of Vote Fault Tolerance [30]. This algorithm does not suit the electricity industry due to its technical approval requirements [33]. We used the Ripple algorithm used in Ripple cryptocurrency, as shown in Figure 2. Ripple is a two-stage algorithm, consisting of an open distributed ledger and a closed one [35]. In our proposed model, every orderer takes the last operating point of the last chain to confirm and register the order package, which can be multiple orders from several anchors. It registers a new microgrid operating point complying with the main system constraints as the new block in the chain. After the block is registered, all orderers should comment on the correctness of this operating point. Orderers receive rewards or penalties for such commenting or not commenting. If fifty-one percent of the companies confirm the block, the block becomes part of the chain and is added to the open ledger. This means confirming the initial order. Each member should confirm whether the main rules of the microgrid are observed in their comment.

The microgrid rules in our design are as follows:

- The new operating point is a correct power flow

$$\bar{S}_i = P_i + jQ_i = V_i e^{j\theta_i} \left[\sum_j Y_{ij} V_j e^{j(\theta_j + \delta_{ij})} \right]^* \tag{5}$$

- The allowable voltage interval of the bus and production range of the energy sources are observed:

$$Vmin_i \leq V_i \leq Vmax_i | Pmin_i \leq P_i \leq Pmax_i \tag{6}$$

- The cost of the rest of the system is declared accurately:

$$Cost_{res} = \sum_t (\lambda_o \, P_o + \lambda_{imp} P_{Imp} - \lambda_{Exp} \, P_{exp} - \tag{7}$$
$$\lambda_{charge} \, P_{charge} \,)$$

Here, $Cost_{rest}$ means the cost of the rest of the system, i.e., the total microgrid cost minus the new load cost. This is based on the internal cost λ_0, P_0, the cost of imported energy λ_{imp}, P_{imp}, the cost of exported energy λ_{exp}, P_{exp}, and the cost of a new order $\lambda_{charge}, P_{charge}$. In the above, P and λ represent energy and price, respectively. Confirming or rejecting the correctness of a block takes seconds. Each company can verify it through a simple algorithm. After the block is initially confirmed, the offering orderer receives fifty percent of the recommended package gas as the reward. In the second phase of consensus certification, all orderers can replace this new operating point block by registering a less costly one.

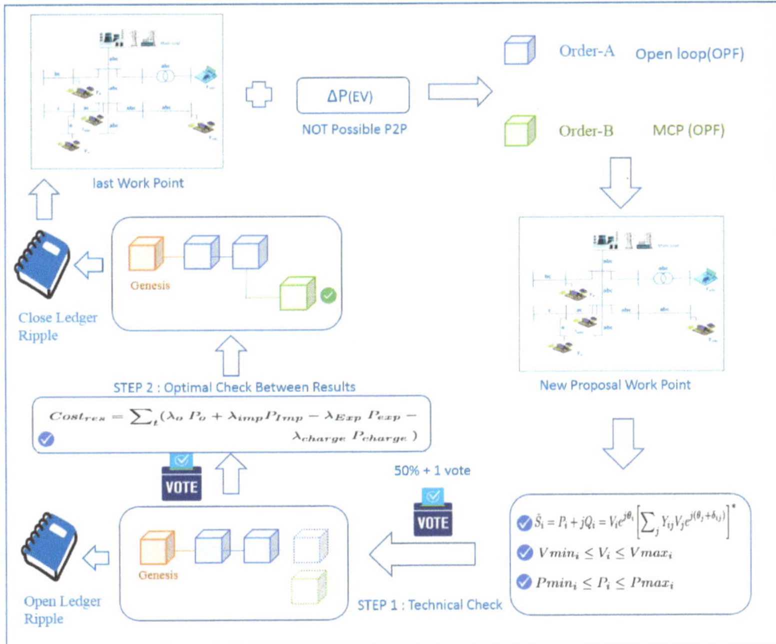

Figure 2. The consensus graph for the proposed decentralized microgrid.

If another orderer manages to do so within the intended interval, the other fifty percent of gas fee is assigned. Otherwise, the first orderer receives the reward at the end of the interval. This can be achieved by combining several blocks into one block. The main chain continues from that block. At the end of the time interval, the last block containing the recommended package is registered in the closed ledger, and the order is finalized. Orders are reviewed and confirmed much faster in this two-stage confirmation. The order is finalized, probably with changes, at the global optimal operating point of the microgrid. Those open ledge blocks which have a utilization cost less than or equal to the last block registered in the closed ledger are automatically sent to the closed ledger, as adding orders on that block has not increased the cost of the system.

4.4. How the Proposed Decentralized Microgrid Works

1. The independent operation loads the genesis block in the daily interval. It indicates an economic dispatch based on the latest load and microgrid generation.
2. Smart load approved by endorsers can connect to the decentralized microgrid and forward their order to it through committers.
3. Committers try to minimize energy costs by offering various offers and smart contracts according to the customers' consumption. They manage the orders and forward them to anchors to be supplied by the microgrid.
4. Anchors manage all internal P2P trades inside the microgrid. They transfer orders that cannot be supplied through local trading to the block pool.
5. Orderers remove the new orders from the block pool according to the last operating point of the microgrid in the main chain using simple and fast, economic dispatch algorithms. They apply such orders to present a new operating point for registering in the main chain subject to the microgrid's technical constraints.
6. The Ripple two-stage consensus mechanism confirms this new operating point's correctness and global optimality. In this consensus model, all the primary microgrid beneficiaries review the correctness of the new operating point according to the approved technical constraints. This review is carried out in a split second using a simple power flow algorithms subject to the microgrid constraints. The block is registered if approved by more than fifty-one percent of the microgrid.
7. The final registering of the block is performed in the second stage of the Ripple consensus algorithm. If any orderer can provide a better operating point, then the previous block becomes a minor one and the new block continues the main chain. The new block can consist of several blocks. However, all previous orders must be considered, and the new block's cost should be less. If a replacement block is not offered and the chain continues, the previous block registration is finalized in the chain.
8. The microgrid is utilized based on the last confirmed operating point in the main chain. The process of the decentralized microgrid is shown in Figure 2.

5. Case Study and Discussion

The proposed decentralized microgrid's performance was tested on standard systems with known centralized management results. We used the IEEE 13-bus test system [36] shown in the microgrid diagram of Figure 3. This is a three-phase power distribution system. A commercial parking lot contains 80 electric vehicles with a 6.6 kWh charging capacity and 36 kW battery capacity. The vehicles can be charged from 6 in the morning and from 9 in the afternoon. The case study parameters are shown in Table 3 [37].

The microgrid has five solar energy sources, each with a 200 kW capacity. The cost of importing and exporting electricity from the slack bus is 0.15 and 0.5 GBP/kWh, and the demand charge is 0.1 GBP/kWh. To compare the random performance of the decentralized microgrid. We planned 30 electric vehicles with the proposed decentralized microgrid. We assume an initial configuration of fifty vehicles has finalized registration. Here, ten vehicles ordered a change of plan, and thirty others ordered a new connection to the microgrid. Ten out of the thirty vehicles have a charge-at-will smart contract. The gas fee for electric vehicles is GBP1, and for modeling the smart contract with committers is GBP2.

Table 3. The parameters of the microgrid adjusted for the case study [37].

Number of electric vehicles	80
Electric vehicle battery sizes	36 kWh
Electric vehicle charger capacities	6.6 kW
PV generation capacities	200 kWp
Phase voltage magnitude limits	0.95 to 1.05 pu
Import price	£0.15/kWh
Export price	£0.05/kWh
Demand charge	£0.10/kW
EMS time-series resolution	30 min
Simulation time-series resolution	5 min

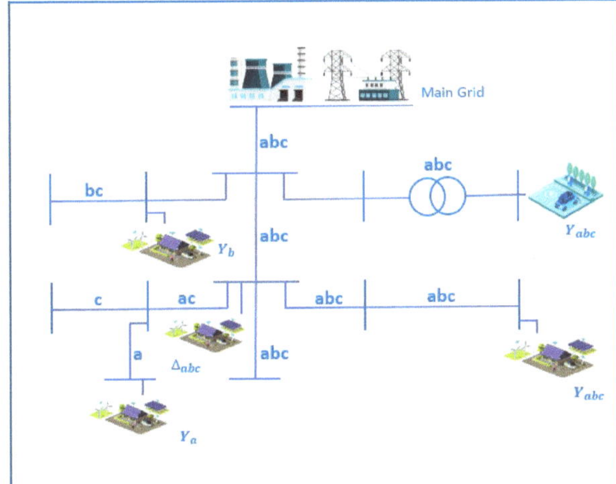

Figure 3. The IEEE 13-bus microgrid adjusted for the electric vehicle case study.

In this paper, the PGOxford of [37] (which is an open-source platform developed in Python) is utilized to model the economic dispatch. The EMS module in this tool forecasts loads using time series and solar power-generating sources. The optimization solvers included open loop solver and MCP solver. Open Loop Solver has shorter optimization times. However, the results in the reference paper indicate that it violates the allowable voltage range. MCP Solver has a longer optimization time but observes all the microgrid constraints in the final configuration.

PGoxford software models the network and distributes the load, including a three-phase model for each line in the network. Three-phase unbalanced distribution systems can be accurately modeled with this software. Using dataframes containing the characteristics of lines, transformers, and capacitor banks, this software method updates the network admittance matrix. We assume a three-phase network with phases a, b, and c, a slack bus voltage and N load buses [37]. In PGoxford software, the solutions were compared with the solutions from OpenDSS [38]. More details about how to model the three phases of the network in this software are provided in detail in the reference appendix C of this software [37], which we refrain from repeating in this article.

In this modeling, we assume the microgrid has five committers, each with six electric vehicle orders. The model has an anchor in charge of P2P transactions and two orderers. Each orderer uses one of the solvers. In each transaction, five vehicles are selected, sorted by the gas fee. Orderer A (OA) uses the Open loop optimization algorithm, and Orderer B (OB) uses the MCP optimization algorithm. They are defined in parallel and different servers and are only connected through the main chain file in the blockchain shown in Figure 4. The objective function and constraints of the microgrid and eclectic vehicle are as in [37].

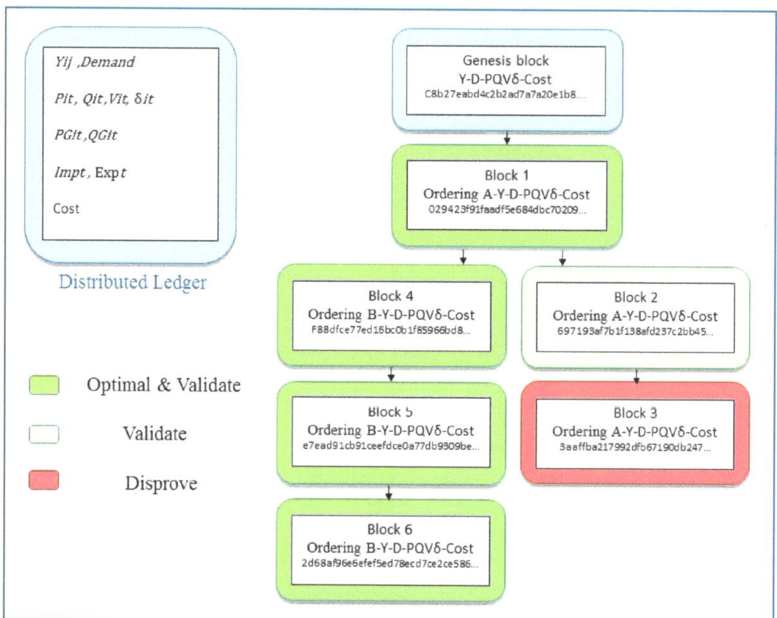

Figure 4. Main chain diagram of decentralized blockchain microgrid.

Simulation Results

Decentralized management enables individuals to solve their problems discretely using simultaneous parallel algorithms. Furthermore, by comparing their results and the consensus network, they can determine which result is more accurate and optimal. This possibility allows various algorithms with different capabilities to be combined, such as for speed and accuracy. Further, the possibility serves to allow faster algorithms for obtaining initial solutions and more accurate algorithms for achieving the final optimal solution. Simulation results indicate that the anchor has P2P replaced all ten vehicles' move orders, i.e., removing their previous order from the initial configuration, with new orders. Seven out of ten orders were vehicles with smart contracts with more gas fees, and three others were other vehicles with timing conflicts. The P2P objective function maximizes the anchor gas fee earnings, and the plan's constraints were the maximum battery charge capacity in the remaining time and time conflict between the orders. The anchor referred the remaining twenty orders to the pool block after settling ten orders. This simulation was carried out connecting two computers together in parallel. Each computer was defined as an order. Figure 4 represents seven blocks registered during the computations: a genesis block, three blocks for OA, and three blocks for OB. However, the end chain consists of one block from OA and three blocks from OB. Block 3 from OA was declared invalid due to a lack of consensus on the correctness of this block, as it violates the bus voltage limitation. Block 2 was replaced by Block 5, As OB offered a block with a lower value cost function. In the end, the orderers registered twenty vehicles. OB received more gas fees by registration, registering fifteen final and ten initial vehicles, compared to OA, which registered five initial request and five final request vehicles. As the first block contains three vehicles with smart contracts, OA and OB's incomes are GBP10.5 and GBP12.5. The results are shown in Table 4.

Table 4. Management of vehicles according to decentralized method.

EV	1	2	3	...	40
State	Charge	Charge	Charge	...	Charge
Smart contract	NO	YES	NO	...	NO
Gas (£)	1	2	1	...	1
Anchor (£)	1	0	0	...	0
Orderer-A (£)	0	2	0	...	0.5
Orderer-B (£)	0	0	1	...	0.5
Initial Block	B1	B1	B5	...	B2
Final Block	B1	B1	B5	...	B4
Chang OPF	NO	YES	YES	...	YES

In this article, the central problem is initially solved using two different algorithms. The results are presented in Figure 5, following reference [37]. The results indicate that the network constraints were violated in an open-loop concentrated algorithm. In addition, a decentralized program was carried out using a couple of two-stage Ripple consensus parallel algorithms. The Ripple consensus algorithm ensures compliance with technical constraints in the first stage, as depicted in Figure 2. In fact, this confidence is achieved by the vote of more than half of the network. The decentralized results are presented in Figure 6, and it is evident that the voltage threshold was not violated. The decentralized network has prevented it by implementing its consensus mechanism. This demonstrates that instability and violations of the algorithms can be prevented by managing algorithms in a decentralized manner and employing a transparent consensus mechanism at the final point of the network.

Comparing the minimum allowable voltage threshold in our results and those of the centralized management in [37] shows that combining several optimization algorithms in a decentralized system eliminates the shortcomings of each model and optimizes the result. It indicates that combining several optimization algorithms in the decentralized network in Figure 6 eliminates the allowable voltage range violation issue of the open loop centralized model in Figure 5 while observing acceptable voltage band constraints. This indicates that a discrete combination of several planning and optimization algorithms provides better results than a single one by combining their speed and accuracy advantages.

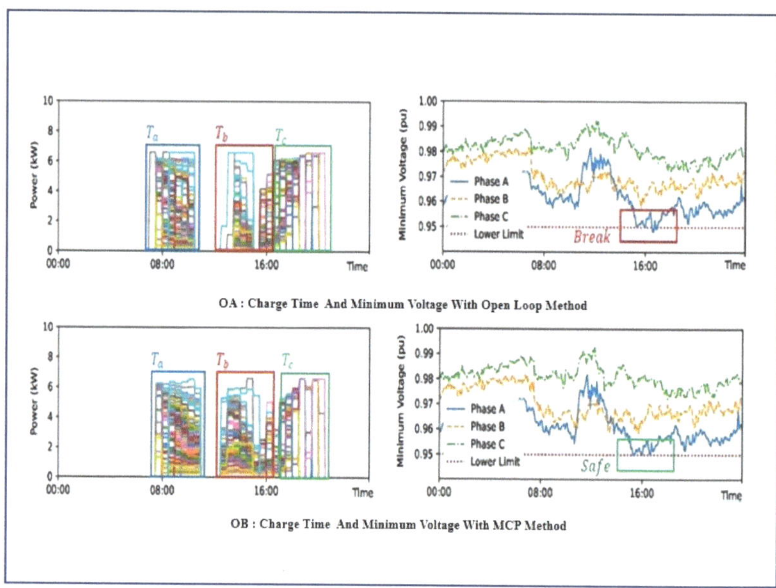

Figure 5. Minimum voltage and charging time of EVs in each phase according to Open Loop and MCP optimization in PGOxford software [37]).

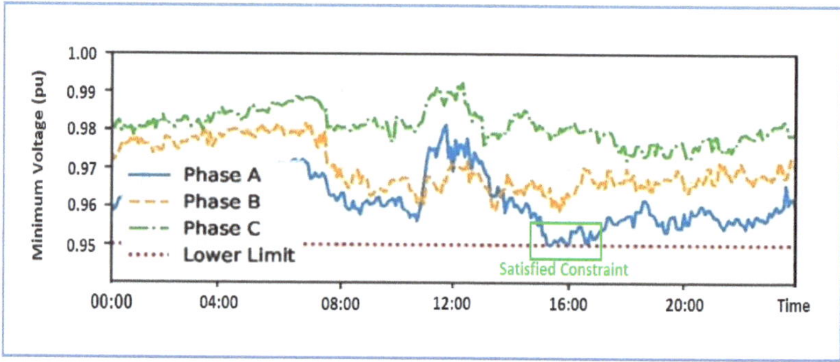

Figure 6. Minimum voltage of each phase in the decentralized optimization.

When charging electric cars, local decentralized P2P exchange networks play a crucial role in providing energy. However, the present article focuses on the non-local provision of decentralized energy, which alters the network's operating point. As observed, by eliminating states that violate technical constraints, the decentralized consensus algorithm optimizes the allocation of charge in the Tb interval. It prevents a breach of the allowable voltage threshold. This algorithm minimizes the cost of network operation by analyzing various states.

Regarding the charging plan of the electric vehicles, Figure 5 shows the timing plan of the electric vehicles in the centralized management system for the models presented in reference [37], while Figure 7 shows the timing obtained from combining the decentralized management of these two models. Dividing the charging interval of the vehicles into three smaller intervals, T_a, T_b, and T_c, shows the most significant change due to decentralized management in the results for interval T_b. This change prevents voltage violation at the end of the interval, which is achieved without increasing the microgrid's cost or violating its other constraints. This means that optimizing the decentralized management was a move in the right direction. The existence of smart contracts for vehicles, planning vehicles charging time using P2P, and presumably differences in the order of receiving orders make these two scenarios inevitably different. Importantly, however, all vehicles have received their intended charge within the due time interval. The average vehicle charge is not changed, and the system's total cost is not increased.

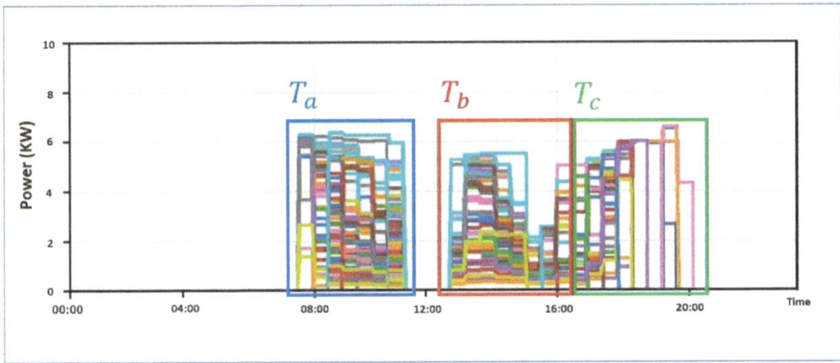

Figure 7. Time for charging EVs in the decentralized optimization.

Finally, examining the results of the decentralized management of power supply shows that microgrid utilization maintains power generation adequacy and observes all system constraints without adding to the total cost of the microgrid. Figure 8 illustrates the method of providing energy using the base load, receiving energy from the primary grid, and the average charging of electric vehicles. The ultimate criterion for verifying each chain is the cost optimization of the system compared to all the states presented in a block. Even if it cannot be proven that the final solution presented in the algorithm is the most optimal global point of the network, it can be claimed that the solution is the most optimal among the proposals of the decentralized network's point of work. In fact, this solution is equal to the most optimal cost of the centralized network. Figure 8 shows no change in the energy imported into the microgrid compared to the value in the reference paper. Considering the difference in the energy price inside and outside the microgrid, the total microgrid cost has not changed.

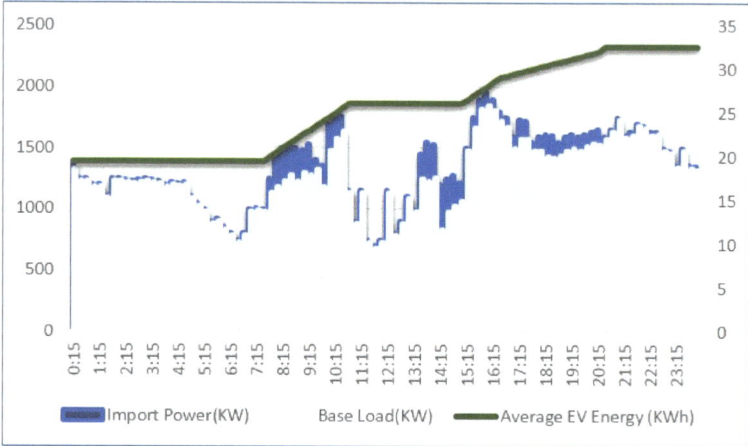

Figure 8. Energy imported into the microgrid.

Comparing the results of modeling the decentralized microgrid management and its centralized peer shows that:

- Decentralized management of a microgrid can utilize it at its optimal operating point while maintaining adequacy and observing its technical constraints.
- A decentralized combination of several optimization models based on Ripple's two-stage consensus algorithm eliminates each model's disadvantages and reaches a globally optimized solution.
- Smart contracts and P2P trades play an essential role in microgrid planning by improving response time, and can be used as an effective tool to direct the microgrid.

The results of comparing the centralized solution in reference [37] with the decentralized solution presented in this paper are provided in Table 5. These results show that the decentralized solution can play a vital role in the rapid and localized network strategy, particularly in electric car charging, by providing an initial solution. Furthermore, this approach ensures the minimum network cost among all available states. As this article is centered on the intra-day period and specifically aimed at addressing load variations from the initial plan, it could serve as a suitable alternative for small-scale intelligent networks, ultimately leading to democratic energy allocation.

Table 5. Comparison between centralized and decentralized modeling.

Method	Centralized	Centralized	Decentralized
Algorithm	MCP	Open Loop	(MCP-OpenLoop)
Initial Result	NO	NO	High-Speed
Final Result	Low-Speed	High-Speed	Medium-Speed
Verify Constraints	Ok	NO	OK
Final Cost System	3345 $	3352 $	3345 $

6. Conclusions

This paper aims to design and propose a blockchain-based decentralized energy market to support the development and establishment of citizen energy communities. The proposed framework is an appropriate solution for small-scale energy transactions sourced by demand-response resources, electric vehicles, and small renewable power sources with high uncertainty. In essence, this article proposes that rather than solving a centralized optimization problem, a decentralized problem can be solved using various algorithms. With this in mind, a decision can be reached on which solution is optimal and most accurate for a given problem by implementing a consensus mechanism. The solution can be selected and announced to all. This perspective can be generally applicable and is not contingent on the power grid or the electrical industry. Comparing the results of the proposed decentralized system with the conventional centralized systems shows that:

- Using several parallel algorithms in a decentralized system makes it much more efficient than a centralized one.
- The two-stage consensus model proposed in this paper ensures adequacy while observing local system constraints and achieves optimality of the final operating point of the system.
- The layered design makes the response time faster than its centralized peer and provides it with more flexibility, particularly in supplying local energy.

This work aims to open up discussion in this direction and pave the way for the future development of the tools and methods required for small-scale energy transactions, including vehicle-to-vehicle, vehicle-to-building, building-to-vehicle, building-to-building, and other forms of peer-to-peer transactions as the main elements of the citizen energy communities of the future.

6.1. Challenges

One of the most critical challenges in implementing decentralized networks in power systems is concern about the network's security and stability. How the network responds in critical moments, resistance against attackers, and alignment with long-term operational strategies are key challenges for this network. However, because this network is an open platform and tools such as smart contracts are used, these challenges cannot pose a severe threat to the development of the networks in the electrical industry.

6.2. Future Work

As discussed in the challenges of decentralized networks, improving reliability, stability, strategic management, and network performance is crucial. Smart contracts play a vital role in controlling and guiding these networks. Thus, by expanding these contracts, we can empower the main network participants to lead the network without exerting direct power and ensure network stability. Therefore, developing smart contracts is a key aspect of network advancement alongside enhancing the decentralized network structure.

Author Contributions: Conceptualization, P.M. and V.V.; Methodology, P.M. and V.V.; Software, P.M.; Validation, P.M., M.S.G. and V.V.; Formal analysis, V.V. and M.S.G.; Investigation, P.M. and V.V.; Resources, V.V.; Data curation, P.M. and V.V.; Writing—original draft, P.M.; Writing—review & editing, V.V. and M.S.G.; Visualization, P.M. and V.V.; Supervision, M.S.G. and V.V.; Project administration, M.S.G. and V.V. All authors have read and agreed to the published version of the manuscript.

Funding: This research received no external funding.

Data Availability Statement: Not applicable.

Conflicts of Interest: The authors declare no conflict of interest.

References

1. Energy Communities. 2022. Available online: Energy.ec.europa.eu/topics/markets-and-consumers/energy-communities (accessed on 22 June 2023).
2. Attaran, M.; Gunasekaran, A. *Applications of Blockchain Technology in Business: Challenges and Opportunities*; Springer: Berlin/Heidelberg, Germany, 2019.
3. Capuano, L. *International Energy Outlook 2018 (IEO2018)*; US Energy Information Administration (EIA): Washington, DC, USA, 2018; Volume 2018, p. 21.
4. Wang, Q.; Ji, T.; Guo, Y.; Yu, L.; Chen, X.; Li, P. TrafficChain: A Blockchain-Based Secure and Privacy-Preserving Traffic Map. *IEEE Access* **2020**, *8*, 60598–60612. [CrossRef]
5. Zhou, Z.; Gong, J.; He, Y.; Zhang, Y. Software defined machine-to-machine communication for smart energy management. *IEEE Commun. Mag.* **2017**, *55*, 52–60. [CrossRef]

6. Xiao, H.; Huimei, Y.; Chen, W.; Hongjun, L. A survey of influence of electrics vehicle charging on power grid. In Proceedings of the 2014 9th IEEE Conference on Industrial Electronics and Applications, Hangzhou, China, 9–11 June 2014; pp. 121–126.

7. Vahidinasab, V.; Mohammadi-Ivatloo, B. *Electric Vehicle Integration via Smart Charging: Technology, Standards, Implementation, and Applications*; Springer: Berlin/Heidelberg, Germany, 2022.

8. Fu, X.; Wang, H.; Wang, Z. Research on block-chain-based intelligent transaction and collaborative scheduling strategies for large grid. *IEEE Access* **2020**, *8*, 151866–151877. [CrossRef]

9. Hassija, V.; Chamola, V.; Garg, S.; Krishna, D.N.G.; Kaddoum, G.; Jayakody, D.N.K. A blockchain-based framework for lightweight data sharing and energy trading in V2G network. *IEEE Trans. Veh. Technol.* **2020**, *69*, 5799–5812. [CrossRef]

10. Liu, H.; Zhang, Y.; Zheng, S.; Li, Y. Electric vehicle power trading mechanism based on blockchain and smart contract in V2G network. *IEEE Access* **2019**, *7*, 160546–160558. [CrossRef]

11. Yahaya, A.S.; Javaid, N.; Javed, M.U.; Shafiq, M.; Khan, W.Z.; Aalsalem, M.Y. Blockchain-based energy trading and load balancing using contract theory and reputation in a smart community. *IEEE Access* **2020**, *8*, 222168–222186. [CrossRef]

12. Cohn, J.M.; Finn, P.G.; Nair, S.P.; Panikkar, S.B.; Pureswaran, V.S. Autonomous Decentralized Peer-to-Peer Telemetry. U.S. Patent 10,257,270, 9 April 2019

13. Afzal, M.; Huang, Q.; Amin, W.; Umer, K.; Raza, A.; Naeem, M. Blockchain enabled distributed demand side management in community energy system with smart homes. *IEEE Access* **2020**, *8*, 37428–37439. [CrossRef]

14. Masaud, T.M.; Warner, J.; El-Saadany, E.F. A blockchain-enabled decentralized energy trading mechanism for islanded networked microgrids. *IEEE Access* **2020**, *8*, 211291–211302. [CrossRef]

15. AlAshery, M.K.; Yi, Z.; Shi, D.; Lu, X.; Xu, C.; Wang, Z.; Qiao, W. A blockchain-enabled multi-settlement quasi-ideal peer-to-peer trading framework. *IEEE Trans. Smart Grid* **2020**, *12*, 885–896. [CrossRef]

16. Hassija, V.; Saxena, V.; Chamola, V.; Yu, F.R. A parking slot allocation framework based on virtual voting and adaptive pricing algorithm. *IEEE Trans. Veh. Technol.* **2020**, *69*, 5945–5957. [CrossRef]

17. Chen, X.; Zhang, X. Secure electricity trading and incentive contract model for electric vehicle based on energy blockchain. *IEEE Access* **2019**, *7*, 178763–178778. [CrossRef]

18. Kang, J.; Yu, R.; Huang, X.; Maharjan, S.; Zhang, Y.; Hossain, E. Enabling localized peer-to-peer electricity trading among plug-in hybrid electric vehicles using consortium blockchains. *IEEE Trans. Ind. Inform.* **2017**, *13*, 3154–3164. [CrossRef]

19. Liu, C.; Chai, K.K.; Zhang, X.; Lau, E.T.; Chen, Y. Adaptive blockchain-based electric vehicle participation scheme in smart grid platform. *IEEE Access* **2018**, *6*, 25657–25665. [CrossRef]

20. Shen, J.; Chen, L.; Zhou, B.; Zhang, M.; Zhou, S.L.; Sun, J.P.; Hou, Z. Intelligent energy scheduling model based on block chain technology. In Proceedings of the 2018 5th International Conference on Information Science and Control Engineering (ICISCE), Zhengzhou, China, 20–22 July 2018; pp. 811–817.

21. Danzi, P.; Angjelichinoski, M.; Stefanović, Č.; Popovski, P. Distributed proportional-fairness control in microgrids via blockchain smart contracts. In Proceedings of the 2017 IEEE International Conference on Smart Grid Communications (SmartGridComm), Dresden, Germany, 23–27 October 2017; pp. 45–51.

22. Duan, B.; Xin, K.; Zhong, Y. Optimal dispatching of electric vehicles based on smart contract and internet of things. *IEEE Access* **2019**, *8*, 9630–9639. [CrossRef]

23. Blockchain Network. 2023. Available online: Hyperledger-fabric.readthedocs.io/en/release-2.2/network/network.html (accessed on 22 June 2023).

24. Cachin, C. Architecture of the hyperledger blockchain fabric. In Proceedings of the Workshop on Distributed Cryptocurrencies and Consensus Ledgers, Chicago, IL, USA, 15 July 2016; Volume 310, pp. 1–4.

25. Blockchain Hyperledger Fabric. 2023. Available online: Cloud.ibm.com/docs/blockchain?topic=blockchain-hyperledger-fabric (accessed on 22 June 2023).

26. Hyperledger Fabric Workflow. 2023. Available online: Techgeek628.medium.com/how-does-hyperledger-fabric-works-d5a4d4 ff6b07 (accessed on 22 June 2023).

27. Shahbazi, M.; Kazem Pourian, S.; Taghva, M. An applied investigation of Consensus Algorithms Used in Blockchain Networks. *Sci. Technol. Policy Lett.* **2020**, *10*, 35–54.

28. Sompolinsky, Y.; Zohar, A. Accelerating Bitcoin's Transaction Processing. Fast Money Grows on Trees, Not Chains. *Cryptology ePrint Archive*. 2013. Available online: https://ia.cr/2013/881 (accessed on 22 June 2023).

29. Yaga, D.; Mell, P.; Roby, N.; Scarfone, K. *arXiv* **2019**, arXiv:1906.11078. [CrossRef]

30. Yang, R.; Yu, F.R.; Si, P.; Yang, Z.; Zhang, Y. Integrated blockchain and edge computing systems: A survey, some research issues and challenges. *IEEE Commun. Surv. Tutor.* **2019**, *21*, 1508–1532. [CrossRef]

31. Vasin, P. Blackcoin's Proof-of-Stake Protocol v2. 2014. Volume 71. Available online: https://blackcoin.Co/blackcoin-PosPdf (accessed on 22 June 2023).

32. Indrakumari, R.; Poongodi, T.; Saini, K.; Balamurugan, B. Consensus algorithms—A survey. In *Blockchain Technology and Applications*; Auerbach Publications: New York, NY, USA, 2020; pp. 65–78.

33. Schwartz, D.; Youngs, N.; Britto, A. The Ripple Protocol Consensus Algorithm. 2014. Available online: https://ripple.com/files/ ripple_consensus_whitepaper.pdf (accessed on 22 June 2023).

34. Armknecht, F.; Karame, G.O.; Mandal, A.; Youssef, F.; Zenner, E. Ripple: Overview and outlook. In Proceedings of the International Conference on Trust and Trustworthy Computing, Heraklion, Greece, 24–26 August 2015; Springer: Berlin/Heidelberg, Germany, 2015; pp. 163–180.
35. Yu, F.R.; Liu, J.; He, Y.; Si, P.; Zhang, Y. Virtualization for distributed ledger technology (vDLT). *IEEE Access* **2018**, *6*, 25019–25028. [CrossRef]
36. Schneider, K.P.; Mather, B.; Pal, B.; Ten, C.W.; Shirek, G.J.; Zhu, H.; Fuller, J.C.; Pereira, J.L.R.; Ochoa, L.F.; de Araujo, L.R.; et al. Analytic considerations and design basis for the IEEE distribution test feeders. *IEEE Trans. Power Syst.* **2017**, *33*, 3181–3188. [CrossRef]
37. Morstyn, T.; Collett, K.A.; Vijay, A.; Deakin, M.; Wheeler, S.; Bhagavathy, S.M.; Fele, F.; McCulloch, M.D. OPEN: An open-source platform for developing smart local energy system applications. *Appl. Energy* **2020**, *275*, 115397. [CrossRef]
38. Dugan, R.C.; McDermott, T.E. An open source platform for collaborating on smart grid research. In Proceedings of the 2011 IEEE Power and Energy Society General Meeting, Detroit, MI, USA, 24–28 July 2011; pp. 1–7.

Article

Robust Control and Active Vibration Suppression in Dynamics of Smart Systems

Amalia Moutsopoulou [1], Georgios E. Stavroulakis [2], Anastasios Pouliezos [2], Markos Petousis [1] and Nectarios Vidakis [1,*

[1] Department of Mechanical Engineering, Hellenic Mediterranean University Estavromenos, 71410 Heraklion, Greece
[2] Department of Production Engineering and Management Technical University of Crete, Kounoupidianna, 73100 Chania, Greece
* Correspondence: vidakis@hmu.gr; Tel.: +30-2810-379-227

Abstract: Challenging issues arise in the design of control strategies for piezoelectric smart structures. Piezoelectric materials have been investigated for use in distributed parameter systems in order to provide active control efficiently and affordably. In the active control of dynamic systems, distributed sensors and actuators can be created using piezoelectric materials. The three fundamental issues that structural control engineers must face when creating robust control laws are structural modeling methodologies, uncertainty modeling, and robustness validation. These issues are reviewed in this article. A smart structure with piezoelectric (PZT) materials is investigated for its active vibration response under dynamic disturbance. Numerical modeling with finite elements is used to achieve that. The vibration for different model values is presented considering the uncertainty of the modeling. A vibration suppression was achieved with a robust controller and with a reduced order controller. Results are presented for the frequency domain and the state space domain. This work cleary demostrated the advantage of robust control in the vibration suppration of smart stuctures.

Keywords: smart structures; robust control; uncertainty; dynamical system

Citation: Moutsopoulou, A.; Stavroulakis, G.E.; Pouliezos, A.; Petousis, M.; Vidakis, N. Robust Control and Active Vibration Suppression in Dynamics of Smart Systems. *Inventions* **2023**, *8*, 47. https://doi.org/10.3390/inventions8010047

Academic Editor: Om P. Malik

Received: 14 December 2022
Revised: 1 February 2023
Accepted: 6 February 2023
Published: 9 February 2023

1. Introduction

A piezoelectric structure with a control strategy has the potential to adapt to both a changing internal environment and a changing external environment, such as stresses or form changes. It includes intelligent actuators that enable controlled modification of system parameters and reactions. Piezoelectric materials (PZT), shape memory alloys, electrostrictive materials, magnetostrictive materials, and fiber optics are only a few examples of the numerous types of actuators and sensors under consideration. We employ piezoelectric material in our paper. In the active control of dynamic systems, piezoelectric materials can be specially adapted to serve as distributed sensors and actuators. The study of intelligent structures has drawn the attention of numerous scholars [1–6]. A smart structure is one that keeps an eye on both its surroundings and itself [7,8].

Robust vibration control of piezoelectric-actuated smart structures has recently attracted a lot of attention. Despite the existence of numerous sources of uncertainty, such control laws are preferred for systems where guaranteed stability or performance are required [9–11].

The later robust controller accounts for the dynamical system's uncertainties as well as the incompleteness of the measured data, which results in the design of smart structures that can be used. To provide a thorough and unitary methodology for designing and validating reliable $H_{infinity}$ (H_{inf}) controllers for active structures, the numerical simulation demonstrates that sufficient vibration suppression can be achieved by using the suggested general methods in a tutorial manner for the case of a piezoelectric smart structure [12–14]. The novelty of the work is that it calculates an $H_{infinity}$ controller with very good results in the frequency domain and the state space, even for different values of the mass and stiffness matrix, considering the uncertainty of the modeling; additionally, good results were acquired with a reduced order $H_{infinity}$ controller. No similar work achieves vibration suppression if there are different values of the mass and the stiffness matrix.

2. Materials and Methods

The approximate discretized variation problem results from using the traditional finite element method. By substituting discretized formulas into the initial variation of kinetic energy and strain energy for a finite element, discrete differential equations are generated [8,15]. The beam element equation of motion is defined in terms of the nodal variable q as follows, integrating over spatial domains and applying Hamilton's principle [8,10]

$$M\ddot{q}(t) + D\dot{q}(t) + Kq(t) = f_m(t) + f_e(t) \tag{1}$$

where K is the global stiffness matrix, D is the viscous damping matrix, M is the global mass matrix, f_e is the global control force vector produced by electromechanical coupling effects, and f_m is the global external loading vector for a beam structure used in this work.

Transversal deflections w_i and rotations ψ_i constitute the independent variable $q(t)$, i.e.,

$$q(t) = \begin{bmatrix} w_1 \\ \psi_1 \\ \vdots \\ w_n \\ \psi_n \end{bmatrix} \tag{2}$$

where in the analysis the number of finite elements used is the n index in the matrix. Vectors w and f_m are upward positive.

Permit state–space representation transformation of control (in the usual manner),

$$x(t) = \begin{bmatrix} q(t) \\ \dot{q}(t) \end{bmatrix} \tag{3}$$

Furthermore, to express $f_e(t)$ as $Bu(t)$ we write it as $F_e^* u$ where F_e^* (of size 2n × n) indicates the voltages on the actuators. The F_e^* (of size 2n × n) matrix also denotes the piezoelectric force for a unit mounted on its corresponding actuator. Lastly, the disturbance vector is designed by the following equation $d(t) = f_m(t)$ [15]. Then,

$$\dot{x}(t) = \begin{bmatrix} 0_{2n \times 2n} & I_{2n \times 2n} \\ -M^{-1}K & -M^{-1}D \end{bmatrix} x(t) + \begin{bmatrix} 0_{2n \times n} \\ M^{-1}F_e^* \end{bmatrix} u(t) + \begin{bmatrix} 0_{2n \times 2n} \\ M^{-1} \end{bmatrix} d(t)$$
$$= Ax(t) + Bu(t) + Gd(t) = Ax(t) + [B\ G]\begin{bmatrix} u(t) \\ d(t) \end{bmatrix} = Ax(t) + \widetilde{B}\widetilde{u}(t) \tag{4}$$

The output equation, as a function of the measured displacements, will help us to strengthen this,

$$y(t) = [x1(t)\ x3(t) \ldots xn - 1(t)]^T = Cx(t) \tag{5}$$

In the equation, the u parameter's matrix size is n × 1 (or smaller), while the d parameter's matrix size is 2n × 1. The units used are Newtons, radians, meters, and seconds.

In the next section, we will examine the behavior of a 32-element cantilever beam containing pairs of elements. The beam's dimensions are $L \times W \times h$. The sensors and actuators have a width and thickness of b_S and b_A, individually. The electromechanical properties of the beam of interest depicted in Figure 1a,b are listed in Table 1.

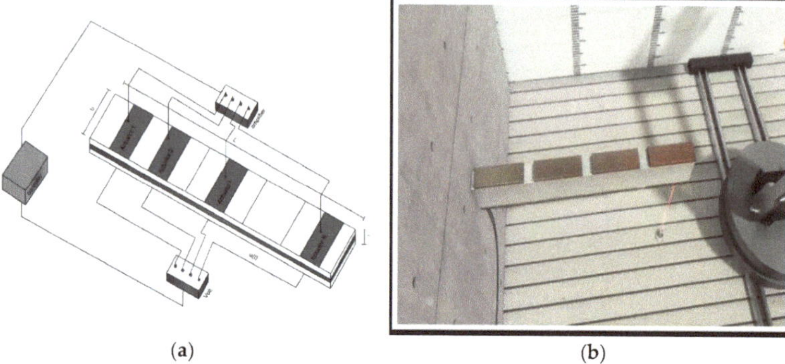

(a)   (b)

Figure 1. (**a**) Piezoelectric smart beam; (**b**) Piezoelectric smart cantilever beam.

Table 1. Parameters of the smart beam.

Parameters	Values
Beam length, L	0.8 m
Beam width, W	0.07 m
Beam thickness, h	0.0095 m
Beam density, ρ	1600 kg/m^3
Young's modulus of the beam, E	1.5×10^{11} N/m^2
Piezoelectric constant, d$_{31}$	254×10^{-12} m/V

2.1. Frequency Domain

In a transfer function matrix, the structured singular value is defined as,

$$
\mu(M) = \begin{cases} \dfrac{1}{\min\limits_{k_m}\left\{ \det(I - k_m M\Delta) = 0, \overline{\sigma}(\Delta) \le 1 \right\}} \\ 0, \text{if no such structured exists} \end{cases}
\tag{6}
$$

This matrix specifies the smallest structured Δ and has $\overline{\sigma}(\Delta)$ as a function (sigma is the structured singular value for the uncertainty modeling), and, as a result, the determinant becomes zero, i.e., $\det(I - M\Delta) = 0$: then $\mu(M) = 1/\overline{\sigma}(\Delta)$. Equation (6) calculates the singular value. The upper and lower limits are visually presented and they should be less than one (1) for the specific Kp (arithmetic parameter for the stiffness matrix) and Km (arithmetic parameter for the scaled mass matrix) values. Following this, it is desired that the μ values are lower than 1, as shown in the results section. The principle followed was the smaller, the better [15–17].

2.2. Design Objectives

Design goals can be divided into two groups:
Nominal performance

1. Small control effort.
2. Attenuation of disturbances with acceptable transient characteristics (overshoot, settling time).
3. Strength of closed loop system (plant + controller).

Robust performance

4. The above criteria (1)–(3) should be satisfied even when noise exists in the modeling procedure.

2.3. System Specifications

To obtain the necessary system specifications, the system should be represented in the (N, Δ) structure to achieve the aforementioned objectives. The conventional diagram is depicted in Figure 2.

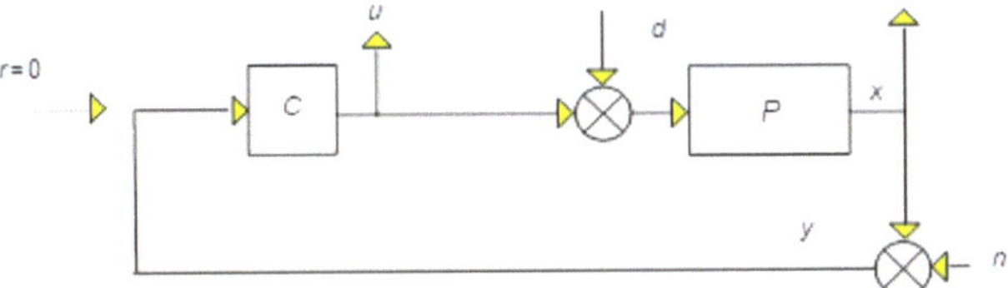

Figure 2. Typical control block graph.

The disturbance vector (mechanical force) d and noise vector n are the diagram's two inputs, and the control vector u and state vector x are the diagram's two outputs. It is expected in what follows that,

$$\left\|\frac{d}{n}\right\|_2 \leq 1, \left\|\frac{u}{x}\right\|_2 \leq 1 \tag{7}$$

If that is not the case, then the original signals can be modified using the right frequency-dependent weights to give the altered signals this feature [9,13].

Rewrite Figure 2 similarly to Figure 3:

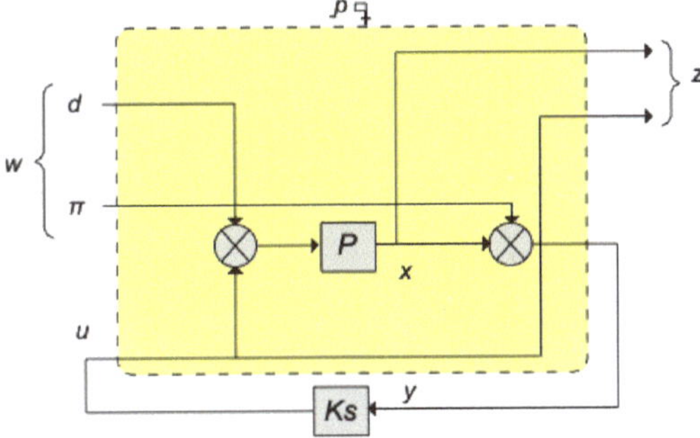

Figure 3. Two-port detailed graph.

Or with fewer details (Figure 4),

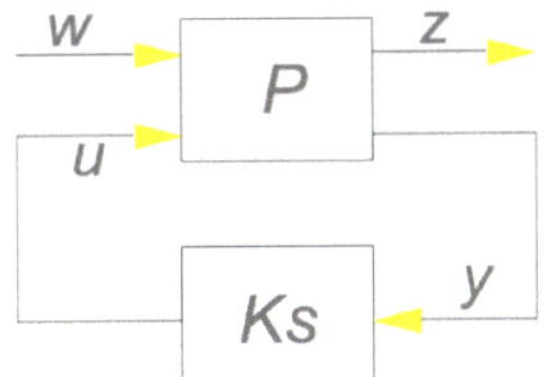

Figure 4. Two-port graph.

with,

$$z = \begin{bmatrix} u \\ x \end{bmatrix}, w = \begin{bmatrix} d \\ n \end{bmatrix} \tag{8}$$

where z is the output (control vector u, and the state vector x) controllable variables as well as exogenous inputs (mechanical disturbances vector and the noise) [12,14,18]. Given that P is composed of two inputs and two outputs, it is typically partitioned as follows,

$$\begin{bmatrix} z(s) \\ y(s) \end{bmatrix} = \begin{bmatrix} P_{zw}(s) & P_{zu}(s) \\ P_{yw}(s) & P_{yu}(s) \end{bmatrix} \begin{bmatrix} w(s) \\ u(s) \end{bmatrix} \overset{op}{=} P(s) \begin{bmatrix} w(s) \\ u(s) \end{bmatrix} \tag{9}$$

Also,

$$u(s) = K_s(s)y(s) \tag{10}$$

The transfer function for a closed loop is obtained by substituting (10) in (9) $N_{zw}(s)$ with $K_s(s)$ the controller of our system,

$$N_{zw}(s) = P_{zw}(s) + P_{zu}(s)K_s(s)(I - P_{yu}(s)K_s(s))^{-1}P_{yw}(s) \tag{11}$$

To determine robustness prerequisites, an additional graph is needed, as shown in Figure 5:

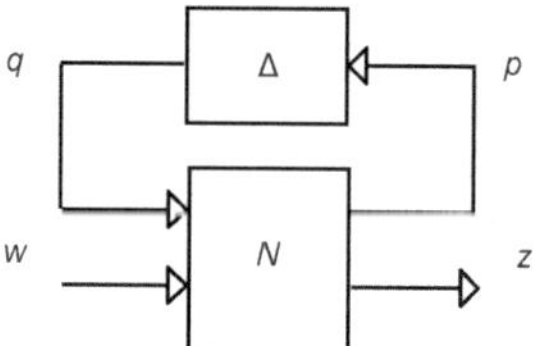

Figure 5. Uncertainty modeling N-Δ structure.

where the N factor is defined by Equation (11) and the uncertainty parameter, which is modeled in Δ, should satisfy the following criterion $||\Delta||\infty \leq 1$ (details later). Where

$$z = \mathcal{F}_u(N, \Delta)w = [N_{22} + N_{21}\Delta(I - N_{11}\Delta)^{-1}N_{12}]w = Fw \tag{12}$$

We can state the following definitions based on this structure, shown in Table 2:

Table 2. Definitions.

Nominal stability (NS) $\Leftrightarrow$	N internally stable
Nominal performance (NP) $\Leftrightarrow$	$||N_{22}(j\omega)||_\infty < 1$, $\forall\omega$ <u>and</u> NS
Robust stability (RS) $\Leftrightarrow$	$F = \Phi u(N, \Delta)$ stable $\forall\Delta$, $||\Delta||_\infty < 1$ <u>and</u> NS
Robust performance (RP) $\Leftrightarrow$	$||F||_\infty < 1$, $\forall\Delta$, $||\Delta||_\infty < 1$ <u>and</u> NS

The following conditions are demonstrated to be true for real or complex block-diagonal perturbations Δ:

I. If M is internally stable, the system is presumably stable;
II. If the system performs about average;
III. If and only if, the system (M, Δ) is robustly stable,

$$\sup_{\omega\in\mathbb{R}}\mu_\Delta(N_{11}(j\omega)) < 1 \tag{13}$$

where the structured singular value of N is the parameter μ_Δ in the criterion, for the structured uncertainty set Δ. This condition is known as the generalized small gain theorem [12–14].

IV. The system (N, Δ) exhibits robust performance if and only if,

$$\sup_{\omega\in\mathbb{R}}\mu_{\Delta_a}(N(j\omega)) < 1 \tag{14}$$

where

$$\Delta_a = \begin{bmatrix} \Delta_p & 0 \\ 0 & \Delta \end{bmatrix} \tag{15}$$

and Δ_p is fully complex and has the same structure as Δ and dimensions corresponding to (w, z). Unfortunately, only bounds on μ can be estimated [19,20].

2.4. Controller Synthesis

All the aforementioned provide solutions to analytical problems and methods for evaluating and contrasting controller performance. A controller that provides a specific performance in terms of the structured singular value may be calculated [12,13].

This is the so-called $(D, G\text{-}K)$ iteration [9], in which finding a μ-optimal controller K_s such that $\mu(\Phi_u(F(j\omega)), K_s(j\omega)) \leq \beta, \forall \omega$, is transformed into the problem of finding transfer function matrices $D(\omega) \in \forall \Delta$ and $G(\omega) \in \Gamma$, such that,

$$\sup_{\omega} \bar{\sigma} \left[\left(\frac{D(\omega)(F_u(F(j\omega), K_s(j\omega))D^{-1}(\omega)}{\gamma} - jG(\omega) \right) \left(I + G^2(\omega) \right)^{-\frac{1}{2}} \right] \leq 1, \forall \omega \tag{16}$$

Unfortunately, even discovering local maxima is not guaranteed by this approach; however, a technique known as D-K iteration is available for complex perturbations (also implemented in MATLAB) [12,13,16]. It combines H_{inf} synthesis and μ-analysis and often produces positive results. An upper limit on μ in terms of the scaled single value serves as the starting point,

$$\mu(N) \leq \min_{D \in \mathcal{D}} \bar{\sigma}(DND^{-1}) \tag{17}$$

It is aimed to determine the controller, which lowers the peak over frequency of its upper limit,

$$\min_{K} \left(\min_{D \in \Delta} \| DN(K_s)D^{-1} \|_{\infty} \right) \tag{18}$$

by alternating between minimizing $\| DN(K_s)D^{-1} \|_{\infty}$ with respect to either K_s or D (while maintaining the other constant) [9].

3. Results and Discussion

Through the relation, the function $f_m(t)$ was produced from the wind velocity data.

$$f_m(t) = \frac{1}{2}\rho C_u V^2(t) \tag{19}$$

where V = velocity, ρ = density, and $C_u = 1.2$.

On one side of the structure, every node is subjected to periodic sinusoidal loading pressure that simulates a severe wind.

The boundaries on the values in the frequency domain are displayed in Figure 6. This results in a deviation of the mass and stiffness matrices M, and K of about 90% from their nominal values.

As can be seen, the system is still stable and performs robustly, because, for all relevant frequencies, the upper bounds of both values remain below 1.

Additionally, we regulate the structure in the state space domain by varying the nominal values of the matrices A and B, stiffness matrix K, and mass matrix M(rel.4). Account factors are considered, such as nonlinearities and system dynamics that the modeling procedure neglects, an insufficient understanding of disturbances, the disturbances caused by the environment's effect, and the decreased accuracy of system sensor data.

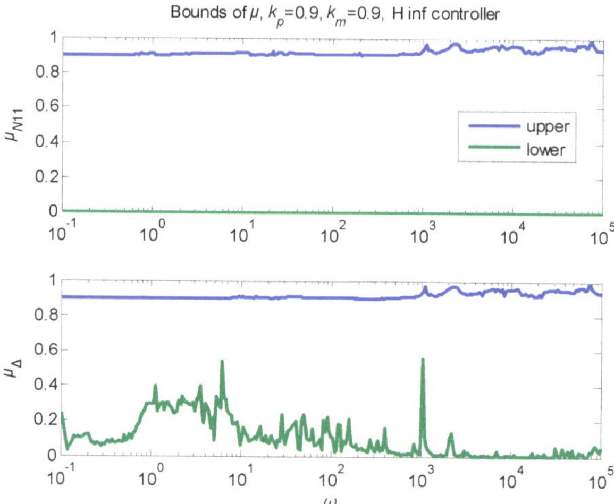

Figure 6. Bounds of the μ value.

The results, as shown in Figure 7, are excellent: oscillations were suppressed even for varying prices of the system's primary matrices A and B; additionally, the oscillations were reduced by differentiating the costs of the mass and stiffness matrices (12) and preserving the piezoelectric components' voltages within their endurance ranges. Figures 7 and 8 show the displacement of the free end of the smart structure when applying $H_{infinity}$ control (close loop with PZT voltages) in the schematic with the blue line. The smart piezoelectric structure almost has no vibrations, and it maintains equilibrium even when the key system matrices (A, B, M, and K) have different prices. In Figure 7 with green, red, light blue and petrol line we can see the displacement of the free end of the beam with different prices of matrices A and B of our system for the open loop that means without PZT voltages. Also in Figure 8 with green, red, light blue and petrol line we can see the displacement of the free end of the beam with different prices of matrices M and K of our system for the open loop that means without PZT voltages. Figures 7 and 8 in the last graph show the changes when the PZT material properties change. The smart piezoelectric structure almost has no vibrations, and it maintains equilibrium even when the key system matrices (A, B, M, and K) have different prices. The initial parameters are the mass, the damping, and the stiffness matrix. In Figures 7 and 8 these parameters change for the open and the closed loop—this means without PZT material and with PZT material. This work focuses on a specific PZT material with its properties shown in Table 1. Figure 7 (last graph) shows the changes when the PZT material properties change.

The discovered $H_{infinity}$ controller is 24 in order. Numerous scientists have proposed algorithms for order reduction as a result of the fact that the order of the controller, which is equal to the order of the system, is substantially higher than the order of conventional controllers such as PI and LQR. The following process will use the most widely used of these algorithms, known as *Hifoo* [21], which has been implemented in the Matlab environment. The main issue is to calculate a reduced-order n < 24 controller that preserves the performance of the $H_{infinity}$ criterion and the behavior of a full-order controller of the given system. As a mechanical input to this controller, 10 KN is taken at the free end of the structure. In Figure 9 we can see the beam-free end displacement with and without control, using *Hifoo* recovery time 0.05 sec (0.03 with $H_{infinity}$), the steady-state error of the order of 10^{-5} m (10^{-6} with $H_{infinity}$) maximum elevation 2.1×10^{-4} (0.3×10^{-4} with $H_{infinity}$) and vibration suppression at 90% (98% with $H_{infinity}$). In Figure 10, we can see the voltages within the piezoelectric limits of 30 volts.

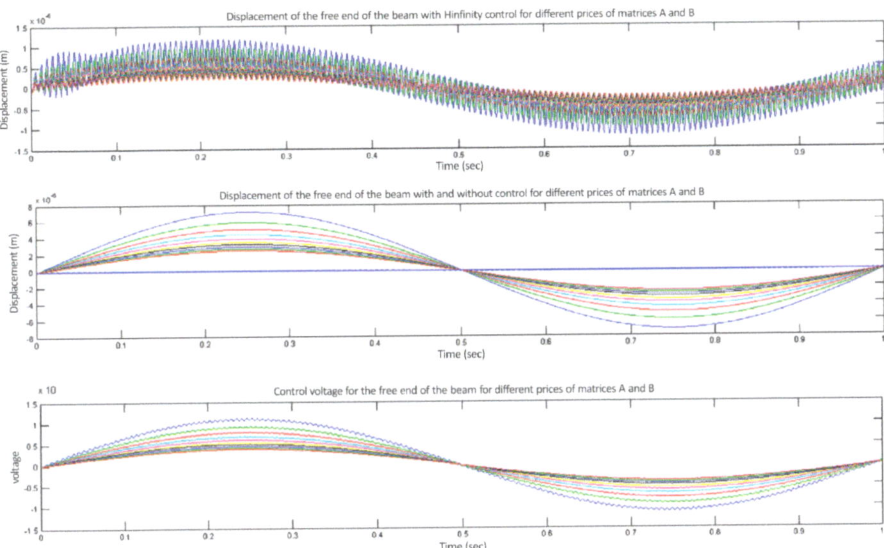

Figure 7. Results for matrices A and B with and without H_{inf}, controlling for sinusoidal external inputs, at various prices.

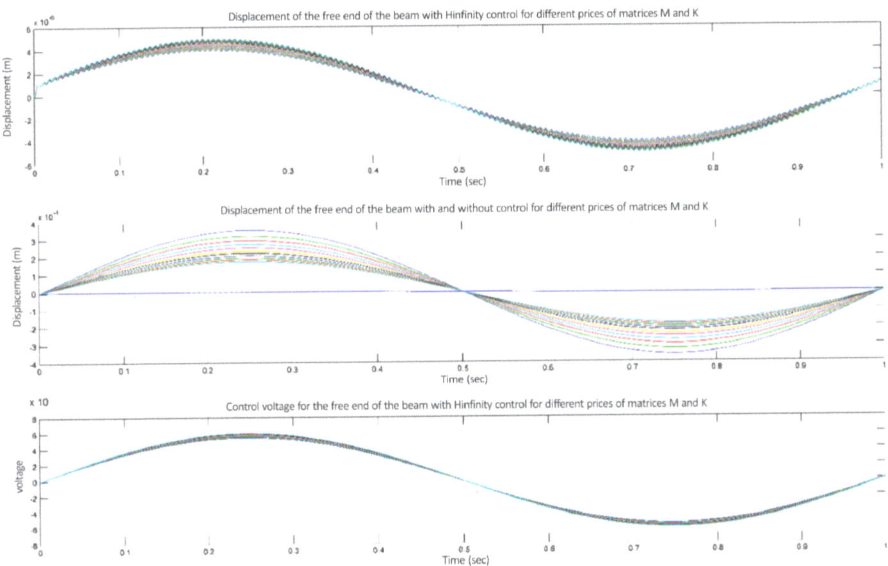

Figure 8. Results for matrices M (mass) and K (stiffness) at various costs, both with and without H_{inf} control of the external sinusoidal inputs.

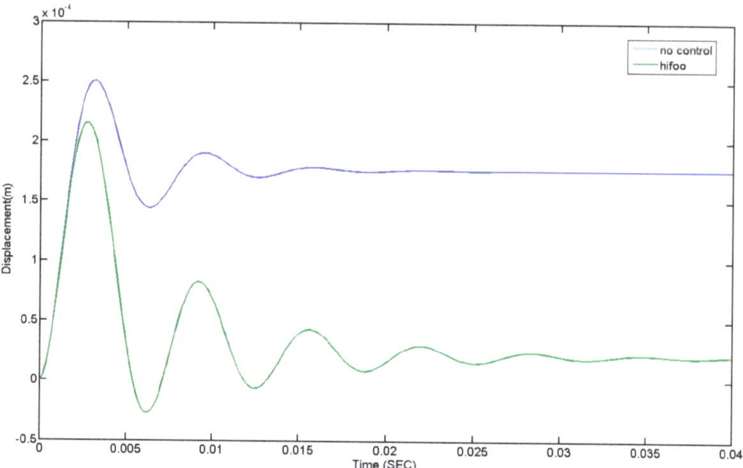

Figure 9. Displacement of the free end of the structure with and without *Hifoo*.

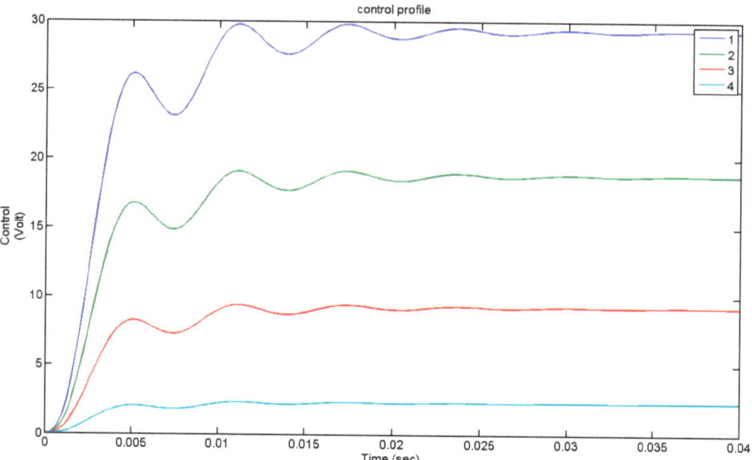

Figure 10. Piezoelectric voltages with reduced order controller for the first four nodes.

The frequency response of the weighting function and matching model is shown in Figure 11. The graph of the function remains below unity so the controller archives robust performance for the given data.

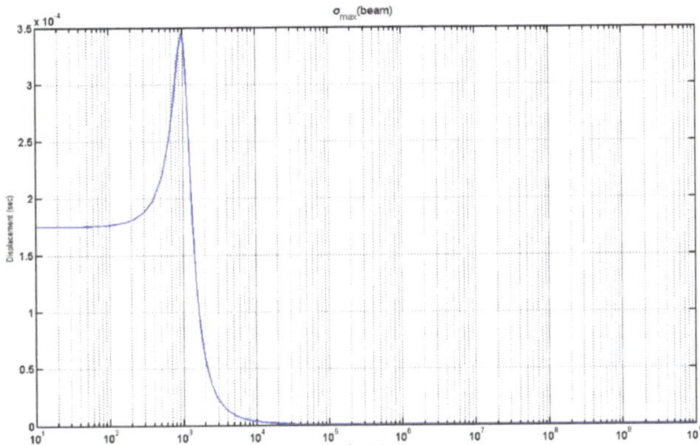

Figure 11. The robust performance criterion measure.

4. Conclusions

The ability of piezoelectric materials to directly transform mechanical energy into electrical energy and vice versa has made them the most desirable functional materials for sensors and actuators in smart constructions. They exhibit outstanding frequency responsiveness and electromechanical coupling properties. In this study, we include active vibration suppression and robust control in the dynamics of a clever piezoelectric system. By using the established vibration control methods on a clever piezoelectric construction, numerical evaluations are performed and analyzed in order to confirm the efficiency of the method. We include modeling uncertainties by accounting for the nonlinearity of the system that was not taken into account in the model, our incomplete understanding of the model's values and parameters, and their physiological fluctuations during the duration of the structures' operation. An $H_{infinity}$-based controller is designed to suppress the vibration of the smart piezoelectric structure under dynamical loading. The robustness of the $H_{infinity}$ controller to parametric uncertainty in vibration suppuration problems is shown. The benefit of robust control and active vibration suppression in the dynamics of smart structures is amply illustrated by this work. $H_{infinity}$ control has certain advantages for the analysis of robust control systems. Unfortunately, relatively complicated modeling and resulting controllers lead to restricted practical applications. These drawbacks will be gradually eliminated due to the availability of cheaper and more powerful electronic components for control implementation. Future research will be focused on experimental verification in this direction.

Author Contributions: G.E.S., methodology; A.M., and M.P., software, writing—review, and editing; N.V., validation; M.P., formal analysis; A.P., investigation, and software. All authors have read and agreed to the published version of the manuscript.

Funding: This research received no external funding.

Data Availability Statement: Not applicable.

Acknowledgments: The authors are grateful for the support from the Hellenic Mediterranean University and the Technical University of Crete.

Conflicts of Interest: The authors declare no conflict of interest.

References

1. Bandyopadhyay, B.; Manjunath, T.C.; Unapathy, M. *Modeling, Control, and Implementation of Smart Structures*; Springer: Berlin, Heidelberg, Germany, 2007; ISBN 10 3-540-48393-4.
2. Burke, J.V.; Henron, D.; Kewis, A.S.; Overton, M.L. Stabilization via Nonsmooth, Non-convex Optimization. *IEEE Trans. Autom. Control* **2006**, *5*, 1760–1769.
3. Doyle, J.C.; Glover, K.; Khargoneker, P.; Francis, B. State space solutions to standard H_2 and H∞ control problems. In Proceedings of the 1988 American Control Conference, Atlanta, GA, USA, 15–17 June 1988; Volume 34, pp. 831–847.
4. Francis, B.A. *A Course on H∞ Control Theory*; Springer: Berlin, Heidelberg, Germany, 1987.
5. Friedman, J.; Kosmatka, K. An improved two node Timoshenko beam finite element. *J. Comput. Struct.* **1993**, *47*, 473–481.
6. Kimura, H. Robust stability for a class of transfer functions. *IEEE Trans. Autom. Control* **1984**, *29*, 788–793.
7. Miara, B.; Stavroulakis, G.; Valente, V. *Topics on Mathematics for Smart Systems, Proceedings of the European Conference, Rome, Italy, 26–28 October 2006*; World Scientific Publishers: Singapore, 2007.
8. Moutsopoulou, A.; Pouliezos, A.; Stavroulakis, G.E. *Modelling with Uncertainty and Robust Control of Smart Beams. Paper 35, Proceedings of the Ninth International Conference on Computational Structures Technology, Athens, Greece, 2–5 September 2008*; Topping, B.H.V., Papa-drakakis, M., Eds.; Civil Comp Press: Edinburgh, UK, 2008.
9. Yang, S.M.; Lee, Y.J. Optimization of non collocated sensor, actuator location and feedback gain and control systems. *Smart Mater. Struct. J.* **1993**, *8*, 96–102.
10. Chandrashekara, K.; Varadarajan, S. Adaptive shape control of composite beams with piezoelectric actuators. *Intell. Mater. Syst. Struct.* **1997**, *8*, 112–124.
11. Lim, Y.H.; Gopinathan, V.S.; Varadhan, V.V.; Varadan, K.V. Finite element simulation of smart structures using an optimal output feedback controller for vibration and noise control. *Int. J. Smart Mater. Struct.* **1999**, *8*, 324–337.
12. Zhang, N.; Kirpitchenko, I. Modelling dynamics of a continuous structure with a piezoelectric sensor/actuator for passive structural control. *J. Sound Vib.* **2002**, *249*, 251–261.
13. Zhang, X.; Shao, C.; Li, S.; Xu, D. Robust H∞ vibration control for flexible linkage mechanism systems with piezoelectric sensors and actuators. *J. Sound Vib.* **2001**, *243*, 145–155. [CrossRef]
14. Kwakernaak, H. Robust control and H∞ optimization. *Tutor. Hper JFAC Autom.* **1993**, *29*, 255–273.
15. Stavroulakis, G.E.; Foutsitzi, G.; Hadjigeorgiou, E.; Marinova, D.; Baniotopoulos, C.C. Design and robust optimal control of smart beams with application on vibrations suppression. *Adv. Eng. Softw.* **2005**, *36*, 806–813. [CrossRef]
16. Packard, A.; Doyle, J.; Balas, G. Linear, multivariable robust control with a μ perspective. *ASME J. Dyn. Syst. Meas. Control 50th Anniv. Issue* **1993**, *115*, 310–319. [CrossRef]
17. Tiersten, H.F. *Linear Piezoelectric Plate Vibrations*; Plenum Press: New York, NY, USA, 1969.
18. Zames, G. Feedback minimax sensitivity and optimal robustness. *IEEE Trans. Autom. Control* **1993**, *28*, 585–601. [CrossRef]
19. Chen, B.M. *Robust and H_{inf} Control*; Springer: London, UK, 2000.
20. Iorga, L.; Baruh, H.; Ursu, I. A Review of H∞ Robust Control of Piezoelectric Smart Structures. *ASME. Appl. Mech. Rev.* **2008**, *61*, 040802. [CrossRef]
21. Millston, M. HIFOO 1.5: Structured Control of Linear Systems with a Non-Trivial Feedthrough. Master's Thesis, Courant Institute of Mathematical Sciences, New York University, New York, NY, USA, 2006.

 *inventions*

Article

Compromised Vibration Isolator of Electric Power Generator Considering Self-Excitation and Basement Input

Young Whan Park, Tae-Wan Kim and Chan-Jung Kim *

School of Mechanical Engineering, Pukyong National University, Busan 48513, Republic of Korea
* Correspondence: cjkim@pknu.ac.kr; Tel.: +82-51-629-6169

Abstract: A previous study proposed an optimal vibration isolator for self-excitation, but the solution results showed a critical drawback for the basement input. Because the plant system is exposed to self-excitation and basement input, the vibration isolator characteristics must meet all the requirements of both excitation cases. Two performance indices of the vibration isolator were introduced to evaluate the vibration control capability over two excitation cases, self-excitation and basement input, using the theoretical linear model of the electric power generator. The compromise strategy was devoted to enhancing the vibration control capability over the basement input, owing to the acceptable margin for self-excitation. The modification of the mechanical properties of the vibration isolator focused on the isolator between the mass block and the surrounding building. Simulation results revealed that an increase in the spring coefficient and a decrease in the damping coefficient of the vibration isolator beneath the mass block could enhance the vibration reduction capability over the basement input.

Keywords: vibration isolator; compromise strategy; theoretical linear model; performance index; electric power generator; basement input; self-excitation

1. Introduction

A vibration isolator is necessary to control the vibration transmissibility between two connected systems. The mechanical property of the vibration isolator is the key design parameter for determining the vibration transfer from one system to another. The passive-type vibration isolator is widespread in industries owing to its low cost and simple installation. Therefore, passive-type vibration isolators are utilized in many applications to control the transmission of harsh excitations to protect supporting systems or preserve quiet environments [1–6]. The active-type mount device also performs well in many applications [7–10]. However, a proper control algorithm should be integrated into the mechanical–electronic system after rigorous identification of the supporting system [11–16]. In addition, the high cost of installation or maintenance is a critical hurdle for the increase in noise and vibration engineering market share.

The mechanism of the vibration isolator plays an essential role in fulfilling the objectives of vibration control over a complicated supporting system. Certain nonlinear factors may prevent the controllability of the supporting system, so novel mount design strategies are used to prevent technical problems [17–19]. A recent study proposed a simplified mount system that included a mass block adjacent to the vibration isolator and applied it to a power plant system [20]. The simplified vibration isolator was compared to the original multilayered isolator, including the mass block, using the measured response accelerations during full-payload operation. A response index was proposed and evaluated by varying the damping coefficient, and the feasibility of the proposed simple vibration isolator was discussed. However, the optimal condition of the vibration isolator was only effective for controlling the excitation from the power plant system and was ineffective for excitation generated from the basement location. The vibration isolator should be equally effective for both excitation cases, the power plant system, and the surrounding building.

Citation: Park, Y.W.; Kim, T.-W.; Kim, C.-J. Compromised Vibration Isolator of Electric Power Generator Considering Self-Excitation and Basement Input. *Inventions* **2023**, *8*, 40. https://doi.org/10.3390/inventions8010040

Academic Editor: Om P. Malik

Received: 15 December 2022
Revised: 19 January 2023
Accepted: 1 February 2023
Published: 2 February 2023

This study selected an electric power generator that uses a combustion engine when an unusual power shortage occurs in a building as the target support system. The electric power generator was manufactured by the DAEHUNG Electric Machinery Company in South Korea, with a maximum capacity of up to 75 kW. The specific production was the same as described in a previous study [20]. Figure 1 shows the electric power generator.

Figure 1. Image of the electric power generator [20].

The combustion engine produces indispensable excitation during electric power generation at a constant rotational speed of 1800 rev/min. Six vibration isolators are used to support the electric power generator to prevent excitation transmission into the surrounding building, as shown in Figure 2.

Figure 2. Image of vibration isolator [20].

The present vibration isolator showed reasonable efficiency in vibration control over self-excitation during operation, but the complex structure prevented quick maintenance service. To overcome this disadvantage, a previous study proposed a novel and simple mount structure without a mass block and compared it using the response index derived from the transmissibility formula for the basement response [20]. However, neither vibration isolator could evaluate any vibration-control capability under basement excitation. If an unexpected basement input is assigned, the supported electric power generator may fail to achieve structural stability owing to external perturbations from the basement. To overcome these shortcomings of the original vibration isolator, a compromise strategy was applied in this study to select mechanical coefficients in the supporting isolators. This study established a theoretical linear model of the electric power generator to simulate the dynamic response or frequency response function (FRF) under several excitation conditions. Two performance indices were derived from the transmissibility between the two responses at different locations and under different excitation conditions. The first performance index represented the capability of the vibration isolator over self-excitation under operation between 30 Hz and 120 Hz, and the second index denoted the vibration control performance of the isolator over the basement input. The dynamic analysis of the electric power generator was focused on the modification of the mechanical properties, both the spring and damper coefficients, of the vibration isolator beneath the mass block. The compromise strategy was to reduce the second performance index while allowing an increase in the first performance index for a frequency range between 1 Hz and 120 Hz. However, because the proposed compromise strategy is only valid for the vibration isolator located at the bottom of the mass block, it may not be valid if the mechanism of the vibration isolators or supported plant system is changed.

2. Theoretical Linear Model of Supported Electric Power Generator

The capability of a vibration isolator can be identified from the dynamic response of the supporting system or attached basements. The performance of a supporting vibration isolator is possible with transmissibility between interesting responses using a theoretical model of an electric power generator. The supported electric power generator (M_p) is supported by a vibration isolator, which is attached to a mass block (M_M), as illustrated in Figure 3. The vibration isolator designed to support the power generator was modeled as linear mechanical elements, and both the stiffness coefficient K_p and damping coefficient C_p, and the connection between the mass block (M_M) and the surrounding building (M_B), were represented by the linear elements, the stiffness (K_M), and the damper (C_M). The responses of the three masses were defined as X_p, X_M, and X_B for the supporting power generator, mass block, and surrounding building, respectively, and X_B could be moved with the virtual spring coefficient (K_B) connected to the ground. The excitation forces from the supporting power plant and surrounding building were defined as F_p and F_B, respectively.

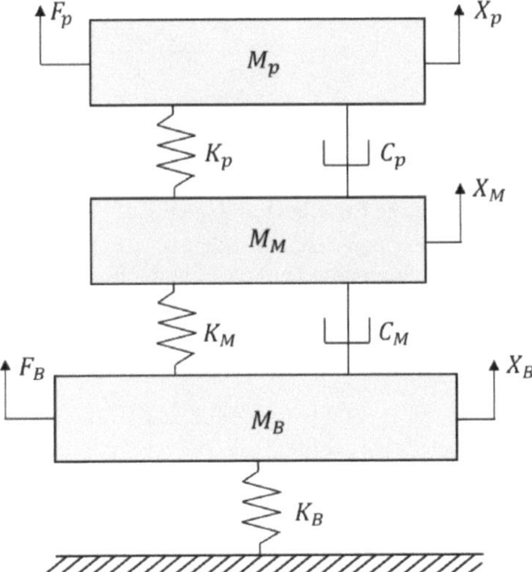

Figure 3. Equivalent electric power generator model supported by vibration isolators.

The dynamics of the supported power generator in Figure 3 can be simulated with theoretical modeling under the linear formulation of connection elements, coefficients of stiffness proportional to the displacement, and a viscous damping coefficient proportional to the velocity. All inertia terms, i.e., the power plant, mass block, and surrounding building, were assumed to be concentrated masses because the theoretical system model was focused on evaluating the vibration isolator alone. The governing equation for each concentrated mass is expressed as:

$$M_p\ddot{X}_p + C_p\dot{X}_p + K_pX_p = C_p\dot{X}_M + K_pX_M + F_p, \tag{1}$$

$$M_M\ddot{X}_M + (C_p + C_M)\dot{X}_M + (K_p + K_M)X_M = C_p\dot{X}_p + K_pX_p + C_M\dot{X}_B + K_MX_B, \tag{2}$$

$$M_B\ddot{X}_B + C_M\dot{X}_B + (K_M + K_B)X_B = C_M\dot{X}_M + K_MX_M + F_B \tag{3}$$

These equations can be expressed using the Laplace transformation in the s-domain to solve the responses under no initial values (both displacements and velocities), as shown in Equations (4)–(6).

$$\left[s^2M_p + sC_p + K_p\right]X_p(s) = \left[sC_p + K_p\right]X_M(s) + F_p(s), \tag{4}$$

$$\left[s^2M_M + s(C_p + C_M) + (K_p + K_M)\right]X_M(s)$$
$$= \left[sC_p + K_p\right]X_M(s) + \left[sC_p + K_p\right]X_M(s), \tag{5}$$

$$\left[s^2M_B + sC_M + (K_M + K_B)\right]X_B(s) = sC_MX_M(s) + K_MX_M + F_B(s) \tag{6}$$

The governing equations can be simplified using the three temporary terms in Equations (7a)–(7c), and the FRF of the supporting power generator can be derived from Equation (8) over the power plant input $F_p(F_B = 0)$ and surrounding building response X_B. In addition, the FRF between the excitation F_B and the response at the power plant X_B can be derived, as shown in Equation (9).

$$\alpha(s) = s^2M_p + sC_p + K_p, \tag{7a}$$

$$\beta(s) = s^2M_M + s(C_p + C_M) + (K_p + K_M), \tag{7b}$$

$$\gamma(s) = s^2M_B + sC_M + (K_M + K_B), \tag{7c}$$

$$\frac{X_B(s)}{F_p(s)} = H_{X_B} = \frac{(sC_M + K_M)(sC_p + K_p)}{\alpha(s)\beta(s)\gamma(s) - \alpha(s)(sC_M + K_M)^2 - \gamma(s)(sC_p + K_p)^2}, \tag{8}$$

$$\frac{X_p(s)}{F_p(s)} = H_{X_p} = \frac{\beta(s)\gamma(s) - (sC_M + K_M)^2}{\alpha(s)\beta(s)\gamma(s) - \alpha(s)(sC_M + K_M)^2 - \gamma(s)(sC_p + K_p)^2} \tag{9}$$

3. Evaluation Indices for Vibration Isolator

If the electric power generator is operated under the rated operational condition, self-excitation can be represented by F_p under no basement force ($F_M = 0$). Two vibration isolators can passively control self-excitation and the transmissibility between X_B and X_p can be formulated into the first performance index (I_1) using Equations (8) and (9). Here, I_1 denotes the vibration transmissibility from the power plant to the surrounding building. This means that the first performance index in Equation (10) represents the vibration isolator's capability to control the power generator's excitation. The smaller the first index value, the better the self-excitation performance of the vibration isolator.

$$I_1(s) = \frac{X_B(s)}{X_p(s)} = \frac{H_{X_B}}{H_{X_p}} = \frac{(sC_M + K_M)(sC_p + K_p)}{\beta(s)\gamma(s) - (sC_M + K_M)^2} \tag{10}$$

The performance of the vibration isolator over the basement input can be derived for the no external force condition ($F_M = F_p = 0$) because the basement input was assigned as the displacement or velocity of basements [21,22]. Under the no external force condition, the ratio between X_p and X_M can be derived using Equations (4) and (6), as shown in Equation (11). The second performance index (I_2) represents the vibration reduction capability of the vibration isolator under basement input conditions. The smaller the second index value, the better the vibration isolator performance for the basement input.

$$I_2(s) = \frac{X_p(s)}{X_B(s)} = \frac{\gamma(s)(sC_p + K_p)}{\alpha(s)(sC_M + K_M)} \tag{11}$$

4. Dynamic Simulation of Supported Power Plant Model

The simulation model of the power generator was studied for the same specifications considered in a previous study [20]. The theoretical model of the power generator is used for the linear formula in Figure 3, and the mechanical properties of the vibration isolator, C_p and K_p, were obtained experimentally using a test machine (835 model/MTS systems, Eden Prairie, Minnesota, United States). Other linear connecting elements were tuned using the response accelerations during full-load operation of the min constant 1800 rev/ internal combustion engine. The parameters defined in Table 1 are verified based on experimental data from a previous study [20].

Table 1. Specification of theoretical power plant model [20].

Variable	Value
M_p(kg)	6070
M_M (kg)	6900
M_B (kg)	$10 \times M_p$
K_p (kN/m)	940 (1 Hz), 1050 (30 Hz), 1245 (60 Hz), 1881 (90 Hz), 4399 (120 Hz)
K_M (kN/m)	$(5 \times 10^6) \times K_p$
K_B (kN/m)	$10^{-1} \times K_p$
C_p (Nsec/m)	603 (1 Hz), 376 (30 Hz), 216 (60 Hz), 184 (90 Hz), 158 (120 Hz)
C_M (Nsec/m)	$(1.5 \times 10^6) * C_p$

The dynamics of the theoretical power plant model can be evaluated via the FRFs formulated using Equations (8) and (9). The frequency of interest was set between 1 Hz and 120 Hz under the operational speed of the combustion engine, and the simulation result is plotted in Figure 4.

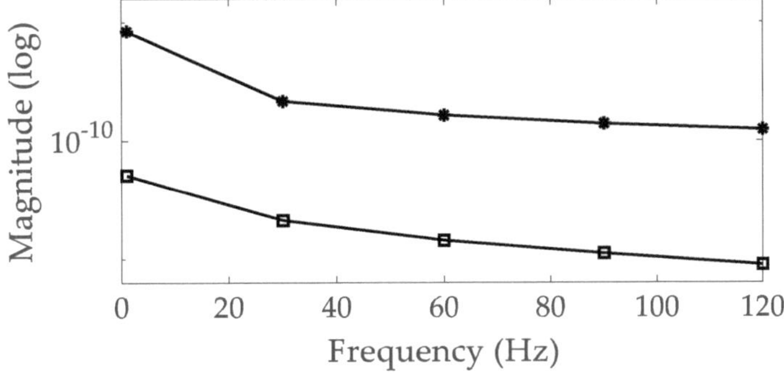

Figure 4. FRF of the theoretical power plant model. ⊟ : H_{X_p}, ✳ : H_{X_B}.

The FRFs showed that the supported electric power generator model had no resonance frequencies for the frequency range of interest, so the original mechanical coefficients of the vibration isolator, C_p, K_p, C_M, and K_M, were suitably selected. The performance of the vibration isolator is also evaluated using the first and second performance indices, plotted in Figures 5 and 6, respectively. Both indices indicated that the transmissibility from the power plant to the surrounding building was very small. In contrast, the transmissibility from the mass block to the power plant was relatively high for the frequency range of interest. Therefore, the controllability of the vibration isolator was reliable when excitation was induced from the supporting power plant side. However, the capability of the vibration isolator to control the excitation from the basement side is not acceptable under the original mechanical property conditions (see Table 1).

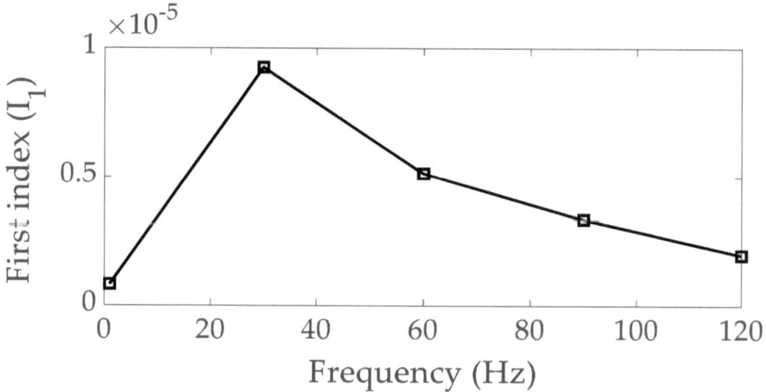

Figure 5. Variation in the first performance index as a function of frequency.

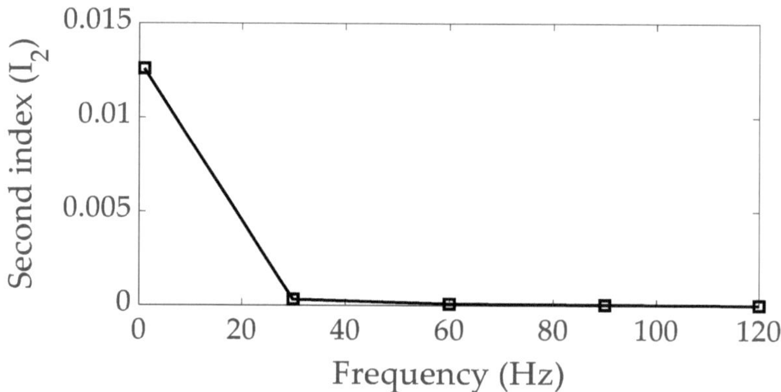

Figure 6. Variation in the second performance index as a function of frequency.

To overcome the poor capability of vibration control over the vibration input from the mass block, revised mechanical properties of the vibration isolator were required for the power plant model. In particular, the mechanical properties of the two vibration isolators cannot be arbitrarily modified from a physical point of view because it is challenging to achieve most mechanical coefficients using isolator specimens, springs, or dampers. Therefore, the modification of coefficients in the vibration isolator should be increased or decreased in proportion to the measured dynamic stiffness and damping coefficient.

Between the two vibration isolators, the vibration isolator beneath the mass block (C_M, K_M) was selected to evaluate the two performance indices in different modification coefficient situations. The vibration isolator (C_p, K_p) to the electric power generator was not considered because the effect of the mass block and the related adjacent joint was considered in a previous study [20] and owing to the high cost of installation or maintenance of the electric power generator. Four vibration isolator cases are selected from the original values, as summarized in Table 2.

Table 2. Modified mechanical properties of vibration isolator.

Case	Value
I	$C_M \div 10$, $K_M \div 10$
II	$C_M \div 10$, $K_M \times 10$
III	$C_M \times 10$, $K_M \div 10$
IV	$C_M \times 10$, $K_M \times 10$

Two indices of the theoretical power generator model were calculated for the selected frequency range in all four cases. The variation in each performance index was calculated as the ratio of the modified index to the original value. The detailed calculation results are summarized in Tables 3 and 4.

Table 3. Ratio of the first performance index for four vibration isolator cases.

Case	Ratio of the First Performance Index (I_1)				
	1 (Hz)	30 (Hz)	60 (Hz)	90 (Hz)	120 (Hz)
I	-	3.2	1.6	1.4	1.6
II	-	3.2	1.7	1.5	1.6
III	-	3.2	1.6	1.4	1.6
IV	-	3.2	1.6	1.4	1.6

Table 4. Ratio of the second performance index for four vibration isolator cases.

Case	Ratio of the Second Performance Index (I_2)				
	1 (Hz)	30 (Hz)	60 (Hz)	90 (Hz)	120 (Hz)
I	1.3	1.0	1.2	1.5	2.0
II	1.3	0.9	0.9	0.8	0.8
III	1.3	1.0	1.0	1.0	1.0
IV	1.3	1.0	1.0	1.0	1.0

The two performance indices exhibited different values in the frequency range of interest by varying the element coefficients. Because the two ratios of performance indicators show the amount of change compared to the reference vibration isolator condition, it can be concluded that the smaller the value, the better the vibration isolator performance under the applied stiffness or damping coefficients. Conversely, when the ratio of index value is greater than 1, it can be determined that the vibration isolator performance is degraded compared to the reference condition. The index ratio at 1 Hz was not considered because self-excitation from the electric power generator was observed only between 30 Hz and 120 Hz. The optimal result is Case II, where the damping coefficient decreases and the stiffness coefficient increases. The other three cases showed similar results for the two performance indices; therefore, the combination of an increase and a decrease in the mechanical coefficients is essential for vibration control from the vibration isolators.

5. Compromise Strategy of Vibration Isolator

The selected target power generator has been reported to have more than ten times the margin at the basement response under full-load operation [20]. However, the capability of vibration control over the basement input was not guaranteed at all. The objective vibration isolator was preliminarily selected for the isolator (C_M, K_M) located beneath the mass block, and the upper vibration isolator (C_p, K_p) was not changed. The simulation results revealed that the best condition for the vibration isolator was an increase in the spring coefficient and a decrease in the damping coefficient. For a ten-fold increase in the spring coefficient and a ten-fold decrease in the damping coefficient, the second performance index was reduced by up to 20% in the high-frequency range; the first performance index was increased up to 320% at 30 Hz. Because the first performance index has a large margin of more than ten times that of the original vibration isolator equipment, the second performance index can be effectively decreased by following the modification guideline with an increase in the stiffness coefficient as well as a decrease in the damping coefficient at the lower vibration isolator. Therefore, it can enhance the vibration control capability of the vibration isolator over a basement input, while decreasing the vibration control performance for self-excitation of the electric power generator.

However, this compromise strategy has some drawbacks. First, the second performance index increased at 1 Hz, which may coincide with the critical frequency for earthquake events. Second, the theoretical linear model was not validated for a low-frequency range of less than 30 Hz. Third, the experiments did not fully verify the feasibility of the design modification of the vibration isolator for Case II. Additional performance enhancement is possible with the modification of the upper vibration isolator (C_p, K_p) in future work.

6. Conclusions

The compromise strategy regarding the vibration isolator for an electric power generator was investigated via simulation of a theoretical linear model. Two vibration isolators (upper and lower) were modeled as linear connecting elements—the spring and damping coefficients. The supporting targets were simplified as concentrated masses. All mechanical specifications followed the system parameters verified in a previous study, and four cases of coefficient modifications were selected for the lower vibration isolator (C_M, K_M). The simulation results revealed that the best case was derived for the combination of isolator coefficients that increases the stiffness coefficient and decreases the damping coefficient, which is expected to decrease the second performance index by 20%. In the same case, the first performance index was increased to 320% at 30 (Hz), so a compromise strategy should be applied to this situation. A previous study verified that the acceptable response margin could be expected to be up to ten times that of the original equipment. Therefore, conditions in Case II were found to be appropriate for the electric power generator.

Author Contributions: Conceptualization, formal analysis, Y.W.P.; software, investigation, T.-W.K.; validation, writing—original draft, visualization, C.-J.K. All authors have read and agreed to the published version of the manuscript.

Funding: This research was supported by the Pukyong National University Development Project Research Fund, 2022.

Data Availability Statement: The data presented in this study are available on request from the corresponding author.

Acknowledgments: This research was supported by Pukyong National University.

Conflicts of Interest: The authors declare no conflict of interest.

References

1. Alujevic, N.; Cakmak, D.; Wolf, H.; Jokic, M. Passive and active vibration isolation system using inerter. *J. Sound Vib.* **2018**, *418*, 163–183. [CrossRef]
2. Siami, A.; Karimi, H.R.; Cigada, A.; Zappa, E.; Sabbioni, E. Parameter optimization of an inerter-based isolator for passive vibration control of Michelangelo's Rondanini Pieta. *Mech. Syst. Signal Process.* **2018**, *98*, 667–683. [CrossRef]
3. Wu, Z.; Jing, X.; Sun, B.; Li, F. A 6DOF passive vibration isolator using X-shape supporting structures. *J. Sound Vib.* **2016**, *380*, 90–111. [CrossRef]
4. Lee, J.; Okwudire, C.E. Reduction of vibrations of passively-isolated ultra-precision manufacturing machines using mode coupling. *Precis. Eng.* **2016**, *43*, 164–177. [CrossRef]
5. Ribeiro, E.A.; Lopes, E.M.O.; Bavastri, C.A. A numerical and experimental study on optimal design of multi-DOF viscoelastic supports for passive vibration control in rotating machinery. *J. Sound Vib.* **2017**, *411*, 346–361. [CrossRef]
6. Oh, H.U.; Lee, K.J.; Jo, M.S. A passive launch and on-orbit vibration isolation system for the spaceborne cryocooler. *Aerosp. Sci. Technol.* **2013**, *28*, 324–331. [CrossRef]
7. Gu, X.; Yu, Y.; Li, J.; Li, Y. Semi-active control of magnetorheological elastomer base isolation system utilizing learning-based inverse model. *J. Sound Vib.* **2017**, *406*, 346–362. [CrossRef]
8. Santos, M.B.; Coelho, H.T.; Neto, F.P.L.; Mafhoud, J. Assessment of semi-active friction dampers. *Mech. Syst. Signal Process.* **2017**, *94*, 33–56. [CrossRef]
9. Oh, H.U.; Choi, Y.J. Enhancement of pointing performance by semi-active variable damping isolator with strategies for attenuating chattering effects. *Sens. Actuators A Phys.* **2011**, *165*, 385–391. [CrossRef]
10. Azadi, M.; Behzadipour, S.; Faulkner, G. Performance analysis of a semi-active mount made by a new variable stiffness spring. *J. Sound Vib.* **2011**, *330*, 2733–2746. [CrossRef]
11. Pingzhang, Z.; Jianbin, D.; Zhenhua, L. Simultaneous topology optimization of supporting structure and loci of isolators in an active vibration isolation system. *Comput. Struct.* **2018**, *194*, 74–85.
12. Beijen, M.A.; Tjepkema, D.; Dijk, J. Two-sensor control in active vibration isolation using hard mounts. *Control Eng. Pract.* **2014**, *26*, 82–90. [CrossRef]
13. Yang, X.L.; Wu, H.T.; Li, Y.; Chen, B. Dynamic isotropic design and decentralized active control of a six-axis vibration isolator via Stewart platform. *Mech. Mach. Theory* **2017**, *117*, 244–252. [CrossRef]
14. Wang, Z.; Mak, C.M. Application of a movable active vibration control system on a floating raft. *J. Sound Vib.* **2018**, *414*, 233–244. [CrossRef]
15. Li, Y.; He, L.; Shuai, C.G.; Wang, C.Y. Improved hybrid isolator with maglev actuator integrated in air spring for active-passive isolation of ship machinery vibration. *J. Sound Vib.* **2017**, *407*, 226–239. [CrossRef]
16. Chi, W.; Cao, D.; Wang, D.; Tang, J.; Nie, Y.; Huang, W. Design and experimental study of a VCM-based stewart parallel mechanism used for active vibration isolation. *Energies* **2015**, *8*, 8001–8019. [CrossRef]
17. Yang, J.; Xiong, Y.P.; Xing, J.T. Vibration power flow and force transmission behavior of a nonlinear isolator mounted on a nonlinear base. *Int. J. Mech. Sci.* **2016**, *115*, 238–252. [CrossRef]
18. Hu, Z.; Zheng, G. A combined dynamic analysis method for geometrically nonlinear vibration isolation with elastic rings. *Mech. Syst. Signal Process.* **2016**, *76*, 634–648. [CrossRef]
19. Yan, L.; Gong, X. Experimental study of vibration isolation characteristics of a geometric anti-spring isolator. *Appl. Sci.* **2017**, *7*, 711. [CrossRef]
20. Kim, C.-J. Design criterion of damper component of passive-type mount module without using base mass-block. *Energies* **2018**, *11*, 1548. [CrossRef]
21. Rao, S.S. *Mechanical Vibration*, 5th ed.; Pearson: Singapore, 2011.
22. Inman, D.J. *Engineering Vibration*, 4th ed.; Pearson: Singapore, 2013.

Article

Organization of Control of the Generalized Power Quality Parameter Using Wald's Sequential Analysis Procedure

Aleksandr Kulikov [1], Pavel Ilyushin [2,3,*], Konstantin Suslov [3,4] and Sergey Filippov [2]

[1] Department of Electroenergetics, Power Supply and Power Electronics, Nizhny Novgorod State Technical University R.E. Alekseev, 603950 Nizhny Novgorod, Russia
[2] Department of Research on the Relationship between Energy and the Economy, Energy Research Institute of the Russian Academy of Sciences, 117186 Moscow, Russia
[3] Department of Hydropower and Renewable Energy, National Research University "Moscow Power Engineering Institute", 111250 Moscow, Russia
[4] Department of Power Supply and Electrical Engineering, Irkutsk National Research Technical University, 664074 Irkutsk, Russia
* Correspondence: ilyushin.pv@mail.ru

Abstract: This paper analyzes the key defining features of modern electric power distribution networks of industrial enterprises. It is shown that the requirements set by industrial enterprises with respect to power quality parameters (PQPs) at the points of their connection to external distribution networks of utilities have been becoming increasingly strict in recent years. This is justified by the high sensitivity of critical electrical loads and distributed generation facilities to distortions of currents and voltages from a pure sine wave. Significant deviations of PQPs lead to significant damage at the consumer end due to the shutdown of electrical equipment by electrical and process protections as a result of overheating and increased wear and tear of individual elements of process lines. This necessitates the implementation of continuous monitoring systems at industrial enterprises, or sampling-based monitoring of PQPs at the boundary bus with an external distribution network. When arranging sampling-based monitoring of PQPs at certain time intervals, only those parameters that are critical for specific electrical loads should be calculated. We provide a rationale for the transition from the monitoring of a set of individual PQPs to a generalized PQP with the arrangement of simultaneous monitoring of several parameters. The joint use of the results of simulation and data from PQP monitoring systems for PQP analysis using the sampling-based procedure produces the desired effect. We present an example of a sequential decision-making process in the analysis of a generalized PQP based on Wald's sequential analysis procedure. This technique makes it possible to adapt the PQP monitoring procedure to the features of a specific power distribution network of an industrial enterprise. We present the structural diagram of the device developed by the authors, which implements the sampling-based monitoring procedure of the generalized PQP. We put forward an approach for determining the average number of sampling data points required to make a decision about the power quality in the implementation of the sequential analysis procedure.

Keywords: modern electric power distribution networks; sampling-based monitoring; generalized power quality parameter; Wald's sequential analysis procedure

Citation: Kulikov, A.; Ilyushin, P.; Suslov, K.; Filippov, S. Organization of Control of the Generalized Power Quality Parameter Using Wald's Sequential Analysis Procedure. *Inventions* **2023**, *8*, 17. https://doi.org/10.3390/inventions8010017

Academic Editor: Om P. Malik

Received: 4 December 2022
Revised: 5 January 2023
Accepted: 9 January 2023
Published: 10 January 2023

1. Introduction

Modern electric power distribution networks (EPDNs) of industrial enterprises are growing increasingly complex due to the integration of various types of generating units (GUs) of distributed generation (DG) facilities, including those based on renewable energy sources (RESs), electricity storage systems (ESSs), and nonlinear loads with power electronic components (e.g., soft starters, variable-frequency drives, uninterruptible power supply units, etc.). The operation of DG and ESS facilities as part of the EPDN makes them active, allowing them to change the mode of electricity consumption in accordance with the signals of the electricity market, which contributes to reducing the cost of power supply [1–3]. The use of RESs—mainly rooftop solar power plants—reduces the carbon footprint by decreasing the value of specific CO_2 emissions per unit of output. The need for this is due to the 2023 introduction of the carbon tax in the European Union, with goods exported to the EU from countries with high levels of atmospheric CO_2 emissions subject to the tax. Thus, the more visible the carbon footprint of an industrial enterprise, the less attractive it will be to consumers and investors.

The main task of the EPDN is to provide a reliable power supply to electrical loads in both steady and transient states [4,5]. However, given the wide adoption of RESs, with their stochastic mode of power generation, as well as in the context of switching high-power electrical loads on/off (as required by specific features of processes), the EPDN records a significant increase in the amplitude of random fluctuations of power flow parameters, with significant deviations of power quality parameters from their standard values [6–8].

The generating units of DG facilities have low values of mechanical inertia constants, as well as low rates of loading when using GUs based on turbocharged internal combustion engines. The rate of electromechanical transients in the EPDN as a result of various disturbances in the island/isolated operating states (e.g., short circuits, significant load surges/shedding, etc.) is 3–10 times higher than in large power systems [9,10].

Simultaneous application of nonlinear loads with power electronic components in the EPDN can lead to prolonged transients and, in some cases, to persistent self-induced oscillations of operating state parameters. When a load of electrical equipment with power electronic components is no more than 30% (for example, uninterruptible power supply units), the total harmonic current distortion (THDi) increases significantly, and the smaller the load, the greater the THDi [11].

The THDi values and the consequences of its growth in the EPDN are as follows:

- THDi < 0.1—normal situation, with no faults in the operation of electrical equipment;
- 0.1 < THDi < 0.5—significant pollution of the EPDN by harmonic components, with the danger of increasing the temperature of electrical equipment, which necessitates the transition to larger cross-sections of cable lines, as well as the use of GUs of DG facilities and backup power supply sources of higher capacity;
- THDi > 0.5—a large degree of pollution of the EPDN by harmonic components, which can lead to failures and shutdowns of electrical equipment due to overheating; requires the installation of filtering and compensation devices.

Significant PQP deviations in the EPDN affect particularly critical electrical loads that are sensitive to deviations from the pure sine wave of currents and voltages, including modern process lines, automatic process control systems, server and network equipment of data centers, etc., causing increased wear of their components and leading to tripping by electrical and process protections [12,13].

The greatest damage to consumers is caused by voltage sags, which are short-term reductions in the RMS voltage on the buses of electrical loads. Some electrical loads are disconnected by protections in case of voltage sags when the RMS value of the voltage drops below 90% of U_{rated} within one or two periods of the utility frequency [14].

Moreover, voltage sags have a negative effect on low- and medium-power GUs of DG facilities, which are disconnected by protection relays with either minimal time delays or no time delays. As a consequence, the balance of generated/consumed active/reactive power in the EPDN is disturbed, which leads to more significant deviations of the operating state parameters and further aggravation of the operating state. Emergency shutdowns of GUs of DG facilities in the EPDN often cause shutdowns of process lines, with corresponding economic damage. Statistical data attest to the fact that emergency disturbances in external distribution networks lead to a cascading development of accidents in the EPDNs of industrial enterprises [15,16].

In Russia, regulation of PQPs at the boundary bus between the utility and the industrial consumer is performed in all states of operation of external distribution networks, except for those containing operating state deviations associated with random events, such as the following:

- Voltage sags (less than 90% of U_{rated} in at least one phase);
- Voltage interruptions (less than 5% of U_{rated} phase voltage in all phases);
- Overvoltages and surge voltages (switching and atmospheric overvoltages).

Hence, the cycles (short circuit, protective relay tripping, operation of automatic reclosing devices or automatic use of backup power, and associated self-starting of motors) are not subject to standardization. At the same time, it is short-term voltage sags and interruptions that cause significant damage at most industrial enterprises [17,18].

In the U.S., the economic losses from power outages due to voltage sags and interruptions for all categories of consumers amount to about USD 79 billion annually. The distribution of damages by consumer type is as follows: commercial—72% (USD 56.8 billion), industrial—26% (USD 20.4 billion), and residential—$\leq$2% (USD 1.5 billion) of total damages. At the same time, two-thirds of the total annual damage—67% (USD 52.3 billion) results from voltage sags due to short circuits in external distribution networks [19].

Under these circumstances, industrial enterprises are forced to install systems of continuous PQP monitoring or sampling-based PQP monitoring in the EPDN to record PQP deviations at the boundary with external distribution networks, in order to make claims to utilities for reimbursement of damages.

The purpose of this article is to justify the expediency of the transition in the EPDN from the monitoring of a set of individual PQPs to a generalized PQP with the arrangement of simultaneous monitoring of several parameters, as well as the effectiveness of the joint use of simulation results and data from PQP monitoring systems to conduct a detailed analysis of PQPs with the use of a sampling-based monitoring procedure.

The systems of continuous monitoring or sampling-based monitoring of PQP implemented by industrial enterprises in the EPDN allow the following tasks to be accomplished:

- To keep the mix and values of regulated PQPs up to date based on periodic review of regulatory requirements and accumulated statistical data, as well as requirements for the efficiency of manufacturing processes and the quality of manufactured products;
- To determine the current operating conditions of the EPDN on-line;
- To carry out continuous automated data collection and processing, analysis, and reporting on PQP deviations from their standard values;
- To create a statistical database on PQPs for information support of the operation of electric power quality management systems;
- To identify critical areas of the EPDN, where deviations of PQPs are the most significant and occur quite often, which requires the arrangement of their continuous monitoring and control;
- To identify (and contain) sources of current and voltage distortions that may cause significant damage, in order to implement measures to change their operating state or power distribution circuit to mitigate the negative impact;
- To make recommendations addressed to duty officers on the implementation of organizational and engineering measures to bring the PQPs to their standard values;
- To determine the most effective ways of load balancing to compensate harmonic components in currents and voltages generated by a nonlinear load with power electronic components;
- To form data packages on PQPs for the system of automatic (automated) control of power quality, allowing the realization of control actions on filtering and compensation devices [20,21].

The introduction of systems of continuous monitoring or sampling-based monitoring of PQPs corresponds to the trend towards making modern EPDNs intelligent, thereby significantly expanding the possibilities of controlling the power flow and power quality, which contributes to improving the reliability of power supply to consumers [22,23].

It is generally believed that the voltage profile in the event of a sag has a rectangular shape, but this is not the case—especially in the EPDNs of industrial enterprises—due to the processes of braking and self-starting of motors after the elimination of a short circuit. In real-world conditions, there is voltage at the terminals of, for example, an induction motor (IM) until the free currents in the rotor have faded and the rotor has stopped. For low-power IMs, the voltage attenuates quickly, but for large IMs it takes a few seconds if there is no static load on the same buses. During a power outage, the static load consumes electricity derived from the kinetic energy of the rotation of the IM rotors, and the losses in the IMs increase proportionally to I^2R. These two factors accelerate the processes of reducing the rotation speed and the values of free currents in the rotors of IMs [24].

Analysis of the results of calculations of transients shows that the duration of self-starting of motors after elimination of a three-phase short circuit is approximately equal to the duration of a short circuit, provided that it does not exceed 0.5–1 s (Figure 1). Therefore, the time during which the voltage is reduced at the terminals of electrical loads is almost twice as long as the duration of the short circuit. When the duration of a short circuit exceeds 1 s (tripping backup protections), self-starting of electric motors becomes either prolonged or completely impossible.

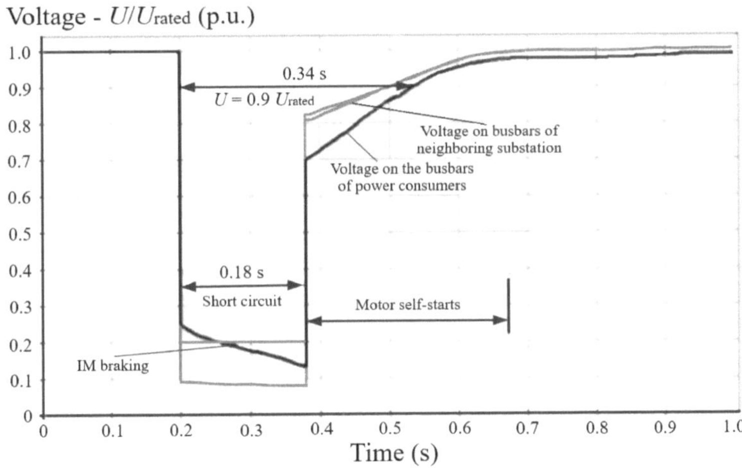

Figure 1. A transient in the EPDN of an industrial enterprise during a near-to-generator three-phase short circuit in the external distribution network.

With a non-rectangular voltage sag shape (Figure 1), the impact on the operation of critical electrical loads can be both underestimated and overestimated. With the different mix of the generating equipment and motors that are switched on, the deviations of the voltage sag shape from the rectangular one may vary. Therefore, the use of conventional methods for estimating voltage sag has its limitations, since the depth and duration of the sags cannot be measured accurately.

Taking into account that the allowable time for decision-making in the EPDNs of industrial enterprises with DG facilities is tens to hundreds of milliseconds, the estimation of voltage parameters at short time intervals by means of digital signal processing has significant errors [25,26]. Consequently, to make decisions about the acceptability of voltage sags for critical electrical loads, and to determine the possible damages, it is advisable to use statistical methods [27].

PQP monitoring systems that are currently in operation and being implemented in EPDNs adopt the following two main approaches:

- Centralized data processing, as a rule, requires a large bandwidth of data transmission channels and significant computing power of the central computing device [28].
- Decentralized data processing requires large sets of circuits in the instruments of PQP monitoring—digital signal processors (DSPs), field-programmable gate arrays (FPGAs), or application-specific integrated circuits (ASICs)—as well as a low-speed communications system [29,30].

To implement the above approaches, algorithms for digital signal processing with a low degree of computational complexity are usually used, allowing the creation of relatively inexpensive equipment for PQP monitoring systems. When creating PQP monitoring systems in the EPDN, both of these approaches are often used in some combination [31,32].

Power quality monitoring systems (PQMSs) use large amounts of various data, so both meters of power quality monitoring and various intelligent electronic devices serve as sources of information for them: phasor measurement units, terminals of digital power system protection, digital loggers of alarm events, digital substation interfacing devices, etc. [33–35].

The PQMS performs the processes of collection, primary processing of data on voltages and currents on the buses and branches of the EPDN, and their transfer to the storage system for analysis and interpretation of data for subsequent decision-making on compliance/non-compliance of PQPs with regulatory requirements (Figure 2).

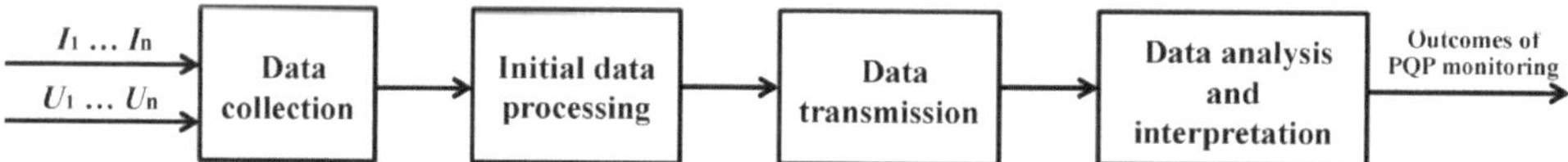

Figure 2. The process of monitoring power quality parameters in the EPDN.

The process of analyzing PQP data was previously performed manually, but the use of digital signal processing techniques and intelligent decision-making techniques has allowed the development and implementation of algorithms for automatic analysis and interpretation of power quality data. The data placed in the storage system can be used to solve the problem of estimating the current PQP values in the EPDN, identifying critical areas of the EPDN—in which PQP deviations are most significant—in order to provide recommendations to duty officers to implement organizational and engineering measures to bring the PQPs to their standard values [36,37].

From the process standpoint, the EPDN should satisfy the electrical loads not only in quantitative terms (i.e., power and amount of electricity), but also in terms of qualitative parameters, e.g., a set of PQPs or a generalized parameter and its variance. The main components of the statistical distribution of the generalized index (i.e., mean value, variance) fully describe both measurable (quantitative) attributes of power quality and non-measurable (alternative) attributes that are qualitative in nature.

2. Materials and Methods

Various distributions of both discrete and continuous random variables are known [38]. Discrete distributions are used to model binary events, relative to which we can conclude that there was or was not a PQP deviation. For example, when arranging PQP monitoring by recording the facts of their deviations from standard values, a random binary event is only the fact of the presence or absence of such deviations. Continuous distributions describe the estimated parameters of currents and voltages associated with the power quality, which are quantitative attributes that can take an arbitrary numerical value in a range of acceptable values [39,40].

When using methods of mathematical statistics in the PQMS, one should distinguish between the tasks related to the distribution of the PQPs and the tasks related to the modeling and application of monitoring techniques. When analyzing PQP distributions, the results of the EPDN's operation are analyzed in terms of random changes in the parameters of currents and voltages. When modeling (i.e., statistically describing) PQP monitoring procedures, the main attention is paid to the mathematical description of the ways of obtaining and processing data to form the current PQPs, along with the issues of efficacy and accuracy of monitoring PQPs, based on which control actions should be implemented in the EPDN in order to bring the PQPs to their standard values. The following distributions of random variables are used when implementing statistical monitoring procedures:

- The hypergeometric distribution determines the numerical value of deviations of PQPs that made it to the aggregate sample, while taking into account the decision-making during monitoring on deviations, as governed by the "acceptable/not acceptable" principle. The hypergeometric distribution from the outset assumes the process of sampling and the execution of the monitoring procedure;
- The binomial distribution is used when estimating the aggregate timed sample of a PQP (a composite parameter) when each instantaneous value has a probabilistic nature and may or may not correspond to the established standard values. The sampling-based monitoring procedure in the analysis of the binary "conform/fails to conform" relationship is modeled by the binomial distribution;
- Poisson distribution can be applied when investigating the distribution of non-conformities, including those of individual PQPs at certain time intervals. The use of this distribution to analyze the results of sampling-based monitoring of PQPs is implemented in order to mathematically simplify the relations of hypergeometric and binomial models of monitoring procedures;
- The normal distribution, as a rule, describes the cumulative result of monitoring with respect to alternative PQPs, as well as modeling the distribution of quantitative PQPs as a source of continuous random variables [27].

It is important to observe the conditions of Mood's theorem, which determines the feasibility of introducing sampling-based monitoring for PQPs, taking into account a given distribution, when implementing the sampling-based monitoring of PQPs.

Suppose that there is some set of N PQP values sampled in certain moments of time. When arranging sampling-based monitoring, a truncated sample with a size of n values is analyzed. Let us introduce a random quantity D characterizing the number of deviations in the aggregate sample of N samples of PQPs, where D takes values in the range $i = 0 \ldots N$. Each numerical value D from the aggregate sample can be assigned a probability $P(D = i)$, where $i \in 0 \ldots N$. Then, there are the expected value $M[D]$ and the variance $\sigma^2[D]$ of the number of identified PQP deviations in the aggregate sample equal to

$$M[D] = \sum_{i=1}^{N} i \times P(D); \; \sigma^2[D] = \sum_{i=1}^{N} (i - M[D])^2 \times P(D).$$

Let us determine the correlation coefficient ρ between the number of deviations d ($d < D$) of PQP in the truncated sample of size n and the number of deviations $(D - d)$ of PQPs in the untested residue.

The correlation coefficient can be one of the following:

- Positive ($\rho > 0$), when $(\sigma^2[D]/\{M[D] \times (1 - M[D]/N)\}) > 1$, or $\sigma^2[D] > \{M[D] \times (1 - M[D]/N)\}$;
- Negative ($\rho < 0$), when $(\sigma^2[D]/\{M[D] \times (1 - M[D]/N)\}) < 1$, or $\sigma^2[D] < \{M[D] \times (1 - M[D]/N)\}$;
- Equal to zero ($\rho = 0$), when $(\sigma^2[D]/\{M[D] \times (1 - M[D]/N)\}) = 1$, or $\sigma^2[D] = \{M[D] \times (1 - M[D]/N)\}$.

If there is no statistical correlation between the number of PQP deviations in the sample and in the untested residue (correlation coefficient $\rho = 0$), or if the correlation is negative ($\rho < 0$), then it is inexpedient to perform sampling-based monitoring, because it fails to provide any additional useful information. Thus, introducing sampling-based monitoring makes sense only when $\rho > 0$ or $\sigma^2[D] > \{M[D] \times (1 - M[D]/N)\}$.

Therefore, statistical analysis and monitoring of power quality should be understood as sampling-based analysis and monitoring of PQPs, as based on the application of methods of mathematical statistics to determine whether they conform to established specifications. In contrast to the manufacturing of industrial products, the results of statistical analysis and monitoring of PQPs in the EPDN can be used not only to determine their compliance at specified intervals with regulatory requirements (i.e., the terms of contractual obligations to supply electricity), but also to implement control actions on electrical equipment to bring the PQPs to their standard values [41,42].

Statistical analysis and monitoring of PQPs can be single-stage, two-stage, multistage, or sequential [43].

In the course of single-step monitoring, the decision on the analyzed PQPs at a given time interval is formed only on the basis of one aggregate sample of these PQPs at a given time interval.

In two-step monitoring, the decision on conformity of PQPs with the regulatory requirements is made based on the results of testing of no more than two aggregate samples, and the selection of the second sample depends on the results of the monitoring of the first one. In this case, the number of identified PQP deviations in the first sample is insufficient for making a decision, and it is made based on the sum of the results of both checks. An advantage of two-step monitoring is that on average it requires less sampling (by 20–30%) than single-step monitoring, but its implementation requires more involved algorithms.

Multistep and sequential monitoring use a number of consecutive samples, and with multistep monitoring the maximum number of samples is limited, whereas with sequential monitoring there are no such constraints. In both cases, the processing of the subsequent sample depends on the results of the previous check. During sequential monitoring, a minimum number of timed samples of PQPs are required to make a decision about their falling within the range of acceptable values. Therefore, PQP monitoring using the sequential analysis procedure is appropriate, especially when high performance of the PQP monitoring system is needed.

3. Results

In the standardization systems of some countries of the world, PQP monitoring is reduced to determining the mix and acceptable ranges of deviations of individual parameters [44,45]. In real-world conditions, there is a complex (integrated) impact of PQPs on the electrical loads of consumers. At the same time, distortions of currents and voltages, as a result of an entire set of PQP deviations that border the area of acceptable values, can cause serious negative consequences and damages for specific consumers.

Therefore, when analyzing power quality, the following should be undertaken:

- To form a generalized PQP, which can be used to estimate the comprehensive impact of a set of deviations of individual PQPs on the operation of electrical loads of a particular consumer;
- To determine the ranges of acceptable deviations of the generalized PQP, within which no damage occurs to specific consumers. This problem can be solved using simulation data for various circuit/operating state situations and the operating conditions of a particular consumer, including in the main maintenance circuits of the external distribution network;
- To develop a procedure for sampling-based monitoring of PQPs on the basis of the generalized PQP for subsequent decision-making on the implementation of organizational and technical measures to bring the generalized PQP into the acceptable range.

Deviations of PQPs at the connection points are subdivided into continuous changes and random events. The latter, as a rule, do not have a significant impact on the operation of the electrical loads of consumers, due to their short duration, and they do not require the implementation of organizational and engineering measures in order to bring the PQPs into the acceptable range.

To estimate the parameters of currents and voltages in various digital devices, including PQP monitoring devices, short time intervals (sliding data window)—for example, constituting one period of the utility frequency—are delineated. The required accuracy of estimation in determining the parameters of distorting harmonic components within such short time intervals cannot be achieved. As a consequence, the results of calculations of some PQPs will not be accurate and will not match the actual situation with distortions of currents and voltages in the EPDN.

Determination of the conformity of PQPs with regulatory requirements should be carried out during control in the process of monitoring of PQPs. Depending on the characteristics of the EPDN, as well as the financial capabilities of industrial enterprises, both continuous and sampling-based monitoring of PQPs can be arranged. With continuous monitoring, measurements and calculations of all PQPs are carried out at every moment of time and at all points of connection of consumers to external distribution networks. Given the economic feasibility factor, such a form of monitoring is unacceptable in most cases. During the sampling-based monitoring, the PQPs are evaluated at individual time intervals and at predetermined monitoring points, with the calculation of only those PQPs that are critical for a particular consumer, taking into account their process-related features.

It is expedient to organize the sampling-based monitoring of PQPs with the use of special sampling procedures of mathematical statistics [43]. In the process of observation within the limited time interval, a conclusion can be formed about the compliance with regulatory requirements within the time period until the next sampling-based monitoring.

As is attested by available experience, it is preferable to arrange the monitoring of PQPs on a quantitative basis. Given such monitoring at the points of connection of electrical loads with respect to the entire set of the calculated PQPs, it is possible to establish the validity of alternative hypotheses of conformity or non-conformity of PQPs with the regulatory requirements [41].

Let us assume that the power quality is characterized by k independent parameters. Then, the result of PQP monitoring is defined by the k-dimensional column vector $x = (x_1, x_2, \ldots, x_j, \ldots, x_k)^T$, in which each component x_j is binary and takes the value of 1—under unacceptable PQP deviation with respect to the j-th index—or 0—under acceptable PQP deviation.

The sampling-based monitoring of the PQP is performed within the interval, including N samples of current (voltage) signals. Let us use m_j $(0 \leq m_j \leq N, j = 1, 2, \ldots, k)$ to denote the number of deviations with respect to the j-th PQP and assign a random k-dimensional vector $m = (m_1, \ldots, m_j, \ldots, m_k)$. Let the component m_j be distributed as governed by the binomial distribution with the parameters n and q_j, where q_j is the probability of the parameter m_j deviating from the acceptable value. Provided that the individual PQPs are independent, the distribution governing the vector m takes the following form:

$$P_n(m) = \prod_{j=1}^{k} C_n^{m_j} \times q_j^{m_j} \cdot (1 - q_j)^{n - m_j}. \tag{1}$$

Estimation of probabilities q_j for a particular EPDN can be obtained from the results of simulation or monitoring of PQPs in various conditions of EPDN operation over a long time interval as related to its circuit and operation.

Let us introduce a generalized (composite) PQP in the form of

$$\xi = \sum_{j=1}^{k} c_j \times m_j \, , \tag{2}$$

where $c = (c_1, \ldots, c_j, \ldots, c_k)^T$ is the column vector of weight coefficients, which determines the ratio of damages from violations of power quality in the event of deviations of individual PQPs.

In [46] it was proposed to group the monitored parameters in the selection of weight coefficients c_j ($j = 1, 2, \ldots, k$) based on the amount of damage associated with the deviation of individual PQPs (a group thereof), with the assignment of the weight c_j. The grouping of PQPs can be carried out using the structured expert judgment technique.

Since in Equation (2) each component of the vector $\boldsymbol{m}$ is distributed as governed by the binomial distribution with the probability independent of n, the random variable ζ, as a linear combination of asymptotically normal quantities m_j ($j = 1, 2, \ldots, k$), also has an asymptotically normal distribution with the expected value m_ζ and variance σ_ζ^2:

$$m_\zeta = n \times \sum_{j=1}^{k} c_j \times q_j; \quad \sigma_\zeta^2 = n \times \sum_{j=1}^{k} c_j \cdot q_j \times \left(1 - q_j\right). \tag{3}$$

The degree to which the distribution ζ approximates the normal distribution depends largely on the vector c and the numerical values of the probabilities q_j.

Let us test the hypothesis that the mean value of the normally distributed PQP value with known variance does not exceed the specified value. Let us assume that ζ is a random PQP value with a time-varying mean value m_ζ and a known variance σ_ζ^2, as determined by the current operating state of the EPDN and the accuracy of the current and voltage measurements by digital signal processing methods. Regarding the chosen generalized PQP ζ, the following statistical problem can be solved: the hypothesis is tested that m_ζ is less than or equal to the specified setpoint value $m_{\zeta\,set}$.

Let there be a set of N consecutive instantaneous sample values of one of the PQPs within the analyzed time interval. It is assumed that the ratio between the analyzed time intervals relative to the sampling interval in the digital processing of current and voltage signals is very large. Deviations of the entire set of PQPs from the regulatory values are recorded by estimating the expected value m_ζ of the random variable ζ. At each point in time, the values of the random variable ζ can generally differ from one another, but the variance of the deviations σ_ζ^2 is a known quantity, and the expected value (mean) m_ζ within the analyzed time interval is unknown.

To illustrate the logic of making a decision based on the generalized PQP, let us stipulate that it is preferable to have a smaller value of m_ζ (e.g., a smaller value of the deviation value from the standard value). Let us specify the setpoint value $m_{\zeta set}$ such that when $m_\zeta < m_{\zeta set}$ the deviations of the generalized PQP will be considered acceptable, and when $m_\zeta > m_{\zeta set}$ the decision on non-compliance of the generalized PQP with the standard value will be made. When $m_\zeta = m_{\zeta set}$, there will be uncertainty in the process of deciding whether or not the PQP conforms to requirements. Furthermore, if m_ζ increases (decreases) in the course of sequential sampling-based monitoring, then the degree of confidence in the power quality for the analyzed EPDN operating state decreases (increases) accordingly.

When performing sequential sampling-based monitoring, values of $m_{\zeta 0}$ and $m_{\zeta 1}$ ($m_{\zeta 0} < m_{\zeta set}$ and $m_{\zeta 1} > m_{\zeta set}$) are specified at which the decision on conformity of the generalized PQP with the standard value is considered in view of damages. If $m_\zeta \leq m_{\zeta 0}$, the wrong decision on the non-compliance of the generalized PQP is characterized by the so-called "supplier [utility] risk", and the decision on the compliance of the generalized PQP if $m_\zeta > m_{\zeta 1}$ is characterized by the "consumer risk". Thus, the area of conformity of the generalized PQP with the standard value is defined by the set of m_ζ values, for which $m_\zeta \leq m_{\zeta 0}$, and the area of non-conformity is defined by the set of m_ζ values, for which $m_\zeta \geq m_{\zeta 1}$. The region for which $m_\zeta \in \left[m_{\zeta 0};\ m_{\zeta 1}\right]$ is the uncertainty region.

The risks inherent in the choices of $m_{\zeta 0}$ and $m_{\zeta 1}$ correspond to the values α and β and are characterized by the probabilities of making wrong decisions. The application of the sequential probability ratio criterion in the implementation of the decision-making procedure leads to the following relations: Let $\zeta_1, \zeta_2, \ldots\ldots, \zeta_m$ be a sequence of instantaneous values of the observed quantity ζ characterizing the power quality. The probability density of the sample $\zeta_1, \zeta_2, \ldots, \zeta_m$, if $m_\zeta = m_{\zeta 0}$ corresponds to the following Equation:

$$p_0(m) = (2\pi\sigma^2)^{-m/2} \times \exp\left\{ -\sum_{i=1}^{m} (\zeta_i - m_{\zeta 0})^2 / (2\sigma^2) \right\}. \tag{4}$$

and if $m_\zeta = m_{\zeta 1}$, to the following Equation:

$$p_1(m) = \left(2\pi\sigma^2\right)^{-m/2} \times \exp\left\{ -\sum_{i=1}^{m} (\zeta_i - m_{\zeta 1})^2 / \left(2\sigma^2\right) \right\}. \tag{5}$$

During the sequential analysis procedure, at each step, the likelihood ratio is calculated, which is defined by the following equality:

$$\eta(m) = \frac{p_1(m)}{p_0(m)} \tag{6}$$

Step-by-step calculations are performed as long as the following conditions are met:

$$B < \eta(m) = \exp\left\{ -\sum_{i=1}^{m} (\zeta_i - m_{\zeta 1})^2 / (2\sigma^2) \right\} / \exp\left\{ -\sum_{i=1}^{m} (\zeta_i - m_{\zeta 0})^2 / (2\sigma^2) \right\} < A. \tag{7}$$

The sequential analysis procedure ends with a decision on the deviation of the generalized PQP from the standard value if

$$\eta(m) = \frac{\exp\left\{ -\sum_{i=1}^{m} \frac{(\zeta_i - m_{\zeta 1})^2}{2\sigma^2} \right\}}{\exp\left\{ -\sum_{i=1}^{m} \frac{(\zeta_i - m_{\zeta 0})^2}{2\sigma^2} \right\}} \geq A. \tag{8}$$

With respect to whether the value of the generalized PQP belongs to the acceptable deviation range, if

$$\eta(m) = \frac{\exp\left\{ -\sum_{i=1}^{m} \frac{(\zeta_i - m_{\zeta 1})^2}{2\sigma^2} \right\}}{\exp\left\{ -\sum_{i=1}^{m} \frac{(\zeta_i - m_{\zeta 0})^2}{2\sigma^2} \right\}} \leq B. \tag{9}$$

The setpoint values of A and B are defined by the following Equations:

$$A = (1 - \beta)/\alpha; \; B = \beta/(1 - \alpha). \tag{10}$$

By taking the logarithm of and transforming Equations (7)–(9), we can obtain

$$\ln \frac{\beta}{1 - \alpha} < \left(\frac{m_{\zeta 1} - m_{\zeta 0}}{\sigma^2} \right) \sum_{i=1}^{m} \zeta_i + \frac{m\left(m_{\zeta 0}^2 - m_{\zeta 1}^2 \right)}{2\sigma^2} < \ln \frac{1 - \beta}{\alpha}, \tag{11}$$

$$\left(\frac{m_{\zeta 1} - m_{\zeta 0}}{\sigma^2} \right) \sum_{i=1}^{m} \zeta_i + \frac{m\left(m_{\zeta 0}^2 - m_{\zeta 1}^2 \right)}{2\sigma^2} \leq \ln \frac{\beta}{1 - \alpha}, \tag{12}$$

$$\left(\frac{m_{\zeta 1} - m_{\zeta 0}}{\sigma^2} \right) \sum_{i=1}^{m} \zeta_i + \frac{m\left(m_{\zeta 0}^2 - m_{\zeta 1}^2 \right)}{2\sigma^2} \geq \ln \frac{1 - \beta}{\alpha}. \tag{13}$$

By adding the summand $m \cdot \left(m_{\zeta 0}^2 - m_{\zeta 1}^2 \right) / (2\sigma^2)$ to both parts of the inequalities and dividing by $(m_{\zeta 1} - m_{\zeta 0})/\sigma^2$, we arrive at the following relations:

$$\left(\frac{\sigma^2}{m_{\xi 1} - m_{\xi 0}}\right) \ln \frac{\beta}{1-\alpha} + \frac{m\left(m_{\xi 0} + m_{\xi 1}\right)}{2} < \sum_{i=1}^{m} \xi_i < \frac{\sigma^2}{m_{\xi 1} - m_{\xi 0}} \ln \frac{1-\beta}{\alpha} + \frac{m\left(m_{\xi 0} + m_{\xi 1}\right)}{2}, \tag{14}$$

$$\sum_{i=1}^{m} \xi_i < \left(\frac{\sigma^2}{m_{\xi 1} - m_{\xi 0}}\right) \ln \frac{\beta}{1-\alpha} + \frac{m\left(m_{\xi 0} + m_{\xi 1}\right)}{2}, \tag{15}$$

$$\sum_{i=1}^{m} \xi_i < \left(\frac{\sigma^2}{m_{\xi 1} - m_{\xi 0}}\right) \ln \frac{1-\beta}{\alpha} + \frac{m\left(m_{\xi 0} + m_{\xi 1}\right)}{2}. \tag{16}$$

In Equalities (14)–(16) make it possible to implement the monitoring of the generalized PQP with the help of "acceptance" numbers. For each step m of the sequential analysis procedure, the "acceptance" number is calculated by the following Equation:

$$a(m) = \left(\frac{\sigma^2}{m_{\xi 1} - m_{\xi 0}}\right) \ln \frac{\beta}{1-\alpha} + \frac{m\left(m_{\xi 0} + m_{\xi 1}\right)}{2}. \tag{17}$$

and the "rejection" number is calculated by the Equation

$$b(m) = \left(\frac{\sigma^2}{m_{\xi 1} - m_{\xi 0}}\right) \ln \frac{1-\beta}{\alpha} + \frac{m\left(m_{\xi 0} + m_{\xi 1}\right)}{2}. \tag{18}$$

The numbers (relations $a(m)$, $b(m)$) are calculated in advance and used as setpoint values. The sequential analysis procedure is performed as long as the following inequalities are satisfied:

$$a(m) < \sum_{i=1}^{m} \xi_i < b(m). \tag{19}$$

When the sum of $\sum_{i=1}^{m} \xi_i$ is outside the interval $[a(m), b(m)]$, a decision is made as to whether the deviation of the generalized PQP from the set value is acceptable or not.

Let us present an example of arranging sampling-based monitoring of the generalized PQP in the EPDN while taking into account the weighting coefficients for individual parameters. Let us assume that, given the weight coefficients of individual parameters of the generalized PQP, the expected values $m_{\xi 0}$ and $m_{\xi 1}$ of the generalized PQP ξ take the values $m_{\xi 0} = 130$ and $m_{\xi 1} = 155$. The value of the variance ξ given the normal distribution is $\sigma^2 = 225$. Let us set $\alpha = 0.01$ and $\beta = 0.03$. In this case, the values of the acceptance and rejection numbers will be determined by the following Equations:

$$a(m) = [225/(155 - 130)]\ln[0.03/(1 - 0.01)] + m(130 + 155)/2 = 142.5m - 87.5;$$

$$b(m) = [225/(155 - 130)]\ln[(1 - 0.03)/0.01] + m(130 + 155)/2 = 142.5m + 114.37$$

Let there be consecutive timed samples of the variable ξ, the values of which are given in Table 1. Table 1 also includes the variables $\sum_{i=1}^{m} \xi_i$, $a(m)$ and $b(m)$, which vary from step to step of the sequential analysis procedure.

Table 1. Values of the variables used in the sequential decision-making procedure regarding the deviations of the generalized PQP.

m	1	2	3	4	5	6	7	8	9
ξ	149	151	154	155	148	160	156	154	150
$\sum_{i=1}^{m} \xi_i$	149	300	454	609	757	917	1123	1287	1437
$a(m)$	55	197.5	340	482.5	625	767.5	910	1052.5	1195
$b(m)$	256.9	399.4	541.9	684.4	826.9	969.4	1111.9	1254.4	1396.9

Figure 3 illustrates the process of sequential decision-making in the analysis of the generalized PQP.

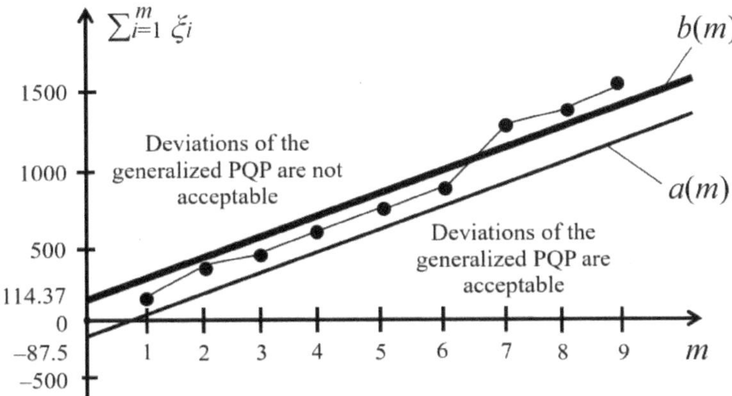

Figure 3. The process of sequential decision-making regarding the deviations of the generalized PQP.

In the graph (Figure 3), points $(m, \sum_{i=1}^{m} \xi_i)$—descriptive of the decision-making process—are plotted. The coefficient s, which determines the slope angle of the setpoint limits $a(m)$ and $b(m)$, corresponds to the following Equation:

$$s = \frac{m_{\xi 0} + m_{\xi 1}}{2}.$$ (20)

The setpoint limits are offset in relation to one another by the value of

$$\left(\frac{\sigma^2}{m_{\xi 1} - m_{\xi 0}} \right) \left(\ln \frac{1 - \beta}{\alpha} - \ln \frac{\beta}{1 - \alpha} \right).$$ (21)

The area between the setpoint limits is the area of uncertainty, which necessitates the continuation of the sampling-based monitoring procedure of the generalized PQP. The analysis of Figure 3 shows that the process of sequential analysis ends at step $m = 7$, when an unambiguous decision is made about the non-conformity of the generalized PQP with the established standard value.

Let us consider an approach to determining the average number of sampling-based monitoring data needed to make a decision about the power quality in the implementation of the sequential analysis procedure by the generalized PQP.

To determine the expected value of the number of sampling monitoring data (i.e., sample size) in the form of values of the generalized PQP, we can use the mathematical derivations obtained in [47]. For the problem under consideration, the probability $P(m_\xi)$ that the sequential analysis procedure will end with a decision on the compliance of the generalized PQP to the established standard value, when m_ξ is the true mean value, is defined by the following equality:

$$P(m_\xi) = \frac{[(1 - \beta)/\alpha]^h - 1}{[(1 - \beta)/\alpha]^h - [\beta/(1 - \alpha)]^h},$$ (22)

where $h = (m_{\xi 1} + m_{\xi 0} - 2m_\xi)/(m_{\xi 1} - m_{\xi 0})$.

The probabilistic relationship $P(m_\xi)$ is called the operational characteristic of Wald's sequential criterion. Since $P(m_\xi)$ is an increasing function with respect to h, and h is a decreasing function of m_ξ, $P(m_\xi)$ is inversely related to m_ξ.

The Equation for determining the average number n of sampling monitoring data points for the sequential analysis procedure associated with the power quality analysis and application of the generalized PQP is presented as the following equality [46]:

$$M(n, \, m_\xi) = \frac{2\sigma^2 \times P(m_\xi) \times \ln\frac{\beta}{1-\alpha} + (1 - P(m_\xi)) \ln\frac{1-\beta}{\alpha}}{m_{\xi 0}^2 - m_{\xi 1}^2 + 2(m_{\xi 1} - m_{\xi 0})m_\xi} \tag{23}$$

It is usually of interest to calculate the values of the operational characteristic $P(m_\xi)$ for special values from the following set:

$$m_\xi = \{-\infty; \, m_{\xi 0}; \, (m_{\xi 0} + m_{\xi 1})/2; \, m_{\xi 1}; +\infty\}$$

where by

$$P(m_\xi = -\infty) = 1;$$

$$P(m_\xi = m_{\xi 0}) = 1 - \alpha;$$

$$P\left(m_\xi = \frac{m_{\xi 0} + m_{\xi 1}}{2}\right) = \frac{\ln[(1 - \beta)/\alpha]}{\ln[(1 - \beta)/\alpha] - \ln[\beta/(1 - \alpha)]};$$

$$P(m_\xi = m_{\xi 1}) = \beta;$$

$$P(m_\xi = +\infty) = 0.$$

When $m_\xi = (m_{\xi 0} + m_{\xi 1})/2$, the right-hand side of Equation (23) corresponds to an uncertainty of $0/0$, and the expected value of the required number of sample data points for the sequential analysis procedure is as follows [47]:

$$M\left(n, \, m_\xi = \frac{m_{\xi 0} + m_{\xi 1}}{2}\right) = \frac{-\ln\frac{\beta}{1-\alpha}\ln\frac{1-\beta}{\alpha}\sigma^2}{(m_{\xi 1} - m_{\xi 0})^2}. \tag{24}$$

Taking into account the previously stated conditions of the problem, let us calculate the value (as per Equation (24)) of the necessary average number of sample data points for the sequential analysis procedure when $m_\xi = (m_{\xi 0} + m_{\xi 1})/2$:

$$M[n] = \{-\ln[0.03/(1 - 0.01)] \times \ln[(1 - 0.03)/0.01] \times 225\}/(155 - 130)^2 \approx 6.$$

Thus, to arrange the sampling-based monitoring with the procedure of sequential analysis of the generalized PQP at the given points of connection of the EPDN of an industrial enterprise to an external power supply network, it is necessary to obtain an average of at least six outcomes of sampling-based PQP monitoring.

4. Discussion

To implement the procedure of sampling-based monitoring of the generalized PQP, a preliminary simulation is performed in order to form a database of acceptable deviations of the generalized PQP in the analyzed points of connection of the EPDN to an external distribution network for different states of its operation [42]. Simulation outcomes are entered into the memory unit of the power quality analysis device (Figure 4). Additionally, the memory unit receives information on possible damages at an industrial enterprise for specific points of connection of the EPDN of an industrial enterprise to an external distribution network, formed either by the results of simulation or by structural expert judgment, taking into account the deviations of each individual PQP [27].

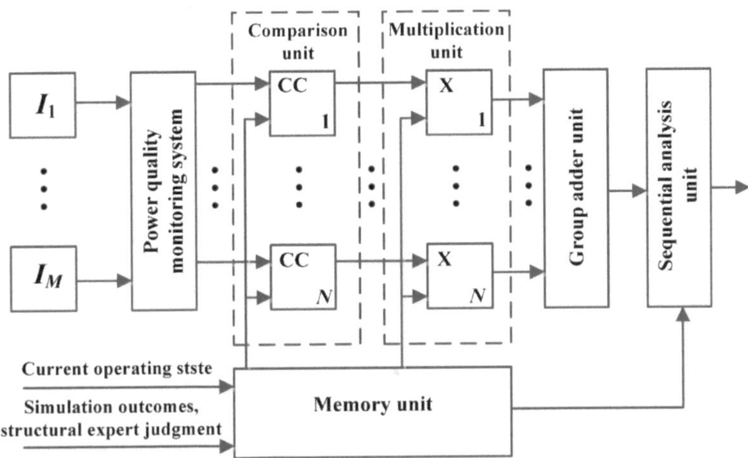

Figure 4. Structural diagram of the device that implements the sampling-based monitoring procedure of the generalized PQP.

Figure 4 shows an example of the structural diagram of the device that implements the sampling-based monitoring procedure of the generalized PQP with the use of sequential analysis.

The device implementing the procedure of sampling-based monitoring of the generalized PQP (Figure 4) includes the following: a power quality monitoring system, which is connected to the power quality monitoring devices ($I_1 \ldots I_M$); a comparison unit, including comparison circuits ($CC_1 \ldots CC_N$) for each of the PCIs; a multiplication unit, consisting of N multiplication units for each of the PCIs; a group adder unit; a sequential analysis unit; and a memory unit.

In the device that implements the sampling-based monitoring procedure of the generalized PQP (Figure 4), the set of processing (calculation) operations is carried out as follows: At each selected moment in time, the calculated PCI values arrive at the inputs of the comparison circuits from the power quality monitoring system. The other inputs of the comparison circuits receive the standardized PQP values from the memory unit as calculated for the current EPDN operating state. According to the results of the comparison performed in the comparison unit, a discrete vector of deviations is formed, whose components are multiplied by the corresponding weight coefficients included in the column vector $c = \left(c_1, \ldots, c_j, \ldots, c_N\right)^T$, defining the damage values for disruptions of power quality with respect to individual parameters.

The group adder unit is designed to form a generalized PQP according to Equation (2), and from its output the sampled values ζ_i are fed to the input of the sequential analysis unit. The other input of the sequential analysis unit receives arrays of acceptance $a(m)$ and rejection $b(m)$ numbers, whose components correspond to the setpoint values for each step of the sequential analysis procedure. Figure 3 illustrates the decision-making process in the sequential analysis aided by a generalized PQP, where the sequential analysis process ends with assuming the hypothesis of an unacceptable deviation of the generalized PQP from the standard value.

The memory unit of the device (Figure 4), which implements the procedure of sampling-based monitoring of the generalized PQP, receives information about the current operating state of the power system (i.e., positions of switching devices, operating state parameters, etc.), where it is recognized, and the state number is assigned. Such information can come, for example, from the SCADA system or the system of the operational information package. In the memory unit, a set of standardized PQP values is selected from the archive for the corresponding operating state number; they are the weight coefficients $c_1, \ldots, c_j, \ldots, c_N$ and the current set of setpoint values $a(m)$, $b(m)$, received from the memory unit outputs by the comparison, multiplication, and sequential analysis units to analyze the power quality at the points of consumer connection to an external distribution network. Along with the information about the current operating state, the simulation data, structural expert judgment, and other information necessary for the operation of the device (Figure 4) are fed to the input of the memory unit before the power quality analysis is performed.

The results of the sampling-based monitoring of the generalized PQP are presented in the form of a discrete signal at the output of the sequential analysis unit. The appearance of "1" at the output of this unit indicates a deviation of the generalized PQP from the standard value (Figure 4), which can lead to incurring damage to the consumer. This occurs when the calculated value of the variables $\sum_{i=1}^{m} \xi_i$ exceeds the upper setpoint limit $b(m)$—for example, for the case under consideration at step $m = 7$ (Figure 3). Consequently, implementation of organizational and engineering measures is required in order to bring the generalized PQP into the acceptable range [48–50].

5. Conclusions

It is advisable to introduce automated systems for monitoring power quality parameters using statistical data processing techniques and the formation of a generalized power quality parameter.

Given that the introduction of continuous monitoring of power quality parameters in most cases is not economically feasible, it is justified to implement sampling-based monitoring at certain time intervals, at predetermined monitoring points, with the calculation of only those parameters that are critical for a particular consumer, taking into account its specific process features.

When the power distribution network operates in quasi-steady states, it is promising to arrange the sampling-based monitoring of the generalized power quality parameter on the basis of Wald's sequential analysis procedure.

The joint use of simulation outcomes, current operating state parameters, and sampling-based monitoring data from the power quality parameter monitoring system using Wald's sequential analysis procedure allows the adaptation of the monitoring procedure to the specific features of a particular power distribution network of an industrial enterprise.

The implementation of the proposed approach, as implemented in the corresponding device, makes it possible to ensure reliable power supply to consumers' electrical loads and prevent damages through the timely implementation of organizational and engineering measures when the generalized power quality parameter deviates from its standardized value.

This technique makes it possible to adapt the PQP monitoring procedure to the features of a specific power distribution network of an industrial enterprise.

Author Contributions: Conceptualization, A.K. and P.I.; methodology, S.F.; software, P.I.; validation, K.S. and S.F.; formal analysis, A.K.; data curation, K.S.; writing—original draft preparation, A.K. and P.I.; writing—review and editing, K.S. and S.F.; visualization, P.I. and K.S.; supervision, A.K.; project administration, S.F. All authors have read and agreed to the published version of the manuscript.

Funding: This research received no external funding.

Data Availability Statement: No new data were created or analyzed in this study. Data sharing is not applicable to this article.

Conflicts of Interest: The funders had no role in the design of the study; in the collection, analyses, or interpretation of data; in the writing of the manuscript; or in the decision to publish the results.

Abbreviations

PQP	Power quality parameter
EPDN	Electric power distribution network
GU	Generating unit
DG	Distributed generation
RES	Renewable energy source
ESS	Electricity storage system
THDi	Total harmonic current distortion
IM	Induction motor
DSP	Digital signal processor
FPGA	Field-programmable gate array
ASIC	Application-specific integrated circuit
PQMS	Power quality monitoring system
SCADA	Supervisory control and data acquisition

References

1. Zeng, B.; Wen, J.; Shi, J.; Zhang, J.; Zhang, Y. A multi-level approach to active distribution system planning for efficient renewable energy harvesting in a deregulated environment. *Energy* **2016**, *96*, 614–624. [CrossRef]
2. Ghadi, M.; Rajabi, A.; Ghavidel, S.; Azizivahed, A.; Li, L.; Zhang, J. From active distribution systems to decentralized microgrids: A review on regulations and planning approaches based on operational factors. *Appl. Energy* **2019**, *253*, 113543. [CrossRef]
3. Begovic, M.M. (Ed.) *Electrical Transmission Systems and Smart Grids: Selected Entries from the Encyclopedia of Sustainability Science and Technology*; Springer Science + Business Media: New York, NY, USA, 2013.
4. Alberto Escalera, A.; Prodanović, M.; Castronuovo, E.D. Analytical methodology for reliability assessment of distribution networks with energy storage in islanded and emergency-tie restoration modes. *Int. J. Electr. Power Energy Syst.* **2019**, *107*, 735–744. [CrossRef]
5. Schwan, M.; Ettinger, A.; Gunaltay, S. Probabilistic reliability assessment in distribution network master plan development and in distribution automation implementation. In Proceedings of the CIGRE, 2012 Session, Paris, France, 26–30 August 2012; p. C4-203.
6. Varma, R.K.; Rahman, S.A.; Vanderheide, T.; Dang, M.D.N. Harmonic impact of a 20-MW PV solar farm on a utility distribution network. *IEEE Power Energy Technol. Syst. J.* **2016**, *3*, 89–98. [CrossRef]
7. Torquato, R.; Freitas, W.; Hax, G.R.T.; Donadon, A.R.; Moya, R. High frequency harmonic distortions measured in a Brazilian solar farm. In Proceedings of the 17th International Conference on Harmonics and Quality of Power (ICHQP), Belo Horizonte, Brazil, 16–19 October 2016; pp. 623–627.
8. *IEEE STD 519-2014*; Recommended practices and requirements for harmonic control in electrical power systems/Transmission and Distribution Committee of the IEEE Power and Energy Society. The Institute of Electrical and Electronics Engineers: New York, NY, USA, 2014.
9. Filippov, S.P.; Dilman, M.D.; Ilyushin, P.V. Distributed Generation of Electricity and Sustainable Regional Growth. *Therm. Eng.* **2019**, *66*, 869–880. [CrossRef]
10. Shushpanov, I.; Suslov, K.; Ilyushin, P.; Sidorov, D. Towards the flexible distribution networks design using the reliability performance metric. *Energies* **2021**, *14*, 6193. [CrossRef]
11. Adineh, B.; Keypour, R.; Sahoo, S.; Davari, P.; Blaabjerg, F. Robust optimization based harmonic mitigation method in islanded microgrids. *Int. J. Electr. Power Energy Syst.* **2022**, *137*, 107631. [CrossRef]
12. Bahrami, M.; Eslami, S.; Zandi, M.; Gavagsaz-ghoachani, R.; Payman, A.; Phattanasak, M.; Nahid-Mobarakeh, B.; Pierfederici, S. Predictive based reliability analysis of electrical hybrid distributed generation. In Proceedings of the International Conference on Science and Technology (TICST), Pathum Thani, Thailand, 4–6 November 2015.
13. Saha, S.; Saleem, M.I.; Roy, T.K. Impact of high penetration of renewable energy sources on grid frequency behavior. *Int. J. Electr. Power Energy Syst.* **2022**, *145*, 108701. [CrossRef]
14. Li, W. *Risk Assessment of Power Systems: Models, Methods and Applications*; John Wiley and Sons: New York, NY, USA, 2005.
15. McDonald, J.D.F.; Pal, B.C. Representing the risk imposed by different strategies of distribution system operation. In Proceedings of the 2006 IEEE PES General Meeting, Montreal, Canada, 18–22 June 2006.
16. Ilyushin, P.V. Emergency and post-emergency control in the formation of micro-grids. *E3S Web Conf.* **2017**, *25*, 02002. [CrossRef]
17. Gurevich, Y.E.; Kabikov, K.V. *Peculiarities of Power Supply, Aimed at Failure-Free Operation of Industrial Consumers*; ELEKS-KM: Moscow, Russia, 2005. (In Russian)
18. Gurevich, Y.E.; Libova, L.E. *Application of Mathematical Models of Electrical Load in Calculation of the Power Systems Stability and Reliability of Power Supply to Industrial Enterprises*; ELEKS-KM: Moscow, Russia, 2008. (In Russian)
19. Lacommare, K.; Eto, J. Cost of power interruptions to electricity consumers in the United States (US). *Energy* **2006**, *31*, 1845–1855. [CrossRef]
20. Suslov, K.V.; Solonina, N.N.; Smirnov, A.S. Distributed power quality monitoring. In Proceedings of the IEEE 16th International Conference on Harmonics and Quality of Power (ICHQP), Bucharest, Romania, 25–28 May 2014; pp. 517–520.

21. Gallego, L.; Torres, H.; Pavas, A.; Urrutia, D.; Cajamarca, G.; Rondón, D. A Methodological proposal for monitoring, analyzing and estimating power quality indices: The case of Bogotá—Colombia. In Proceedings of the IEEE Power Tech 2005, St Peterburg, Russia, 27–30 June 2005.
22. Gavrilov, D.; Gouzman, M.; Luryi, S. Monitoring large-scale power distribution grids. *Solid-State Electron.* **2019**, *155*, 57–64. [CrossRef]
23. Byk, F.L.; Myshkina, L.S.; Khokhlova, K. Power supply reliability indexes. In Proceedings of the International Conference on Actual Issues of Mechanical Engineering, Tomsk, Russia, 27–29 July 2017; pp. 525–530.
24. Bollen, M.H.J.; Dirix, P.M.E. Simple model for post-fault motor behaviours for reliability/power quality assessment of industrial power systems. *IEE Proc. Gener. Transm. Distrib.* **1996**, *143*, 56–60. [CrossRef]
25. Ribeiro, P.F.; Duque, C.A.; Silveira, P.M.; Cerqueira, A.S. *Power Systems Signal Processing for Smart Grids*; John Wiley & Sons Ltd.: Hoboken, NJ, USA, 2014.
26. Kulikov, A.L.; Ilyushin, P.V.; Suslov, K.V.; Karamov, D.N. Coherence of digital processing of current and voltage signals at decimation for power systems with a large share of renewable power stations. *Energy Rep.* **2022**, *8*, 1464–1478. [CrossRef]
27. Knowler, L.A.; Howell, J.M.; Gold, B.K.; Coleman, E.P.; Moan, O.B.; Knowler, W.S. *Statistical Methods of Product Quality Control*; McGraw-Hill Book Company: New York, NY, USA, 1989.
28. Won, D.-J.; Chung, I.-Y.; Kim, J.-M.; Moon, S.-I.; Seo, J.-C.; Choe, J.-W. Development of power quality monitoring system with central processing scheme. In Proceedings of the IEEE Power Engineering Society Summer Meeting, Chicago, IL, USA, 21–25 July 2002; Volume 2, pp. 915–919.
29. Marple, S.L., Jr. *Digital Spectrum Analysis with Applications*; Prentice Hall: Hoboken, NJ, USA, 1987.
30. Won, D.-J.; Chung, I.-Y.; Kim, J.-M.; Moon, S.-I.; Seo, J.-C.; Choe, J.-W. A new algorithm to locate power-quality event source with improved realization of distributed monitoring scheme. *IEEE Trans. Power Deliv.* **2006**, *21*, 1641–1647. [CrossRef]
31. Valsan, S.P.; Swarup, K.S. Wavelet transform based digital protection for transmission lines. *Int. J. Electr. Power Energy Syst.* **2009**, *31*, 379–388. [CrossRef]
32. McGranaghan, M.F.; Santoso, S. Challenges and trends in analysis of electric power quality measurement data. *EURASIP J. Adv. Signal Process.* **2007**, *2007*, 057985. [CrossRef]
33. Kulikov, A.L.; Misrikhanov, M.S. *Introduction to Methods of Digital Relay Protection of High-Voltage Transmission Lines*; Energoatomizdat: Moscow, Russia, 2007. (In Russian)
34. Kezunovic, M.; Meliopoulos, S.; Venkatasubramanian, V.; Vittal, V. *Application of Time-Synchronized Measurements in Power System Transmission Networks*; Springer: New York, NY, USA, 2014.
35. Ilyushin, P.V.; Shepovalova, O.V.; Filippov, S.P.; Nekrasov, A.A. Calculating the Sequence of Stationary Modes in Power Distribution Networks of Russia for Wide-Scale Integration of Renewable Energy Based Installations. *Energy Rep.* **2021**, *7*, 308–327. [CrossRef]
36. NASPI. *Synchrophasor Monitoring for Distribution Systems: Technical Foundations and Applications*; A White Paper by the NASPI Distribution Task Team NASPI-2018-TR-001; NASPI: Tempe, AZ, USA, 2018.
37. Edomah, N. Effects of voltage sags, swell and other disturbances on electrical equipment and their economic implications. In Proceedings of the 20th International Conference and Exhibition on Electricity Distribution, Prague, Czech Republic, 8–11 June 2009; pp. 1–4.
38. Ilyushin, P.V.; Sukhanov, O.A. The Structure of Emergency-Management Systems of Distribution Networks in Large Cities. *Russ. Electr. Eng.* **2014**, *85*, 133–137. [CrossRef]
39. Wentzel, E.S. *Probability Theory (First Steps)*; Mir: Moscow, Russia, 1986.
40. Won, D.J.; Chung, I.Y.; Kim, J.M.; Ahn, S.J.; Moon, S.I. A modified voltage sag duration for power quality diagnosis. *Proc. IFAC* **2003**, *36*, 739–743. [CrossRef]
41. Won, D.J.; Ahn, S.J.; Moon, S.I. A modified sag characterization using voltage tolerance curve for power quality diagnosis. *IEEE Trans. Power Deliv.* **2005**, *20*, 2638–2643. [CrossRef]
42. Cowden, D.J. *Statistical Methods in Quality Control*; Prentice-Hall, Inc.: Englewood Cliffs, NJ, USA, 1957.
43. Shor, Y.B. *Statistical Methods of Analysis and Quality Control and Reliability*; Soviet Radio: Moscow, Russia, 1962. (In Russian)
44. Belyaev, Y.K. *Probabilistic Sampling Methods*; Nauka: Moscow, Russia, 1975. (In Russian)
45. *EN 50160:2019*; Voltage Characteristics of Electricity Supplied by Public Electricity Networks. European Committee for Standards: Brussels, Belgium, 2019; 34p.
46. *ANSI C84.1-2020*; Electric Power Systems and Equipment—Voltage Ratings (60 Hz). American National Standards Institute: Washington, DC, USA, 2020; 21p.
47. Wald, A. *Sequential Analysis*; John Wiley and Sons: New York, NY, USA, 1947.
48. Ayvazyan, S.A.; Mkhitaryan, V.S. *Applied Statistics and Fundamentals of Econometrics: A Textbook for High Schools*; UNITY: Moscow, Russia, 1998; 1022p.

49. Suslov, K.; Shushpanov, I.; Buryanina, N.; Ilyushin, P. Flexible power distribution networks: New opportunities and applications. In Proceedings of the 9th International Conference on Smart Cities and Green ICT Systems, Prague, Czech Republic, 2–4 May 2020; Volume 1, pp. 57–64.
50. Ilyushin, P.V.; Suslov, K.V. Operation of automatic transfer switches in the networks with distributed generation. In Proceedings of the 2019 IEEE Milan PowerTech, Milan, Italy, 23–27 June 2019. [CrossRef]

Research of Static and Dynamic Properties of Power Semiconductor Diodes at Low and Cryogenic Temperatures

Mikhail Ostapchuk *, Dmitry Shishov, Daniil Shevtsov and Sergey Zanegin

Moscow Aviation Institute, National Research University, 4, Volokolamskoe Shosse, 125993 Moscow, Russia
* Correspondence: ostapchukma@mai.ru

Abstract: Systems with high-temperature superconductors (HTSC) impose new requirements on power conversions, since the main part of the losses in such systems is induced in the semiconductors of the converters. Within the framework of this study, the possibility of improving the static and dynamic characteristics of power semiconductor diodes using cryogenic cooling was confirmed; in some cases, a loss reduction of up to 30% was achieved.

Keywords: HTS generators; power electronics; cryogenic devices; cryoelectronics; cryogenic cooling; power diodes

Citation: Ostapchuk, M.; Shishov, D.; Shevtsov, D.; Zanegin, S. Research of Static and Dynamic Properties of Power Semiconductor Diodes at Low and Cryogenic Temperatures. *Inventions* 2022, 7, 96. https://doi.org/10.3390/inventions7040096

Academic Editor: Om P. Malik

Received: 23 September 2022
Accepted: 13 October 2022
Published: 18 October 2022

Publisher's Note: MDPI stays neutral with regard to jurisdictional claims in published maps and institutional affiliations.

1. Introduction

To improve the weight and size parameters of generation and electric propulsion systems, electric machines and cable lines based on high-temperature superconducting materials of the second generation (HTSC-2) can be used [1]. In such systems, it is impossible to consider an electric machine in isolation from other elements, especially electronic converters. At present, in the vast majority of cases, an electric machine works in conjunction with a semiconductor energy converter: a power factor corrector, a rectifier, an electric motor controller, etc. These devices, which are directly connected to the windings of the electrical machine, affect the harmonic composition of the phase currents. The harmonic composition of the current is a very important parameter for HTSC windings since the losses in them strongly depend on it [2]. Studies conducted at the Department 310 MAI have shown that when a non-linear load is supplied from a generator with HTSC windings, an active correction of the shape of the phase currents using a power electronic converter is necessary. Otherwise, the losses from the higher current harmonics can be so great that the cryosupport system can no longer cope with heat removal and the super-conductor will go into a normal state.

Due to the fact that superconducting electrical machines require the presence of a cryogenic cooling circuit, it seems appropriate to investigate the possibility of integrating power semiconductor converters into this circuit. Cooling the heatsinks of power semiconductor devices with liquid nitrogen can also have a positive effect on the weight and size parameters of the system. It is proposed to use indirect cryogenic cooling of converters by pumping liquid nitrogen through tubes built into the radiator (the so-called cold plate). In this regard, it must be emphasized that the temperature of semiconductor components can fluctuate over a wide range with this method of cooling [3]. It also means that the lowest achievable temperature of the component will be higher than the boiling point of liquid nitrogen.

The main problem of the development of cryogenic cooling of power semiconductors is the unexplored effect of cryogenic cooling of components on the operation of the device. Since temperature and thermal processes are the main factor determining reliability, this issue is a priority. In particular, this follows from the various expansion coefficients of materials that can not only destroy the internal structure of the component, but also completely destroy the printed circuit board of the device [4].

There are a fairly large number of studies devoted to determining the characteristics of electronic semiconductor components at the boiling point of liquid nitrogen [5–8]. However, there are very few works in which power semiconductor devices are tested in the entire temperature range from +25 °C to −195.75 °C. Among them, we can highlight as the most outstanding, for example, [9], some works where the diode model of promising structures was corrected with experimental confirmation; and [10], where the static and dynamic characteristics of semiconductors were considered in a wide temperature range. However, the information presented in these publications is not universal, since the behavior of the characteristics of each particular semiconductor component is unique. This article is devoted to the experimental determination of the static characteristics of some power semiconductor diodes in a given temperature range and the analysis of the obtained dependencies.

2. Materials and Methods

To determine the dependence of the characteristics of power semiconductor devices on temperature in the range from −195.75 °C to +25 °C, a special test bench was developed and manufactured.

The device under study is mounted on a radiator through the tubes from which liquid nitrogen is pumped. The temperature sensor is attached to the heatsink as close as possible to the component under test. To control the temperature, the flow of liquid nitrogen is controlled using a manual valve. To prevent the formation of frost on the surface of the component under study, the space surrounding the component is evacuated.

When the valve is opened, the flow of liquid nitrogen under the action of pressure injected in the first Dewar vessel is pushed out through the tube towards the radiator. For a sufficiently long time, nitrogen passes through the tubes in a gaseous state. Once the radiator is cool enough, nitrogen begins to flow through the tubes in a liquid state and drains into a second open Dewar vessel. Cooling occurred at a rate of approximately 2–5 K per minute, which made it possible to measure the static and dynamic parameters of the diodes at a certain temperature, set with an accuracy of one degree. Measurements were made in steps of 10 K from 293 K to the minimum achievable temperature, which was 100 K with this method of cooling. The block diagram of the test bench is shown in Figure 1.

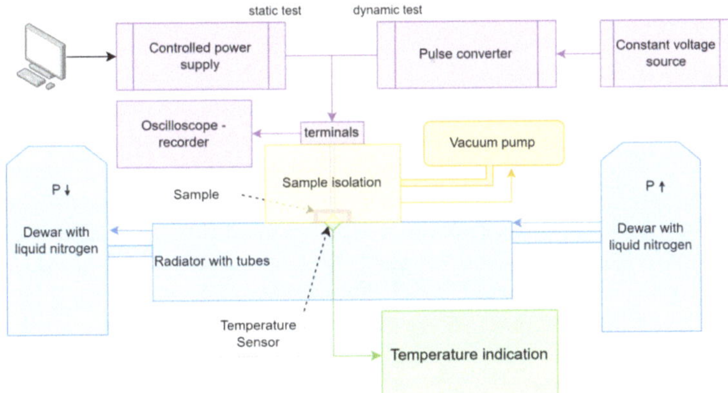

Figure 1. Structural diagram of the test bench.

As part of the tests for temperature measurement, an OVEN 2TRM1 thermostat was used as an indicator and a DTS324-100P.VZ.41/0.5 sensor compatible with it. The absolute error of temperature measurement is the sum of the error of the temperature controller and the temperature sensor, and in total is 3.35–3.5 K. The appearance of the test bench is shown in Figure 2.

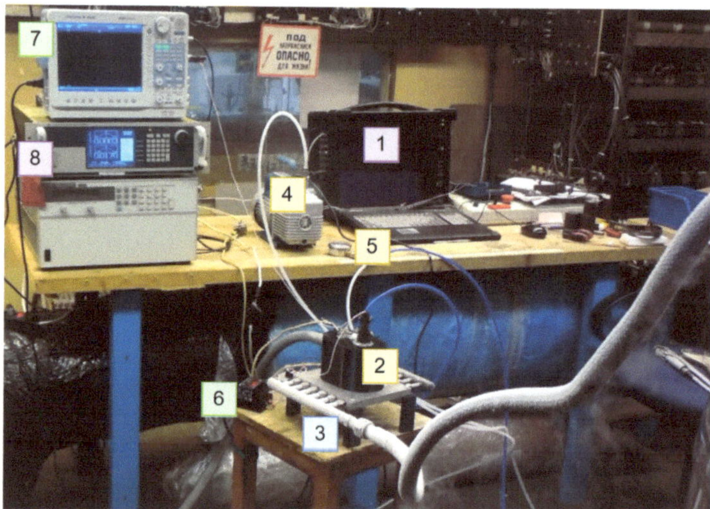

Figure 2. Photo of the test bench. 1—computer that controls the source of test signals, 2—vacuum casing, 3—radiator with tubes, 4—vacuum pump, 5—pressure sensor, 6—temperature indicator, 7—oscilloscope recorder, and 8—source of test signals.

For static tests, a Kepco BOP 10–100 MG controlled current source was used. The signal was controlled using PC with NI SignalExpress 2014 and NI SCB-68a. During dynamic tests, two TEC 14 power supplies connected to a pulsed transistor power amplifier were used to form a bipolar signal. As a result of static tests, a current–voltage characteristic (CVC) was obtained.

A Yokogawa DL850E oscilloscope–recorder was used as a means of measuring electrical parameters and recording data. In the study of static characteristics, a Yokogawa 720,250 measuring module was used to record the parameters, and a Yokogawa 720,211 measuring module was used for dynamic characteristics. Both modules make it possible to measure voltage with an error of 0.5% of the maximum value of the selected range [11]. The stand measurement errors are presented in Table 1.

Table 1. Characteristics of the measured signals.

Value	Dynamic Tests	Static Tests
Measured current range, A	$-0.2\ldots+0.2$	$0\ldots200$
Absolute current measurement error, A	±0.002	±1
Max. rev. current values, A	$-0.05\ldots+0.05$	100
Measured voltage range, V	$-20\ldots+20$	$0\ldots5$
Absolute voltage measurement error, V	±2	±0.025
Max. rev. voltage values, V	$-15\ldots+15$	3

The current values during static tests were calculated based on the voltage values on the measuring shunt made using low-inductance resistors LVR03, with an accuracy of 1%. The sensor contains 10 LVR03 resistors connected in parallel and has a resistance of 0.001 ohm.

To study the dynamic characteristics of a diode, a test signal of a special shape is required. This is due to the high rate of reverse recovery processes of modern diodes. In the case of a test signal with a low switching rate, the reverse recovery processes may be hardly noticeable, or even absent. To form the corresponding signals, a pulsed transistor power amplifier was developed, the simulation computer model of which is shown in Figure 3.

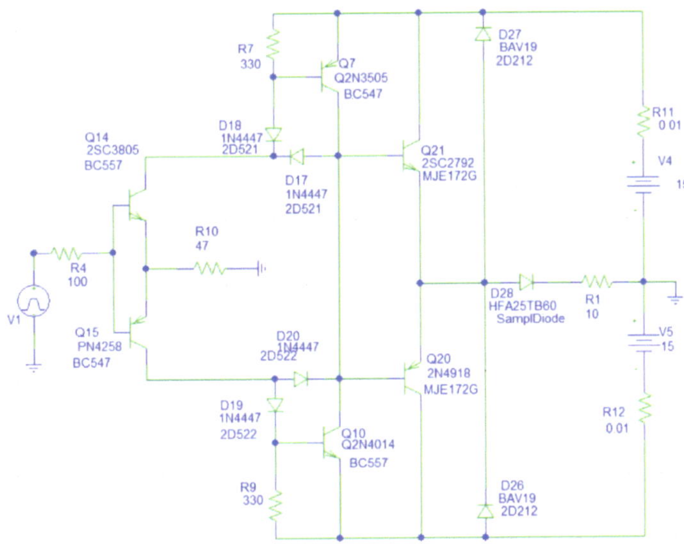

Figure 3. Simulation model of a pulsed amplifier in the OrCad program.

The amplifier is assembled using high-frequency complementary bipolar transistors. To prevent the saturation of transistors, non-linear Baker circuits on diodes D17–D20 are used. The amplifier is controlled using the Aktakom signal generator, which in Figure 4 is represented by a source of pulsed voltage V1. The signal generator ensures the rise and fall time of the pulse is no more than 20 ns.

A problem with both static and dynamic testing is the heating of the semiconductor component crystal by the test signal. This means that during the test pulse, as the current increases, the temperature difference between the sensor and the crystal also increases. This causes an additional error in estimating the temperature of the component. This error can be estimated using either analytical calculations (assuming that the heating process is adiabatic) or finite element analysis. In both cases, it is necessary to be aware of the design of the semiconductor device and its thermophysical properties, including the heat capacities of all the structural elements. Obtaining this information is a very difficult task. In this regard, in this work, the temperature of the radiator of the studied component was measured de facto. In addition, it should be noted that the test signal is a small series of short pulses with low energy, which allows us to assume that the temperature of the crystal of the device under study will slightly differ from the temperature of the radiator.

For processing and outputting graphs in the .wdf format, the Mathworks MatLab program version 2022a with the WDF Access add-on was used.

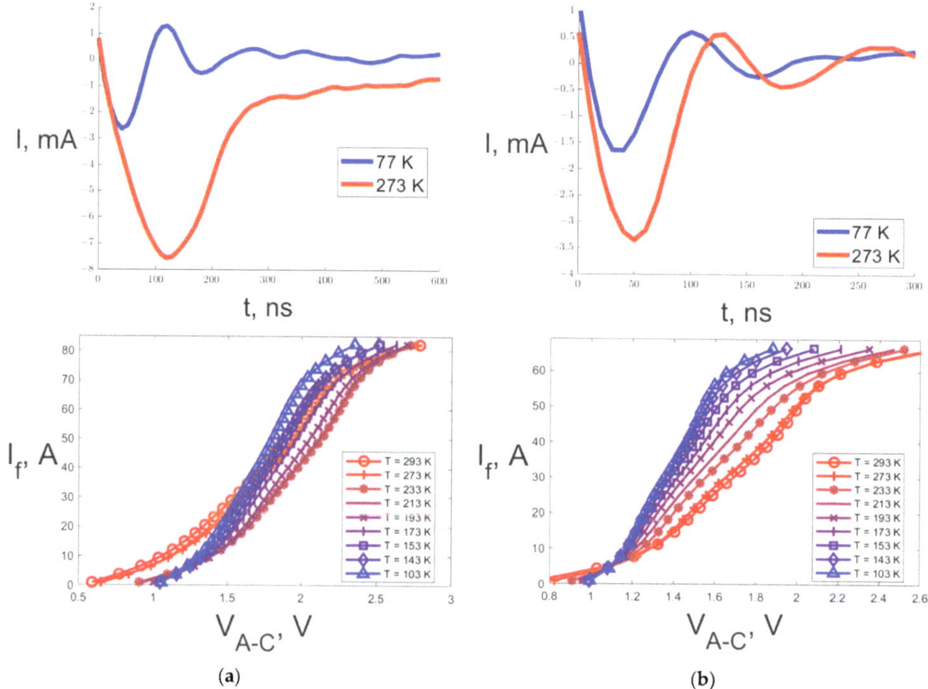

Figure 4. Graphs of the reverse recovery process (top) of the samples and the forward branch of the CVC (bottom) for samples 1 (**a**) and 2 (**b**).

3. Results

During the tests, the dynamic and static characteristics of power silicon diodes were obtained. The components used by the authors in current developments were selected as samples. The list of components is presented in Table 2.

Table 2. Characteristics of the studied samples.

Sample №	Models Name	Rev. Voltage, V	Avg. Forward Current, A	Rev. Recov. Time, ns	Corps	Dimens., mm	Mass, g
1	SKR47F17	1700	47	120	TO-218	$38 \times 15 \times 4$	5
2	HFA30PA60C	600	15	42	TO-247	$34 \times 16 \times 5$	6
3	VUO160-12NO7	1300	175	5000	PWS-E	$94 \times 54 \times 30$	284
4	VS-HFA120	1200	145	14	SOT-227	$38 \times 26 \times 12$	30

Figures 4 and 5 show the graphs of the reverse recovery process obtained as a result of experiments and the direct branch of the CVC. For samples 1 and 2, due to some circumstances, the dynamic characteristics were investigated only for two temperatures.

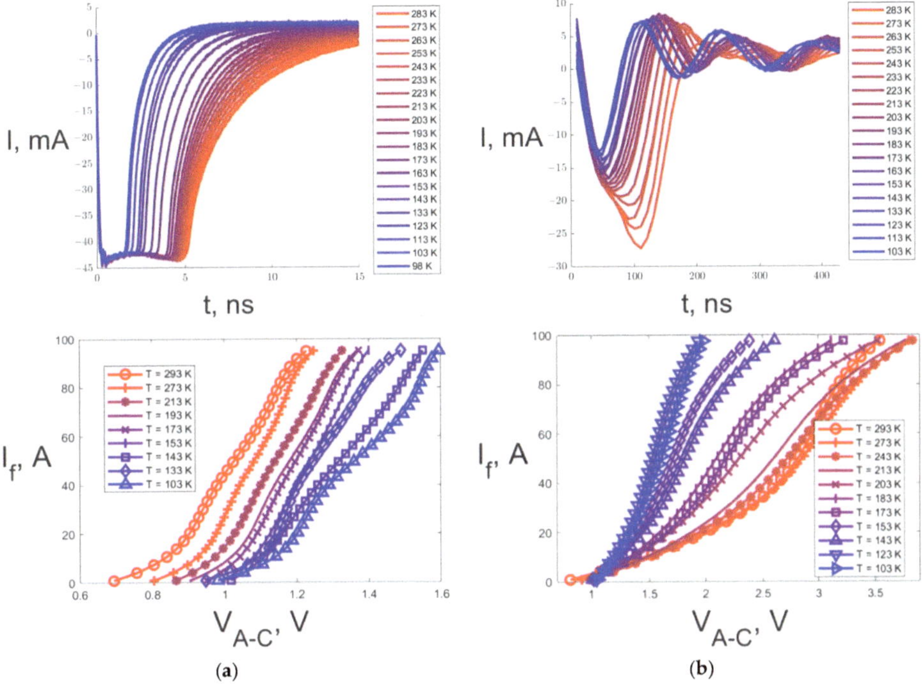

Figure 5. Graphs of the reverse recovery process (top) of the samples and the forward branch of the CVC (bottom) for samples 3 (**a**) and 4 (**b**).

The dynamic characteristics showed a uniform change in the form of a decrease in dynamic losses. This is also associated with a decrease in the reverse recovery time of the diode. If we pay attention only to the characteristics of sample 4 (Figure 5a), then we can assume that the reverse recovery time decreases almost linearly with decreasing temperature. However, the nature of the change in the reverse recovery process of sample 3 suggests that, in fact, there is a reduction in the reverse recovery stage with the most intense current strength (Figure 5b). This is confirmed by the non-linearity of the transition point from the amplitude value of the reverse recovery current to the value of the leakage current.

In addition, it should be noted that the recovery process of samples 1, 2, and 4 has an oscillatory character. This may be due to the influence of the parasitic inductance of the design and capacitance of the pn junction of the diode at low reverse recovery times (samples 1,2, and 4 belong to the class of "fast" diodes).

The change in static characteristics with temperature is less unambiguous. The threshold voltage of the diodes in all cases increased with decreasing temperature. This can be seen from the transfer of the minimum current at a lower temperature in the direction of increasing voltage for each diode.

It is important to note a significant indicator of the change in static losses—the voltage drop across the diode for a given forward current. Since the region of high currents was chosen, then the studied components will reveal three options for changing this indicator. In the first case, there was an increase in voltage with decreasing temperature; in the second—a decrease; and in the third—a non-linear nature of the changes. The non-linear nature of the changes in the voltage drop at a given current was revealed in sample 1, which at first showed a tendency to increase the voltage in the region of high currents with a decrease in temperature, and after a certain temperature, the reverse process began. The CVC of sample 3 can be attributed to the first type, and samples 2 and 4, to the second. The CVC of samples 2 and 4 suggest a significant energy benefit from the use of cryogenic cooling in the case of prevailing static losses during the normal operation of the converter.

4. Discussion

To obtain an estimate of the change in static losses according to the CVC, the dependencies shown in Figure 6 were built. They reflect the ratio of the power of static losses at different temperatures to the maximum heat release, which for different devices falls at different temperatures. Thus, the resulting set of such energy characteristics is approximately in the same range and a numerical assessment of the change in static losses for each characteristic is possible. Given the equal sampling frequency of the oscilloscope, the losses were calculated as the average instantaneous power at positive currents and voltages of the test signal.

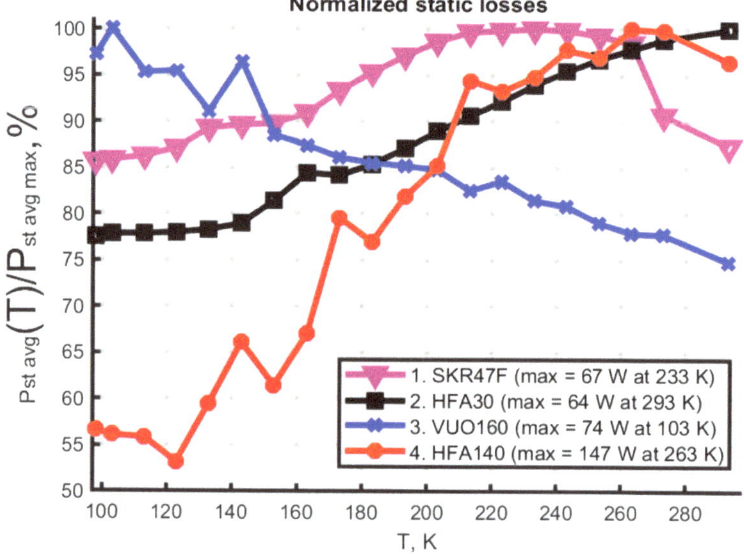

Figure 6. Normalized static losses.

Based on the characteristics obtained, it can be concluded that the voltage drop at a given forward current has a strong influence on the resulting static losses. A decrease in this indicator corresponds to a decrease in static losses. The following conclusion can be made on the fact that it is energetically more profitable to use diodes in the region of higher currents in cryogenic cooling.

The dynamic loss during diode reverse recovery is calculated as the average of the power dissipated. The decrease in dynamic losses is explained by a decrease in the charge that is accumulated by the diode in the off state as the temperature decreases. This leads to both a decrease in the reverse recovery time and a decrease in the amplitude of the reverse recovery current in the case of samples 2–4. Characteristics normalized by the maximum value are shown in Figures 7 and 8. In contrast to static losses, the maximum losses in the reverse recovery process are recorded at a maximum temperature of 293 K. This point corresponds to the maximum reverse recovery time for all the samples.

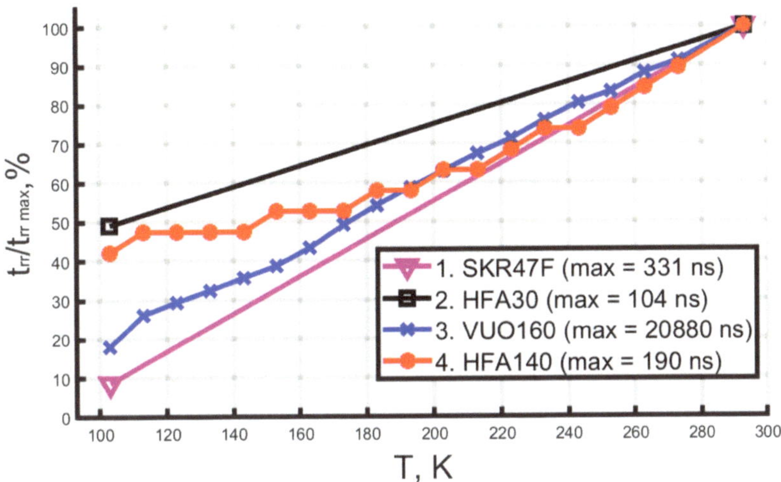

Figure 7. Dependencies of the normalized values of the reverse recovery time on temperature.

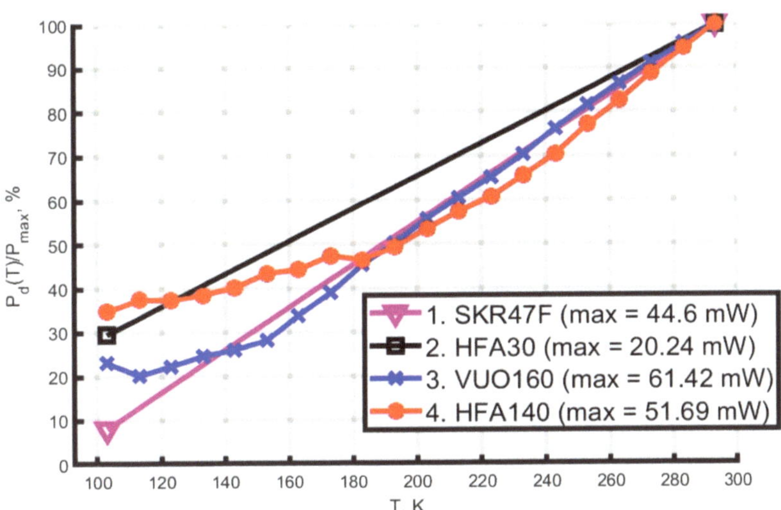

Figure 8. Dependencies of the normalized losses in the process of reverse recovery on temperature.

Dynamic losses are an indicator that depends on the actual mode of operation of the diode. In the case of a single reverse recovery process, this only suggests an idea of how the dynamic losses will change, not allowing you to accurately calculate how the dynamic losses will change during the normal operation of the proposed device. However, this indicator allows one to obtain an estimate of heat dissipation using oscillographic measurements in a real circuit.

If it is advisable to cool electronics with liquid nitrogen, this will improve the efficiency of all devices where liquid nitrogen cooling is already used (for example, in MRI). In addition, this effect can be used with other refrigerants used as fuel—liquefied natural gas or liquid hydrogen.

5. Conclusions

In the course of studying the influence of low and cryogenic temperatures on the static and dynamic characteristics of power silicon semiconductor diodes, the dependencies of the current–voltage characteristics and the reverse recovery time on temperature were obtained for several samples in the range from $+25\ ^{\circ}C$ to $-195.75\ ^{\circ}C$.

Based on the results of the study, it can be concluded that further research is promising in the field of using cryogenic cooling of power stages of semiconductor converters operating in the same system with superconducting electrical machines. This conclusion is based on the fact that both static and dynamic parameters of power semiconductor diodes demonstrate such a tendency with decreasing crystal temperature, which should lead to a decrease in both types of losses.

In order to confirm the rationality of such a solution, further studies of the effect of cryogenic temperatures on the static and dynamic characteristics of power semiconductor components, including power transistors, are required. The final conclusion can be made only after testing a real power semiconductor converter with experimental determination of losses in its elements.

Separately, it should be emphasized that for the practical application of cryogenic cooling in power electronics, it is necessary to study the effect of thermal cycling on the electrical parameters of semiconductor components and their design. This may be the subject of further research.

Author Contributions: Conceptualization, D.S. (Daniil Shevtsov) and D.S. (Dmitry Shishov); methodology, D.S. (Dmitry Shishov); software, M.O.; validation, S.Z., D.S. (Daniil Shevtsov), and D.S. (Dmitry Shishov); formal analysis, M.O.; investigation, M.O.; resources, D.S. (Dmitry Shishov); data curation, S.Z.; writing—original draft preparation, M.O.; writing—review and editing, D.S. (Dmitry Shishov) and S.Z.; visualization, M.O.; supervision, D.S. (Daniil Shevtsov); project administration, D.S. (Dmitry Shishov); funding acquisition, D.S. (Dmitry Shishov). All authors have read and agreed to the published version of the manuscript.

Funding: The study was carried out with the financial support of a project by the Russian Federation represented by the Ministry of Education and Science of the Russian Federation, agreement No. 075-15-2020-770.

Institutional Review Board Statement: Not applicable.

Informed Consent Statement: Not applicable.

Data Availability Statement: Not applicable.

Conflicts of Interest: The authors declare no conflict of interest.

References

1. Kovalev, K.L.; Ivanov, N.S.; Penkin, V.T.; Development of HTSC electrical machines and devices at MAI. Status and prospects for the development of electrical and thermal technology (Benardos readings). In Proceedings of the Materials International (XX All-Russian) Scientific and Technical Conference, Ivanovo, Russia, 29–31 May 2019; Ivanovo State Power Engineering University named after V.I. Lenin: Ivanovo, Russia, 2019; pp. 153–156.
2. Zanegin, S.Y.; Ivanov, N.S.; Shishov, D.M.; Shishov, I.M.; Kovalev, K.L.; Zubko, V.V. AC losses test of HTS racetrack coils for HTS motor winding. *J. Phys. Conf. Ser.* **2020**, *1559*, 012142. [CrossRef]
3. Wang, F.; Chen, R.; Gui, H.; Niu, J.; Tolbert, L.; Costinett, D.; Blalock, B.; Liu, S.; Hull, J.; Williams, J.; et al. MW-Class Cryogenically-Cooled Inverter for Electric-Aircraft Applications. In Proceedings of the 2019 AIAA/IEEE Electric Aircraft Technologies Symposium (EATS), Indianapolis, IN, USA, 22–24 August 2019; pp. 1–9. [CrossRef]
4. Korobkov, M.; Vasilyev, F.; Mozharov, V. A Comparative Analysis of Printed Circuit Boards with Surface-Mounted and Embedded Components under Natural and Forced Convection. *Micromachines* **2022**, *17*, 634. [CrossRef] [PubMed]
5. Elwakeel, A.; Feng, Z.; McNeill, N.; Zhang, M.; Williams, B.; Yuan, W. Study of power devices for use in phase-leg at cryogenic temperature. *IEEE Trans. Appl. Supercond.* **2021**, *31*, 1–5. [CrossRef]

6. Pereira, P.; Valtchev, S.; Pina, J.; Gonçalves, A.; Neves, M.V.; Rodrigues, A.L. Power electronics performance in cryogenic environment: Evaluation for use in HTS power devices. *J. Phys. Conf. Ser.* **2008**, *1*, 012219. [CrossRef]
7. Chen, S.; Cai, C.; Wang, T.; Guo, Q.; Sheng, K. Cryogenic and high temperature performance of 4H-SiC power MOSFETs. In Proceedings of the 2013 Twenty-Eighth Annual IEEE Applied Power Electronics Conference and Exp osition (APEC), Long Beach, CA, USA, 17–21 March 2013; pp. 207–210. [CrossRef]
8. Singh, R.; Baliga, B.J. Cryogenic operation of PiN power rectifiers. *Solid-State Electron.* **1994**, *37*, 1833–1839. [CrossRef]
9. Chen, J.; Zhu, M.; Lu, X.; Zou, X. Electrical characterization of GaN Schottky barrier diode at cryogenic temperatures. *Appl. Phys. Lett.* **2020**, *116*, 062102. [CrossRef]
10. Wei, Y.; Hossain, M.M.; Mantooth, H.A. Static and Dynamic Cryogenic Characterizations of Commercial High Performance GaN HEMTs for More Electric Aircraft. In Proceedings of the 2022 International Power Electronics Conference (IPEC-Himeji 2022-ECCE Asia), Himeji, Japan, 15–19 May 2022; pp. 2300–2306.
11. YOKOGAWA TEST & MEASUREMENT CORPORATION. Plug-in Modules Specifications; ScopeCorder Series DL950/DL850E/ DL850EV/DL350/SL1000; 2021 Printed in Japan. Available online: https://cdn.tmi.yokogawa.com/1/8934/files/BUDL950-0 2EN.pdf (accessed on 13 July 2022).

Article

Energy-Saving Load Control of Induction Electric Motors for Drives of Working Machines to Reduce Thermal Wear

Tareq M. A. Al-Quraan [1], Oleksandr Vovk [2], Serhii Halko [2], Serhii Kvitka [2], Olena Suprun [3], Oleksandr Miroshnyk [4], Vitalii Nitsenko [5,6,*], Nurul Mohammad Zayed [7] and K. M. Anwarul Islam [8]

[1] Technical Sciences Department, Ma'an College, Al-Balqa Applied University, Al-Balqa 19117, Jordan
[2] Department of Electrical Engineering and Electromechanics Named after Prof. V.V. Ovharov, Dmytro Motornyi Tavria State Agrotechnological University, 72312 Melitopol, Ukraine
[3] Department of Foreign Languages, Dmytro Motornyi Tavria State Agrotechnological University, 72312 Melitopol, Ukraine
[4] Department of Electricity Supply and Energy Management, State Biotechnological University, 61052 Kharkiv, Ukraine
[5] SCIRE Foundation, 00867 Warsaw, Poland
[6] Department of Entrepreneurship and Marketing, Institute of Economics and Management, Ivano-Frankivsk National Technical Oil and Gas University, 76019 Ivano-Frankivsk, Ukraine
[7] Department of Business Administration, Daffodil International University, Dhaka 1341, Bangladesh
[8] Department of Business Administration, The Millennium University, Dhaka 1217, Bangladesh
* Correspondence: vitaliinitsenko@onu.edu.ua

Citation: Al-Quraan, T.M.A.; Vovk, O.; Halko, S.; Kvitka, S.; Suprun, O.; Miroshnyk, O.; Nitsenko, V.; Zayed, N.M.; Islam, K.M.A. Energy-Saving Load Control of Induction Electric Motors for Drives of Working Machines to Reduce Thermal Wear. *Inventions* 2022, 7, 92. https://doi.org/10.3390/inventions7040092

Academic Editor: Om P. Malik

Received: 31 July 2022
Accepted: 3 October 2022
Published: 12 October 2022

Publisher's Note: MDPI stays neutral with regard to jurisdictional claims in published maps and institutional affiliations.

Abstract: The influence of reduced voltage on the service life of an induction motor is considered in this article. An algorithm for calculating the rate of thermal wear of induction motor insulation under reduced supply voltage depending on the load and the mechanical characteristics of the working machine has been developed. It determines the change in the rate of thermal wear under alternating external effects on the motor (supply voltage and load) and allows forecasting its service life under these conditions. The dependency graphs of the rate of insulation thermal wear on the motor load for various levels of supply voltage and various mechanical characteristics of working machines are provided in the work. It was determined that the rate of thermal wear of the induction motor insulation increases significantly when the voltage is reduced compared to its nominal value with nominal load on the motor. The authors propose to consider this fact for resource-saving control of the motor. The paper presents the results of experimental verification of the obtained rule for "Asynchronous Interelectro" (AI) series electric motors that confirm its accuracy. Based on the obtained correlation, the rule of voltage regulation in energy-saving operation mode has been derived. The proposed rule takes into account the thermal impact on the electric motor running in energy-saving mode and enables saving its resource, which, in turn, results in extending its service life. The research does not consider additional effects on the electric motor except the thermal one.

Keywords: induction motor; reduced voltage; working machine; motor load; insulation heating; thermal wear of insulation; extending service life; rule for voltage regulation

1. Introduction

Currently, about 40% of electricity produced in the world is consumed by induction motors, the number of which exceeds 300 million units [1,2]. They are mostly used in various industrial production processes, consuming up to 80% of electricity in this sector of the economy [3,4]. A significant share of that electricity is used by pumping systems, which consume about 22% of the mentioned amount [5].

Such a wide distribution of induction motors can be explained by their high reliability and relatively low manufacturing cost [6,7]. At the same time, the operational reliability of induction motors in all sectors of the European economy is insufficient, as evidenced by the following: the annual cost of repairs and maintenance of these motors amounts to 8% of the annual cash flow of any branch of economic activity [8]. Up to 4% of induction motors fail annually [9]. The primary reason for such low operational reliability of these motors is external exposure to various factors on the part of the mains power supply and the peculiarities of working machines. This becomes especially relevant with the increasing use of alternative energy sources for power supply from local networks (such factors include changes in wind speed, solar insolation, etc.) [10–12]. The thermal ageing of insulation and reduced supply voltage at the terminals of induction motors have a significant impact on their operational reliability [13–19].

It is known that even a slight decrease in the quality of the supply voltage leads to negative consequences associated with the deterioration of insulation and a decrease in the energy efficiency and power of induction motors [20]. When the voltage in the network deviates from the nominal value, the active power on the motor shaft remains almost constant, while the active power loss changes [21,22]. This causes a change in the heating of the induction motor and, as a consequence, a change in the rate of the thermal wear of its insulation. Thermal wear is one of the main factors influencing the deterioration of insulation [23–25].

For many induction motors, energy-saving control systems are currently used. One of the regulating factors of these systems is the supply voltage, which is usually reduced in relation to the nominal value. Many studies are concerned with this problem, and it has been analyzed in sufficient detail in review works [26–30]. The number of publications on this topic continues to increase, which indicates the urgency of the issue. In [28], energy-saving control methods for induction motors are divided into the following groups: (1) methods concerning the condition of the motor; (2) methods based on the model of power loss in the motor; and (3) methods using direct optimization.

The first group includes methods in which the regulation of the speed of electric motors is performed by applying voltage or frequency or both [26–30]. Direct or indirect control of the current (or the square of the current strength) of the stator winding is performed. When its value is minimal at the given speed of the shaft, it is believed that the power consumption of the motor will also be minimal. However, only variable power losses depend on the current of the stator winding; constant power losses are independent and are determined by the magnetic flux and the speed of rotation. Therefore, the disadvantages of this group of methods include the fact that the mains deviation, the load of the motor, the mechanical characteristics of the working machine, the heating of its active parts, and the rate of thermal wear of the motor insulation are not taken into account.

The second group of methods is based on the dependence of the combined power loss on the magnetizing current [26–30]. By finding the first derivative of this dependence and setting it to zero, an equation for controlling the speed of the motor is obtained. The methods differ by types of dependencies and control algorithms, and the load of the motor is usually taken into account by considering the current consumption. The disadvantages of these methods include significant linearization of the motor parameters and neglecting the mains deviations. Further, the mechanical characteristics of the working machine and the thermal wear of the motor insulation are not taken into account.

The third group includes methods using artificial neural networks or fuzzy logic devices for the optimization of power consumption [26–30]. These methods include the approximation of non-linear dependence, for example, between its magnetizing current and torque, speed, and power loss. The optimum value of the magnetizing current is then determined. The disadvantages of these methods include the relatively long optimization time and neglecting the mains deviation. The mechanical characteristics of the working machine and the thermal wear of the motor insulation are not taken into account.

In [31], a method for increasing the efficiency of the induction motor is proposed, provided that the temperature of its stator winding increases. It is proposed to compensate the voltage drop in the stator winding, which is caused by additional heating. For this purpose, voltage is increased in direct proportion to the compensation factor. This voltage compensation does not take into account its phase, and therefore the current in its stator windings increases. This results in its overheating and accelerated wear of insulation.

In [32], it is proposed to run the induction motor with minimum power loss by adjusting the voltage. For this purpose, the author obtained a voltage dependency of the motor, for which the power loss is minimal. However, mechanical losses in electric motors that may vary due to even minor bearing damage are not taken into account. Further, the author disregards the fact that the motor slip is a function of the applied voltage and load of the electric motor. The mentioned deficiencies will result in reduced voltage. This will cause motor overload, which will lead to the overheating and wear of the stator winding insulation.

In [33], it is proposed to run the induction motor at minimum energy loss by adjusting the voltage. The authors developed the field-oriented control system of the motor speed that provides the motor phase currents control and the rotation speed of its shaft. These values are used to determine the voltage, under which the motor will consume a minimum of power under the given operating conditions. This method assumes that the parameters of the motor are constant. However, when the motor is running, these parameters change due to the magnetization reversal of the magnetic circuit and the heating of its active parts. Unless this is taken into account, the optimal voltage will be too low, which will cause motor overload and result in the overheating and accelerated wear of its insulation.

Thus, all of the mentioned methods of energy-saving control disregard the relationship between the voltage of the motor and the rate of its insulation wear, which results in its low operational reliability.

Let us consider the influence of voltage reduction on the speed of the insulation wear of the motor and its operational reliability based on [6,34–37]. Reducing the voltage of running induction motors leads to an increase in their slip and, as a consequence, to an increase in current consumption and current overload. The latter causes vibration, high electrodynamic forces between the conductors of the winding, low resistance of the turn insulation, and overheating of the turn insulation, the ground insulation, and the core, resulting in its thermal wear and leading to local defects of the insulation in its slot and coil end parts. These defects cause incomplete breakdowns of the turn insulation, followed by complete inter-turn short circuits, which lead to motor failures.

At the same time, it is essential that the change in the load on the motor and the change in the temperature of the surroundings are considered. Increasing the load on the working machines leads to an increase in current consumption, overload of the motors, overheating of the turn insulation, its thermal wear, local defects of turn insulation, incomplete break-downs, and eventually, to complete inter-turn short circuits and failures of the motor. The temperature of the surroundings rise, leading to the same consequences.

Thus, the influence of voltage reduction on the rate of the insulation wear of the motor and its operational reliability will be as follows (Figure 1).

As follows from Figure 1, the main manifestation of the unacceptable voltage reduction of induction motors is an increase in their heating, which leads to thermal wear of the insulation. The rate of thermal wear of the insulation is calculated as follows [38,39]:

$$\varepsilon = \varepsilon_n \cdot exp\left[B\left(\frac{1}{\tau_{1n} + \vartheta_{sur.n} + 273} - \frac{1}{\tau_{1st} + \vartheta_{sur} + 273} \right) \right], \tag{1}$$

where ε—rate of thermal wear of insulation, *bas.h/h.* (base hours per hour);
ε_n—nominal rate of thermal wear of insulation, *bas.h/h.*;
B—insulation class, K;
τ_{1n}—nominal steady-state temperature excess of insulation, °C;
$\vartheta_{sur.n}$—nominal temperature of the surroundings, °C;
τ_{1st}—current steady-state temperature excess of insulation, °C;
$\vartheta_{sur.}$—current temperature of the surroundings, °C.

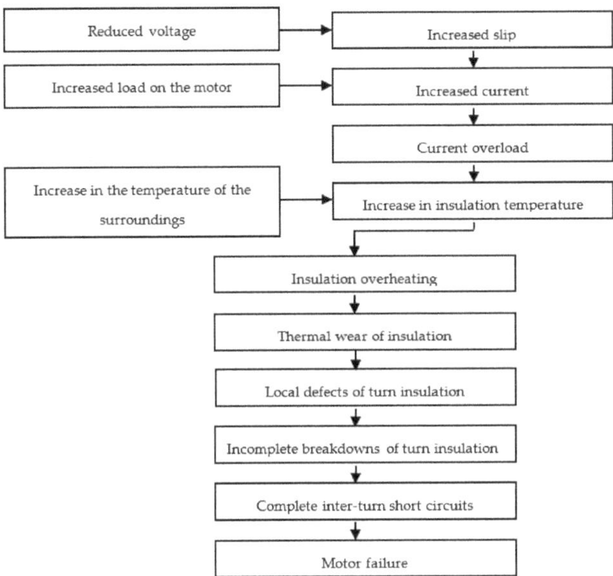

Figure 1. The influence of voltage reduction on the rate of insulation wear of the motor and its operational reliability.

Thus, the rate of thermal wear of insulation depends on the nominal rate of thermal wear of insulation ε_n, the parameter characterizing the insulation class, B, nominal steady-state temperature excess of insulation, τ_{1n}, nominal temperature of the surroundings, $\vartheta_{sur.n}$, current steady-state temperature excess of insulation, τ_{1st}, and current temperature of the surroundings, ϑ_{amb}, that is $\varepsilon = f\,(\varepsilon_n,\, B,\, \tau_{1n},\, \vartheta_{sur.n},\, \tau_{1st},\, \vartheta_{sur.})$. The values of $\varepsilon_n,\, B,\, \tau_{1n},\, \vartheta_{sur.n}$ depend only on the design of the induction motor, the value of $\vartheta_{sur.}$ determines the effect of temperature of the surroundings on the rate of thermal wear of insulation, and the value of τ_{1st} determines the influence of the operation mode on the rate of thermal wear of insulation, which is characterized by the load voltage, the load level, and the mechanical characteristics of the working machine. These factors affect the slip, power losses, and other performance indicators of the motor, causing changes in its thermal condition. The latter is characterized by changes in steady-state temperature excess of its winding, leading to a change in the rate of thermal wear of its insulation. Therefore, the purpose of the present work is to establish the resource-saving rule of the voltage regulation of the motor in energy-saving operation mode. To achieve the purpose of the study, the following problems must be solved:

- To establish how the steady-state temperature excess of the stator insulation depends on the losses of active power in its structural components;
- To find out how the active power losses in the structural components of the motor depend on the voltage and the load;
- To establish the dependence between the current slip of the motor and the voltage, the load of the motor, and the type of mechanical characteristics of the working machine;
- To perform analytical modelling of the dependence of the rate of thermal wear of insulation on voltage and load, taking into account the mechanical characteristics of the working machines.

2. Materials and Methods

- To solve the above problems, the following methods were used. The first problem of the research was solved by analyzing the thermal processes in the induction motor according to its thermal equivalent circuit, which contains three elements (stator winding, rotor winding, and steel). The main focus was on the heating of the motor in the steady operation mode under the uncertainty of its thermal conductivity. The coefficient of the influence of the power loss of the motor on its heating was calculated. The analysis of the steady-state thermal condition of the motor was carried out in nominal operation mode, idling and short circuit.
- The second objective of the study was achieved by analyzing the process of electromechanical energy conversion in an induction motor according to its L-shaped equivalent circuit. This conversion of energy was analyzed for the steady-state operation mode of the motor. Special attention was paid to the power loss in the bodies of the thermal equivalent circuit of the motor. These losses were analyzed with nominal and non-nominal voltage and load.
- The third objective of the research was achieved by analyzing the process of electromechanical energy conversion in the motor according to its mechanical characteristics, the mechanical characteristics of the working machine, and its empirical equation. This energy conversion was studied for the steady-state operating mode of the motor in the operating range of loads. The linearization of the listed characteristics and the main provisions of analytical geometry on the plane were applied. The slip analysis was performed at nominal and non-nominal voltage and load taking into account the type of mechanical characteristics of the working machine.

3. Results and Discussion

3.1. Dependence of the Steady-State Temperature Excess of the Stator Insulation on the Active Power Losses in Its Structural Components

To establish this dependence, we apply the thermal equivalent circuit of the induction motor. There are two types of such circuits: two-element and multi-element, with different interpretations. The two-element thermal circuit implies that power losses take place in the copper and steel of the motor [40]. It requires few data for determining the steady-state temperature excess of the stator winding, all of which are available. The calculations are performed quickly, but their accuracy is far from perfect. In addition, the results of calculations are applied for all electric motors of this type at the same load and voltage, not taking into account possible changes in their structural components caused by operation (deterioration of cooling conditions, malfunctions, etc.). The multi-element thermal circuit distributes the structural components of the motor into sections, and the power losses in motor units are distributed into separate components [41,42]. In this case, calculations have high accuracy, but are quite bulky and require a considerable amount of input data: geometrical data of the electric motor, characteristics of materials, boundary conditions, etc., which are not always available for a specific motor. Further, the results of the calculations are applied for all motors of this type at the same load and voltage, not taking into account possible changes in the structural components.

To avoid the above-mentioned disadvantages, it is necessary to get a new algorithm for determining the steady-state temperature excess of the stator winding, which will take into account the thermal condition of a particular motor and will have sufficient accuracy. To establish this dependence, let us consider the induction motor as a system of three bodies: (1) stator winding; (2) rotor winding; (3) steel (core, mechanical part, and housing) and assume that the heat capacity of the environment around the motor is equal to infinity and the temperature of the surroundings is nominal and constant (Figure 2). This is sufficient for the analysis of the operational processes in it, more accurate than [40], and does not require a large amount of output data as opposed to [41,42].

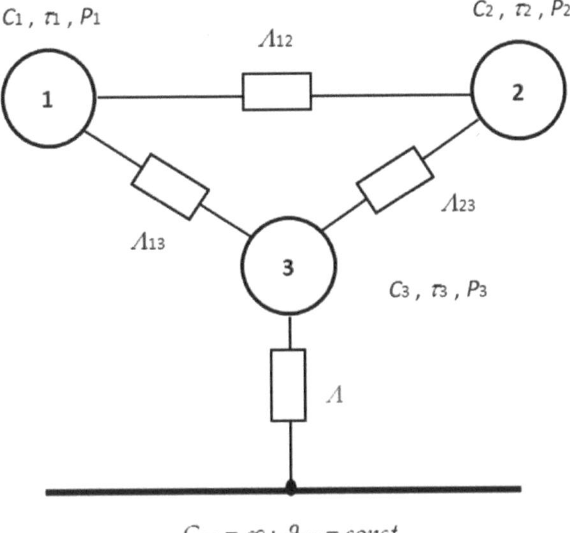

Figure 2. Thermal equivalent circuit of the motor.

The diagram (Figure 2) shows the following symbols:

C_1, C_2, C_3—heat capacity of the respective bodies, J/°C;

τ_1, τ_2, τ_3—excess temperature of the respective bodies over the temperature of the surroundings, °C;

P_1, P_2, P_3—loss of active power in the respective bodies, W;

Λ_{12}, Λ_{13}, Λ_{23}—thermal conductivity between the respective bodies, W/°C;

Λ—thermal conductivity between the third body and the surroundings, W/°C;

C_{sur}—heat capacity of the surroundings, W/°C;

ϑ_{sur}—temperature of the surroundings, °C.

The system of heat balance equations for the circuit shown in Figure 2 is as follows:

$$\left.\begin{array}{l}
C_1 d\tau_1 + \Lambda_{13}(\tau_1 - \tau_3)dt - \Lambda_{12}(\tau_2 - \tau_1)dt = P_1 dt; \\
C_2 d\tau_2 + \Lambda_{23}(\tau_2 - \tau_3)dt + \Lambda_{12}(\tau_2 - \tau_1)dt = P_2 dt; \\
C_3 d\tau_3 + \Lambda\tau_3 dt - \Lambda_{13}(\tau_1 - \tau_3)dt - \Lambda_{23}(\tau_2 - \tau_3)dt = P_3 dt.
\end{array}\right\} \tag{2}$$

The solution of the system of Equation (2) for τ_1 in the steady-state mode of the operation of the motor (τ_{1st}) is as follows:

$$\tau_{1st} = a \cdot P_1 + b \cdot P_2 + c \cdot P_3, \tag{3}$$

where a—impact coefficient of losses P_1 on the heating of the stator insulation, °C/W; b—impact coefficient of losses P_2 on the heating of the stator insulation, °C/W; c—impact coefficient of losses P_3 on the heating of the stator insulation, °C/W.

The coefficients obtained after solving (2) are equal to

$$\left.\begin{array}{l}
a = \dfrac{\Lambda_{12}\Lambda_{23} + (\Lambda + \Lambda_{13})(\Lambda_{12} + \Lambda_{23})}{m}, \\
b = \dfrac{\Lambda_{13}\Lambda_{23} + \Lambda_{12}(\Lambda + \Lambda_{13} + \Lambda_{23})}{m}; \\
c = \dfrac{\Lambda_{13}(2\Lambda_{12} + \Lambda_{23})}{m},
\end{array}\right\} \tag{4}$$

where

$$m = (\Lambda + \Lambda_{13} + \Lambda_{23})(\Lambda_{12}\Lambda_{23} + \Lambda_{13}(\Lambda_{12} + \Lambda_{23})) -$$
$$-\Lambda_{23}(\Lambda_{12}\Lambda_{23} + \Lambda_{13}(\Lambda_{23} - 2\Lambda_{12} - \Lambda_{13})) - \Lambda_{12}\Lambda_{13}^2. \tag{5}$$

Thus, the steady-state temperature excess depends on the coefficients of the impact of active power losses in the motor units on the heating of the stator insulation and is a function of these losses (P_1, P_2, P_3).

In turn, coefficients a, b, c represent a complex function of thermal conductivity. These coefficients allow avoiding thermal conductivity of the electric motor, in contrast to [41,42]. In order to determine these coefficients, the excess of the stator insulation temperature over the temperature of the surroundings was studied under short circuit and idling. Based on these parameters, a system of three equations was derived, each of which represents an Equation (3) for the tests of nominal load, short circuit, and idling. It is assumed that these coefficients are the same in all of these modes. The system of equations is as follows:

$$\left.\begin{array}{l} a \cdot P_{1n} + b \cdot P_{2n} + c \cdot P_{3n} = \tau_{1n}; \\ a \cdot P_{1sh} + b \cdot P_{2sh} + c \cdot P_{sh} = \tau_{1sh}; \\ a \cdot P_{1id} + b \cdot P_{2id} + c \cdot P_{3id} = \tau_{1id}, \end{array}\right\} \tag{6}$$

where τ_{1n}, τ_{1sh}, τ_{1id}—steady-state excess of the stator insulation temperature in experiments of nominal load, short circuit, and idling, respectively, °C;

P_{1n}, P_{2n}, P_{3n}—loss of active power in the respective bodies in the nominal load test, W;

P_{1sh}, P_{2sh}, P_{3sh}—loss of active power in the respective bodies in the short circuit test, W;

P_{1id}, P_{2id}, P_{3id}—loss of active power in the respective bodies in the idling test, W.

In the short circuit test, the mechanical losses were equal to zero, and it was performed at nominal current in the stator insulation, so it is assumed that $P_{3sh} \approx 0$, and $P_{1sh} \approx P_{1n}$ and $P_{2sh} \approx P_{2n}$. In the idling test, the voltage at the motor terminals was nominal, and the slip was $s \approx 0$, so it is assumed that $P_{3id} \approx P_{3n}$ and $P_{2id} \approx 0$. Taking into account these assumptions, the system of Equation (6) with coefficients a, b, c is solved as follows:

$$\left.\begin{array}{l} a = \dfrac{\tau_{1sh} + \tau_{1id} - \tau_{1n}}{P_{1id}}; \\[2ex] b = \dfrac{\tau_{1sh} - \dfrac{P_{1n}}{P_{1id}} \cdot (\tau_{1sh} + \tau_{1id} - \tau_{1n})}{P_{2n}}; \\[2ex] c = \dfrac{\tau_{1n} - \tau_{1sh}}{P_{3n}}. \end{array}\right\} \tag{7}$$

The obtained Equations (3) and (7) allow the calculation of the steady-state temperature excess of the induction motor at any rate of power loss in the bodies of its thermal equivalent circuit due to the impact of these losses on the heating of the stator winding. These coefficients are determined by standard operational tests (idling and short circuit) and allow identification of the thermal condition of a particular motor, taking into account the current condition of its active parts. This increases the accuracy of the proposed method, which distinguishes it from the ones proposed in [40–42]. The value of the steady-state temperature excess, calculated according to the methods described in [40–42], is applied for all motors of a certain type and size, and according to [40–42], it does not matter whether the motor is new or has already been in operation for some time. However, while the motor is in service, changes may occur in its active parts (deterioration of cooling conditions, malfunctions, etc.), which will lead to changes in its thermal conductivity. The latter will cause a change in the steady-state temperature excess of the motor, which is not taken into account in [40–42]. The method of determining the steady-state temperature excess, which is proposed in the present work, has no such deficiencies, which is conditioned by the added coefficients of the active power loss impact on the heating of the stator windings, which are experimentally determined for each electric motor, instead of using the nameplate data.

3.2. Dependence of Active Power Losses of the Motor on the Voltage and Load

Proceeding from (3), it is necessary to establish the dependence of power losses, which occur in the bodies of thermal equivalent circuit of the induction motor (Figure 2), on the voltage taking into account the load of the motor. In [43], such correlation is indicated, but it is assumed that the load is constant. In [44], the correlation of losses is partially indicated without taking into account the change in the load. This correlation is described in [20,32], but only between two components, permanent and variable losses. Thus, the existing publications do not solve the problem in full.

To establish this dependence, let us consider the L-shaped equivalent circuit of a three-phase induction motor shown in Figure 3.

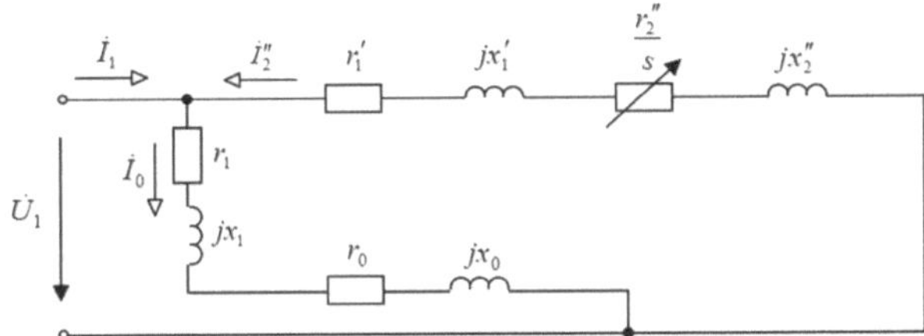

Figure 3. L-shaped equivalent circuit of a three-phase induction motor.

The diagram (Figure 3) shows the following: $\dot{U}_1$—complex of effective phase voltage; $\dot{I}_1$—complex of effective phase current; $\dot{I}_2''$—complex of effective current in the load branch; $\dot{I}_0$—complex of effective current in the magnetizing branch; s—slip of the motor; r_1', x_1', r_2'', x_2'', r_1, x_1, r_0 —circuit parameters.

According to this circuit, the loss of active power in the first body of the thermal equivalent circuit can be represented as

$$P_1 = 3 \cdot r_1' \cdot \left(I_2''\right)^2 = 3 \cdot \frac{r_1'}{\left(r_1' + \frac{r_2''}{s}\right)^2 + \left(x_1' + x_2''\right)^2} \cdot U_1^2. \tag{8}$$

At nominal voltage U_{1n} and nominal load (that is nominal slip s_n), Equation (8) will take the following form:

$$P_{1n} = 3 \cdot \frac{r_1'}{\left(r_1' + \frac{r_2''}{s_n}\right)^2 + \left(x_1' + x_2''\right)^2} \cdot U_{1n}^2. \tag{9}$$

By dividing (8) by (9), we obtain:

$$P_1 = P_{1n} \cdot \frac{\left(r_1' + \frac{r_2''}{s_n}\right)^2 + \left(x_1' + x_2''\right)^2}{\left(r_1' + \frac{r_2''}{s}\right)^2 + \left(x_1' + x_2''\right)^2} \cdot k_u^2, \tag{10}$$

where $k_u = U_1/U_{1n}$—reduction of voltage coefficient.

The loss of active power in the second body of the thermal equivalent circuit can be represented as follows:

$$P_2 = 3 \cdot r_2'' \cdot \left(I_2''\right)^2 = 3 \cdot \frac{r_2''}{\left(r_1' + \frac{r_2''}{s}\right)^2 + \left(x_1' + x_2''\right)^2} \cdot U_1^2. \tag{11}$$

At nominal voltage U_{1n} and nominal load (that is nominal slip s_n), Equation (11) will take the following form:

$$P_{2n} = 3 \cdot \frac{r_2''}{\left(r_1' + \frac{r_2''}{s_n}\right)^2 + \left(x_1' + x_2''\right)^2} \cdot U_{1n}^2.$$ (12)

By dividing (11) by (12), we obtain:

$$P_2 = P_{2n} \cdot \frac{\left(r_1' + \frac{r_2''}{s_n}\right)^2 + \left(x_1' + x_2''\right)^2}{\left(r_1' + \frac{r_2''}{s}\right)^2 + \left(x_1' + x_2''\right)^2} \cdot k_u^2.$$ (13)

The loss of active power in the third body of the thermal equivalent circuit is calculated as follows:

$$P_3 = 3 \cdot (r_1 + r_0) \cdot I_0^2 = 3 \cdot \frac{(r_1 + r_0)}{(r_1 + r_0)^2 + (x_1 + x_0)^2} \cdot U_1^2.$$ (14)

At nominal voltage U_{1n}, equation (14) will take the following form:

$$P_{3n} = 3 \cdot \frac{(r_1 + r_0)}{(r_1 + r_0)^2 + (x_1 + x_0)^2} \cdot U_{1n}^2.$$ (15)

By dividing (14) by (15), we obtain:

$$P_3 = P_{3n} \cdot \left(\frac{U_1}{U_{1n}}\right)^2 = P_{3n} \cdot k_u^2.$$ (16)

The derived Equations (10), (13) and (16) show the dependence of power losses, which occur in the bodies of the proposed thermal equivalent circuit (Figure 2), on the voltage and load of the motor according to its L-shaped equivalent circuit. These equations allow the determination of constant and variable power losses divided into two parts, which, in contrast to [20,32], completely satisfy the conditions of the objective of the present study. Compared to [43,44], the above expressions take into account the operating mode of the motor more accurately. When the motor is controlled with voltage, the load can change, which will require clarification of the value of voltage. Mathematical models of power loss given in [43,44] do not take this into account and therefore lead to a lower optimum voltage. The latter causes the switching of the electric motor to the overload mode, its overheating, and acceleration of the thermal wear of its insulation. The mathematical model of power loss proposed in the present work is devoid of such deficiencies as it takes into account simultaneous changes of the voltage and the load of the motor. The mains deviation is taken into account through coefficient k_u, and the load of the motor —through the slip s. The additional losses in electric motors are not taken into account; however, they amount to no more than 1% of the power consumed, so this disadvantage will not affect the results of calculations significantly.

Therefore, taking into account (10), (13), and (16), Equation (3) will take the following form:

$$\tau_{1st} = \left((a \cdot P_{1n} + b \cdot P_{2n}) \cdot \frac{\left(r_1' + \frac{r_2''}{s_n}\right)^2 + \left(x_1' + x_2''\right)^2}{\left(r_1' + \frac{r_2''}{s}\right)^2 + \left(x_1' + x_2''\right)^2} + c \cdot P_{3n} \right) \cdot k_u^2.$$ (17)

3.3. Dependence of the Slip of the Motor on Voltage and Load

In Equation (17), the slip is the indicator of the operation mode of the motor, in addition to the multiplicity of the voltage, which also depends on the voltage. It is known that the slip of the induction motor at constant load depends on the square of the voltage [45,46], but it is not clear how the simultaneous change of voltage and load on the motor shaft affects the slip. Moreover, it is unclear whether it is necessary to take into account the mechanical characteristics of the working machine driven by the electric motor. Therefore, let us analyze the effect of reducing the voltage on the slip of the motor taking into account its load coefficient and the mechanical characteristics of the working machine that it drives. For this purpose, let us use the mechanical characteristics of the three-phase induction motor and the working machine (Figure 4).

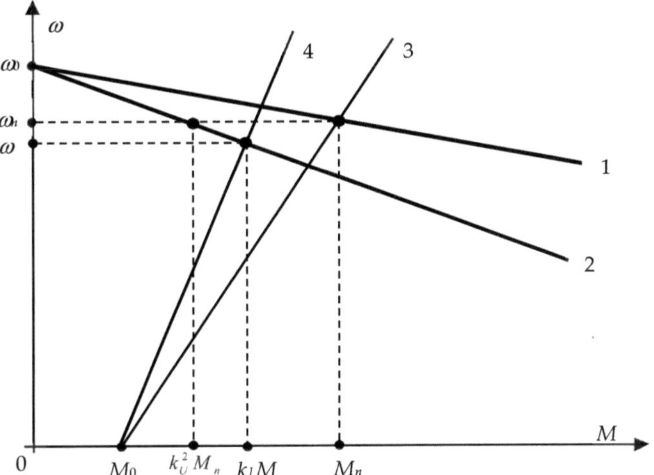

Figure 4. Mechanical characteristics of the induction motor and the working machine.

Figure 4 shows the following:

1, 2—linearized working sections of the mechanical characteristics of the induction motor, at nominal and reduced voltage, respectively;
3, 4—mechanical characteristics of the working machine, at nominal and reduced load, respectively;
ω_0, ω_n, ω—synchronous, nominal and current angular velocity of the induction motor, respectively;
M_0—initial resisting torque of the working machine;
M_n, M—nominal and current torques on the motor shaft, respectively;
k_l—load coefficient of the motor;
k_u—voltage reduction coefficient.

Based on Figure 4 and the triangle similarity principle,

$$\frac{\omega_0 - \omega_n}{\omega_0 - \omega} = \frac{k_l \cdot M}{k_u^2 \cdot M_n}. \tag{18}$$

After transformations of (18), we obtain:

$$M = \frac{s}{s_n} \cdot \frac{k_u^2}{k_l} \cdot M_n. \tag{19}$$

The slip of the motor determines the effective current in the stator and rotor insulations, the square of which determines the amount of heat released during motor operation. Therefore, let us investigate how the motor slip changes depending on the load and voltage reduction coefficients, taking into account the mechanical characteristics of the working machine. To do this, let us use the empirical equation of the mechanical characteristics of the working machine [32]:

$$M_r = M_0 + (M_{r.n} - M_0) \cdot \left(\frac{\omega}{\omega_n}\right)^x, \tag{20}$$

where M_r, $M_{r.n}$—current and nominal values of the resisting torque of the working machine, respectively, $N \cdot m$; x—degree indicator characterizing the change in the static torque of the working machine when its speed changes.

If we apply (19) to (20) under the condition that $M = M_r$, after transformations we obtain:

$$\frac{s}{s_n} \cdot \frac{k_u^2}{k_l} = M_{0*} + (1 - M_{0*}) \cdot \left(\frac{1 - s}{1 - s_n}\right)^x, \tag{21}$$

where $M_{0*} = M_0 / M_{r.n}$—relative value of the initial resisting torque of the working machine.

Having applied (21), we found the expressions of motor slip depending on the load and voltage reduction factors, taking into account the mechanical characteristics of the working machine (i.e., the value of the degree indicator characterizing the change in the static torque of the working machine when its speed changes).

For a working machine with a mechanical characteristic that does not depend on the speed ($x = 0$), this dependence will be as follows:

$$s = \frac{k_l}{k_u^2} \cdot s_n. \tag{22}$$

For a working machine with a ramp mechanical characteristic ($x = 1$), this dependence will be as follows:

$$s = \frac{M_{0*} + \dfrac{1 - M_{0*}}{1 - s_n}}{\dfrac{k_u^2}{k_l \cdot s_n} + \dfrac{1 - M_{0*}}{1 - s_n}}. \tag{23}$$

For a working machine with a parabolic mechanical characteristic ($x = 2$), this dependence will be as follows:

$$s = 1 + \frac{(1 - s_n)^2}{2 \cdot (1 - M_{0*})} \cdot \left(\frac{k_u^2}{k_l \cdot s_n} - \sqrt{\frac{k_u^2}{k_l \cdot s_n} \cdot \left(\frac{k_u^2}{k_l \cdot s_n} + \frac{4 \cdot (1 - M_{0*})}{(1 - s_n)^2}\right)} - \frac{4 \cdot (1 - M_{0*})}{M_{0*} \cdot (1 - s_n)^2}\right). \tag{24}$$

For a working machine with a hyperbolic mechanical characteristic ($x = -1$), this dependence will be as follows:

$$s = \frac{k_l \cdot s_n}{2 \cdot k_u^2} \left(\frac{k_l \cdot s_n}{k_u^2} + M_{0*} - \sqrt{\left(\frac{k_l \cdot s_n}{k_u^2} + M_{0*}\right)^2 - \frac{4 \cdot k_u^2}{k_l \cdot s_n}((1 - M_{0*})(1 - s_n) + M_{0*})}\right). \tag{25}$$

Compared to [45,46], in Equations (22)–(25), the operating mode of the motor is taken into account more accurately. When controlling the motor, the use of voltage can change the load of the motor, which will require clarification of the voltage value. The mathematical models of the slip given in [45,46] do not take this into account and therefore lead to a lower value of optimum voltage. The latter causes the switching of the motor to the overload mode, its overheating, and accelerated thermal wear of its insulation. The mathematical model of the slip proposed in the present work does not have such a drawback because it takes into account simultaneous changes in the voltage and load of the motor. The mains deviation is taken into account through coefficient k_u, and the load of the motor—through coefficient k_l; the mechanical characteristics of the working machine are also taken into account. Thus, the mathematical model of the slip obtained in the work is more universal compared to that in [45,46]. The disadvantage of this model is its limited application restricted only by the operating range of the slip; however, it is sufficient for the analysis of the steady-state operating mode of the motor.

3.4. Modelling of the Dependence of the Insulation Thermal Wear Rate of the Induction Motor in the Function of Voltage and Load Taking into Account the Mechanical Characteristics of the Working Machines

Analytical studies allow us to conclude that the rate of thermal wear of insulation is a function of the design and operational performance of the motor and the working machine. The structural parameters of the electric motor are determined by both its general design and its insulation design, and the design parameters of the working machine—by its mechanical characteristics. Performance indicators depend on the mode of operation and take into account the voltage reduction of the motor, its load coefficient, and the temperature of the surroundings. That is, $\varepsilon = f\,(\varepsilon_n,\, B,\, \tau_{1n},\, \tau_{1sh},\, \tau_{1id},\, \vartheta_{sur.n},\, r_1',\, r_2'',\, x_1',\, x_2'',\, P_{1n},\, P_{2n},\, P_{3n},\, s_n,\, M_{0*},\, k_u,\, k_l,\, \vartheta_{amb})$. The obtained dependence allows us to calculate the rate of thermal wear of insulation using Equations (1), (7), (17) and (22)–(25). These equations, written in the appropriate sequence, represent an algorithm for calculating ε in relation to the specified design and performance indicators.

Using the described calculation algorithm, the change in the rate of thermal wear of insulation of the induction motor AIR132S4 (nominal motor power 7.5 kW) was modelled depending on performance indicators: voltage reduction coefficient and load coefficient.

It was assumed that the load coefficient of the motor varies from 0 to 1, the voltage reduction coefficient has the following values: 1.0; 0.95; 0.9; and 0.85, and the temperature of the surroundings is 40 °C.

The results of modelling the dependences $\varepsilon = f\,(k_l)$ for electric motors of working machines with different mechanical characteristics are presented in Figures 5–8. The figures show: 1—dependence at $k_u = 1$; 2—dependence at $k_u = 0.95$; 3—dependence at $k_u = 0.9$; 4—dependence at $k_u = 0.85$.

As can be seen from Figures 5–8, the dependences are of the same type, which is determined by the exponential dependence of the rate of thermal wear of insulation shown in (1). However, the numerical values of the rate of thermal wear of insulation with the same electric motor driving working machines with different mechanical characteristics differ when the voltage is reduced. This can be seen from the results shown in Tables 1–4. The tables show examples of the dependence $\varepsilon = f\,(k_l)$ at $k_u = 0.9$ for AIR132S4 induction motor driving working machines with different types of mechanical characteristics.

The above results show that when the motor drives different types of working machines, the rate of thermal wear of its insulation differs at loads close to nominal. This is clearly seen at nominal load, at which it differs by 4 to 10 base hours per hour. In addition, the above results show that when the voltage decreases, the rate of thermal wear of insulation increases sharply. Therefore, for example, at a decrease in voltage of 10% in relation to the nominal voltage and with the electric motor working at nominal load, the rate of thermal wear of insulation depending on the type of the working machine increases by 30–45 times compared to the nominal value. The analysis of the obtained dependencies, shown in Figures 5–8, revealed the following regularities between the voltage reduction coefficient and the load coefficient. Regardless of the type of working machine, the rate of thermal wear of the insulation of the specified motor will be nominal under such conditions: $k_u = 1.0$ and $k_l = 1.0$; $k_u = 0.95$ and $k_l = 0.9025$; $k_u = 0.9$ and $k_l = 0.81$; $k_u = 0{,}85$ and $k_l = 0.7225$. A comparison between these values shows that the load factor equals the square of the voltage deviation coefficient, i.e.,

$$k_l = k_u^2. \tag{26}$$

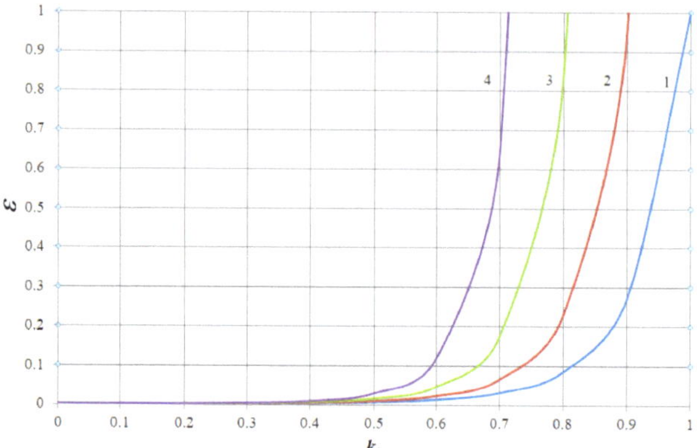

Figure 5. Dependence $\varepsilon = f(k_l)$ at $x = 0$.

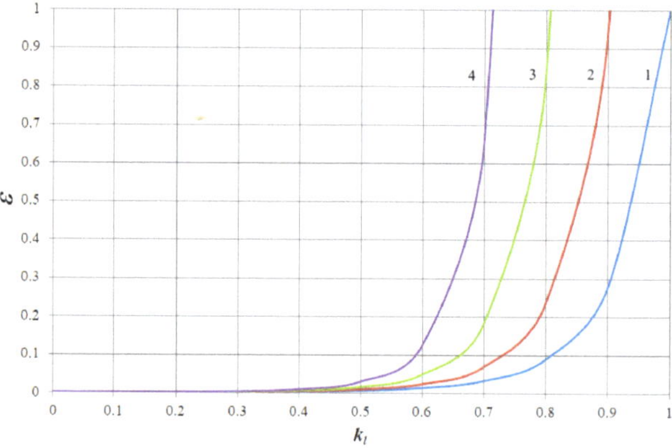

Figure 6. Dependence $\varepsilon = f(k_l)$ at $x = 1$.

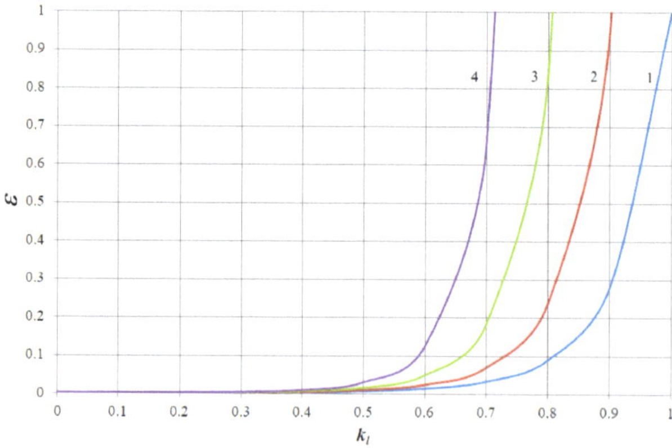

Figure 7. Dependence $\varepsilon = f(k_l)$ at $x = 2$.

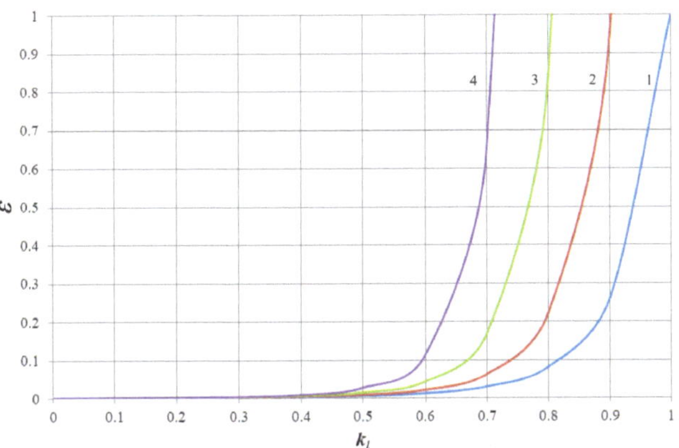

Figure 8. Dependence $\varepsilon = f(k_l)$ at $x = -1$.

Table 1. Dependence $\varepsilon = f(k_l)$ at $x = 0$, $k_u = 0.9$.

k_l	0.4	0.5	0.6	0.7	0.8	0.9	1.0
ε, bas.h/h	0.006	0.015	0.046	0.176	0.844	5.10	38.9

Table 2. Dependence $\varepsilon = f(k_l)$ at $x = 1$, $k_u = 0.9$.

k_l	0.4	0.5	0.6	0.7	0.8	0.9	1.0
ε, bas.h/h	0.006	0.016	0.048	0.183	0.848	4.84	34.1

Table 3. Dependence $\varepsilon = f(k_l)$ at $x = 2$, $k_u = 0.9$.

k_l	0.4	0.5	0.6	0.7	0.8	0.9	1.0
ε, bas.h/h	0.006	0.016	0.051	0.189	0.852	4.61	30.1

Table 4. Dependence $\varepsilon = f(k_l)$ at $x = -1$, $k_u = 0.9$.

k_l	0.4	0.5	0.6	0.7	0.8	0.9	1.0
ε, bas.h/h	0.006	0.014	0.043	0.169	0.840	5.37	45.0

Similar results were obtained when modelling the rate of thermal wear of insulation of most electric motors in this series.

In order to experimentally test (26) for the AIR132S4 induction motor, a heating test was performed. It was outlined as follows: sinusoidal voltage of a certain value was applied to the motor, and a load of a certain value was applied to the shaft. An adjustable shutter fan was used as the load. The motor was running for 6 h at each voltage value, every 1000 s its insulation temperature was measured using a thermistor, and the temperature of the surroundings was measured with an electronic thermometer. According to the measurement results, the excess of the insulation temperature over the temperature of the surroundings was calculated. The experiments were performed in the following sequence:

1. Nominal voltage was applied to the motor terminal, and nominal load was applied to the shaft ($k_l = k_u = 1.0$);
2. Reduced voltage was applied to the motor terminal (90% of nominal voltage), and nominal load was applied to the shaft;
3. Reduced voltage was applied to the motor terminal (90% of nominal voltage), and reduced load was applied to the shaft ($k_l = k_u = 0.9$);
4. Reduced voltage was applied to the motor terminal (90% of nominal voltage), and reduced load was applied to the shaft ($k_l = k_u^2 = 0.81$);
5. Reduced voltage was applied to the motor terminal (90% of nominal voltage), and reduced load was applied to the shaft ($k_l = k_u^3 = 0.73$).

Cold start of the motor was used, i.e., before each experiment the motor was cooled to the temperature of the surroundings, which remained unchanged during the experimental test. The results of the experimental test are shown in Figure 9.

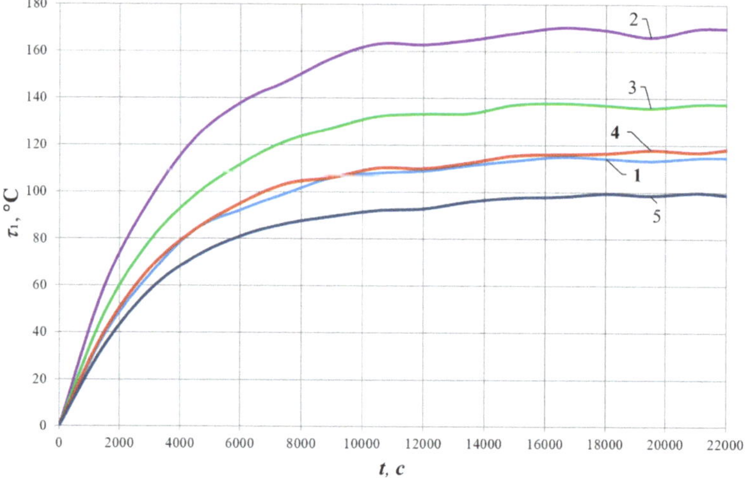

Figure 9. Dependence $\tau = f(t)$ at different values of k_l and k_u.

Figure 9 shows that the dependences obtained from $k_l = k_u = 1$ and $k_l = k_u^2 = 0.81$ almost coincide. This indicates the accuracy of the obtained Equation (26).

Let us compare the results with the information given in [31–33]. For example, consider the case when the motor drives a working machine with a parabolic mechanical characteristic at the load of 80% of nominal. The results obtained in the work show that according to (26), the rate of thermal wear of insulation will be nominal when the voltage is at least 89.5% of nominal and does not exceed it. Having applied the results of [31–33] to the modelling conditions of this work, we will get the following. According to [31], at the specified motor load, we propose to apply a voltage equal to 86% of nominal. This results in the rate of thermal wear of insulation equal to 3.27 bas.h/h. According to [32], at the specified motor load, it is proposed to apply voltage equal to 87.2% of nominal. This results in the rate of thermal wear of insulation equal to 2.15 bas.h/h. According to [33], at the specified motor load, we apply voltage equal to 88.6% of nominal. This results in the rate of thermal wear motor insulation equal to 1.63 bas.h/h.

The comparison shown indicates that all of the control methods presented in works [31–33] result in low voltage. This causes the switching of the motor to overload mode and accelerated thermal wear of its insulation. The increased rate of thermal wear of insulation leads to the deterioration of its characteristics and premature failure of the motor. To prevent this from happening, it is necessary to take into account the relationship between the motor load and its voltage by including this relationship into the control algorithm, which is not provided in [31–33]. This will ensure the nominal rate of thermal wear of insulation and will prevent the premature failure of the motor. Consideration of the mentioned relationship in the control algorithm is possible if the algorithm includes an additional rule for voltage control. This rule based on (26) has the following form:

$$\sqrt{k_3} \leq k_u \leq 1. \tag{27}$$

The fulfilment of this condition will allow saving the resource of the motor while controlling it by voltage. For example, if the AIR132S4 electric motor drives by a pump unit for 100 h with a load of 80% of nominal, then according to the described voltage control methods, additional wear of its insulation will be observed. According to [31], it will additionally use $(3.27 - 1) \times 100 = 227$ bas.h/h of the insulation resource. According to [32], it will additionally use $(2.15 - 1) \times 100 = 115$ bas.h/h of the insulation resource. According to [33], it will additionally use $(1.63 - 1) \times 100 = 63$ bas.h/h of the insulation resource. By applying the control rule (27), presented in this work, additional wear of insulation is excluded.

4. Conclusions

As a result of the study, by using a three-element thermal equivalent circuit of the induction motor, which is more accurate than that in [40], the dependence of the steady-state temperature excess of the stator winding on the active power loss in the elements of the circuit under uncertainty of its thermal conductivity has been obtained. This distinguishes the dependence from [41,42], which require determining the thermal conductivity. The dependence obtained in the work helps to avoid this by introducing coefficients of power loss impact on the electric motor heating. They have been determined by standard operational tests (idling and short circuit). The power loss coefficient makes it possible to identify the thermal condition of a particular electric motor taking into account the current technical condition of its active parts, which are different in each electric motor. As a result, the obtained dependence is more accurate for a particular electric motor compared to [40–42], assuming that all electric motors of a certain size have the same technical condition of active parts.

According to the L-shaped equivalent circuit, the dependence of losses in the three-element thermal equivalent circuit on voltage and load has been obtained. This dependence takes into account the joint effect of voltage deviation and a change in the load of the motor. This approach differs from the existing ones, which take into account either voltage deviation or load change.

The dependence of induction motor slip on voltage and load by imposing the linearized workable range of the mechanical characteristic of the motor and the working machine has been determined for the main types of mechanical characteristics of working machines, which varies for each type of mechanical characteristics of the working machine. This approach, unlike [45,46] taking into account only voltage deviations, is more universal. It allows taking into account the main factors of the mode-based nature, which influence the motor slip.

By applying the established dependencies, the modelling of the rate of thermal wear of insulation with various mechanical characteristics at various reduction coefficients of voltage and load (dependence for AIR132S4 electric motor is included in the work) has been performed. The results of modelling have shown that at reduced voltage, the rate of thermal wear of insulation sharply increases. At a decrease in voltage of 10% compared to the nominal value and operation of the motor with nominal load, the rate of thermal wear increases by 30–45 times in relation to the nominal value depending on the type of the working machine.

The dependency of the nominal rate of thermal wear of insulation on voltage has been established. It is as follows: the rate of thermal wear of insulation of the motor is nominal if its load factor equals the square of the voltage reduction coefficient. Verification of this dependency for the temperature of the stator winding showed that the difference between modelling results and the test does not exceed 5%.

By comparing the dependency of the nominal speed of thermal wear of insulation with control methods presented in [31–33], the following was obtained. If the electric motor drives a working machine with parabolic mechanical characteristics and is loaded by 80%, according to the obtained dependency, voltage must be no less than 89.5% of nominal and must not exceed it. This results in the nominal rate of thermal wear of insulation, i.e., 1 bas.h/h. Methods described in [31–33] not taking into account this regularity lead to undervoltage, which causes an increase in the rate of thermal wear of insulation.

It has been proved that in order to maintain the nominal speed of thermal wear of insulation while controlling it by energy-saving voltage regulation, it is necessary to include an additional rule of voltage regulation in the control algorithm, which states that the coefficient of the voltage reduction of an electric motor must be no less than the square root of its load coefficient and no more than 1. The existing control methods [31–33] result in additional wear of insulation. The disadvantage of this rule is the lack of consideration of other influences on the electric motor, except the thermal one.

The obtained rule belongs to the category of algorithmic support of control systems of induction motors and allows extending the service life of electric motors when controlling them by voltage. It is applied in control systems as an additional condition to those involved in them.

Author Contributions: Conceptualization, T.M.A.A.-Q., O.V. and S.H.; methodology, O.V. and V.N.; software, S.H.; validation, O.V., S.K. and O.S.; formal analysis, O.M., T.M.A.A.-Q. and N.M.Z.; investigation, O.M. and V.N.; data curation, S.H., S.K. and N.M.Z.; writing—original draft preparation, T.M.A.A.-Q., O.M. and K.M.A.I.; writing—review and editing, V.N. and O.S.; visualization, S.K., O.V. and K.M.A.I. All authors have read and agreed to the published version of the manuscript.

Funding: This research received no external funding.

Institutional Review Board Statement: Not applicable.

Informed Consent Statement: Not applicable.

Data Availability Statement: Data sharing not applicable.

Conflicts of Interest: The authors declare no conflict of interest.

References

1. Brunner, C.U.; Arquit Niederberger, A.; de Almeida, A.T.; Keulenaer, H. Standards for efficient electric motor systems SEEEM building a worldwide community of practice. In Proceedings of the ECEEE 2007 Summer Study on Energy Efficiency: Saving energy—Just Do It, La Colle sur Loup, France, 4–9 June 2007; Volume 3, pp. 1443–1455.
2. Lu, S.-M. A review of high-efficiency motors: Specification, policy and technology. *Renew. Sustain. Energy Rev.* **2016**, *59*, 1–12. [CrossRef]
3. Waide, P.; Brunner, C.U. Energy-Efficiency Policy Opportunities for Electric Motor-Driven Systems. In *IEA Energy Papers*; OECD Publishing: Paris, France, 2011; p. 132. [CrossRef]
4. Ferreira, F.J.T.E.; de Almeida, A.T. Overview on energy saving opportunities in electric motor driven systems—Part 1: System efficiency improvement. In Proceedings of the 2016 IEEE/IAS 52nd Industrial and Commercial Power Systems Technical Conference (ICPS), Detroit, MI, USA, 1–5 May 2016; pp. 1–8. [CrossRef]
5. Oshurbekov, S.; Kazakbaev, V.; Prakht, V.; Dmitrievskii, V.; Gevorkov, L. Energy Consumption Comparison of a Single Variable-Speed Pump and a System of Two Pumps: Variable-Speed and Fixed-Speed. *Appl. Sci.* **2020**, *10*, 8820. [CrossRef]
6. Terron-Santiago, C.; Martinez-Roman, J.; Puche-Panadero, R.; Sapena-Bano, A.A. Review of Techniques Used for Induction Machine Fault Modelling. *Sensors* **2021**, *21*, 4855. [CrossRef]
7. Boyko, A.; Volianskaya, Y. Synthesis of the System for Minimizing Losses in Asynchronous Motor with a Function for Current Symmetrization. *East.-Eur. J. Enterp. Technol.* **2017**, *4–5*, 50–58. [CrossRef]
8. Sudhakar, I.; Adi Narayana, S.; Anil Prakash, M. Condition Monitoring of a 3-Ø Induction Motor by Vibration Spectrum anaylsis using Fft Analyser—A Case Study. *Mater. Today Proc.* **2017**, *4*, 1099–1105. [CrossRef]
9. Ferreira, F.J.T.E.; Ge, B.; de Almeida, A.T. Reliability and Operation of High-Efficiency Induction Motors. *IEEE Trans. Ind. Appl.* **2016**, *52*, 4628–4637. [CrossRef]

10. Qawaqzeh, M.Z.; Szafraniec, A.; Halko, S.; Miroshnyk, O.; Zharkov, A. Modelling of a household electricity supply system based on a wind power plant. *Prz. Elektrotech.* **2020**, *96*, 36–40. [CrossRef]
11. Halko, S.; Suprun, O.; Miroshnyk, O. Influence of Temperature on Energy Performance Indicators of Hybrid Solar Panels Using Cylindrical Cogeneration Photovoltaic Modules. In Proceedings of the 2021 IEEE 2nd KhPI Week on Advanced Technology (KhPIWeek), Kharkiv, Ukraine, 13–17 September 2021; pp. 132–136. [CrossRef]
12. Bazurto, A.J.; Quispe, E.Q.; Mendoza, R.M. Causes and Failures Classification of Industrial Electric Motor. In Proceedings of the 2016 IEEE ANDESCON, Arequipa, Peru, 19–21 October 2016; pp. 1–4. [CrossRef]
13. Sumislawska, M.; Gyftakis, K.N.; Kavanagh, D.F.; McCulloch, M.; Burnham, K.J.; Howey, D.A. The Impact of Thermal Degradation on Electrical Machine Insulation Insulation. In Proceedings of the 2015 IEEE 10th International Symposium on Diagnostics for Electrical Machines, Power Electronics and Drives (SDEMPED), Guarda, Portugal, 1–4 September 2015; pp. 1–8. [CrossRef]
14. Höpner, V.N.; Wilhelm, V.E. Insulation Life Span of Low-Voltage Electric Motors—A Survey. *Energies* **2021**, *14*, 1738. [CrossRef]
15. Ghassemi, M. Accelerated insulation aging due to fast, repetitive voltages: A review identifying challenges and future research needs. *IEEE Trans. Dielectr. Electr. Insul.* **2019**, *26*, 1558–1568. [CrossRef]
16. Liu, T.; Jiang, M.; Zhang, D.; Zhao, H.; Shuang, F. Effect of Symmetrical Voltage Sag on Induction Motor Considering Phase-angle Factors Based on A New 2-D Multi-Slice Time-Stepping Finite Element Method. *IEEE Access* **2020**, *8*, 75946–75956. [CrossRef]
17. Gnaciński, P. Insulations Temperature and Loss of Life of an Induction Machine Under Voltage Unbalance Combined With Over- or Undervoltages. *IEEE Trans. Energy Convers.* **2008**, *23*, 363–371. [CrossRef]
18. Al-Issa, H.A.; Qawaqzeh, M.; Kurashkin, S.; Halko, S.; Kvitka, S.; Vovk, O.; Miroshnyk, O. Monitoring of power transformers using thermal model and permission time of overload. *Int. J. Electr. Comput. Eng. (IJECE)* **2022**, *12*, 2323–2334. [CrossRef]
19. Zhao, K.; Cheng, L.; Zhang, C.; Nie, D.; Cai, W. Induction motors lifetime expectancy analysis subject to regular voltage fluctuations. In Proceedings of the IEEE Electrical Power and Energy Conference (EPEC), Saskatoon, SK, Canada, 22–25 October 2017. [CrossRef]
20. Kostic, M. Effects of Voltage Quality on Induction Motors' Efficient Energy Usage. In *Induction Motors—Modelling and Control*; Araújo, R.E., Ed.; IntechOpen: London, UK, 2012; pp. 127–156. [CrossRef]
21. Petronijević, M.; Mitrović, N.; Kostić, V.; Banković, B. An Improved Scheme for Voltage Sag Override in Direct Torque Controlled Induction Motor Drives. *Energies* **2017**, *10*, 663. [CrossRef]
22. Rachev, S.; Dimitrov, L.; Koeva, D. Study of the Influence of Supply Voltage on the Dynamic Behavior of Induction Motor Low Voltage Drive. *Int. Sci. J. Mach. Technol. Mater.* **2017**, *11*, 206–209.
23. Masoum, M. Lifetime Reduction of Transformers and Induction Machines. In *Power Quality in Power Systems and Electrical Machines*; Fuchs, E.F., Masoum, M.A.S., Eds.; Academic Press: Cambridge, MA, USA, 2008; pp. 227–260. [CrossRef]
24. Zyuzev, A.M.; Metel'kov, V.P. Toward the evaluation of the thermal state of an induction motor in the recursive short-term mode. *Russ. Electr. Eng.* **2014**, *85*, 554–558. [CrossRef]
25. Gundabattinia, E.; Kuppanb, R.; Gnanaraj Solomona, D.; Akhtar Kalamc, A.; Kotharid, D.P.; Abu Bakar, R. A review on methods of finding losses and cooling methods to increaseefficiency of electric machines. *Ain Shams Eng. J.* **2021**, *12*, 497–505. [CrossRef]
26. Marchesoni, M. Energy Efficiency in Electric Motors, Drives, Power Converters and Related Systems. In *Energies*; MDPI: Basel, Switzerland, 2020; 248p. [CrossRef]
27. Saidur, R. A review on electrical motors energy use and energy savings. *Renew. Sustain. Energy Rev.* **2010**, *14*, 877–898. [CrossRef]
28. Diachenko, G.G.; Aziukovskyi, O.O. Review of methods for energy-efficiency improvement in induction machines. *Nauk. Visn. Nats. Hirn. Univ.* **2020**, *1*, 80–88. [CrossRef]
29. Iegorov, O.; Iegorova, O.; Miroshnyk, O.; Savchenko, O. Improving the accuracy of determining the parameters of induction motors in transient starting modes. *Energetika* **2020**, *66*, 15–23. [CrossRef]
30. Vovk, O.; Kvitka, S.; Halko, S.; Strebkov, O. Energy-Saving Control of Asynchronous Electric Motors for Driving Working Machines. In *Modern Development Paths of Agricultural Production*; Springer: Cham, Switzerland, 2019; pp. 415–423. [CrossRef]
31. Jirasuwankul, N. Simulation of Energy Efficiency Improvement in Induction Motor Drive by Fuzzy Logic Based Temperature Compensation. *Energy Procedia* **2017**, *107*, 291–296. [CrossRef]
32. Guo, C. Voltage regulation and power-saving method of asynchronous motor based on fuzzy control theory. *Open Phys.* **2022**, *20*, 334–341. [CrossRef]
33. Resa, J.; Cortes, D.; Marquez-Rubio, J.F.; Navarro, D. Reduction of Induction Motor Energy Consumption via Variable Velocity and Flux References. *Electronics* **2019**, *8*, 740. [CrossRef]
34. Chulines, E.; Rodríguez, M.A.; Duran, I.; Sánchez, R. Simplified Model of a Three-Phase Induction Motor for Fault Diagnostic Using the Synchronous Reference Frame DQ and Parity Equations. *IFAC-Papers* **2018**, *51*, 662–667. [CrossRef]
35. Simion, A.; Livadaru, L.; Munteanu, A. Mathematical Model of the Three-Phase Induction Machine for the Study of Steady-State and Transient Duty under Balanced and Unbalanced States. In *Induction Motors—Modelling and Control*; IntechOpen: London, UK, 2012; pp. 3–44. [CrossRef]
36. Bonnett, A.H. The impact that voltage and frequency variations have on AC induction motor performance and life in accordance with NEMA MG-1 standards. In Proceedings of the IEEE Conference Record of Annual Pulp and Paper Industry Technical Conference, Seattle, WA, USA, 21–25 June 1999; pp. 1–16. [CrossRef]
37. Iegorov, O.; Iegorova, O.; Miroshnyk, O.; Cherniuk, A. A calculated determination and experimental refinement of the optimal value of the single-phase induction motor transformation ratio. *Energetika* **2021**, *67*, 13–19. [CrossRef]

38. Vovk, O. Steady-state process of induction motor heating. *TDATA Rev.* **2002**, *5*, 62–66.
39. Metelkov, V.P.; Esaulkova, D.V.; Kondakov, K.A. Method for Estimating a Residual Resource of Induction Motor Insulation Insulation Based on Capacitive Leakage Currents. In Proceedings of the 2021 XVIII International Scientific Technical Conference Alternating Current Electric Drives (ACED), Ekaterinburg, Russia, 24–27 May 2021. [CrossRef]
40. Duran, M.J.; Iraizoz Fernández, J.M. Lumped-Parameter Thermal Model for Induction Machines. *IEEE Trans. Energy Convers.* **2005**, *19*, 791–792. [CrossRef]
41. Boglietti, A.; Cavagnino, A.; Staton, D.; Shanel, M.; Mueller, M.; Mejuto, C. Evolution and Modern Approaches for Thermal Analysis of Electrical Machines. *IEEE Trans. Ind. Electron.* **2009**, *56*, 871–882. [CrossRef]
42. Meneceur, N.; Boulahrouz, S.; Boukhari, A.; Meneceur, R. Thermal Behavior Simulation of An Externally Cooled Electric Motor. *Eng. Technol. J.* **2022**, *40*, 81–90. [CrossRef]
43. Szychta, E.; Szychta, L. Collective Losses of Low Power Cage Induction Motors—A New Approach. *Energies* **2021**, *14*, 1749. [CrossRef]
44. Zhong, B.; Ma, L.; Dong, H. Principle of Optimal Voltage Regulation and Energy-Saving for Induction Motor With Unknown Constant-Torque Working Condition. *IEEE Access* **2020**, *8*, 187307–187316. [CrossRef]
45. Kuznetsov, V.V.; Nikolenko, A.V. Models of operating asynchronous engines at poor-quality electricity. *East.-Eur. J. Enterp. Technol.* **2015**, *1*, 37–42. [CrossRef]
46. Wu, H.; Dobson, J. Cascading Stall of Many Induction Motors in a Simple System. *IEEE Trans. Power Appar. Syst.* **2012**, *27*, 2116–2126. [CrossRef]

MDPI AG
Grosspeteranlage 5
4052 Basel
Switzerland
Tel.: +41 61 683 77 34

Inventions Editorial Office
E-mail: inventions@mdpi.com
www.mdpi.com/journal/inventions